Engineering Materials

This series provides topical information on innovative, structural and functional materials and composites with applications in optical, electrical, mechanical, civil, aeronautical, medical, bio- and nano-engineering. The individual volumes are complete, comprehensive monographs covering the structure, properties, manufacturing process and applications of these materials. This multidisciplinary series is devoted to professionals, students and all those interested in the latest developments in the Materials Science field.

More information about this series at http://www.springer.com/series/4288

Keisuke Fujisaki
Editor

Magnetic Material for Motor Drive Systems

Fusion Technology of Electromagnetic Fields

Springer

Editor
Keisuke Fujisaki
Department of Engineering
Toyota Technological Institute
Nagoya, Japan

ISSN 1612-1317 ISSN 1868-1212 (electronic)
Engineering Materials
ISBN 978-981-32-9905-4 ISBN 978-981-32-9906-1 (eBook)
https://doi.org/10.1007/978-981-32-9906-1

This Springer imprint is published by the registered company Springer Nature Singapore Pte Ltd.
The registered company address is: 152 Beach Road, #21-01/04 Gateway East, Singapore 189721, Singapore

Preface

With a commercialization of a hybrid car, the hybrid technology of an internal combustion engine and an electric motor is applied to the overall means of transportation such as a ship, an airplane, and a car. The strong orientation that most of the transportation body is driven by an electrical motor is shown. This is regarded as a revolution of the means of transportation from the drive of the internal combustion engine to the drive of the electric motor. Even if a problem of a accumulation technology of the electric energy such as the battery still remains essentially, this flow to the electrical motor which is high efficiency and high responsibility depends on the power electronics technology basically, and it may be said that the technology will enlarge the spread more and more in future.

The electric motor itself was developed and used for more than 100 years, and it was one leading role of the secondary industry revolution until now. However, the motor drive system causing a moving revolution in this time leaves from the inherent technique, and two kinds of new technical requests come out. One is a variable velocity technology by the power electronics which did not exist in those days, and it introduces a new requirement called a change of an operating point of the motor and a change of operating environment of the motor. The other is that the setting place of the motor is changed from an on-ground operation which was conventional to an onboard operation being put on the means of transportation, and a further request of a downsizing and high-efficiency technology for the motor drive system comes out.

The high-efficiency electrical motor has been researched and developed widely until now in the combination with the social request of the energy-saving technique. It may be said that the considerable downsizing and high-efficiency technology has been realized at present. However, it may be said that the expansion of the further application of the onboard motor drive system by the transportation revolution requires more-better-than downsizing and high-efficiency technology for the motor drive system. Among loss reduction technologies of the electrical motor, a study of iron loss characteristics for the motor drive system due to a magnetic material was not seen by technical difficulty of understanding or so. However, nowadays, when most of the loss reduction technologies are basically solved, the study of iron loss technologies based on the magnetism characteristic for the motor drive system is required to be an important technique problem to be solved in earnest from now on.

Therefore, this book describes the high efficiency of the motor drive system from a point of view of the magnetic material. There is a combination of different technologies such as the motor, the power electronics, and the magnetic material for the iron loss reduction. The high-efficiency research and development of each elemental technology goes for each area deeply. However, the most important issue to be apt to overlook here is whether the high-efficiency motor drive system is provided to when the motor drive system is looked at as a whole system. It may be said that their electromagnetic field phenomenon is mutually combined between the motor and the power electronics circuit through supplied voltage and electrical current. So those fusion technology and science are strongly demanded.

It is important for a technical fusion to understand each technical essence mutually well, and then, an influence of the interaction can be discussed after the mutual understanding. Therefore, this book tried to write with the goal of understanding the technical fusion of the motor drive system about the magnetic material widely for the engineer and the researcher of the different fields such as the motor, the power electronics, and the magnetic material related to the motor drive system.

The motor drive system with large and small motors is now applied to a home, a car, an office, and a production factory, and is related to the various and wide-ranged people who are engaged in not only the research and development of the motor drive system technology but also the production, setting, operation, maintenance, and management. They research, develop, produce, manage, and maintain the motor drive system for their application. Mutual technical discussions are indispensable in order that various people varying in the situation and the position judge it definitely and accomplish it in each application of the motor drive system of the intense technical progress.

Therefore, this book is decided to list the minimum contents about the motor drive system. So the conceptual description as well as the considerably "rough" discussion may be carried out. The details are left to each technical book, and the coarseness of the discussion may be said to stand out according to each expert, but please forgive it due to the purpose mentioned above. The authors will be happy if each reader can reach the heartstrings.

Finally, gratitude is expressed to all the authors who had willingly consented to the writing of this book and Springer Nature which had allowed the publication of this book.

Nagoya, Japan Keisuke Fujisaki

Contents

Motor Drive System and Magnetic Material: Contents of This Book

Keisuke Fujisaki

Abstract Motor drive system is now widely applied to a transportation system, and its high efficient and downsizing are required more and more. So magnetic material becomes a key technology for power electronics excitation especially. The relation among electrical motor, power electronics, and magnetic material is described in detail. The contents of this book are shown in briefly.

Keywords Iron loss · Electrical motor · Power electronics · Magnetic material

1 Motor and Power Electronics and Magnetic Material [1–8]

An electric motor has been widely used in the society, industry, and home as an apparatus changing electrical energy of alternate current (AC) or direct current (DC) to mechanical energy of torque and rotational speed for more than 100 years. It was one leading role of the secondary industry revolution following a steam engine.

When electric power system networks of AC at the commercial frequency of 50 or 60 Hz spread, an electrical motor based on uniformity frequency, phase, and a constant voltage has been used as a power source of air or water movement technology such as fan or blower. Precise and detailed control technology of positioning, speed, or torque was not so used, and the downsizing and high-efficient technology of the motor was not a necessarily primary problem there because the electric motor was used on the ground. Enough space and enough energy supply are basically possible in on-ground operation. As for the magnetic material, because it was used for a core of the motor, transformer, and reactor at the commercial frequency operation without time-harmonic components, evaluation methods of the magnetic material established in JIS and IEC were enough by sine wave excitation not to include time-harmonic components.

K. Fujisaki (✉)
Toyota Technological Institute, Nagoya, Japan
e-mail: fujisaki@toyota-ti.ac.jp

K. Fujisaki (ed.), *Magnetic Material for Motor Drive Systems*,
Engineering Materials, https://doi.org/10.1007/978-981-32-9906-1_1

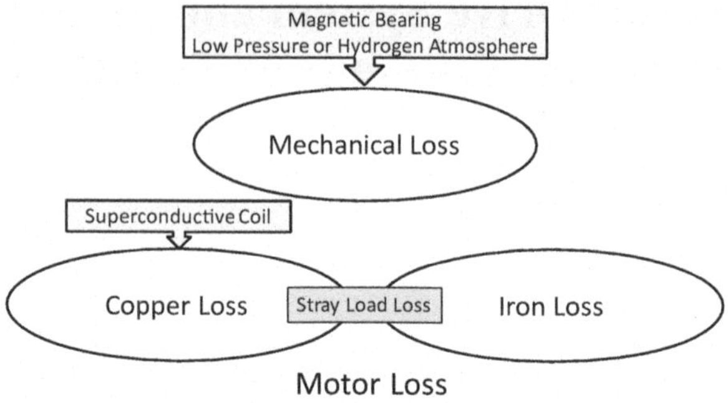

Fig. 1 Three major loss components of electrical motor

In contrast, a power electronics technology that enables variable voltage and variable frequency by switching operation of power semiconductor makes the variable speed drive of the electrical motor particularly AC motor. High efficient and high responsible variable velocity drive system enables the application of the electrical motor to the means of transportation. So the drive system of the electrical motor is now considered and put to practical use in not only a car but also all the means of transportation such as a ship, a locomotive and an airplane.

As transportation system is driven by a limited energy source on it when a motor drive system comes to be put on overall means of transportation, the first demand specifications to the drive system are higher efficiency, lower loss and more downsizing than ever. This is required not only for the field of power electronics that many studies are accomplished until now but also for the electrical motor itself.

The loss of the electrical motor consists of mechanical loss, copper loss, and iron loss like Fig. 1. The mechanical loss is windage loss occurring in rotating a rotor and friction loss with the bearing supporting a rotor mechanically.

The copper loss is Joule loss occurring in flowing an exciting current in a copper coil and eddy current in electrical conductive material. The iron loss is a loss occurring in magnetic material such as a stator or a rotor by a flow of AC magnetic flux density in it and the generation of magnetic hysteresis curve as B–H curve (B: magnetic flux density, H: magnetic field).

Stray load loss is often picked up as motor loss other than the three components described in the above. It is the Joule loss of the eddy current flowing in a conductive material by slot harmonics and time harmonics with the inverter. However, it is left out here, if it is considered to include three components described in the above as it is considered to be the copper loss when the eddy current flows in the copper coil and it is the iron loss when the eddy current flows in the magnetic material.

Among such the motor loss, the mechanical loss can become a zero theoretically, as windage loss can become a zero by making low pressure or a hydrogen atmosphere and as the friction loss can become a zero by using the non-contact support with the magnetic bearing to which electromagnetic suspension control technology is applied. The copper loss is theoretically possible to be a zero if superconductive coil is applied

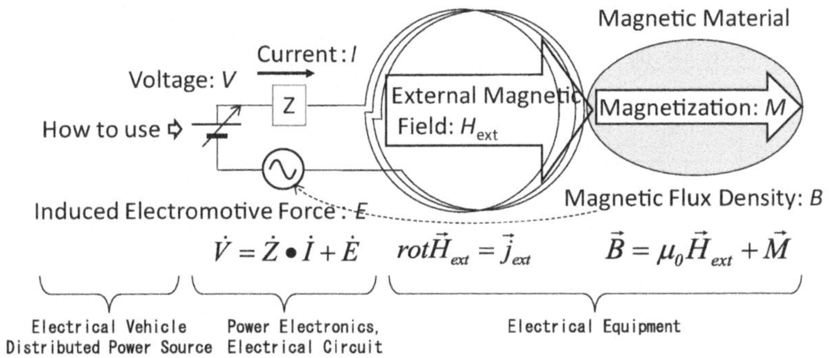

Fig. 2 Relation between electrical engineering and magnetic engineering [2–6, 8–10]. Adapted from [4, 10] with permission ©2014, 2017 The Magnetics Society of Japan

to the coil in the motor. However, about the reduction plan of the iron loss, it may be said that it is theoretically the present conditions not to reach the place called a zero.

Furthermore, the magnetic material producing the iron loss has complicated characteristics such as nonlinearity, magnetic domain structure by the saturated magnetization, magnetic anisotropy, stress sensitivity, and magnetism hysteresis characteristic. In other words, among the three major elements of the motor loss, it is the iron loss that reduction is theoretically difficult, and understanding is difficult. So it may be also said that the clarification of its characteristic is the last problem remained in the motor study. Therefore, this book decides to pay the attention to the magnetic material which is not enough elucidated among the motor losses.

The magnetic material is magnetized by external magnetic field H_{ext} as shown in Fig. 2 (outbreak of M, M: magnetization), and then, a big magnetic flux density B can be obtained as the next equation.

$$B = \mu_0(H_{\text{ext}} + M) \tag{1}$$

The big magnetic flux density realizes the small size and the high output of the motor, the transformer, and the reactor described as below. The external magnetic field occurs based on the Ampere law by an electric current flowing in an electrical circuit, and the electric current occurs by the electrical circuit with an applied voltage, an impedance of the circuit, and an induced electromotive force. And an electric circuit used for the variable velocity drive of the motor is power electronics.

As the power electronics equipment such as an inverter operates by switching of a power semiconductor in the circuit, time-harmonic components will occur essentially. The switching operation makes a time series voltage of square shape in the output voltage. Therefore, magnetic material used for a motor drive system or so is an excitation by the power electronics equipment including the time-harmonic components, that is, a power electronics excitation.

The switching operation of the power electronics can realize the variable voltage and the variable frequency in high efficiency and in high responsibility at the same

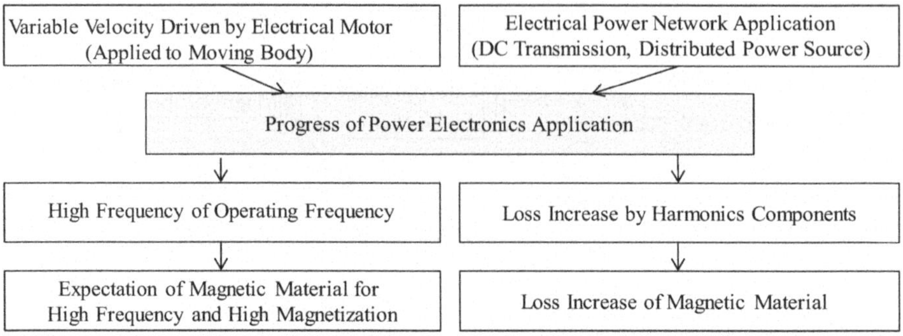

Fig. 3 New magnetic material requirement by a progress of power electronics technology

time only by changing a switching timing of the power semiconductor described as bellows. Therefore, an application of the power electronics is unfolded for not only the motor drive but also the electric power network system like Fig. 3 and pushes forward the progress of the power electronics technology.

As the high-frequency trend of the operating frequency generally leads to the downsizing and weight reduction of the equipment, high-frequency operation is also required to magnetic material. As a cooldown method becomes a large scale when the equipment becomes a high-frequency and large-capacity operation, the requirement for low loss and high magnetization is extremely high. Furthermore, the switching operation of the power semiconductor causes time-harmonic components, and the magnetic material excited by it causes the loss increase.

In other words, the electrical motor gets big magnetic flux density by using a magnetic material and the reduction of size and weight can be realized, but the motor drive system is a power electronics excitation by the switching operation, and then, a new problem called the core loss increase of the whole motor drive system will occur by the high-frequency operation and the outbreak of the time-harmonic components. A power electronics technology is an outbreak source of the problem of the core loss increase, and the evaluation should be performed with the electrical motor like Fig. 4. The magnetic material is an origin of outbreak of the loss, and a study and a solution for new magnetic material should be performed.

2 Contents of This Book

This book describes the basics and the application of the high efficient and downsizing motor drive system from a point of view of the magnetic material. Five parts are divided, and then, the details are shown as follows.

At first, Part I describes the background of the requirement for new magnetic material by the progress of power electronics and motor drive system as general remarks. The minimum fundamental concepts of magnetic material, electrical motor, and

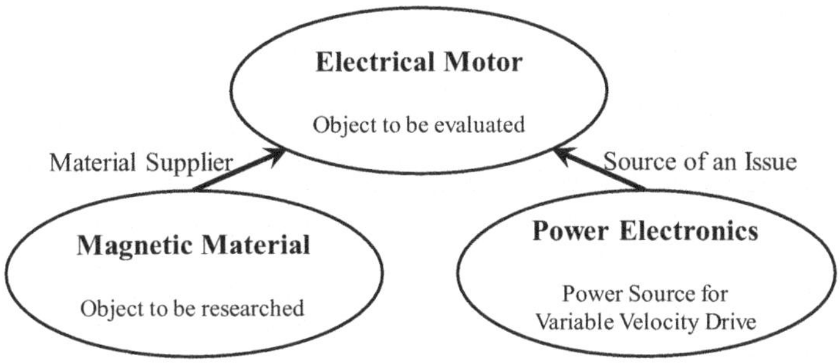

Fig. 4 Magnetic material inflection technology of the motor drive system in the power electronics excitation [11]

power electronics, which are three elemental technologies of the motor drive system, are explained for the different academic engineers and researchers to understand them each other. Particularly, as the big technical and academic wall called the electrical engineering and the magnetic engineering is assumed between motor, power electronics, and magnetism, an important concept for each field is specialized and is written down. Finally, the electromagnetic field fusion science and technology of equipment, application, and materials are spoken.

In the second, Part II describes an inflection technology of the magnetic material, where power electronics excitation properties and the motor inflection of the magnetic material are shown. At first as for the power electronics excitation properties, the loss increase phenomenon of the magnetic material by inverter excitation, and the magnetic phenomenon there, and then the core loss characteristic with the motor are described. Next as for an inflection technology of the magnetic material, the motor application of commercially available low iron loss materials and an iron loss reduction method is described.

In Part III, the foundations of magnetic material and the modeling of the magnetized characteristic are described in order to have the person concerned of the motor drive system understand a magnetic material more deeply. At first, the magnetization phenomenon in the wide meaning is arrested in the viewpoint of the magnetic body multi-scale more deeply such as expression of the magnetization with the electronic spin, magnetic domain structure, and the polycrystalline magnetic body. And then, the modeling of the magnetic hysteresis phenomenon, the modeling of high-frequency magnetism properties, the modeling of the analysis to connect with the semiconductor properties in the power electronics circuit, and the magnetic characteristic are mentioned. Finally, a vector property of magnetism characteristics is spoken.

Furthermore, future magnetic material is described in Part IV. At first, the summarization of so far and the future of the magnetic material is carried out, and the soft magnetic material for low loss is shown. The sintered permanent magnet (PM) and the bond PM which greatly contributed to the downsizing and high efficiency of the motor including the rare earth problems are described there. Finally, high-frequency magnetism to become an issue of power electronics excitation is spoken.

Finally, the application of magnetic material is shown in Part V. The research situation about the motor application and the reactor application of the magnetic material is described and the application of electrical vehicle and the linear motor as an application of the electric apparatus using a magnetic material is shown.

References

1. Fujisaki, K.: Future magnetic material property installed in and driven by power electronics technology. In: IUMRS-ICAM 2015 (14th International Union of Materials Research Societies—International Conference on Advanced Materials), IV-1Th3F1-1 (IS), Korea, 29 Oct 2015
2. Investigating R&D Committee of Magnetic Material for Electromagnetic Actuator: Magnetic material and its evaluation technology for electromagnetic actuator. Technical report of IEEJ, No. 1397 (2017) (in Japanese)
3. Fujisaki, K.: Advanced magnetic material requirement for higher efficient electrical motor design. In: The 38th Annual Conference on Magnetics in Japan, Symposium "Challenge of Magnetics to Improve Energy Efficiency", 4aB-2, Sept 2014
4. Fujisaki, K.: Future magnetic material based on power electronics. In: The 202nd Topical Symposium of the Magnetics Society of Japan, "Current Trends and Future Developments of Energy-Related Magnetic Materials", 202-1, Tokyo, Japan, 26 May 2015, pp. 1–6. ISSN 1882-2940 (in Japanese)
5. Fujisaki, K.: Magnetization of magnetic material excited by power electronics equipment. Jpn. Inst. Power Electron. **42**, 3–6 (2016). JIPE-42-01 (in Japanese)
6. Fujisaki, K.: Necessity and problem of magnetic material for electromagnetic actuator system. In: The 2015 Annual Meeting of the Institute of Electrical Engineering of Japan, 5-S22-1, Mar 2015, pp. S22(1)–S22(4) (in Japanese)
7. Fujisaki, K.: Necessity and problem of magnetic material for electromagnetic actuator system. In: The 2017 Annual Meeting of the Institute of Electrical Engineering of Japan, 5-s13-1, Toyama, Mar 2017 (in Japanese)
8. Fujisaki, K.: Magnetic property of magnetic body necessary for power electronics development. In: Technical Meeting on Magnetics of the Institute of Electrical Engineering of Japan, MAG-13-149, Tokyo, Dec 2013 (in Japanese)
9. Fujisaki, K.: Magnetic material property for high efficient motor drive system. Mater. Stage **16**(4), 12–13 (2016) (in Japanese)
10. Fujisaki, K.: Future magnetic material excited by power electronics equipment. Technical report of 213, Magnetics Society of Japan (2017) (in Japanese)
11. Fujisaki, K.: New magnetic material expectation required on power electronics excitation. Monthly EMC, No. 349, May 2017 (in Japanese)

General (Background of New Magnetic Material Requirement for Power Electronics Technology)

Technical Requirement to Magnetic Material in Motor Drive System

Keisuke Fujisaki

Abstract Due to an expansion of electrical motor drive system application for transportation means such as a car, a ship, and an airplane, its high efficient and downsizing technology required more and more. Nowadays, when studies of power semiconductor material as GaN and SiC are going ahead, and 80% of electrical energy is expected to be used through power electronics technology in 2030 such as a realization of distributed power supply system from mW level to GW level, it is said to be that the bottleneck technology has become magnetic material in a viewpoint of downsizing, cost, and efficiency. So a new magnetic material for high power and high frequency is required strongly. One of the solutions for the requirement is thought to be a mutual consideration of magnetic material and power semiconductor. Then, future trend of electrical motor is considered to be motor physics as microscopic direction as well as motor integration as macroscopic direction.

Keywords Movement · On-machine · Distributed power supply · Motor integration · Motor physics

1 Introduction

A hybrid car driven by an internal combustion engine and an electrical motor was commercialized in 1997 for the first time, and the application of the means of transportation with the electric motor was in use in earnest. The drive use of the electrical motor realizes the fuel economy improvement and the environmental load reduction by the high controllability of the motor drive system. It not only made the big trend of the car drive system, but also was with the big movement to spread in all the means of transportation such as a ship [1], a locomotive [2], and an airplane [3] like Fig. 1.

The application of electrical motor to the ship gives us an added value to keep the weight balance of the ship by locating the heavy generator and the heavy motor drive in front and back in the ship and to obtain the quiet sea voyage to a passenger.

K. Fujisaki (✉)
Toyota Technological Institute, Nagoya, Japan
e-mail: fujisaki@toyota-ti.ac.jp

K. Fujisaki (ed.), *Magnetic Material for Motor Drive Systems*,
Engineering Materials, https://doi.org/10.1007/978-981-32-9906-1_2

(a) Car (b) Ship

(c) Locomotive (d) Airplane

Fig. 1 Expansion of the means of transportation driven by electrical motor [1–3]. **b**, **c**, **d** are reprinted from 1–3, respectively. Images courtesy of **b** BAN PR inc., **c** Japan Freight Railway Company, and **d** All Nippon Airways Co., Ltd

The electrical motor application contributes the improvement of driving property to the locomotive, and it is also in advance for the airplane.

The feasibility study to a commercial large plane is also performed like Fig. 2 [4]. As about half of the domestic airline flight fuel is said to be consumed for a transportation in an airport, an application of the generator and the motor of the MW class is considered for the efficiency improvement of the run there. According to the survey of Ministry of Economy, Trade, and Industry, the market size of the airplane is said that there is global market of approximately 30,000 planes, approximately 4 trillion dollars mainly on market expansion of the Asia-Pacific region for the next

Fig. 2 Feasibility study of electrical motor drive of airplane [4] [basic design in the specification of Boeing 777-200LR (300 passengers airplane), turboelectric distribution propulsion (30 MW generator × 2, 4 MW electrical motor × 14)]

20 years, and it may be said that a use opportunity of the motor drive in the field of airplane increases more and more [5].

In this way, an electric motor was used generally in the big domain where the means of transportation used gist 30% degree of domestic energy. Therefore, about a technical background and a technical problem that happened there are considered.

2 Conventional Motor and Coming Power Electronics Excitation Motor [6–8]

The electric motor played a big role in a secondary industry revolution and has been used as clean power source until now. Its power source is supplied in high efficiency and stability by using the electric power network which spread as a power supply in those days there.

The generator to rotate a shaft using waterpower or heat power to make electric power becomes high efficiency by centralizing and making a big size. Since the accumulation of electric energy is theoretically difficult, the generated electric power must be always the same with the consumption in the load. Upsizing of a synchronization generator also aims at the stabilization of electric power network. The load includes lighting equipment, electronic intelligence apparatus other than the motor, and it has been demanded to have stability, high efficiency, and low cost.

Because the electric power system network connects various kinds of load at the same time, the common specifications to connect them are an important issue. In other words, it is important that a voltage and a frequency, and a phase should be always constant in the network. Therefore, the power supply for the electric motor premises that the frequency is a commercial one (50/60 Hz) and the voltage is the constant one such as 100 or 6600 V [9].

The conventional style motor was mainly used to send water or air given constant pressure (electromagnetic torque) in sewage or an air conditioner, and it was set on ground. In other words, it can be said to be an era of the constant-speed motor in the constant frequency and the constant voltage. Figure 3 is an induction motor which has been actually used as a blower to send the chill or warm air [10].

In contrast, the progress of the power electronics technology enabled a variable velocity drive system.

As an example, a synchronous motor with two poles is considered here. Basic equations of torque T, armature voltage V_a, rotational angular frequency ω_m of the motor are expressed as follows [11].

$$T = r\Phi_f I_a \tag{1}$$

$$V_a = -\frac{d\Phi_f}{dt} \tag{2}$$

Fig. 3 30 kW, 200 V, 3-phase AC 60 Hz induction motor (a blower in Toyota Technological Institute) [10]

$$\omega_e = 2\pi f \tag{3}$$

$$\omega_m = \omega_e \tag{4}$$

Here, r: radius of rotor, Φ_f: magnetic flux of the field magnet, I_a: armature electric current, ω_e: electric synchronous angular velocity, f: power supply frequency. Equation (1) is an expression of Lorenz force, and Eq. (2) is an expression of electromotive force, and Eq. (3) is an expression of the angular velocity of the rotating magnetic field in the motor and the power supply frequency, and Eq. (4) is an expression indicating that the angular velocity of the rotating magnetic field is synchronizing the angular rotational velocity of a rotating rotor.

As the change of the magnetic flux in Eq. (2) becomes a rotary speed in the case of a synchronous machine, the armature voltage is proportional to the rotary speed of the motor. In addition, from Eqs. (3) and (4), the power supply frequency is proportional to the rotary speed of the motor. In other words, the voltage and the frequency of the motor have to be changed when the rotation velocity of the motor is changed. In this case, an armature electric current will flow depending on the torque that the output torque is needed from Eq. (1).

As a movement is a variable velocity as being described later, the motor also should have a variable velocity characteristic, when a motor is used as a drive source. For a variable velocity, the frequency of the power supply must be changed from Eq. (3) (variable frequency), and the output voltage of the power supply also must be changed from Eq. (2) (the variable voltage). Furthermore, it is necessary to change the variable frequency and the variable voltage with a change of the speed in real-time operation, and the conversion efficiency of the electricity must be not only high efficiency but also a high reply.

3 What Is a Movement

Here, a "movement" carried out in a car, a ship, and an airplane is considered. The movement is that a moving body such as a person or an object moves from a point A of the moving beginning to a point B of the moving end like Fig. 4 [10, 12]. The moving body must be a stationary state of speed zero by all means in point A and point B, and it is necessary to let the moving body give some velocity for the moving action between point A and point B.

In other words, an acceleration to give some velocity and a slowdown to become a stationary state at a destination are required in the movement by all means, and a coasting driving such as constant velocity exists meanwhile. Some combinations of these are used in the real movement, but the acceleration and the slowdown exist for a movement by all means, and it may be said that it is a "variable velocity" state that velocity changes in every time. When a driving equipment is applied to a means of transportation, variable velocity system should be realized in high efficiency and high reply.

As mentioned above, the movement requires a variable velocity, and when it is realized by electrical motor, it requires an electrical energy conversion system of the variable voltage and variable frequency in high efficiency and high responsibility. The electrical energy conversion technology of the variable voltage and variable frequency in high efficiency and high reply came to be realized only after the appearance of the power electronics technology.

The power electronics technology can realize a high reply and high-efficiency electrical energy conversion system of variable voltage and variable frequency by the switching operation of the power semiconductor as described in Chap. 5. In other words, it may be said that the application to the means of transportation of the motor was enabled only after a motor drive system like Fig. 5 that an electric motor was driven and excited by a power electronics equipment [7].

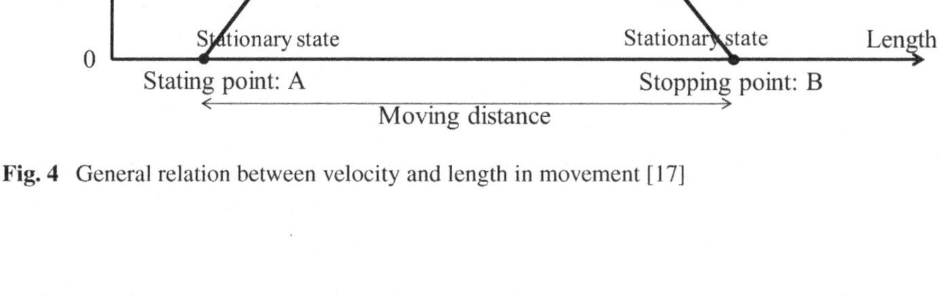

Fig. 4 General relation between velocity and length in movement [17]

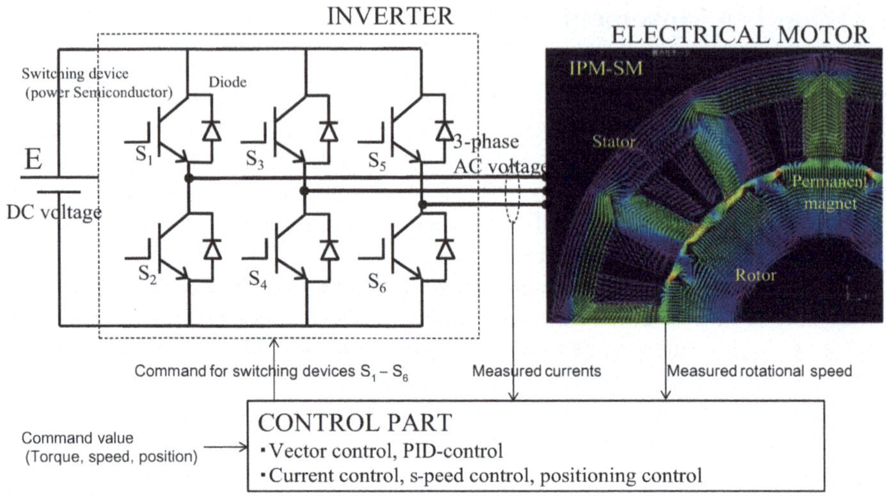

Fig. 5 Motor drive system [11, 17–20]. Adapted from [11, 18–20] with permission ©2015–2017 The Magnetics Society of Japan

An application of the electrical motor having been enabled to overall means of transportation such as not only a car but also a ship and an airplane may be said at least to be a premise of power electronics technology.

As the motor drive system requires a real-time change of the voltage and the frequency according to the rotational velocity of the motor, a feedback control to measure the rotational velocity of the motor and to make appointed ON/OFF signals to switching devices of power semiconductors in the inverter is with requisiteness. In other words, the motor drive system is comprised of the electric motor, inverter, and control part.

Various kinds of control theories are used in the control part. The vector control theory is the one that can treat the AC motor which is hard to control but is in high performance like the DC motor which is easy to control and realizes a high reply for an appointed order level such as a position or a speed immediately.

The control system has a multiplex loop structure of the current control, velocity control, and positioning control, and PI control is often used for simplicity. An output voltage order of the inverter will be given by the inverter control, and it introduces a gate signal order (ON/OFF order) of each power semiconductor in the inverter.

4 Electrical Energy and Power Electronics Technology

Energy exists with some forms such as chemical energy of oil or coal, machine energy of heat, pressure, power, position, electric energy, light/radiant energy, bioenergy, and nuclear energy. The electric energy is said to be secondary energy, since it does not

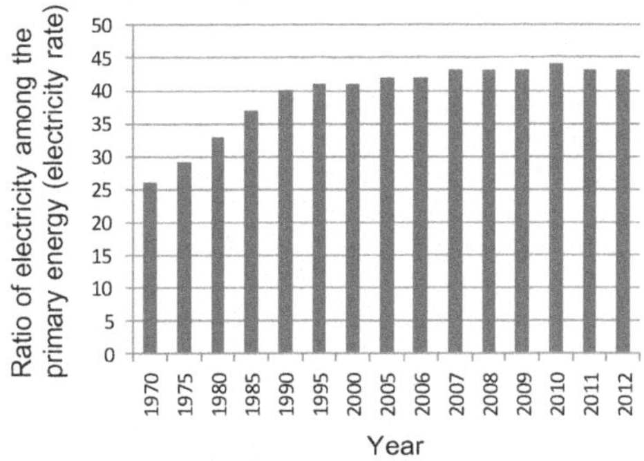

Fig. 6 Annual change of the ratio to be used as electric energy in primary energy to be supplied [21]

exist in itself but is provided by energy conversion from chemical energy, potential energy, and nuclear energy or so.

However, electrical energy can be converted directly into a lot of kinds of physics quantity that is necessary for the human life such as force, velocity, position, light, heat, temperature, information, and a new application to organic or inorganic matter and metal refinement. The electrical energy has superior characteristics to controllability, responsiveness, safety, cleanness, reversibility and, besides, can support the variety of resources as the renewable energy. In this way, the ratio to be used as electric energy increases year by year like Fig. 6 because the electric energy is superior and useful to the other energy forms [5, 13].

In late year, electrical energy seems to have two trends. The one is the moving revolution described as so far, and the other is the spread of renewable energy and dispersion power systems.

As for the renewable energy such as a wind generation or a photovoltaic power generation, the natural energy occurring in the natural world converts electrical energy. So the generated electrical energy changes every time, and the voltage and the frequency generated by renewable energy are not always constant. On the other hand, in order that the renewable energy is optionally used in the other fields which need electrical energy, it connects to the common electric power system, where the voltage, the frequency, and the phase are always strictly managed to be constant. Therefore, it is necessary to convert electric energy in high reply and high efficiency in the meantime, and a power electronics technology is used there.

The large and public electric power system was mainly used to connect the generation to the load (consumption) until now. However, when the generated or storage electric energy can convert any voltage, any frequency, and any phase freely and in low cost by means of power electronics technology, a distributed power supply system like Fig. 7, which is not always connected to the large electrical power system, can be realized in any electrical power range as from mW degree to GW or higher degree.

The distributed power supply system in the small power supply capacity is a cell phone, a smartphone class of the mW degree, and the next larger one is a personal computer (PC) class of W degree and the next one is an electric car, or a home

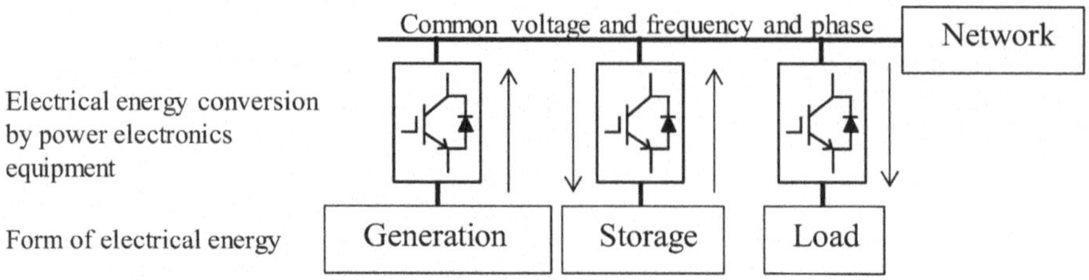

Fig. 7 One of the basic structure of distributed power supply system

electronics class of the kW degree and the next is a factory, or a hospital class of the MW degree, and the next is a country level of GW degree. It is comprised of three elements as the generation, the accumulation, and the load. The generation is considered to include receiving electric power from the electric power system basically. Since their voltage and frequency are not always the same and they fluctuate in terms of time, the electricity conversion technology by use of the power electronics technology is extremely useful to construct a common electric system.

The use of electrical energy, which is superior in availability stated in the above, spreads rapidly from mW class to GW class by electric energy conversion technology among the generation, the accumulation, and the load.

Such a distributed power supply system varies in a spread of techniques level depending on the power supply capacity like Fig. 8. A cell phone and smartphone classes have already spread widely and go to the further advance of downsizing technology now. Most of PC classes spread with notebook type, too.

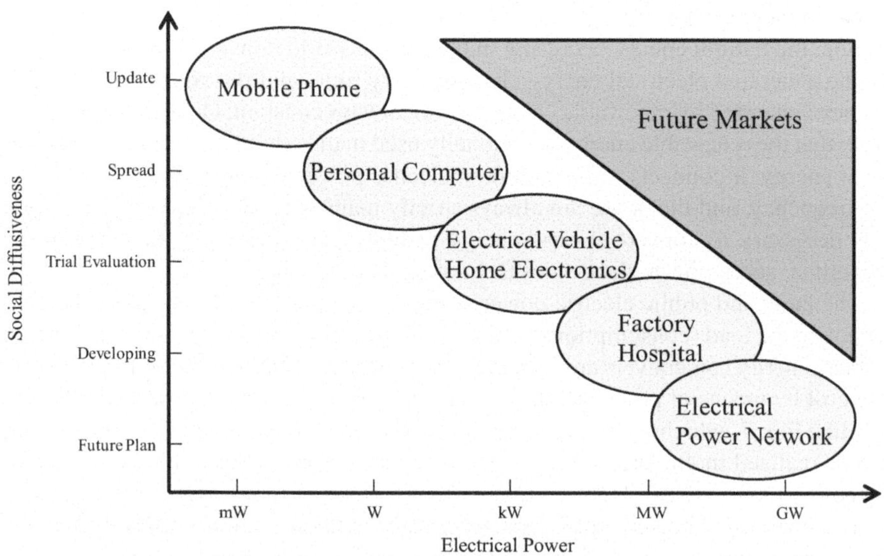

Fig. 8 Electric capacity and social spread for distributed power supply system

The electric car and the home electronics class are partially in use as HV (hybrid vehicle), but most of them are in a trial manufacturing level generally, and the full-scale spread has just begun. The factory, the hospital class, and the country level as GW class are now in developing or a design level for the future.

In other words, it may be said that the application development of the power electronics technology largely increases in considering the moving revolution with the electrical motor and the application to electrical network system business.

For example, the 30% of the electric energy is said to be used through power electronics technology in the USA in 2005, but a story that the ratio reaches the 80% in 2030 is discussed [14]. Therefore, a technical problem in the power electronics technology spread is considered here.

5 High-Frequency Requirement and Magnetic Material in Power Electronics

The boost chopper used in electric cars like Fig. 9 is considered here as an applied example of the power electronics. When the power semiconductor in the electrical circuit of Fig. 9a is operated as ON/OFF in the condition that switching period $T_0 = T_{ON} + T_{OFF} = 1/f_c$ (f_c: switching frequency) where T_{ON} is ON time of the semiconductor and T_{OFF} is OFF time, time series of electrical current in the reactor i_L and its terminal voltage v_L is shown as Fig. 9b.

When the semiconductor is in ON time, the reactive coil voltage equals to the battery power supply voltage as $v_L = E_d$ and then the reactor current i_L increases.

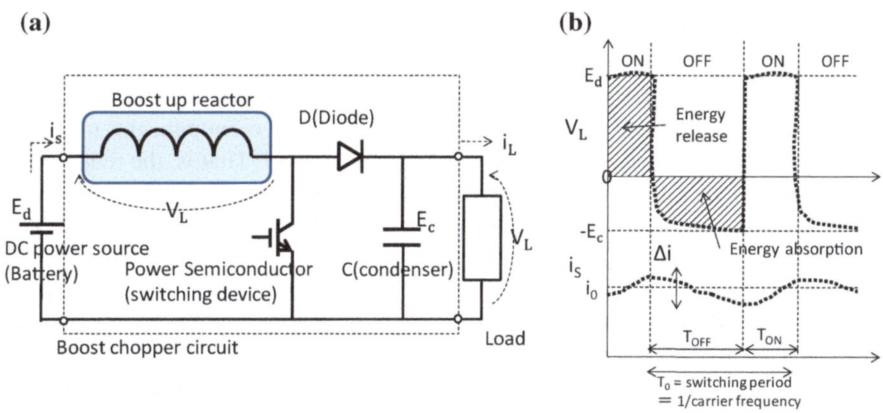

Fig. 9 Boost chopper circuit (**a**) [22] and time series of voltage and current in reactor coil (**b**). **a** Adapted from [22] with permission ©2013 The Magnetics Society of Japan

When it is assumed that the increase in the reactive coil electric current in ON time interval is Δi_L and a switching period is small enough, the next equation is introduced from $v_L = L\frac{di_L}{dt}$.

$$E_d T_{ON} = L \Delta i_L \quad (5)$$

When the switching operation of the semiconductor is assumed to be a quasi-stationary state, the reactor electric current increases in ON time and it decreases in OFF time, and the one at the first time of the current increase must be the same as the one at the last time of the current decrease. Therefore, the reactive coil voltage in OFF time became the minus number, and if its value is assigned as $-V_c$, the next equation is introduced.

$$-V_c T_{OFF} = -L \Delta i_L \quad (6)$$

Then, the equation $E_d T_{ON} = V_c T_{OFF}$ is introduced, and then, the next equation is introduced.

$$E_d i_0 T_{ON} = V_c i_0 T_{OFF} \quad (7)$$

Here, i_0 assumes to be an average electric current flowing of the reactor coil in ON/OFF operation constantly. The left side of Eq. (7) is energy of the reactive coil in ON time, and it may be said to be magnetic energy of the reactor coil which is transmitted from the battery to the reactor coil. In addition, the right side of Eq. (7) is energy of the reactor coil in OFF time, and it may be said to be magnetic energy which is transmitted from the reactor coil to the load. In other words, the reactor coil may be said to save the electric energy of the battery temporarily during a period of ON time and to emit the energy in OFF time through the switching operation of the power semiconductor.

As T_{ON} which is a time to save the energy becomes small so that the switching period T_0 is small and the switching frequency f_c is large enough, the magnetic energy to be accumulated in the reactor coil becomes small. That is, the reactor coil can become small. The reactor coil using the ferromagnetic comes to have a big energy density, but there is the upper limit in an energy density in the reactor coil since saturated magnetization exists.

In other words, a higher carrier frequency operation of the power semiconductor downsizes the shape of the reactor coil. As the reactor coil accounts for a certain ratio in an electrical circuit, the downsizing of the reactor coil shape by a high-frequency operation is related to the downsizing of circuit itself. As the downsizing demand for driving means in the moving body which includes an electric car is quite big, the demand to the high-frequency switching operation of the power semiconductor increases. This means that the high-frequency operation is required for the magnetic material which is the main materials of the reactor coil. The problem here is whether the magnetic material itself follows the research and development trend for the high-frequency operation.

Cost-effectiveness of the power semiconductor element, which was expensive conventionally, improves like Moore's Law by the recent progress of the power electronics technology year by year. In contrast, it is hard to say that a magnetic material necessarily follows it. Figure 10 shows a cost analysis of each element for 20 kVA UPS (uninterruptive power supply) [15]. The ratio of power semiconductor which would account for a considerable ratio conventionally is shown to fall down to 18%. A capacitor to accumulate electricity and the cooling element are less than 10%. The most important issue to be paid attention to in Fig. 10 is that the 44% of the 20 kVA UPS cost is occupied by inductors for the filter and it is a big ratio in that. In other words, it should be understood that a magnetic material accounts for most of the cost.

The weight ratio of the magnetic material in an electrical conversion system as the inverter is shown in Fig. 11a as an example [16]. In the large-capacity inverter of 1.5 MVA, a magnetic material accounts for a big ratio as more than 30% of the weigh. Figure 11b shows the efficiency in a vertical axis and the weight of the power supply per 1 MVA unit in a horizontal axis [14]. As the operating frequency becomes large from 20 kHz or less to 40 kHz or more, the power supply equipment

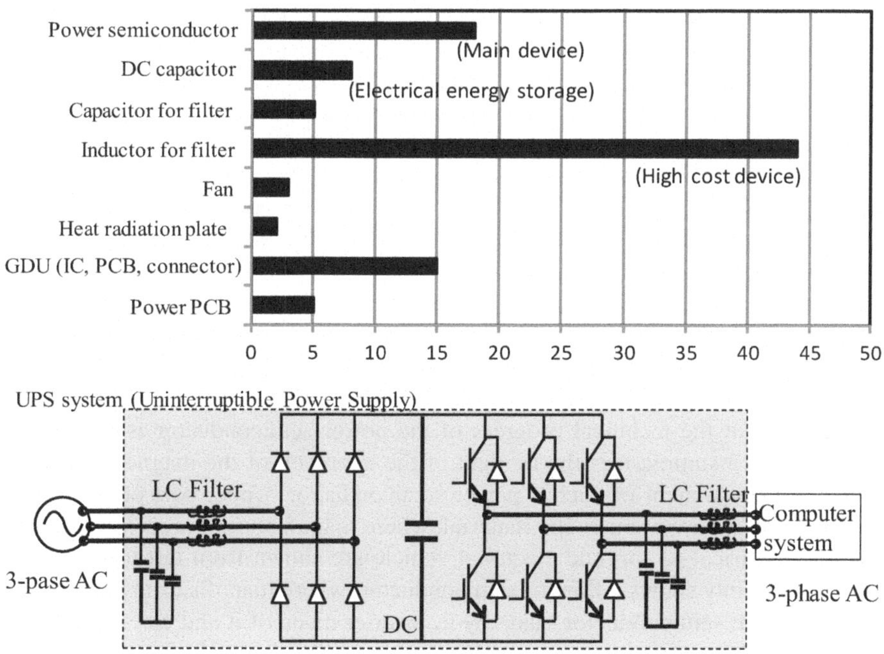

Fig. 10 Cost analysis of 20 kVA UPMS (upper) and its considerable electrical circuit system (lower) [7, 8, 18, 20, 23]. Adapted from [18, 20] with permission ©2015, 2017 The Magnetics Society of Japan

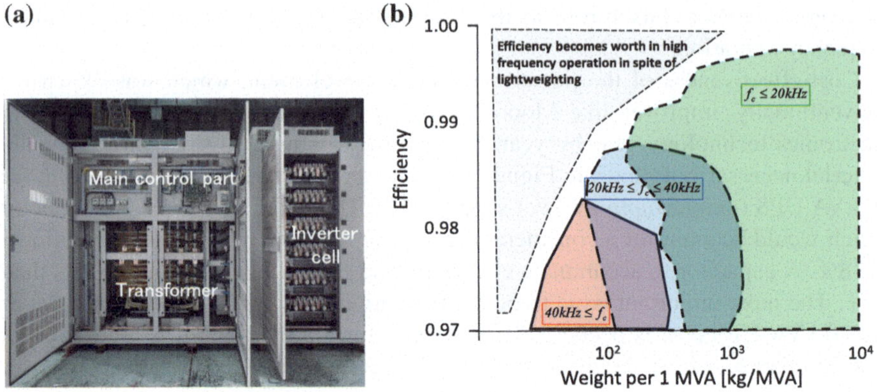

Fig. 11 Weight ratio of magnetic material in power electronics equipment (**a**) [16], and downsizing and low efficiency in high-frequency operation (**b**) [14]. **a** Image courtesy of Fuji Electric Co., Ltd

size becomes small, but the efficiency decreases. It may be said that the loss of the magnetic material in the high-frequency domain is one of the problems.

The power semiconductor realizes an ability of low price and high performance by constantly advancing progress, and as a result, the operating frequency rises, and the downsizing of the equipment is carried out. However, the problem from the point of view of a price, a volume, and a loss is found out to be a magnetic material. In other words, the appearance of a magnetic material, which can tolerate in high-frequency operation and has a high saturated magnetization and is low-priced and low loss, is being strongly expected nowadays.

Si is used as power semiconductor materials now, but a new power semiconductor material as GaN or SiC which is superior to Si materials like Fig. 12 at the point of breakdown voltage, electrical conductivity, and high energy band is studied energetically in domestic and abroad [12]. This research trend will introduce a realization of a new power electronics device in high-voltage and high-frequency operation. Therefore, an appearance of new magnetic material in high frequency, high saturated magnetization, low price, and low loss is required more and more.

Now the reason why the research and development of the magnetic material does not advance for the technical progress of the power semiconductor is considered here. Figure 13 summarizes the thought of the approach of the magnetic material for the new movement of such a power semiconductor. Upper part of this figure mainly shows a flow of magnetic material, where manufacturing process, magnetic material, electrical motor, and electrical vehicle are shown from left to right. The lower part mainly shows a flow of semiconductor, where manufacturing process of semiconductor, semiconductor made by it, inverter made of it and electrical power network are shown from left to right. Final user of electrical vehicle and electrical power network is a human being.

It is a motor drive system that the upper flow is connected to the lower flow mutually. To make a motor drive system, magnetic material and semiconductor are

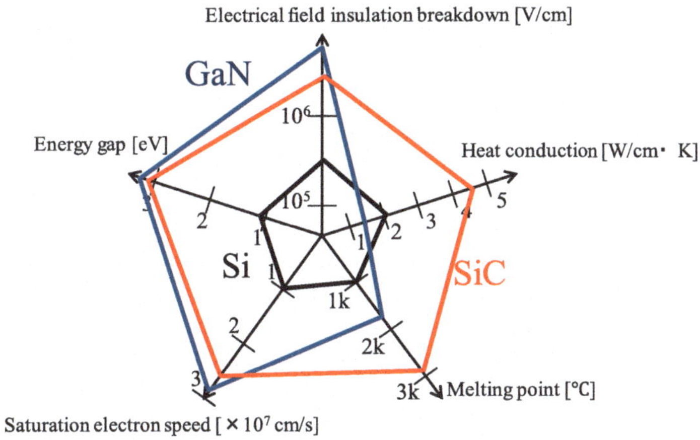

Fig. 12 Material constants comparison of new power semiconductor SiC and GaN [21, 22]. Adapted from [22] with permission ©2013 The Magnetics Society of Japan

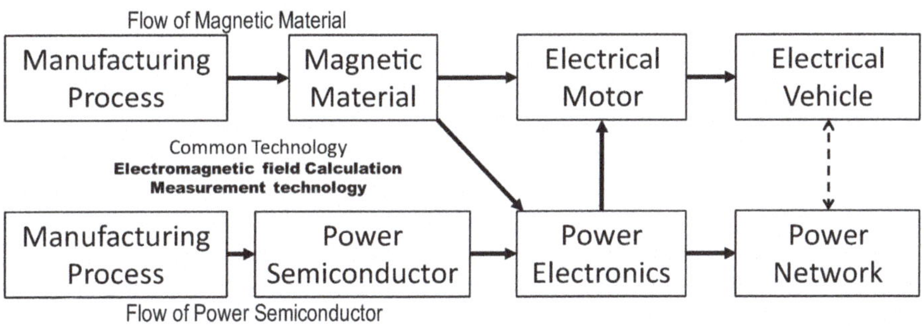

Fig. 13 Relation between magnetic material flow and semiconductor flow [7–9, 11, 18–21, 23–25]. Adapted from [11, 18–20] with permission ©2015–2017 The Magnetics Society of Japan

strongly combined, but in just looking at their front and back parts mutually their flows seem to have no relation to each other. However, the appearance of the power semiconductor materials for high efficiency as GaN and SiC requires a strong change for a magnetic material. In other words, it may be said that just looking at one's technical domain and its front and back is insufficient.

In considering each element technology of the two flows as the magnetic material and the semiconductor in Fig. 13, the different academic departments and subjects such as physics, metallurgy, materials, chemical engineering, electrical and electronics engineering, and a mechanical enginieering are found out to be constructed there [12, 13]. It is strongly required to understand and to grasp all these subjects in order to consider their relations with the magnetic material and the semiconductor. The need for electromagnetic field fusion science and technology comes out here.

6 Magnetic Material for Electrical Energy Application

The use of the magnetic material can mainly be divided into two systems. The one is an information, communication system such as hard disks, and the other is an electrical energy system such as the motor, the transformer, and the reactor. A hard magnetic material and a soft magnetic material are used in the electrical energy system to take up here mainly.

A permanent magnet of hard magnetic material is widely used for the electrical motor by an appearance of NdFeB which has a big residual magnetic flux density B_r and is comparatively easy to obtain the raw materials. "The field magnet" giving constant magnetic flux is necessary for a synchronization motor and a DC (direct current) motor, and an electromagnet has been used until now for it. However, a conventional electromagnet type has a problem of efficiency, maintenance, and downsizing, as it was necessary to be connected to the external electrical power source and to generate the copper loss due to the current flow. The permanent magnet can solve the problem, but some problems are also observed. Speed control in the high velocity is difficult due to a high electromotive force, and drug force loss generates in the time of the coasting driving.

On the other hand, the soft magnetic material is widely used because large magnetic flux density (B) is provided by small external magnetic field (H_{ext}) due to the magnetization of magnetic material (M) as the next equation.

$$\vec{B} = \mu_0(\vec{H} + \vec{M}) \tag{8}$$

The large magnetic flux density derived by the magnetization is used for three kinds of electrical equipment classified by the main electromagnetic principle as follows.

The first one is a transformer using electromagnetic induction theory based on the next equation.

$$\text{rot}\vec{E} = -\frac{\partial \vec{B}}{\partial t} \tag{9}$$

Electromotive force generates in the terminal of the secondary electromagnetic coil which is crossed by the magnetic flux generated in primary coil, and the current flows by connecting the terminal to a load. The secondary terminal voltage can be changed by changing the number of the winding of the primary coil and the secondary coil, and it may be said that the transformer is a conversion equipment of the electrical energy in quasi-static condition, not dynamics.

The second one is an electrical motor that uses an electromagnetic force based on Maxwell stress expressed by the next equation.

$$\vec{F} = \oiint_{\text{Thin Steel Plate}} [T_{\text{m}}] \cdot \vec{n} dS \tag{10}$$

$$[T_{\mathrm{m}}] = \frac{1}{\mu_0}\begin{bmatrix} \frac{1}{2}(B_x^2 - B_y^2 - B_z^2) & B_x B_y & B_x B_z \\ B_y B_x & \frac{1}{2}(B_y^2 - B_z^2 - B_x^2) & B_y B_z \\ B_z B_x & B_z B_y & \frac{1}{2}(B_z^2 - B_x^2 - B_y^2) \end{bmatrix} \quad (11)$$

An electromagnetic force is introduced to act to an object (a rotor of electrical motor) due to a surface integration of Eq. (10) based on Maxwell tensor of Eq. (11) on the closed plane around the object. Maxell tensor of Eq. (11) shows that the electromagnetic force per unit surface is approximately proportion to square of magnetic flux density. A soft magnetic material is used to get big magnetic flux density. However, the electromagnetic force per unit surface has an upper limitation because of the saturated magnetization of magnetic material due to Eq. (11). When rotational speed of electrical motor increases due to an acceleration of the electromagnetic force, speed electromotive force generates and the terminal voltage of the electromagnetic coil increases because the magnetic flux with rotational speed crosses the coil. So in high-speed operation, the supplied voltage should increase. This electrical machine becomes a motor since the energy flows from electrical power source to the machine when supplied voltage of the power source is larger than the speed electromotive force, but it becomes a generator when the speed electromotive force is larger than the supplied voltage.

The magnetic energy is defined as the next equation, and it may be said that it is a reactor to use an accumulation of the magnetic energy.

$$E = \int \vec{H} \cdot d\vec{B} \quad (12)$$

The reactor is used as an element of the electric circuit. An electric current flows in supplying a voltage and magnetic field generates due to Ampere's law and big magnetic flux density occurs due to the magnetization of magnetic material, and then, magnetic energy occurs in the reactor. The magnetic energy flows from the one element in the electrical circuit and is saved in some interval of one period. In the other interval, the saved energy is released to the other element of the circuit.

7 Future Research of Electrical Motor

The motor drive system realizing the moving method in high efficiency, high reply is used in not only each industry but also each home or each car, and it is thought with "rice (bread)" of industry and/or society now due to the cheapness and the high performance. It may be said to be greatly important for a future technical trend to think about the research directionality of electrical motor when the motor application progresses more and more.

As most of variable velocity motors are controlled by power electronics, plural motors are possible to be unified, and then, they are highly organized and integrated.

Since electric is used for both electrical motor and information technology (IT), the affinity of the electrical motor with the IT is high. The motor drive system can be controlled through an IT network when the control part of the power electronics circuit is connected to the network. Then, a huge system integration of motor drive system is enabled. In other words, when the future of the motor is considered, the one directionality of electrical motor is considered to be a system integration of motor drive system which should be called a motor integration based on mathematics studies.

At the same time, the demand for high efficiency, downsizing, and high responsivity for the electrical motor increases more and more due to the on-machine operation of the electrical motor drive system. For its realization for further efficiency and downsizing, it will be insufficient only in treating the optimization of the motor shape and the control characteristics anymore. Then, the research and development that sailed up the manufacturing method of the motor material and an atom and/or a spin level are expected. In other words, the other directionality of electrical motor is considered to be a physics integration of motor drive system as multi-physics and multi-scale. The physical integration treats wide range of physical phenomena from spintronics or atomic behavior to electrical vehicle application and is synthesized in the reconstruction to obtain a high-performance motor drive system. It should be called motor physics based on physical studies.

When both directions change a viewpoint like Fig. 14, two directions as the horizontal axis of Fig. 14 are considered to be macroscopic and microscopic, and there is an electrical motor size in the center. The macroscopic is to argue with km degree size or more such as a society system connecting some motors by network, and the microscopic is to treat a motor of the cm degree size back to micro- or nanometer

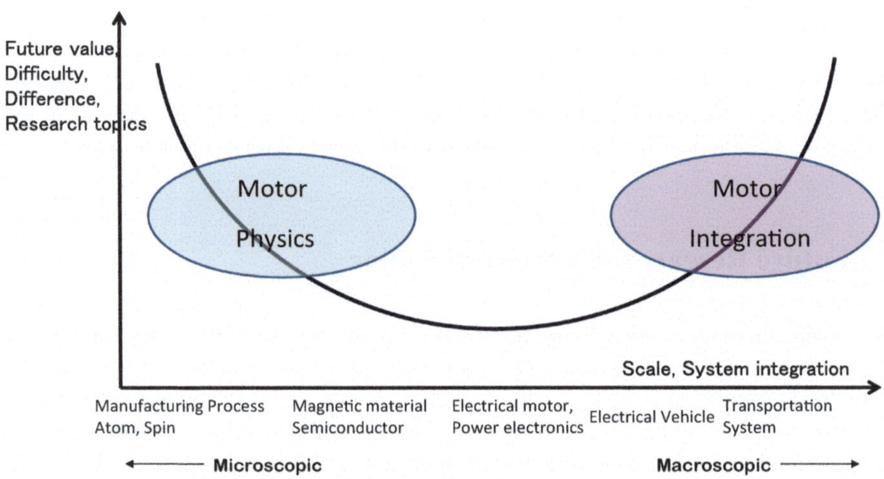

Fig. 14 Future direction of electrical motor research

size. The expansion of the new application of the means such as transportation needs the new approach in the research field.

References

1. AMANOGAWA by BAN PR inc. http://osaka-mizubebar2012.seesaa.net/category/14537690-3.html. Accessed 5 Dec 2017
2. Hybrid Locomotive by Japan Freight Railway Company. http://www.jrfreight.co.jp/transport/improvement/development.html. Accessed 5 Dec 2017
3. 787 Airplane by All Nippon Airways Co., Ltd. https://www.ana.co.jp/ja/jp/promotion/b787_9/. Accessed 5 Dec 2017
4. NASA N3-X with Turboelectric Distributed Propulsion. https://ntrs.nasa.gov/archive/nasa/casi.ntrs.nasa.gov/20150002081.pdf. Accessed 5 Dec 2017
5. Pamphlet of Ministry of Economy, Trade and Industry. http://www.meti.go.jp/committee/summary/0001640/pdf/059_h02_00.pdf. Accessed 5 Dec 2017
6. Fujisaki, K.: Future magnetic material property installed in and driven by power electronics technology. In: IUMRS-ICAM 2015 (14th International Union of Materials Research Societies—International Conference on Advanced Materials), IV-1Th3F1-1 (IS), Korea, 29 Oct 2015
7. Fujisaki, K.: Necessity and problem of magnetic material for electromagnetic actuator system. In: The 2015 Annual Meeting of the Institute of Electrical Engineering of Japan, 5-S22-1, Mar 2015, pp. S22(1)–S22(4) (in Japanese)
8. Investigating R&D Committee of Magnetic Material for Electromagnetic Actuator: Magnetic material and its evaluation technology for electromagnetic actuator. Technical report of IEEJ, No. 1397 (2017) (in Japanese)
9. Steinmetz, C.P.: Theoretical investigation of some oscillations of extremely high potential in alternating high potential transmissions. Trans. Am. Inst. Electr. Eng. **XVIII**, 383–405 (1901)
10. Fujisaki, K.: Required magnetic property for magnetic material in use of future electrical energy. In: The Japan Association of Bonded Magnetic Materials, No. 86, JABM Autumn Symposium, Tokyo, Sept 2014 (in Japanese)
11. Fujisaki, K.: Permanent magnets and their applications, VI cutting edge of permanent magnets. Magn. Jpn. **11**(1), 34–41 (2016) (in Japanese)
12. Fujisaki, K.: Motor core property with high performance magnetic material and new trend of electrical engineering. In: The Japan Association of Bonded Magnetic Materials, No. 90, JABM Autumn Symposium, Tokyo, Sept 2016 (in Japanese)
13. Figure Set of Energy. http://www.fepc.or.jp/library/pamphlet/zumenshu/pdf/all01.pdf. Accessed 5 Dec 2017
14. Heidel, T.: ARPA-E initiatives in high efficiency power conversion. In: APEC (Applied Power Electronics Conference and Exposition) 2014, Plenary Session Presentations, Fort Worth (2014). http://www.apec-conf.org/Portals/0/Plenary%20Presentations/Speaker%204%20Heidel%20PRES.pdf. Accessed 5 Dec 2017
15. Kolar, J.W., Krismer, F., Nee, H.P.: What are the "Big CHALLENGES" in power electronics? Presentation for the 8th International Conference of Integrated Power Electronics Systems (CIPS 2014), Nuremberg, Germany, 25–27 Feb 2014. https://www.pes-publications.ee.ethz.ch/uploads/tx_ethpublications/CIPS_2014_Kolar_Challenges_Power_Electronics_Video.pdf. Accessed 5 Jan 2018
16. Catalogue of Fuji Electric: Fuji Medium-voltage IGBT Inverters FRENIC4600FM5e, 2014-9(2014a/G2007)/KO-D/CTP5EP
17. Fujisaki, K.: In: Annual Meeting of IEEJ, 5-s13-1, 2-s2-4 (2017) (in Japanese)
18. Fujisaki, K.: Technical report of 202, Magnetics Society of Japan (2015) (in Japanese)
19. Fujisaki, K.: In: 9aA-3, Symposium of Magnetics Society of Japan (2015) (in Japanese)

20. Fujisaki, K.: Technical report of 213, Magnetics Society of Japan (2017) (in Japanese)
21. Fujisaki, K.: Technical report of IEEJ, MAG-13-149 (2013) (in Japanese)
22. Fujisaki, K.: In: 6pB-2, Symposium of Magnetics Society of Japan (2013) (in Japanese)
23. Fujisaki, K.: Monthly EMC, No. 349, pp. 61–64 (2017) (in Japanese)
24. Fujisaki, K.: Power Electron. **42**, JIPS-42-01 (2017) (in Japanese)
25. Fujisaki, K.: Mater. Stage **16**(4), 12–13 (2016) (in Japanese)

Fundamental Concept of Magnetic Material for Electrical Engineer

Keisuke Fujisaki

Abstract When external magnetic field is applied to magnetic material, a magnetization phenomenon is caused and big magnetic flux density can be obtained. The big magnetic flux density plays an extremely important role in a motor, a generator, a transformer, and a reactor which are the applied apparatus of the magnetic material in electrical engineering. Therefore, a magnetization phenomenon to obtain the big magnetic flux density and an iron loss produced in magnetization process in AC supplied are important material indexes for magnetic material. Here, an outline about the magnetization process and the iron loss is mainly described on an electrical steel sheet.

Keywords Magnetic domain · Magnetization · Recrystallization · Iron loss · Hysteresis curve · Mechanical stress · Magnetic anisotropy · Two-dimensional magnetic property · Power magnetics

1 Multi-scale of Magnetic Material

The magnetic material causes different physical phenomena governed by different equations which depend on the size to deal with.

A motor drive system is used in a drive source of an electrical vehicle which is a meter (m) size. As a general rule, the driving properties of the electrical vehicle are mainly influenced by Newton dynamics. Since soft magnetic material can get big magnetic flux density by a small electrical current to obtain a big electromagnetic torque, it is used in an electrical motor which has a centimeter (cm) size.

The electromagnetic torque generated in the electrical motor is governed by electromagnetic equations of Maxwell. The electrical steel used as soft magnetic material includes several percent of Si to increase electrical resistance and to reduce eddy current loss. It also has a texture of the polycrystalline having a lot of crystal grains whose size is from some micrometers (μm) to several millimeters (mm). Each crystal grain has several magnetic domain structures of the micrometer (μm) size that is smaller

K. Fujisaki (✉)
Toyota Technological Institute, Nagoya, Japan
e-mail: fujisaki@toyota-ti.ac.jp

K. Fujisaki (ed.), *Magnetic Material for Motor Drive Systems*,
Engineering Materials, https://doi.org/10.1007/978-981-32-9906-1_3

than the grain size. Magnetization process advances due to an external magnetic field, so that magnetic wall between the magnetic domain structures moves and then magnetization rotates.

The magnetic domain structure is governed by Landau–Lifshits–Gilbert (LLG) equation [1, 2]. All the atomic size within the magnetic domain structure has a saturated magnetization vector, which is generated by an electronic spin of the nanometer (nm) size. The behavior of the spin is governed by the Schrodinger equation of the quantum theory.

For magnetic characteristic comprehension and improvement, it is necessary to understand the magnetic phenomenon and the governed equation on each scale. The individual detailed contents will be described in Part III. Here, only their general characteristics are shown for the help of the magnetic phenomenon understanding.

1.1 Appearance of Magnetization

Appearance of magnetization is caused by an electronic spin, and it is governed by a quantum theory of many bodies. Magnetization is derived from each atomic structure or molecular structure, and saturated magnetization exists as material constants.

Saturated magnetization is different depending on atomic number and the alloy, and the saturated magnetization is provided as Slater–Pauling curve experimentally [3, 4].

In the normal temperature and the ordinary pressure, Fe, Ni, Co, and their alloy are ferromagnetic with big magnetization. As iron has relatively big saturated magnetization and is manufactured cheaply among them, it is usually used as a core of a motor or a transformer.

Electrical steel has a body-centered cubic (bcc) structure at the normal temperature, and its magnetic characteristic varies according to the crystal orientation like Fig. 1. (100) direction becomes the easy magnetization axis and has the best magnetic characteristic [5]. (110) direction has the second, and (111) direction has the worst one.

It is a grain-oriented (GO) steel material that prepares the crystal orientation of steel so that easy magnetization direction (100) is in rolling direction and steel surface is on [011] of GOSS side like Fig. 2 and then has a strong magnetic anisotropy. GO steel is superior magnetic materials, if all the magnetic flux in the GO material flows to the easy magnetization direction [6]. The crystal grain size of GO materials is around some tens mm or so, and it accumulates crystal orientation of easy magnetization direction where most of the crystal grains orientation directs to easy magnetization direction within around several degrees.

On the contrary, in case of the polycrystalline in which the crystallographic orientation is arranged at random, its magnetic characteristic has a weak anisotropy and then isotropic. It is called non-oriented (NO) steel) materials and is used for an electrical motor [7].

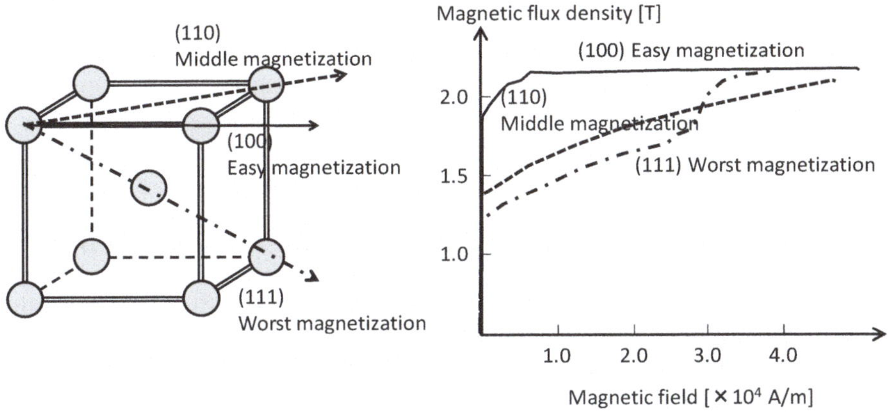

Fig. 1 bcc-structure of iron and its crystal orientation (left) and its magnetization (right)

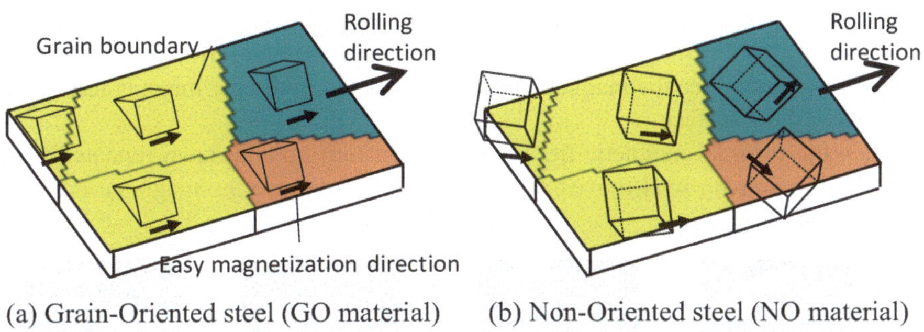

Fig. 2 Electrical steel with texture in some orientations of crystals

1.2 Magnetic Structure

Weiss is said to show that saturated magnetization of atomic level appears in overcoming heat disturbance, and Landau–Lifshits are said to study magnetic domain structure theoretically [1]. There are several magnetic domain structures with the same saturated magnetization vector in magnetic material, and it has the magnetic domain structure whose direction and shape are decided so that total magnetic energy is minimized.

The boundary line or surface where the direction of the saturated magnetization vector changes is called as magnetic wall in a crystal grain and the interval between one magnetic domain and the next magnetic domain is called as magnetic domain width. Magnetic domain structure is proofed to exist as Bitter observed the magnetic domain. Magnetic energy is comprised of static magnetic energy, anisotropic energy, exchange energy, demagnetizing energy, external stress energy, and magnetostriction energy.

The minimized condition of magnetic energy introduces the LLG equation shown in the next equation, which is the relational expression between magnetic field vector $\vec{H}$ and saturated magnetization vector $\vec{M}$ [1, 2].

$$\frac{d\vec{M}}{dx} = -\gamma \vec{M} \times \vec{H} + \frac{\alpha}{M_s} \vec{M} \times \frac{d\vec{M}}{dt} \quad (1)$$

Here, γ is the electronic gyromagnetic ratio, effective magnetic field H is the magnetic field that added the revision of quantum mechanics to an external magnetic field and an internal magnetic field, α is the fixed number of no dimension quantity called the Gilbert constant number, and M_s is the saturation magnetization.

When the external magnetic field excites the ferromagnetic, the domain wall moves, and the magnetic pole appears, and the magnetization phenomenon emerges in macro scale. Figure 3 shows the calculation result of the magnetic domain structure based on LLG equation in a certain moment of the ferromagnetic. In red part, a saturated magnetization vector directs to upper direction, and in blue part, a saturation magnetization vector directs to the lower direction. The arrow-formed magnetic domain called Lancet magnetic domain often exists in the real ferromagnetic as GO steel, in order to reduce the magnetic domain width by producing the magnetic pole on the steel sheet surface.

When the external magnetic field is excited in turn up and down repeatedly as AC field, the domain wall moves as shown in Fig. 3, where the magnetic domain

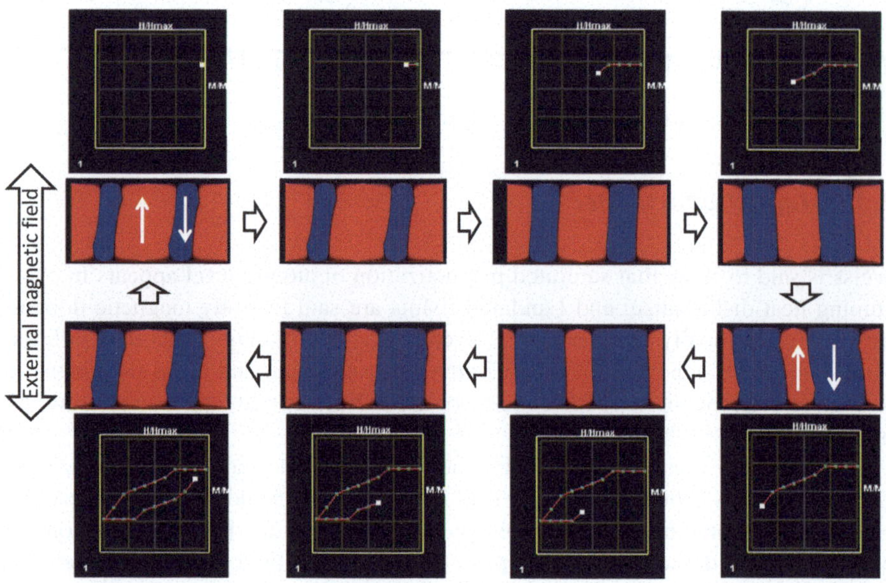

Fig. 3 Magnetic domain structure calculated by LLG equation (external magnetic field is excited in up and down directions)

width lengthens and shrinks. In other words, external magnetic field introduces different situations: the one is that a blue downward saturated magnetization vector becomes dominant and the other is that a red ascending saturation magnetization vector becomes dominant. These magnetic phenomena show in micro, and a magnetization vector appears in the macroscopic view by averaging the microphenomena.

The magnetization properties introduced by averaging the calculated magnetic domains (*M*–*H* curve) are also shown in up or down part of the magnetic domain structure in each time. The situation that domain wall moves by external magnetic field and a magnetization process is going ahead is expressed in this numerical analysis results.

1.3 Polycrystalline Body

Electrical steel has a texture structure of the polycrystalline body. GO steel has crystal grains the size of which is around some tens of mm as Fig. 4a and NO steel has crystal grains the size of which is around some tens of μm as Fig. 4b.

The GO materials become the materials which integrate the degree of the GOSS direction within some degrees, and magnetic anisotropy is strong. It is mainly used as a core of transformer. It is divided and arranged so as that the main magnetic flux in the transformer moves to its easy magnetization direction in order to realize low iron loss.

In contrast, NO steel prevents for the direction of each crystal grain to direct to a certain specific direction, and then NO steel has magnetism isotropy in which the magnetism characteristic of any direction has almost the same magnetism characteristic. Since magnetic flux vector directs to any direction on a steel core in the electrical motor depending on rotating of motor and excitation method of three-phase

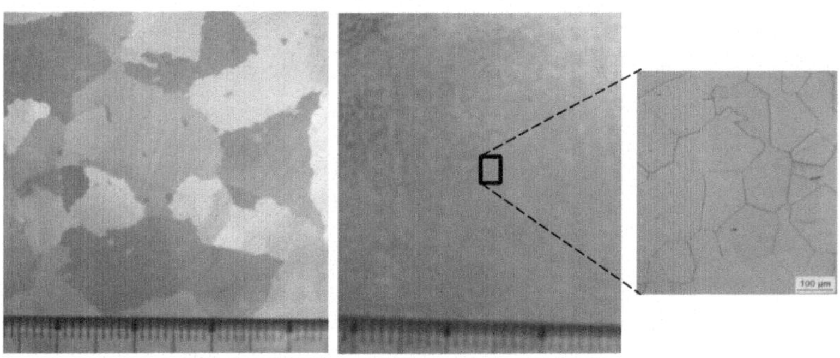

(a) GO steel (b) NO steel

Fig. 4 Polycrystalline body of GO steel (**a**) and NO steel (**b**) (after having torn off an insulation film) current, NO steel of electrical sheet with magnetic isotropy is easy to be used for electrical motor core because it is punched in one body and molded.

In the grain boundary that is the border between the crystal grains, magnetic domain structure changes as crystallographic orientation changes, and the magnetic pole is easy to appear and, therefore, the domain wall movement is disturbed. The grain boundary per unit surface area decreases as a crystal grain is big, but, on the other hand, the magnetic domain width becomes large and the outbreak of the eddy current by the domain wall movement (abnormal eddy current) increases. The size of the crystal grain will affect the magnetism properties of materials in this way.

1.4 Crystal Grain Control

Crystal grain size can be controlled to some extent, because the crystal grain size depends on annealing time as shown in Fig. 5 [8]. Since the materials are usually rolled and are made plastic deformation, they have a lot of distortion energy by the lattice defect. If annealing time becomes long, a release phenomenon of the distortion energy is gradually caused, and a recovery phenomenon happens, and a crystal grain grows big little by little.

After the crystal grain does not include a defect in more than a certain time, it is formed to be stable and the other domain with defects is encroached. So, the primary recrystallization that the distortion energy with the defects becomes extinct at one sweep happens, and particle size grows big rapidly. In this way, the distortion energy becomes extinct, but the re-crystal grain has a relatively small size and grain boundary per unit becomes large.

Therefore, when annealing process continues in longer time, the crystal grain growth whose drive source is grain boundary energy, in other words, the normalcy crystal grain growth continuously happens. When some conditions are satisfied in

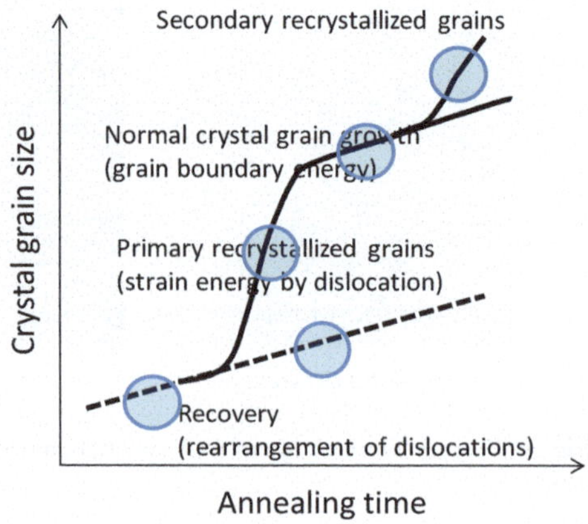

Fig. 5 Annealing time and crystal grain size [8]

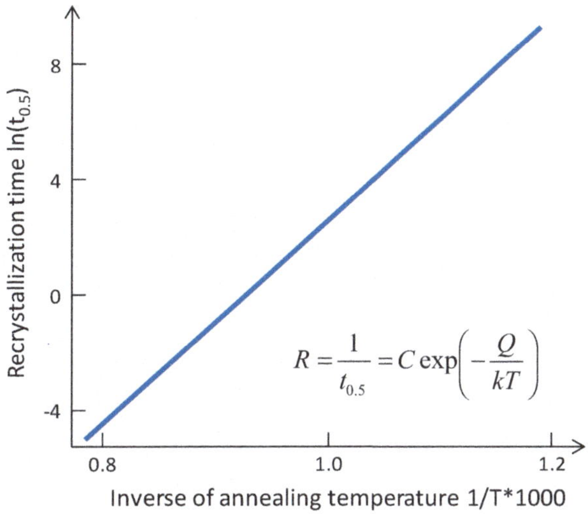

Fig. 6 Annealing temperature (T) and recrystallization time $t_{0.5}$ [8]

this process, the phenomenon that the specific few crystal grains in the re-crystal grain grow up more rapidly than the other crystal grains, that is, the second recrystallization happens.

The recrystallization time is expressed as Arrhenius equation as Fig. 6. It depends on the temperature at that time. In other words, recrystallization time shortens rapidly so that annealing temperature is high.

2 Magnetization Process

A magnetization phenomenon is considered here, when external magnetic field is applied to ferromagnetic. As shown in Fig. 7, the magnetization phenomenon progresses by domain wall movement at first so as that saturated magnetization vectors direct to the external magnetic field as much as possible. However, when domain wall movement is finished and the direction of the external magnetic field vector is different from the direction of the saturated magnetization vector derived from the crystallographic orientation, a magnetization rotation happens to rotate the magnetization vector so that the saturation magnetization vector agrees with the direction of the external magnetic field. Finally, the magnetization process is finished when the direction of the external magnetic field agrees with the direction of the saturated magnetization vector. As a magnetization rotation needs greater energy than domain wall movement, big external magnetic field is required in the last magnetization process.

The magnetization process of Fig. 7 is said to be an initial magnetization curve generally, but a magnetic hysteresis phenomenon like Fig. 8 comes out when external AC magnetic field is actually applied to ferromagnetic repeatedly.

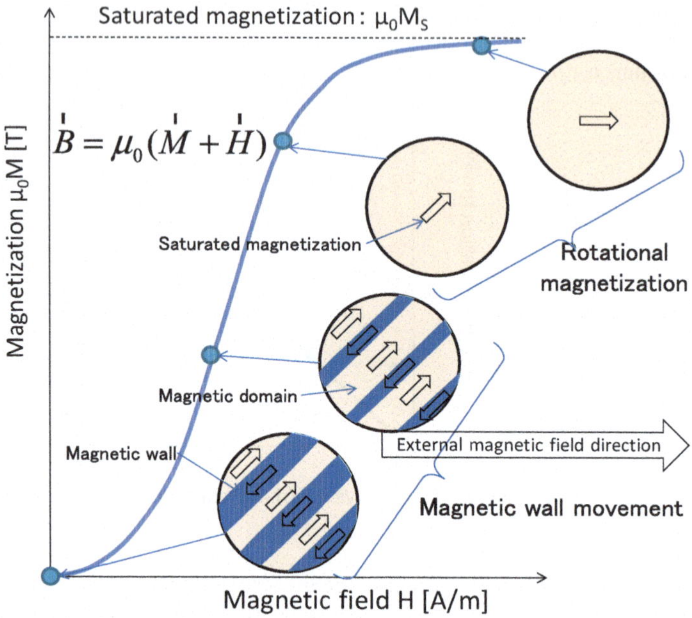

Fig. 7 Magnetic domain structure and magnetization process

Fig. 8 Magnetic hysteresis curve

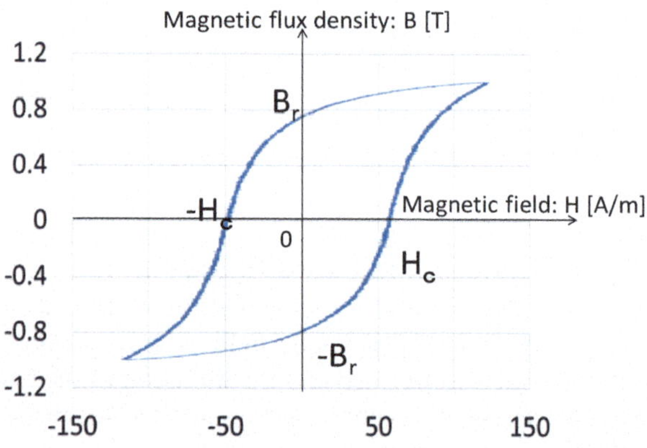

As the internal area of the magnetic hysteresis curve is a magnetic energy, an integrated value in one period defined as the next equation becomes a loss (heat generation) caused by the excitation of a steel sheet being applied to by external magnetic field and is called an iron loss.

$$W_{Fe} = \frac{f}{\rho} \int_{t}^{T+t} H dB \tag{2}$$

With a magnetic hysteresis curve of Fig. 8, the intercept with magnetic flux density B is called a residual magnetic flux density B_r, and the intercept with magnetic field H is called a coercive field H_c. When B_r and H_c are small, as the iron loss of one period excitation is small, it is called a soft magnetic material. It is used for a core material of a motor and a transformer and a reactor in order to obtain high magnetic flux density. In contrast, when B_r and H_c are big, as almost constant magnetic flux density of residual magnetic flux density B_r is obtained even if there is no external magnetic field and is small, it is called a hard magnetic material, in other words, a permanent magnet. It is used for a magnet field part of a synchronous motor or a direct current motor.

3 Iron Loss

The iron loss is the internal area of the magnetism hysteresis curve in Fig. 8, and it is generally divided into hysteresis loss and eddy current loss as Fig. 9.

The hysteresis loss occurs by direct current excitation and is derived from the obstruction of domain wall movement such as a pinning effect in a magnetization process. It is caused by the impurities or deposit within the magnetic material, or by the residual internal stress to remain in the magnetic material or by the existence of the crystal grain boundary where the crystallographic orientation varies.

The eddy current loss occurs only by AC excitation and is divided into classic eddy current loss and abnormality eddy current loss. The classical eddy current loss is a Joule loss derived from the electromagnetic phenomena that an induced electromotive force is introduced to disturb the time change of the AC magnetic flux to flow in a steel sheet, and, as a result, an eddy current occurs in a steel sheet. Therefore, it can be small when electric resistivity becomes big and the thickness of the steel sheet is thin. In contrast, the abnormal eddy current loss occurs by domain wall movement of the magnetic domain structure. As it might be said that domain wall movement excited by the AC external magnetic field is a change of the magnetic flux in the magnetic materials, induced electromotive force is introduced and then the eddy current occurs. It is also a Joule loss. Therefore, the abnormal eddy current comes to have a small one where magnetic domain width is narrow in addition to small electric resistivity.

Among three components of the iron loss, an influence of the hysteresis loss is big in the low frequency such as the commercial frequency, and an influence of the classical eddy current grows bigger when it becomes the kHz degree.

As the hysteresis loss occurs by direct current excitation, it is proportional to the frequency of the excited magnetic field, but the eddy current loss is almost proportional to the square of the frequency, as the eddy current itself is proportional to the frequency and Joule loss is proportional to the square of the eddy current. Therefore, the rough sketch of iron loss characteristics is shown in Fig. 10, when the value that divided iron loss by a frequency is taken in a vertical axis and a frequency is taken in horizontal axis. The intercept of this line is hysteresis loss and

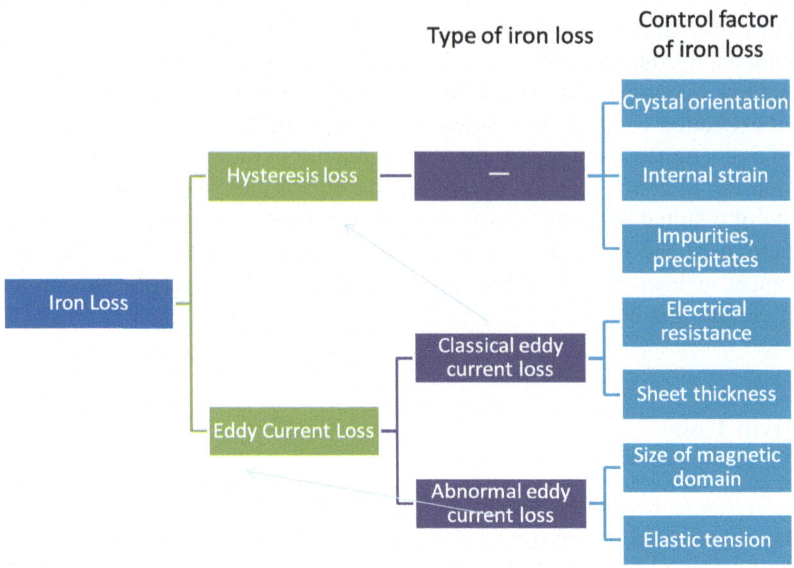

(a) Components of iron loss

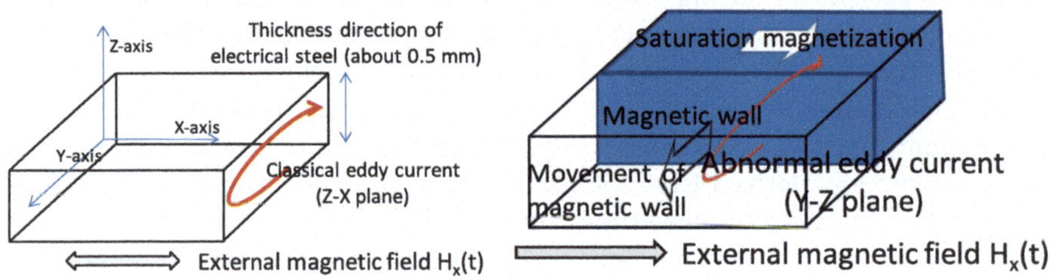

(b) Outbreak factor of iron loss as classical eddy current (left) and abnormal eddy current (right)

Fig. 9 Components of iron loss (upper) and its outbreak factor of iron loss as classical eddy current (lower left) and abnormal eddy current (lower right)

the vertical axis increment as much as frequencies increase is eddy current loss. The quantifiability of this tendency is established in the relatively near frequencies, but it is not in the large different frequency domain. Particularly, a special attention should be paid when an electromagnetic penetration depth begins to become smaller than steel sheet thickness.

It is a Steinmetz equation as shown in the next that formulated this qualitative discussion.

$$W_{\text{Fe}} = k_{\text{h}} f B^{\alpha} + k_{\text{e}} f^2 B^2 \tag{3}$$

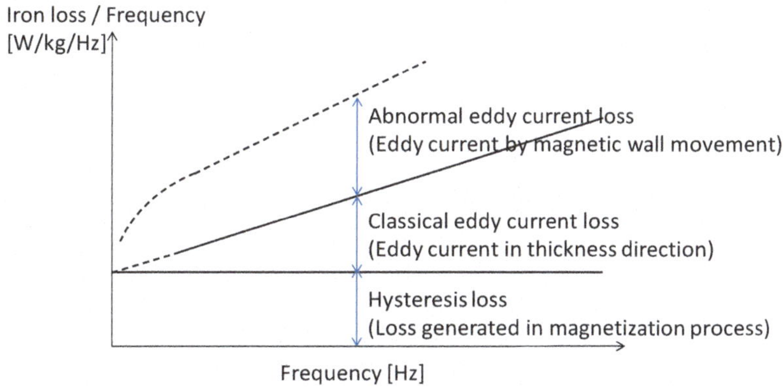

Fig. 10 Frequency characteristics of iron loss and its components [9]

Here, k_h, k_e, f, B is a hysteresis coefficient, an eddy current coefficient, an excitation frequency, maximum magnetic flux density in one period, respectively. In addition, α is a coefficient which takes about from 1.6 to 2. This equation was provided by Mr. Steinmetz from an experimental data approximately 100 years ago [9].

4 High-Frequency Magnetization

The conventional magnetic material has been used in a commercial frequency such as 50/60 Hz mainly, but the low iron loss magnetic material for the high frequency is expected when an operating frequency becomes high and an influence of the harmonic content comes out by the spread of power electronics technologies.

High-frequency operation requires the thin thickness of the steel sheet and the increase of electrical resistivity in order to reduce the iron loss. In the iron material including the electromagnetic steel sheet, electrical resistivity changes like Fig. 11 depending on the ingredient containing [10].

The real electrical steel contains Si. In general, steel including Si stiffens and is hard to roll in manufacturing process. Most of electrical steel contains around 3% of Si. When Si density becomes 6.5%, the magnetostriction becomes zero and the stress sensitivity becomes low.

5 Mechanical Stress Influence

The residual mechanical stress generates in the electrical steel, when it is punched and cut off to make a core shape and the cut-off steel is fixed as motor core such as a shrink fitting. Figure 12 shows a cross-sectional view in cutting the steel and

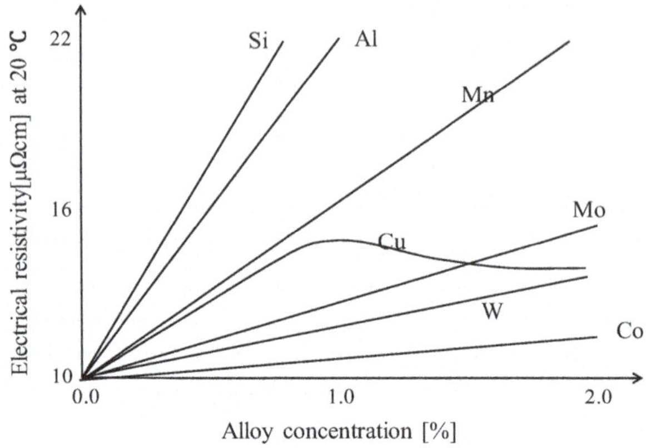

Fig. 11 Alloy concentration in steel sheet and electrical resistivity [10]

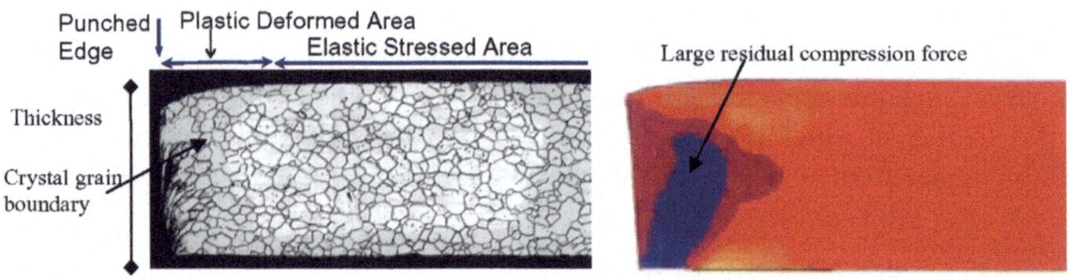

Fig. 12 Residual mechanical stress in cutting off the electrical steel (NO material). 2007 IEEE Reprinted, with permission, from [11]

its numerical analysis distribution of the residual stress at that time. There is a plastic deformation domain from the punched edge to the place of the thickness degree of the steel sheet, where residual mechanical stress is observed. In addition, the other part becomes an elastic deformation domain, where mechanical stress remains [11, 12].

Figure 13 shows magnetic characteristics when compression stress ($\sigma < 0$) and tensile stress ($\sigma > 0$) are applied to the direction that the steel sheet is excited to. Magnetization properties of Fig. 13a show that the magnetic flux density becomes big with the tensile stress in the same magnetic field and the magnetization characteristic is improved slightly, but it becomes small with the compression stress in the same magnetic field, and the magnetization characteristic turns worse. Iron loss properties of Fig. 13b show that iron loss becomes small in the same magnetic induction with the tensile stress a little, but it increases with the compression stress. The influence of the stress on steel sheet depends on hysteresis loss.

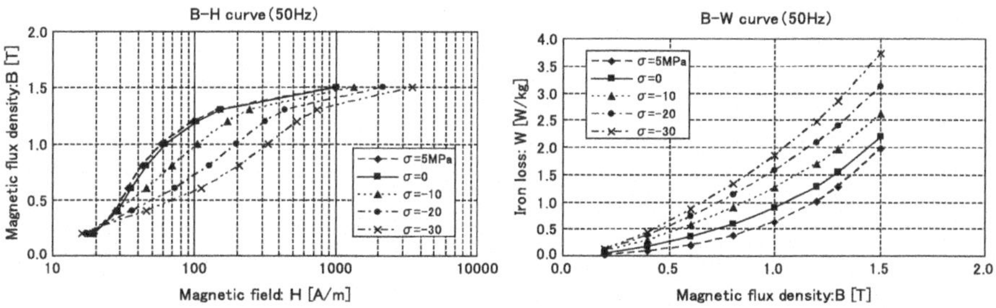

Fig. 13 Magnetic properties in mechanical stress supplied (NO material) **a** magnetization properties, **b** iron loss properties. 2007 IEEE reprinted, with permission, from [12]

6 Magnetic Anisotropy

Generally, the magnetic material has the magnetic anisotropy that magnetism properties vary according to a direction to excite, and it has some kinds of the magnetic anisotropy.

One of them is the one due to the bcc-structure of iron atom as shown in Fig. 1 described as before. As it is derived from the crystallographic orientation, it is called a crystal magnetic anisotropy. A shape magnetic anisotropy is caused by particle shape when ferromagnetic is shaped in particle. An induced magnetic anisotropy is caused by heat treatment under magnetic field supplied and is weak. Magnetic property becomes better in the direction to which mechanical tensile stress is supplied. The GO materials which aligned the crystallographic orientation have a crystal magnetic anisotropy. NO materials include some anisotropy which is observed when they measure magnetically.

When magnetic anisotropy appears, the magnetic field vector exciting a steel sheet and the magnetic flux density vector magnetized by it in the steel sheet usually do not have the same direction. Then the magnetism properties of the steel sheet essentially have a vector characteristic in the two-dimensional plane of the steel sheet. Therefore, a two-dimensional vector magnetism measurement as shown in Fig. 14 is used to measure the vector characteristics of magnetic field and magnetic flux density [13, 14]. The details are shown in Chap. 17, but an example of the magnetism characteristic of two-dimensional vector magnetism is explained here using Fig. 15 as it may be said to be much useful to know the magnetism properties of the real steel sheet.

Figure 15 shows the magnetic characteristic and the phase difference θ_{BH} of a magnetic field vector and a magnetic flux density vector when a magnetic flux density vector is rotated around 360° in the condition that its size is constant as 1 T in the steel.

In a magnetic measure equipment of Fig. 14, the central steel sheet is excited to *X*-direction and the *Y*-direction independently. After measuring the magnetic flux density of *X*-direction and *Y*-direction in the steel sheet, it is controlled and excited so

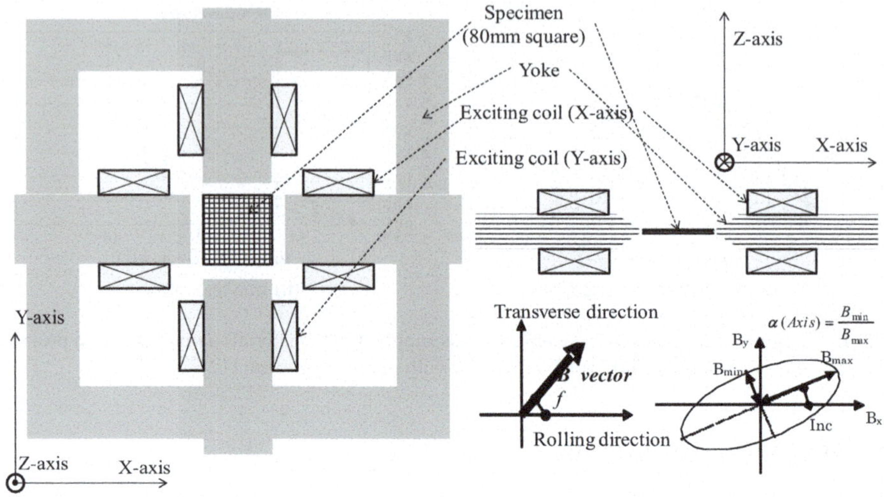

Fig. 14 Measurement equipment of two-dimensional vector magnetic property [13, 14]. Image courtesy of Nippon Steel & Sumitomo Metal Corporation

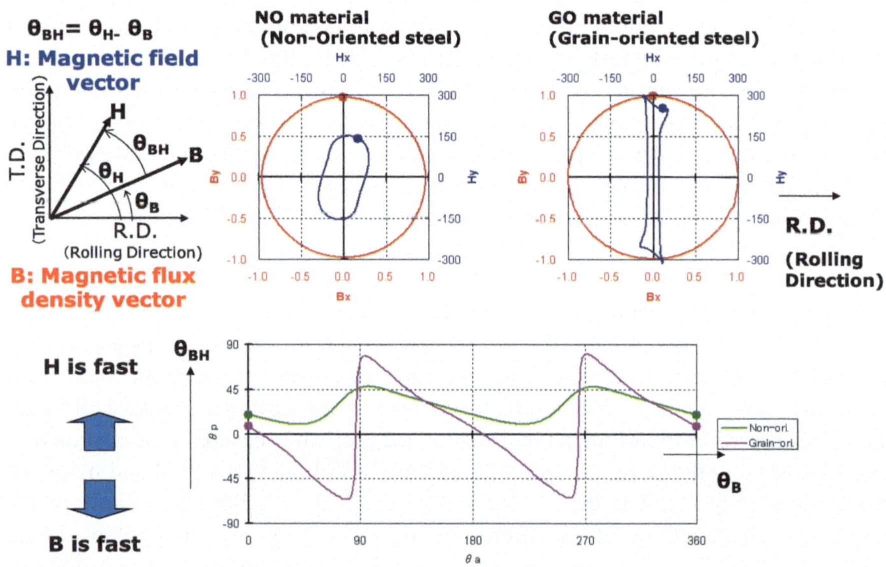

Fig. 15 Magnetic property of GO steel and NO steel in rotational magnetic field and angle difference between magnetic field vector and magnetic flux density vector θ_{BH}. Image courtesy of Nippon Steel & Sumitomo Metal Corporation

that magnetic flux density vector is to be a decided characteristic. Here, the circular rotary magnetic flux density vector is rotating to the counterclockwise direction from the magnetization easy axial direction so as that its norm becomes 1 T.

The exciting voltage of the *X*-, *Y*-directions is changed by feedback control and excites the steel specimen so that the magnetic flux density vector of *X*-direction and the *Y*-direction becomes the appointed rotating magnetic field. At the same time, a magnetic measurement coil to detect the one of *X*-direction and the *Y*-direction on the steel sheet is arranged and the magnetic field vector is measured.

When GO steel is excited by two-dimensional vector magnetic field in order to get a circular rotary magnetic flux density the norm of which is 1 T, the shape of the magnetic field vector trajectory deforms very much, and then the magnetic property of rolling direction and transverse direction is found to be different. In the case of NO steel, the shape of the magnetic field vector trajectory deforms to some extent, and then the magnetic anisotropy is found to be observed.

In the phase difference θ_{BH} properties in the lower figure of Fig. 15, the horizontal axis is an angle difference between the magnetic flux density vector and the rolling direction of the steel. θ_{BH} property shows that it does not become constant for not only the GO materials but also NO materials. It means that the magnetic field vector is not parallel to the magnetic flux density vector and the relative magnetic permeability in one period is not constant with directionality. Not only GO material but also the NO material is found to have magnetic anisotropy from this property.

7 Magnetic Measurement

The magnetism properties of the magnetic material are provided by a magnetic measurement. Because a magnetic field vector and a magnetic flux density vector are not always parallel and they have vector characteristics like statement above, the magnetic characteristics should be discussed by a two-dimensional vector magnetism measurement. However, since it takes a great time and measurement technique, a simpler magnetic measurement method is usually used. Typical magnetic measurement methods are the Epstein method, single sheet tester (SST) method, and the ring sample method as shown in Fig. 16.

Epstein method is an international magnetism measurement method established in International Electrotechnical Commission (IEC) or so [15–17]. A steel sheet sample of 30 mm width and 280 mm length is laminated and they are built in parallel crosses. In the Epstein method, steel sheets are excited directly. Magnetic field is introduced by measured exciting current and magnetic path length, and magnetic flux density is introduced by a time-integral of secondary voltage of electromotive force as shown in Fig. 16a. The excitation power supply excites it in a sine wave not to include time harmonics components, and steel sheet is measured in the stress-free condition.

Epstein method requires some quality of steel sheet and it is often difficult to get it. So to cope with the small amount of steel whose situation often happens in trial manufacturing or so, a single sheet measurement method is also considered as single

(a) (b) (c)

Fig. 16 Magnetic measurement equipment. **a** Epstein [15, 17]. **b** SST [16]. **c** Ring core [18]

sheet tester (SST) which is also established in IEC. One of SST methods is shown in Fig. 16b, where around 5 mm width and 100 mm length sheet is excited by a U-shaped excitation coil (under this photograph). Magnetic flux density is provided by *B*-coil being wound up to a steel sheet and introduced by time-integral of the voltage of the both ends. Magnetic field is introduced by time-integral of the both ends voltage of the *H*-coil located on the steel sheet. In the ring method, a steel sheet is cut in ring shape and laminated and excited in direct. The details are described in Chap. 7.

8 Information Magnetics and Power Magnetics

The application of the magnetic material is mainly considered to be an information system and an energy system like Fig. 17; the former is a hard disk application, and latter is a motor or transformer application.

In the information system, the data occurring in central processor unit (CPU) or MM (main memory (MM) are accumulated in hard disk (HD) and taken from HD when the data are required. Electromagnetic physical phenomena such as E (electric field), B (magnetic induction), D (electric displacement), and H (magnetic field) occur in a magnetic material in HD. However, the interaction of the electromagnetic physical phenomena such as E, B, D, H does not exist between the HD and the CPU or the MM, because the data between the HD and the CPU or the MM are accumulated in or taken out with chronological order one by one. Therefore, the research and development of the magnetic material of HD can be performed independently without considering the interaction with the peripheral device too much.

In contrast, in the energy system, the voltage produced by an inverter circuit is applied to an electrical motor, and electrical current flows between the electric circuit and the motor with magnetic material. Electromagnetic physical phenomena such as H (magnetic field), E (electric field), B (magnetic induction), and D (electric displacement) occur at the same time as the voltage generated in the inverter circuit and the electrical current flows in the motor.

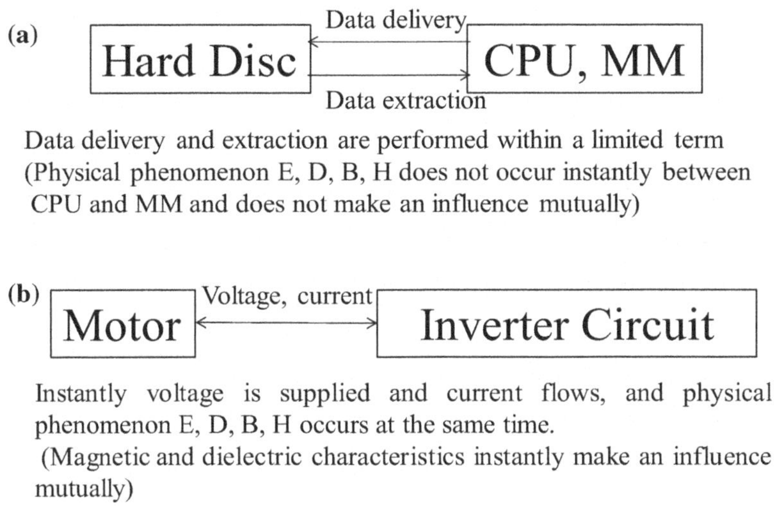

Fig. 17 How to use the magnetic material. **a** Information magnetics. **b** Power magnetics

Therefore, an interaction of electromagnetic physical phenomenon occurs between them, as it happens through the voltage and the electric current between the inverter circuit and the motor instantly. Usually, the inverter circuit and the motor are designed independently as far as they are operated in a linear operating domain and a single frequency operation.

However, an interaction often occurs between them when a nonlinearity of the magnetic material occurs or time harmonics components of the inverter generates. Therefore, it is difficult only to perform the research and development of the magnetic material in the energy system independently. The magnetic material research and development should be evaluated in the operating conditions as inverter excitation.

When a magnetic material is used in high magnetic flux density operation for reducing the size and weight or a spark voltage or current generates in the electric current, the nonlinearities of the magnetism characteristic occur, and then the operating point may be different from the designed operating point which was designed in linear characteristics.

When such a phenomenon occurs, it is assumed that it leads to the damage and the combustion of the electromagnetic apparatus. This phenomenon in the energy system is quite different from the information system in the process of research and development of the magnetic material.

For the future research of energy magnetic material, a new technology which is researched in information system such as spintronics is strongly expected and implements in energy system for fusion technology of electromagnetic field.

References

1. Landau, L.D., Lifshitz, L.M.: On the theory of the dispersion of magnetic permeability in ferromagnetic bodies. Phys. Z. Sowjetunion **8**, 153–169 (1935)
2. Gilbert, T.L.: A Lagrangian formulation of the gyromagnetic equation of the magnetic field. Phys. Rev. **100**, 1243 (1955)
3. Chikazumi, S.: Physics of magnetism. In: Chikazumi, S., Charap, H. (eds.) Wiley Series on the Science and Technology of Materials. Wiley (1978)
4. Kaya, S.: Ferromagnetics. Iwanami-zensho, vol. 158 (1952) (in Japanese)
5. Honda, K., Kaya, S.: Sci. Rep. Tohoku Imp. Univ. **15**, 721 (1926)
6. Goss, N.P.: Electrical sheet and method and apparatus for its manufacture and test. U.S. Patent 1,965,559, 1934
7. Hubert, A., Schaefer, R.: Magnetic Domains. The Analysis of Magnetic Microstructure. Springer (2000)
8. Barret, C.R., Nix, W.D., Tetelma, A.S.: The Principles of Engineering Materials. Prentice-Hall, Inc. (1973)
9. Steinmetz, C.P.: On the law of hysteresis. Trans. Am. Inst. Electr. Eng. **IX**(1), 3–64 (1892)
10. Bozorth, R.M.: Ferromagnetism, p. 40. Wiley-IEEE Press (1993)
11. Fujisaki, K., Hirayama, R., Kawachi, T., Satou, S., Kaidou, C., Yabumoto, M., Kubota, T.: Motor core iron loss analysis evaluating shrink fitting and stamping by finite-element method. IEEE Trans. Magn. **43**(5), 1950–1954 (2007)
12. Fujisaki, K., Satoh, S.: Numerical calculations of electromagnetic fields in silicon steel under mechanical stress. IEEE Trans. Magn. **40**(4), 1820–1825 (2004)
13. Enokizono, M.: Two-dimensional magnetic property. IEEJ-A **115**(1), 1–8 (1998)
14. Fujisaki, K., Nemoto, Y., Sato, S., Enokizono, M., Shimoji, H.: 2-D vector magnetic method in comparison with conventional method. In: Sievert, J. (ed.) 7th International Workshop on 1&2-Dimensional Magnetic Measurement and Testing. Proceedings (PTB-E-81), pp. 159–166 (2002)
15. Japanese Industrial Standard, C 2550-1 (2011)
16. Japanese Industrial Standard, C 2556 (2015)
17. International Electrotechnical Commission, 60404-3, 2nd edn. (1992)
18. Fujisaki, K., Yamada, R., Kusakabe, T.: Difference in iron loss and magnetic characteristics for magnetic excitation by PWM inverter and linear amplifier. Trans. Inst. Electr. Eng. Jpn. D **133**(1), 69–76 (2013) (in Japanese)

Fundamental Concept of Electrical Motor for Magnetic Researcher

Keisuke Fujisaki

Abstract Electrical motor itself is well known for most of the electrical engineers who have already studied in universities or so. However, it is not always well-known topics for researchers or engineers related to magnetic material or so. In order to research or design the magnetic material of an electrical motor or the related technologies, the magnetic researcher or engineer at least should know the fundamental concept of electrical motor. So, the minimum contents of the summary from the fundamental theory of the electric motor, which are considered to be important for magnetic researcher, are extracted and are explained here. Fundamental theory of electrical motor is explained based on Maxwell equations. As a basic concept of electrical engineering, three-phase AC theory and traveling magnetic field are shown. AC motor theory and PM motor characteristics are finally explained since they are mainly used in EV drive system.

Keywords Electrical motor · Traveling magnetic field · Permanent magnet motor · Synchronous motor · Induction motor

1 Fundament Theory and Structure of Electrical Motor

An electrical motor is an energy conversion equipment from an electric energy to a mechanical energy. The electrical energy is the product of the voltage and the electric current, and the mechanical energy is the product of the velocity and the electromagnetic force based on electromagnetic phenomena. The detailed electromagnetic phenomena are shown in Fig. 1. The electromagnetic force is introduced in the electrical motor, and then the speed and the position are provided by its time integration. This figure is shown in order that it leads to the comprehension of the torque occurrence to break an electromagnetism phenomenon to be caused at the same time into some electromagnetic principles.

K. Fujisaki (✉)
Toyota Technological Institute, Nagoya, Japan
e-mail: fujisaki@toyota-ti.ac.jp

K. Fujisaki (ed.), *Magnetic Material for Motor Drive Systems*,
Engineering Materials, https://doi.org/10.1007/978-981-32-9906-1_4

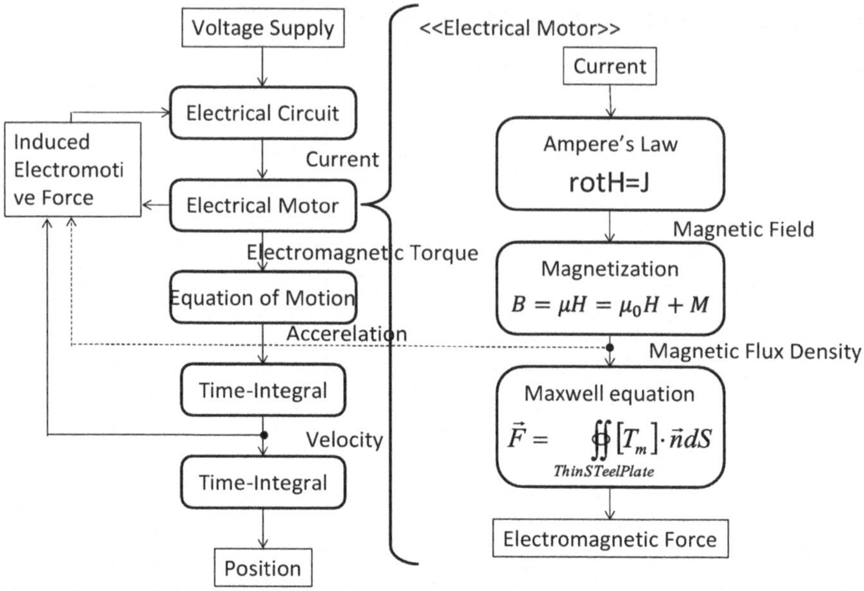

(a) Physical phenomena of motor drive system (b) Torque generation in electrical motor

Fig. 1 Fundamental electromagnetic and motion theory of electrical motor obtain a velocity or positioning in voltage supply

The voltage is applied to the electrical motor in connecting with power supply or inverter circuit. An electric current flows by a circuit equation with induced electromotive force based on Faraday's law. In the electrical motor, a magnetic field occurs based on Ampere's law from the electric current. In supplying the magnetic field to ferromagnetic material in the electrical motor, a magnetization phenomenon is caused and then large magnetic flux density can be introduced. The magnetic flux goes and comes between the stator of the standstill in the motor and the rotor of the reel and interlinks across each electromagnetic coil. The time variation of magnetic flux produces an induced electromotive force in the electromagnetic coil, which becomes one term of the electrical circuit mentioned in the above. Electromagnetic torque is provided by applying a Maxwell stress theory to the magnetic flux density distribution in the air gap between the stator and the rotor. The angular acceleration is derived by dividing the electromagnetic torque by the moment of inertia based on an equation of motion, and the angular velocity is provided by time integral of the angular acceleration, and the angle is provided by time integral of the angular velocity. The speed and the position are introduced by multiplying a radius of the motor in the angular velocity and the angle, respectively.

Since these electromagnetic phenomena show that the electrical motor is energy conversion equipment, the input electrical energy is equal to the output mechanical

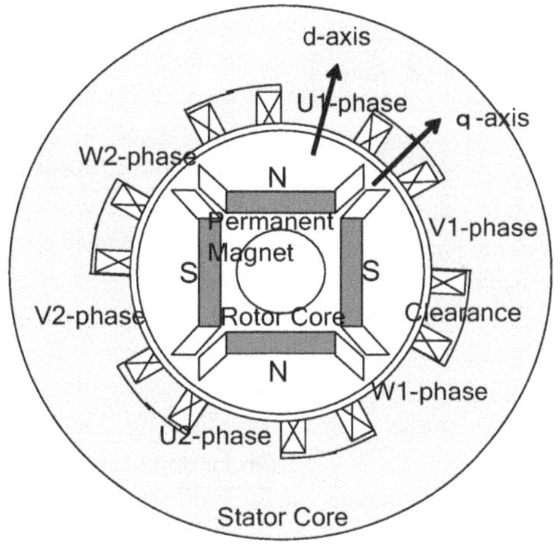

Fig. 2 Motor structure and stator, rotor (IPMSM, central winding, 4 poles)

energy if motor loss is considered to be zero. Electrical energy is a product of voltage and current, and mechanical energy is a product of electromagnetic torque and angular velocity. Electrical motor loss is considered as mechanical loss, copper loss, and iron loss.

Electrical motor is a body of rotation not to speak of, but it is a body of rotation for the thing standing still. Therefore, the basic constitution of the motor is comprised of a standstill body (stator) and a body of rotation (rotor) like Fig. 2. As the stator is installed on the ground part and the power supply is often supplied from the ground, the power supply is often supplied to the stator. The stator makes an electromagnetic field and it is supplied to the rotor. So, an electromagnetic phenomenon acts on the stator and the rotor mutually, and an electromagnetic torque occurs.

The real electrical motor is divided into some kinds depending on the kind of the power supply, the mechanical structure of the motor, outbreak theory of the electromagnetic torque. Figure 3 shows them. A power supply includes DC power supply and an AC power supply.

A direct current (DC) motor is comprised of an armature and a magnetic field. The armature connected with the DC voltage and DC current flow in it. Magnetic flux is derived from the magnetic field. Then, the electromagnetic torque occurs by a Lorentz force of the electric current in the armature and the magnetic flux. The armature connects to the DC voltage power source via a commutator, which changes a direction of electric current flowing in the armature so that the current direction is always the same at the magnetic field position in rotating. Since the direction of armature current and magnetic field is always the same, an electromagnetic torque has always the same direction. The commutator makes a role of switching operation. However, the commutator has a mechanical contact point where the current flows from the static power source to the rotating armature. So, it causes the generation of an arc, and then the deterioration advances and needs periodical maintenance, and a large-scale DC motor is impossible to provide. Though DC motor is easy to understanding,

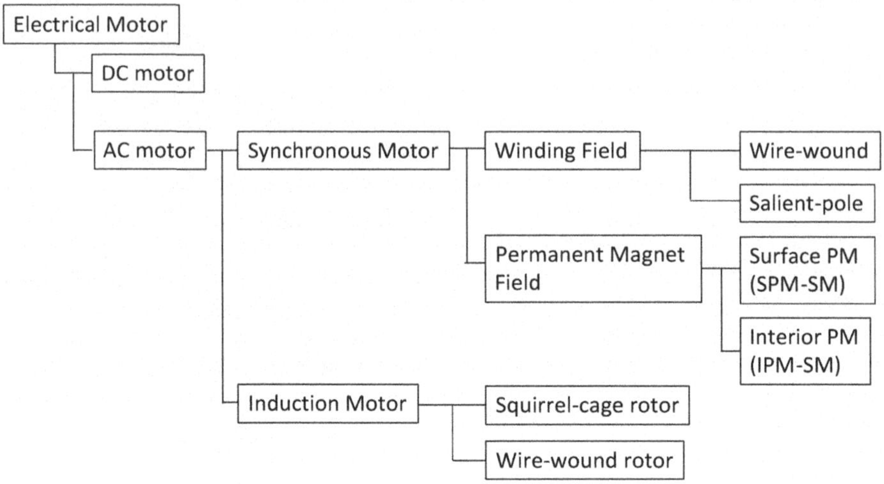

Fig. 3 Type of electrical motor

AC motor is now widely used by changing the mechanical commutator operation to an inverter in which power semiconductors are used as a switching operation in mechanical contact.

Therefore, to understand the characteristics of AC motor that will also be used in future, the fundamental concept of AC theory is shown at first and then the main property of AC motor is shown.

2 Three-Phase Alternating Current and Traveling Magnetic Field

Because alternating current (AC) is easily realized to change voltage by a transformer, and current cut-off is relatively easier than direct current (DC), large-scale transmission is possible by AC, not DC. Large-scale generator and transmission of electrical power make the system high efficient, so AC is widely used in electrical power transmission in second industrial revolution.

The basic concept of AC technology is considered to be a single phase, and it is necessary to grasp the relation between the voltage and the current in single-phase. The relation between them should be discussed in an electrical circuit with AC power source and its load, and the load is expressed by an impedance that is complex number. Complex number notation premises that only a single frequency is treated and the load has only lumped constant elements with linear property. The basics of complex number are Euler's formula expressed as follow.

$$e^{j\theta} = \cos\theta + j\sin\theta, \qquad j = \sqrt{-1} \tag{1}$$

This equation makes it possible to discuss a phase difference in single frequency: *f*, by putting it with $\theta = \omega t,\ \omega = 2\pi f$, where *t*: time.

Now AC current is considered to be expressed by the next complex number. The variable with dot on it means that it is a complex number, and the variable with no dot on it means that it has only a real part.

$$\dot{I} = I_0 e^{j\omega t} \tag{2}$$

From this equation, since the current which we can measure is a real part of the complex current, it is expressed as follows.

$$I = \mathrm{Re}[I_0 e^{j\omega t}] = I_0 \cos \omega t \tag{3}$$

Here, $\mathrm{Re}[\dot{I}]$ means a real part of $\dot{I}$.

Now the load connected to the electrical power source is expressed as lumped constants of resistance *R* and inductance *L*. As an electrical motor usually treats in comparative low frequencies less than MHz, and displacement current term in electromagnetic field is generally ignored, this load expression is enough. Then, the impedance is expressed as follows with Fig. 4.

$$\dot{Z} = R + j\omega L = Z_0 e^{j\varphi} \tag{4}$$

$$Z_0 = \sqrt{R^2 + (\omega L)^2}, \quad \varphi = \tan^{-1} \frac{\omega L}{R} \tag{5}$$

Then, the voltage expressed by complex number is as follows.

$$\dot{V} = \dot{Z} \cdot \dot{I} = Z_0 e^{j\varphi} \cdot I_0 e^{j\omega t} = Z_0 \cdot I_0 e^{j(\omega t + \varphi)} = e^{j\omega t}(Z_0 I_0 e^{j\varphi}) \tag{6}$$

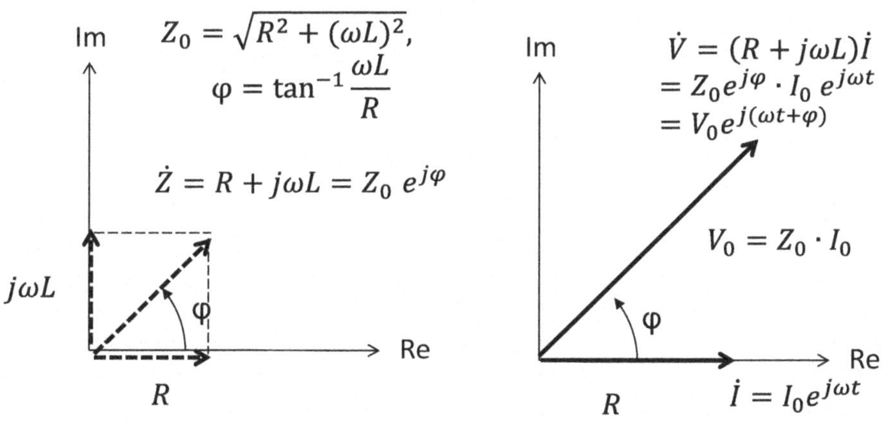

Fig. 4 Impedance and complex expression of voltage and current

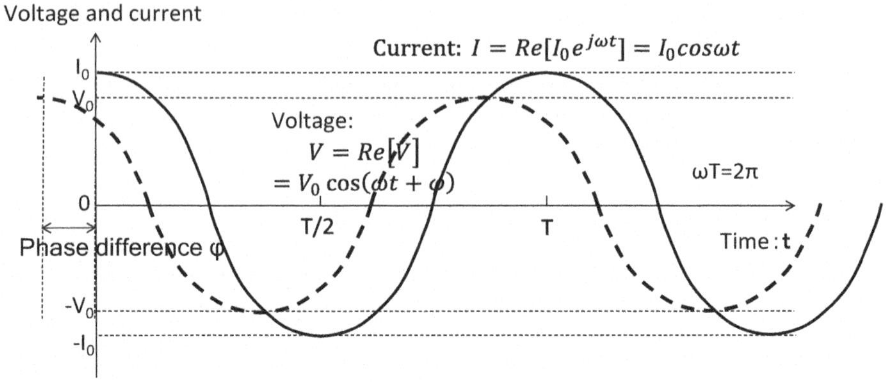

Fig. 5 Time-series waveform of voltage and current and phase difference

The voltage to be measured becomes:

$$V = \mathrm{Re}[\dot{V}] = Z_0 I_0 \cos(\omega t + \varphi) = V_0 \cos(\omega t + \varphi), \quad V_0 = Z_0 I_0 \tag{7}$$

Here, the voltage and the current have the same frequency as f, and there is a phase difference φ between them as shown in Fig. 4. Time series of waveforms of the current and the voltage are shown in Fig. 5.

The (active) power, in this case, is considered here. Since the definition of electrical power is a product of current and voltage, instantaneous power is expressed as follows.

$$P(t) = IV = \frac{1}{2} I_0 V_0 (\cos(2\omega t + \varphi) + \cos \varphi) \tag{8}$$

This equation means that the instantaneous power of single-phase is not constant. It has double frequency component and variable. Average power in one period of instantaneous power is introduced as followed by the time integral of $P(t)$.

$$P_0 = \frac{1}{T} \int_t^{t+T} P \mathrm{d}t = \frac{1}{2} I_0 V_0 \cos \varphi \tag{9}$$

As I_0 and V_0 are the peak values of an electric current and voltage, the value that divided them by $\sqrt{2}$ is an effective value. Then, $\frac{1}{2} I_0 V_0$ is the product of the effective value of electric current and voltage, and is called an apparent power. P_0 is also called an active power. Then, cos φ is a ratio of the active power and the apparatus power and is called a power factor. The power factor occurs by a phase difference between the electric current and the voltage and occurs by an inductance ingredient of the load. In addition, a value defined in the next expression is called a reactive power.

$$Q_0 = \frac{I_0 V_0}{2} \sin \varphi \tag{10}$$

The cross-area of the copper cable which an electric current flows in is decided by the value of the electric current which they want to apply, and the electricity insulation characteristics are decided by the value of the voltage that they want to apply. In other words, the electric equipment of copper cable and electricity insulation characteristics are decided by the apparatus power. However, effective electrical power for the facilities actually is an active power. Therefore, the power factor may be said that it is an index to show how the electric equipment is effectively used. It is an important issue to increase a power factor in using the electric equipment effectively.

It is a big problem that instantaneous power is not constant. Motor and generator are considered to be energy conversion equipment between the electric energy and the mechanical energy. Then, the electric energy IV and the mechanical energy $T\omega$ (T: torque, ω: an angular frequency) should be equal as the next equation as far as it is an ideal situation as no loss.

$$IV = T\omega \tag{11}$$

This equation means that if electric energy is variable in time, the mechanical energy also should be variable in time. When one-phase electrical power is used, since instantaneous power is variable in time with double frequency component, mechanical energy also must be variable with the double frequency component in torque or angular velocity. The variable mechanical energy is not realized without mechanical troubles because of mechanical strength, especially in high power system. So, one-phase motor or generator is not usually used in high power operation.

It is a three-phase AC to solve the variable problem of electrical power. Three-phase AC is defined as u-phase, v-phase, and w-phase, and each phase has a phase difference as $\frac{2\pi}{3}$. So, the current of each phase is expressed as follows.

$$I_u = I_0 \cos \omega t, \quad I_v = I_0 \cos\left(\omega t - \frac{2\pi}{3}\right), \quad I_w = I_0 \cos\left(\omega t - \frac{4\pi}{3}\right) \tag{12}$$

When the load connected to three-phase AC is considered to be in the state of three-phase equilibrium that each phase load is electrically the same, as there is a phase difference φ between electric current and voltage, the voltage of each phase is expressed as follows.

$$\begin{aligned} V_u &= V_0 \cos(\omega t + \varphi), \\ V_u &= V_0 \cos\left(\omega t - \frac{2\pi}{3} + \varphi\right), \\ V_w &= V_0 \cos\left(\omega t - \frac{4\pi}{3} + \varphi\right) \end{aligned} \tag{13}$$

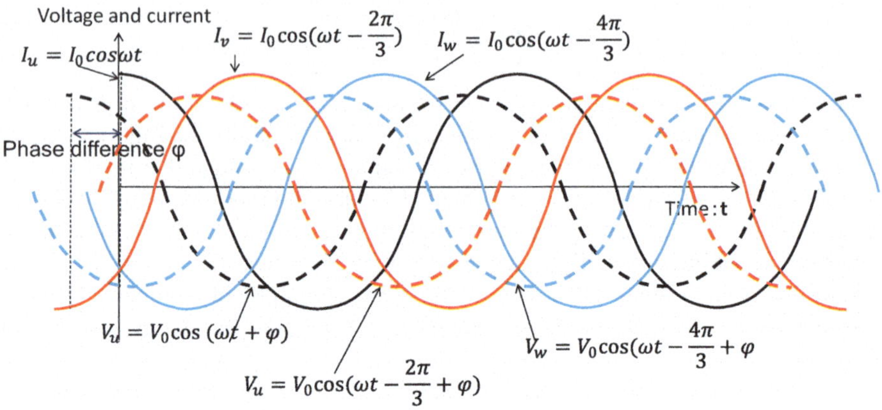

Fig. 6 Time series of current and voltage of three-phase AC

Time-series wave pattern of each phase current and voltage is shown in Fig. 6. Then, the instantaneous power P_3 is given in the next equation.

$$\begin{aligned} P_3 &= I_u V_u + I_v V_v + I_w V_w \\ &= I_0 V_0 \left(\cos(\omega t) \cos(\omega t + \varphi) + \cos\left(\omega t - \frac{2\pi}{3}\right) \cos\left(\omega t - \frac{2\pi}{3} + \varphi\right) \right. \\ &\quad \left. + \cos\left(\omega t - \frac{4\pi}{3}\right) \cos\left(\omega t - \frac{4\pi}{3} + \varphi\right)\right) \\ &= \frac{3}{2} I_0 V_0 \cos\varphi \end{aligned} \tag{14}$$

This equation shows that instantaneous power P_3 of three-phase AC is not variable in time with no time-variable component. Then, electrical power with three-phase equilibrium load is always constant. Therefore, large energy conversion equipment as large motor and large generator uses three-phase AC.

Another important concept of three-phase AC in electrical motor and generator is traveling magnetic field (rotating magnetic field). In fact, rotating magnetic field has two meanings in electrical machine. The one is that the magnetic poles rotate (travel) along the outer of rotor or the inner of stator, which is described here. The other is that a trace of magnet field vector in one period at a certain fixed point such as on the stator or rotor becomes a circle or an oval or so. They are a different electromagnetic phenomenon but are often called the same words as "rotating magnetic field". To distinguish them, the former is called as traveling magnetic field and the latter is called as rotating magnetic field here. The detailed description of rotating magnetic field is done in Chap. 17.

The direction of the electrical motor is considered to be presented so as that the circumference direction along the inside diameter of stator with the single-layered winding is X-direction and the radial direction is in Y-direction as shown in Fig. 7 top.

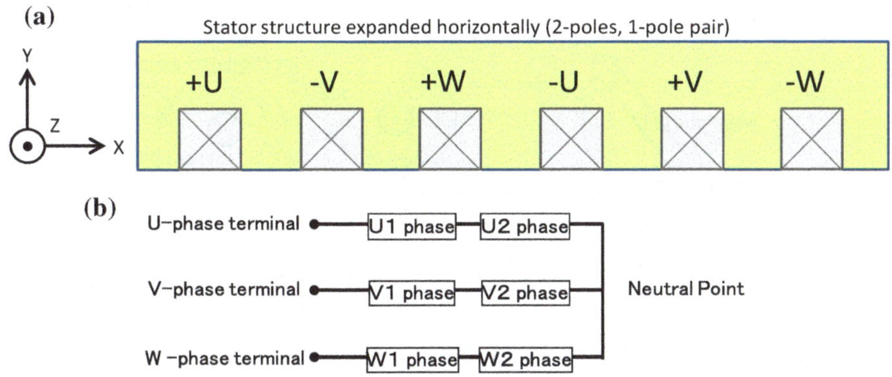

Fig. 7 Stator structure expanded horizontally (2 poles, 1 pole pair) (**a**) and *Y*-connection (**b**)

It is considered to be that three-phase coils are connected as *Y*-connection as Fig. 7 bottom and three-phase AC flow in the coils. For a three-phase current flowing as Fig. 8, an electric current at each time as $\omega t = 0, 60, 120, 180, 240, 300, 360$-degree

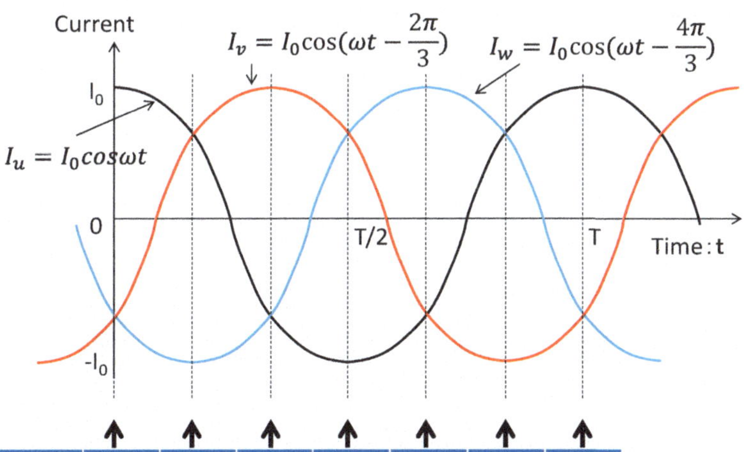

ωt	0	60	120	180	240	300	360
I_u	I_0	$\frac{1}{2}I_0$	$-\frac{1}{2}I_0$	$-I_0$	$-\frac{1}{2}I_0$	$\frac{1}{2}I_0$	I_0
I_v	$-\frac{1}{2}I_0$	$\frac{1}{2}I_0$	I_0	$\frac{1}{2}I_0$	$-\frac{1}{2}I_0$	$-I_0$	$-\frac{1}{2}I_0$
I_w	$-\frac{1}{2}I_0$	$-I_0$	$-\frac{1}{2}I_0$	$\frac{1}{2}I_0$	I_0	$\frac{1}{2}I_0$	$-\frac{1}{2}I_0$

Fig. 8 Three-phase current AC in time series

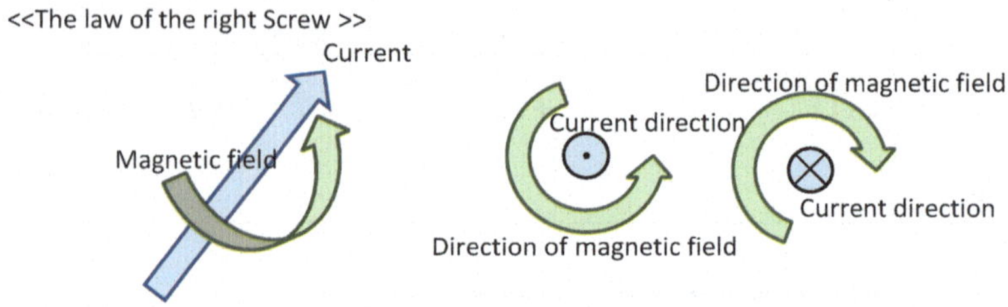

Fig. 9 Current direction and magnetic field direction based on right-handed screw rule

is considered. As a magnetic field generates by the right-handed rule like Fig. 9, depending on the direction of the electric current, the direction of the electric current and the direction of the magnetic field in each time are considered here.

For an example, a time of $\omega t = 0$ is took up. Then, each phase current becomes as followed by Fig. 8.

$$I_u = I_0, \quad I_v = -\frac{1}{2} I_0, \quad I_w = -\frac{1}{2} I_0 \tag{15}$$

So, the current distribution on the stator becomes as Fig. 10a. The spatial distribution of Y-direction magnetic field along the X-direction at that time is introduced by the right-handed screw rule as shown in Fig. 10b.

The current distribution and the magnetic field distribution derived from the current are shown along X-direction in each time as $\omega t = 0°, 60°, 120°, 180°, 240°, 300°$,

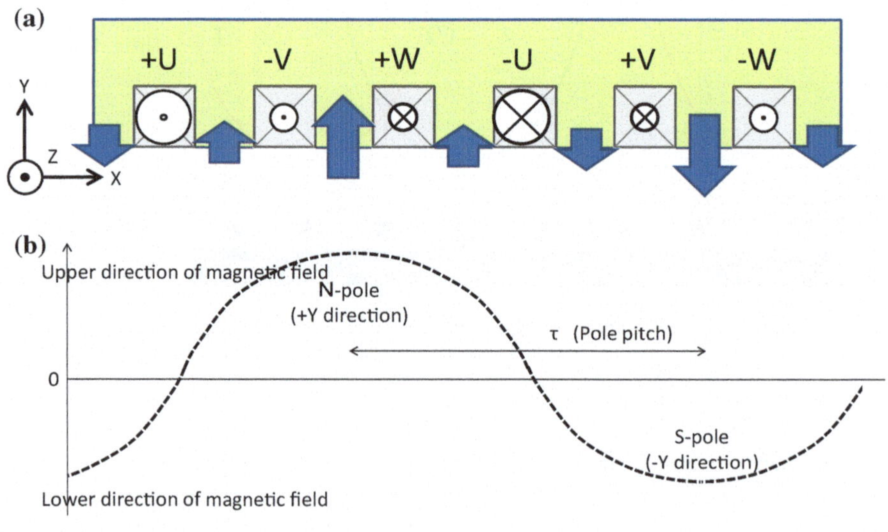

Fig. 10 Current distribution in $\omega t = 0$ (**a**) and the magnetic field distribution directed in Y-direction at that time (**b**)

360° as Fig. 11. Though the magnetic field directed in Y-direction is a simple harmonic oscillation in each position, a magnetic field moves to the X-direction as time is passing, and then traveling magnetic field is observed, when three-phase current is arranged in an appropriate connection. With one period, a magnetic field moves for two poles (1 pole pair). Therefore, when pole pitch is put with τ (distance between N-pole and the S-pole), the traveling speed v_0 of the magnetic field is expressed as follows.

$$v_0 = 2\tau f \tag{16}$$

Traveling angle of rotation frequency of the magnetic pole is $\omega_0 = rv_0$ in the stator when the radius of the rotor is assumed to be r. These are also called as a synchronous speed and a synchronous angular velocity.

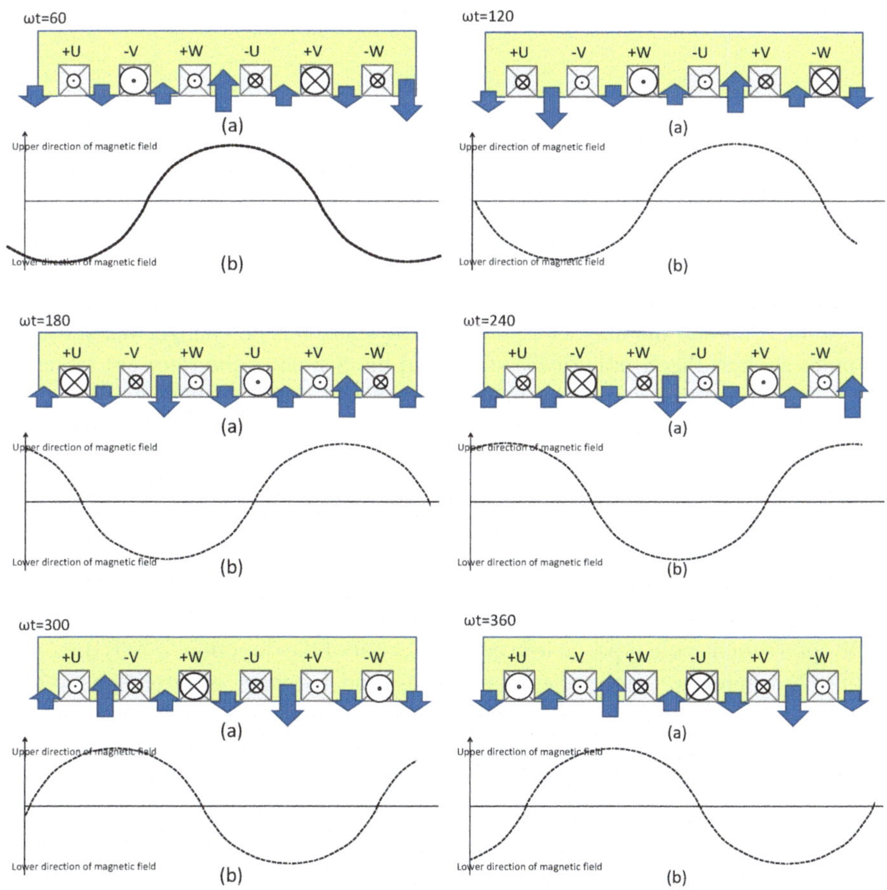

Fig. 11 Current distribution in each time of one period and the magnetic field distribution at that time

3 AC Motor [1–4]

When electric current of three-phase AC is applied to a stator coil connected appropriately in the electrical motor, magnetic pole such as S-pole and the N-pole generates, and a traveling magnetic field occurs in the stator, where the magnetic pole in a stator travels spatially in traveling angle of rotation frequency ω_0 (synchronous angular velocity). In this case, two kinds of the electrical motor are considered whether ω_0 is the same as angle of rotation speed ω_m of the rotor, which is called as mechanical angular velocity. Here, it is assumed that the pole number of motor is 2, pole pair is 1 for convenience.

When

$$\omega_m = \omega_0 \tag{17}$$

the motor is called as a synchronous motor, and when

$$\omega_m \neq \omega_0 \tag{18}$$

the motor is called as an asynchronous motor or an induction motor.

At first, a synchronous motor is thought about. The magnetic pole of the pole number of stators, which is the same number of stator pole, is set up and produced in the rotor using an electromagnet or a permanent magnet or a ferromagnetic body. Because the traveling angle of rotation frequency is the same as the angle of rotation speed of rotor, the rotor rotates so as that the relative position between the magnetic pole of the rotor and the magnetic pole with the stator has no change. When different poles such as the N-pole of the stator and the S-pole of the rotor are arranged here, absorption force caused by magnetic power is produced between the stator and the rotor. This force is an outbreak source of the electromagnetic torque in the synchronization motor.

Rotor structure to produce the magnetic pole includes an electromagnet type and a permanent magnet type. It is necessary to apply an electric current to the electromagnetic coil of the rotor directly for the electromagnet type. A slip ring is required to connect the rotor core to on-site DC power supply. The electromagnet structure includes a salient-pole type and a cylindrical type.

As for a salient-pole type, an iron core sticks out of the direction (d-axis direction of Fig. 2) where the magnetic pole produces, and a copper electromagnetic coil is wound up in the orthogonal direction to d-axis (q-axis direction of Fig. 2), and there is no iron core. Therefore, the reluctance ingredients with d-axis and q-axis are different, and the reluctance torque derived from the difference occurs newly, and the electromagnetic torque becomes large in some condition. However, because the salient-pole type is not symmetric for a rotational body structurally, it is unsuitable for high-speed rotation and upsizing.

As for the cylindrical type, the slot structure of the stator is placed to the circumference direction of the rotor uniformly, and a copper coil is wounded there, and a direct current is applied, and the magnetic pole is produced. So, the reluctance ingredients of d-axis and the q-axis become the same, and the reluctance torque does not occur, and the electromagnetic torque does not so increase. However, since it is a rotary symmetry, it is suitable for high-speed rotation and upsizing.

When a rotor rotates in synchronous speed, the relative position relations with the magnetic pole in the rotor and the magnetic pole in the stator do not change, and an angle difference δ exists between both magnetic poles. This is called as a load angle in the synchronous machine. As for the rotatory direction of the rotor, if the magnetic pole (e.g., N-pole) position of the rotor advances from the magnetic pole (e.g., S-pole) position of the rotating magnetic field of the stator ($\delta > 0$), since the magnetic pole of the rotor is considered to be accompanied by the magnetic pole of the stator, this synchronous machine in this condition is arrested to be a synchronous generator which converts the mechanical energy into the electrical energy. On the contrary, if the magnetic pole (e.g., N-pole) position of the rotor is behind the magnetic pole (e.g., S-pole) position of the rotating magnetic field of the stator ($\delta < 0$), since the magnetic pole of the stator is considered to be accompanied by the magnetic pole of the rotor, this synchronous machine in this condition is arrested to be a synchronous motor which converts the electrical energy into the mechanical energy. Even if a synchronous motor has the same structure of stator and rotor, its function to convert the energy depends on the sign of the load angle (δ); it becomes a generator when $\delta > 0$ and it becomes a motor when $\delta < 0$.

As the magnetic pole of the stator seems to stand still judging from a rotor, as a general rule, the eddy current does not occur to a rotor basically. However, since stator has usually teeth and slots and slot part has large magnetic flux and teeth part has small magnetic flux, rotor has large and small magnetic flux which depends on the slot and teeth part. So, the harmonic components occur in the rotor and eddy current are introduced which depend on the slot and teeth shape. In the case of inverter excitation, higher harmonic components add with the slot harmonics by the switching operation of power semiconductor in the inverter.

In contrast, as for an instruction motor, since the rotational speed of traveling magnetic field in the stator is faster than the rotational mechanical speed of rotor, a rotational speed of magnetic pole in the stator is observing from the rotor is observed as the next equation.

$$\omega_s = \omega_0 - \omega_m \tag{19}$$

This ω_s is defined as a slip angular velocity. Here, a slip s is defined as the next equation.

$$s = \frac{\omega_0 - \omega_m}{\omega_0} = \frac{\omega_s}{\omega_0} \tag{20}$$

Since the magnetic field in the rotor is traveling in the slip angular velocity ω_s, an induced electromotive force occurs by the electromagnetic induction operation. When a rotor has a structure so as that an eddy current easily flows by the induced electromotive force, the eddy current occurs in the rotor. Such a rotor structure is considered as a squirrel-cage structure to shorten the coil end in the rotor called an end ring and a cylindrical structure which is the same rotor structure as the cylindrical synchronous motor. The former structure is simple, solid, and cheap. The latter structure is possible to connect the outer power source or electrical circuit through a slip ring, and then it can change the secondary resistance of the rotor.

The total with the magnetic field occurring by an exciting current applied to a stator and the one by an eddy current induced in the rotor becomes the magnetic field in the clearance between the stator and the rotor. Lorentz force by the product with the eddy current in the rotor and the total magnetic field becomes the electromagnetic torque produced in the induction motor.

When a rotor is thought to be in stationary state as slip $s = 1$, here, an eddy current occurs in a rotor, and then a certain electromagnetic torque occurs. When a rotor rotates and a slip decreases in the condition that supplied voltage is constant, the electromagnetic torque increases. However, when the rotary speed of the rotor becomes the same as the synchronous speed and a slip becomes 0 ($s = 0$), since no eddy current is induced in the rotor, the electromagnetic torque becomes zero.

An induction motor (IM) has a little complicated theory of electromagnetic phenomena, but a solid structure. Fundamentally, IM has larger loss and lower efficiency than a permanent magnet synchronous motor (PMSM) in making an electromagnetic torque because eddy current flows in IM's rotor but PMSM has no current basically. However, IM has no induced electromotive force in no-current-flow condition, and these situations often occur when the electrical vehicle drives in highway or so. In addition, IM has no drag force loss at that condition. However, PMSM produces induced electromotive force in its coil end as well as drag force loss. Therefore, IM is, sometimes, applied to an electrical vehicle also.

4 Permanent Magnet Synchronous Motor (IPMSM, SPMSM) [5]

A permanent magnet synchronous motor is used in air conditioner or motor drive system of electrical vehicle, and its field part is made of permanent magnet (PM). PM is widely applied to the motor, nowadays, because of the appearance of the permanent magnet as NdFeB which has large residual magnetization in relatively inexpensive. Figure 2 shows a rotor structure of the permanent magnet motor.

When permanent magnet is used as magnetic field, a slip ring is not necessary, and no extra electric power is required to generate magnetic field for the rotor. Therefore, higher efficient and smaller size motor is realized than a conventional motor.

The theory of the permanent magnet motor is almost as same as the electromagnet type SM, but one point is different as for the motor torque characteristics. Relative magnetic permeability of the permanent magnet is approximately one.

Permanent magnet SM has two types as shown in Fig. 12. The one is a surface

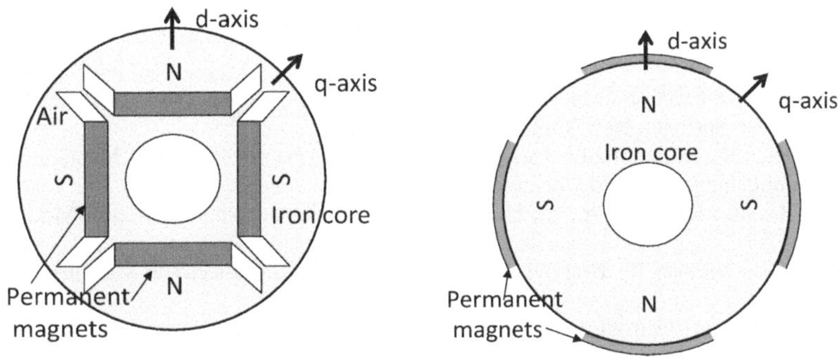

(a) Interior permanent magnet – Synchronous motor

(b) Surface permanent magnet – synchronous motor

Fig. 12 Permanent magnet arrangement in rotor [6]. Adapted from [6] with permission ©2016 The Magnetics Society of Japan

permanent magnet synchronous motor (SPMSM) in which the PM is arranged on the surface of iron rotor. The other is an interior permanent magnet SM (IPMSM) in which the PM is buried into the rotor.

Since the relative permeability of permanent magnet is almost one or so, a reluctance X_d in d-axis and a reluctance X_q in q-axis become the same in SPMSM, and then the reluctance torque does not occur such as a cylindrical SM. However, in the case of IPMSM, reluctance torque occurs like a salient-pole SM as a reluctance X_q in q-axis is different from a reluctance X_d in d-axis. However, as $X_d < X_q$ because the relative permeability of permanent magnet is almost one, reluctance torque characteristics of IPMS-SM is different from the one of salient-pole SM, in which $X_d > X_q$. Therefore, a maximum torque of IPMS-SM is observed in $\pi/2 < \delta < \pi$, where the one of salient-pole SM is observed in $0 < \delta < \pi/2$.

As an induced electromotive force is produced between the terminals of armature coil, when PMSM is rotating in some angular velocity, further power voltage more than that is required to be supplied to maintain the angular velocity. Generally, for example, the supplied voltage of the power source has some upper limit as battery capacity. Therefore, it is necessary to decrease the induced electromotive force in order to obtain the further angular velocity. To decrease it, the magnetic flux density of the field in the rotor should decrease. In the case of using the electromagnetic magnet in the field, the magnetic flux can be decreased by lowering an electric current flowing through a slip ring. However, the permanent magnet is usually difficult to decrease its magnetic flux density which it produces. So, a field-weakening method is used for PMSM to get higher velocity. Magnetic field from permanent magnet becomes smaller by changing the load angle δ. Magnetic field of PM can be suppressed by one of the stator current.

References

1. Vukosavic, S.N.: Electrical Machines. Part of the Power Electronics and Power Systems Book Series (PEPS). Springer, New York (2013)
2. Hendershot, J.R., Miller, T.J.E.: Design of Brushless Permanent – Magnet Machines. Magna Physics Publishing & Oxford University Press (1994)
3. Fitzgerald, A.E., Kingsley, Jr., C., Umans, S.D.: Electric Machinery, 4th edn. McGraw Hill (1983)
4. Pyrhoenen, J., Jokinen, T., Hrabovcova, V.: Design of Rotating Electrical Machines, 2nd edn. Wiley (2014)
5. Bose, B.K.: Power Electronics and AC Drives. Prentice Hall (1986)
6. Fujisaki, K.: Permanent magnets and their applications, VI cutting edge of permanent magnets. Magn. Jpn. **11**(1), 34–41 (2016) (in Japanese)

Fundamental Concept of Power Electronics for Magnetic Researcher

Keisuke Fujisaki

Abstract Power electronics technology has been studied in earnest for more than 40 years, and its application was limited to some electric energy. However, the positioning is going to turn big when 80% of the electric energy comes to be used through a power electronics technology from now on in 10–20 years or so. Therefore, the power electronics technology may be said to be the minimum required knowledge to be learned by not only the electric person concerned but also the person concerned with magnetism. The summary of the power electronics technology is described here, and the detailed explanation is carried out about one-phase inverter dynamic behavior which is said to be one of the most important fundamental concepts in power electronics technology.

Keywords Power electronics · Inverter · Power semiconductor · Switching operation · Dead time

1 Summary of Power Electronics Technology [1, 2]

Power electronics is a dynamic electrical energy conversion technology between Alternating Current (AC) and Direct Current (DC) by using a power semiconductor as switching operation. The electrical energy conversion only changing a timing of the switching of ON/OFF of the power semiconductor realizes a variable frequency and a variable voltage in high efficiency and in high responsibility. A command value of the frequency and voltage is decided by the control theory of the motor, and that value is realized by the power electronics technology. So finally, predetermined position, speed or torque comes to be provided. In other words, the component of the power electronics technology is said to be a compound technology that is comprised of a power semiconductor, an electrical circuit, and control as shown in Fig. 1.

High efficient and high responsible exchange technology to a certain frequency and voltage realizes to obtain a variable velocity motor control economically, and then electrical motor drive system with electrical motor, power electronics and control is

K. Fujisaki (✉)
Toyota Technological Institute, Nagoya, Japan
e-mail: fujisaki@toyota-ti.ac.jp

K. Fujisaki (ed.), *Magnetic Material for Motor Drive Systems*,
Engineering Materials, https://doi.org/10.1007/978-981-32-9906-1_5

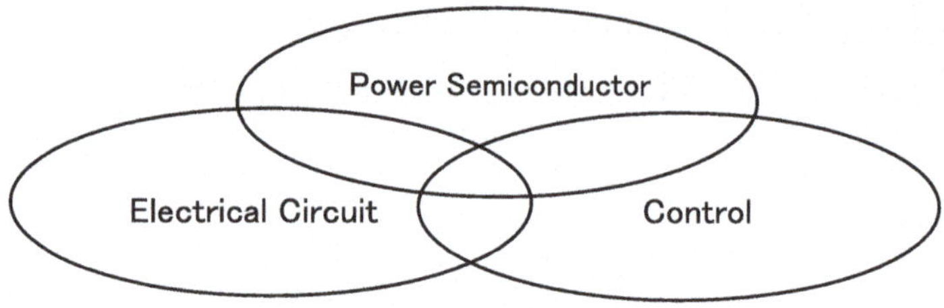

Fig. 1 Three elements of power electronics technology

now widely applied to the most of the transportation equipment. So, it may be said to be that a transportation revolution is now going ahead from inner combustion engine to electrical motor drive system.

There are two kinds of power semiconductor as 2-terminal element and 3-terminal element. The former selects ON/OFF operation by the direction of the electric current flowing in the element, and the latter selects ON/OFF operation by the outside signal of the third terminal. The 2-terminal element includes a diode, and the 3-terminal element includes a thyristor, bipolar transistor, Metal-Oxide-Semiconductor Field-Effect Transistor (MOSFET), Gate Turn Off thyristor (GTO), and Insulated Gate Bipolar Transistor (IGBT). The element of power semiconductor is used depending on electrical capacity or responsiveness.

As the form of the electrical energy is divided into Alternating Current (AC) and Direct Current DC, the way of the electrical energy conversion is thought about like Fig. 2. A representative electrical circuit name exists depending on a way of the electrical energy conversion.

A control system of electrical motor is composed of the multiplex loop consisting of three minor loops as a current control, velocity control and positioning control. Vector control theory is applied to an AC motor to get a high responsibility and a simplification. Furthermore, the outside of the loop includes methods to control some motors like a robot.

As mentioned above, power electronics is a compound technique comprised of some technical elements, and it is not a purpose of this book to explain all of them. It is left to other technical books. Instead of it, for magnetic researcher or engineer to consider and research magnetic material and magnetism in motor drive system, fundamental concept of power electronics technology is focused on such as a switching

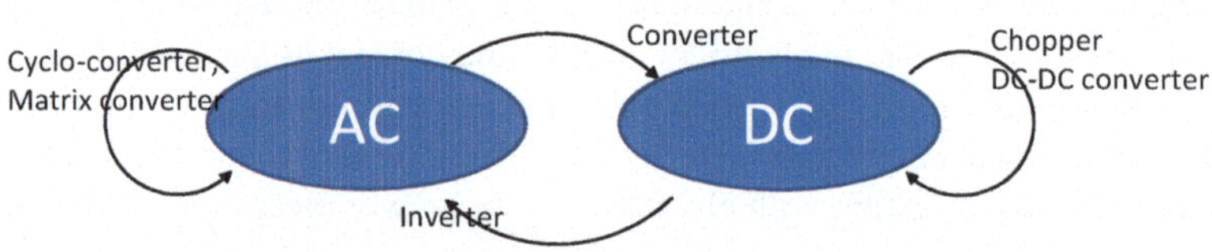

Fig. 2 Power electronics functions and their electrical circuits (electrical energy conversion technology)

operation of power semiconductor and operating theory of inverter circuit which is a main circuit of AC motor in electrical motor.

2 Switching Operation of Power Semiconductor

The power electronics circuit uses a power semiconductor in switching operation.

When it is assumed that an electric current flowing in a power semiconductor is I_s and the supplied voltage to it is V_s, electrical power P_s becomes the next equation as shown in Fig. 3.

$$P_s = I_s V_s \tag{1}$$

Like a right figure of Fig. 3, it is considered that a power semiconductor becomes an ON-state from an OFF-state now. An ideal state is considered here that no current flows in OFF-state, no voltage is supplied in ON-state, namely the power semiconductor is in short-circuit state, and the power semiconductor can change instantly from the OFF-state to the ON-state or from the ON-state to the OFF-state. Then in the OFF-state, since the electric current does not flow ($I_s = 0$), the loss of the power semiconductor becomes zero ($P_s = 0$). In the ON-state, since the voltage is not supplied in the power semiconductor ($V_s = 0$), the loss of the power semiconductor becomes zero ($P_s = 0$).

In other words, the loss of the semiconductor is "0" as there is only either OFF-state or ON-state if the power semiconductor is an ideal state.

Main electrical circuit of power electronics such as the inverter is comprised of the power semiconductor where the electrical current flows. Detailed behavior will be shown in later. So, the loss of the electrical energy conversion by the power electronics circuit becomes zero as far as the power semiconductor operates in an

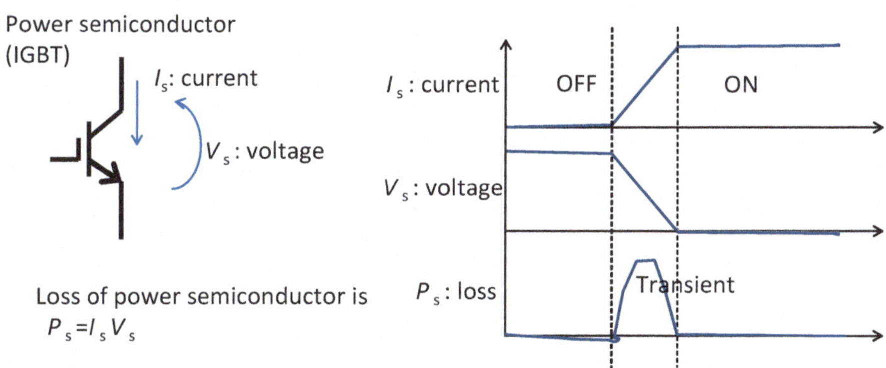

Fig. 3 Switching operation of power semiconductor [7, 8]. Adapted from [8] with permission ©2016 The Magnetics Society of Japan

ideal switching operation. In addition, it may be said to be a high responsibility as the change of the switching between ON/OFF can be performed instantly. This is a reason enabling the power electronics equipment to operate in the high responsible and high efficient.

However, a power semiconductor operation has a transient phenomenon which is the change from the ON-state to the OFF-state or from the OFF-state to the ON-state. A rising time or a falling down time of voltage or current is usually observed, and it is a μs order for Si semiconductor case. So the switching loss generates in the transient phenomena within the power semiconductor as shown in Fig. 3 right. In addition, Joule loss of the power semiconductor occurs in the ON-state and the OFF-state as the electrical resistance of the semiconductor is not zero in the ON-state, and as the leak electric current flows in the OFF-state to some extent. Therefore, the efficiency of the electrical conversion of the power electronics device does not always become 100%. However, such a loss is not such generally big.

3 Inverter Circuit and Its Operation [3]

Several kinds of power electronics circuits are considered which depends on electrical energy conversion as shown in Fig. 2, and operation explanation for all the circuit is not our purpose of this book. So as a typical example of power electronics circuit, one-phase inverter circuit is picked up. It converts the electrical energy from DC to AC as shown in Fig. 4. The reason is that most of future electrical vehicle is expected to be driven by AC motor excited by the inverter. Four power semiconductors (IGBT in the circuit) S_1–S_4 with switching operation and four diodes (they are said to be a flywheel diode) make a pair with the semiconductor, and they constitute H-bridge. The input is a DC voltage V_{DC}, and the output is connected to load and assumes to be a voltage V.

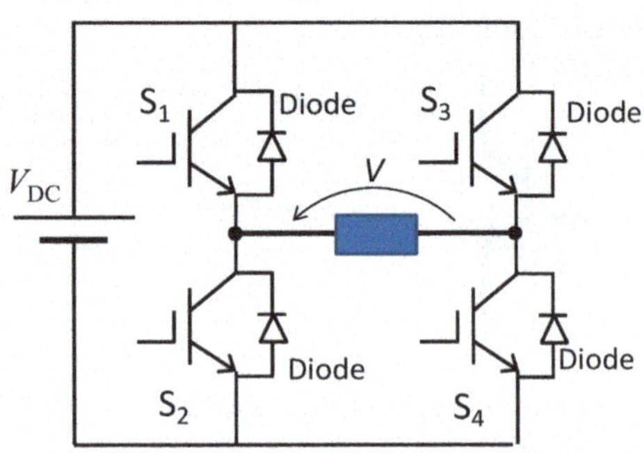

Fig. 4 One-phase full-bridge inverter circuit [7]

Table 1 Switching operation of power semiconductor and its mode

	S_1	S_2	S_3	S_4
Mode 1	ON(○)	OFF(×)	OFF(×)	ON(○)
Mode 2	ON(○)	OFF(×)	ON(○)	OFF(×)
Mode 3	OFF(×)	ON(○)	ON(○)	OFF(×)
Mode 4	OFF(×)	ON(○)	ON(○)	OFF(×)

When the semiconductors in the same leg (a pair of S_1 and S_2 or a pair of S_3 and S_4 in Fig. 4) in the inverter circuit are turned on at the same time, large amount of short-circuit electric current flows from the DC power supply, and then the semiconductor is destroyed due to heat up. So, the next equation is necessary to avoid the heat destruction.

$$S_1 = \bar{S}_2, \quad S_3 = \bar{S}_4 \tag{2}$$

The bar on the parameter means an inversion of ON/OFF here. Then four kinds of combinations of ON/OFF patterns of four power semiconductor S_1–S_4 are considered as four modes like Table 1. Electrical circuit for each mode and its output voltage at the four modes are shown in Fig. 5. They show that the output voltage V of the

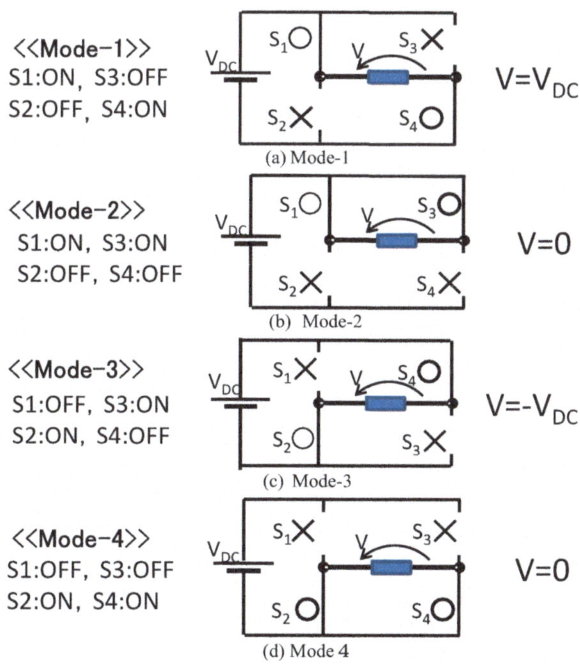

Fig. 5 Four modes in inverter circuit and its output voltage [7]

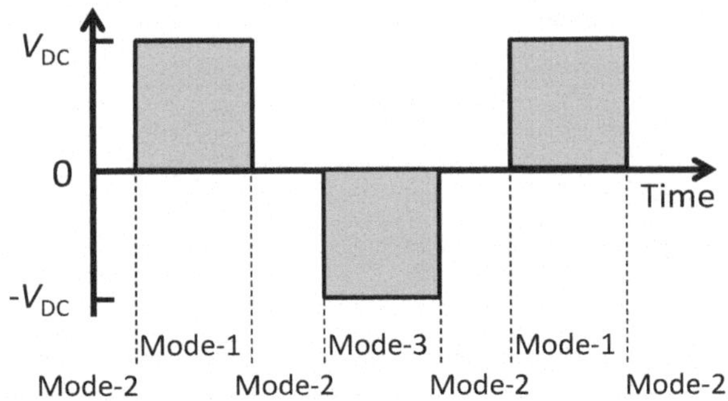

Fig. 6 Output voltage waveform of one-phase inverter

inverter can be selected as $-V_{\mathrm{DC}}$, 0, V_{DC} according to the switching ON/OFF pattern of power semiconductor S_1–S_4.

Now time-series of switching mode are considered to be selected and operated as the next chronological order for example

$$\begin{aligned}&\text{Mode 2} \rightarrow \text{Mode 1} \rightarrow \text{Mode 2} \rightarrow \text{Mode 3}\\ &\rightarrow \text{Mode 2} \rightarrow \text{Mode 1} \rightarrow \text{Mode 2} \rightarrow \cdots.\end{aligned}$$

Then, the time-series of output voltage are obtained like Fig. 6. An output voltage wave pattern becomes + voltage and − voltage in turn there periodically. As a definition of AC is that + voltage and − voltage repeat themselves in turn in a constant period, it may be said that this voltage wave pattern is AC. In other words, electrical power conversion from DC input to AC output would be realized by the switching operation of the power semiconductor in one-phase inverter of Fig. 4.

As the circuit of Fig. 4 shows only a main part, and the inverter circuit is comprised only of the power semiconductors mainly, the main voltage is applied in this circuit and the electric current flow there. Therefore, the inverter circuit of Fig. 3 is said to convert electrical power from DC to AC with loss "0", as the power consumption of the power semiconductor is "0" ideally, as far as the power semiconductor operates in ideal switching. Besides, the inverter can operate in high responsibility as it operates only in the switching operation of the semiconductor. In other words, it may be said that the inverter circuit is the electricity converter with high efficiency and high responsibility.

A voltage wave pattern of Fig. 6 is an AC wave, but it is different from a normal sine wave. The difference is derived from the step waveform with the rinsing up and the falling down of the output voltage, which is caused by the switching operation of the power semiconductor. The output voltage of the step waveform has a lot of harmonic components by Fourier transform. Therefore, the wave pattern of the inverter output has essentially the harmonic components other than the fundamental frequency component.

Now, a voltage wave V of Fig. 6 is assumed to have a periodical wave pattern of a fundamental frequency f_{o} as the next equation

$$V = V(\omega t), \tag{3}$$

here $\omega = 2\pi f_{\mathrm{o}}$. So, it can be transformed into Fourier series present as follows,

$$V(\omega t) = \sum_{i=1}^{\infty} v_i \sin(i\omega t + \varphi_i). \tag{4}$$

Then the effective voltage V_{rms} of overall is defined as follow.

$$V_{\mathrm{rms}} = \sqrt{\frac{\int_t^{t+T} V(\omega t)^2 \mathrm{d}t}{T}} \tag{5}$$

Here, $T = 1/f_{\mathrm{o}}$. On the other hand, an effective voltage of fundamental frequency V_{o} is defined as follows.

$$V_{\mathrm{o}} = v_1 \tag{6}$$

So, a representative voltage of inverter output voltage has two values as V_{rms} and V_{o}.

Usually, only a fundamental component makes an electromagnetic torque in electrical motor, but loss as copper loss should be evaluated in effective value. In general,

$$V_{\mathrm{o}} \neq V_{\mathrm{rms}} \tag{7}$$

Next, how to change the output voltage and the output frequency of the inverter is considered here. The output voltage pattern of Fig. 6 is decided by the switching pattern as mode 1–4. So, the output voltage can be changed by changing the switching pattern. The output voltage becomes large when width of the square wave of the output voltage is wider, and the output voltage becomes small when the width becomes narrower as shown in Fig. 7b. As an interval between the square waves of the output voltage is a period of the fundamental frequency, the frequency becomes low when the interval becomes widely, and the frequency becomes high when the interval becomes narrower as Fig. 7c.

The width and the period of the square wave of the voltage wave pattern can be changed like Fig. 7 by changing the timing from each mode of Table 1 to the other, namely by changing the timing of ON/OFF of switching element S_1–S_4. In other words, it may be said that how to change the output voltage and frequency can be realized by changing the timing of the change of ON/OFF of switching element S_1–S_4.

However, the harmonic components occur necessarily because of the switching operation of power semiconductor. The harmonic components deteriorate electromagnetic circumstance generally and increase the iron loss in the load. Therefore, it is necessary to reduce the harmonic components.

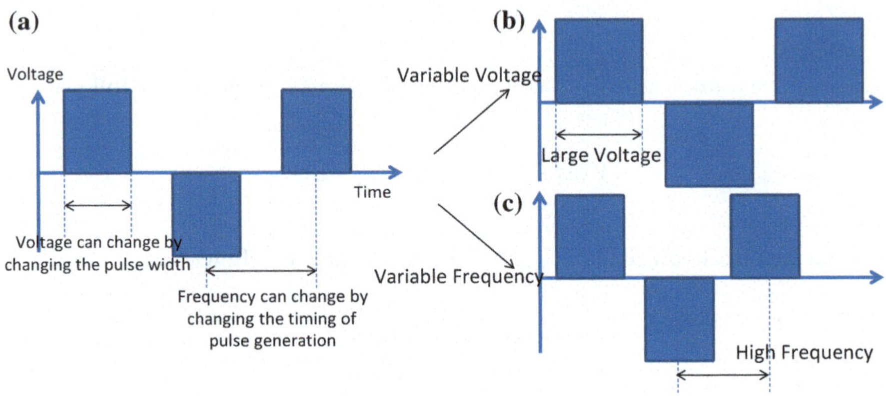

Fig. 7 Changing method of output voltage and frequency in inverter circuit [8]. Adapted from [8] with permission ©2016 The Magnetics Society of Japan

Several methods are considered to reduce the harmonic components of the inverter output wave pattern, but a sine-wave Pulse-Width-Modulation (PWM) inverter method is used well which uses triangular waves with the carrier frequency that is a higher frequency than the fundamental frequency.

Sine wave with a fundamental frequency f_0 and triangle wave with a carrier frequency f_c are considered, and the ON/OFF switching pattern of four power semiconductors is decided as Fig. 8 in the inverter. The sine wave has a wave height level of m (it is also said to be a modulation index), and the triangle wave has a wave height level of 1. The switching pattern is decided by which is higher or smaller.

Then, the wave pattern of the sine wave and the triangular wave are shown in upper part of Fig. 9, and the ON/OFF patterns of four power semiconductors

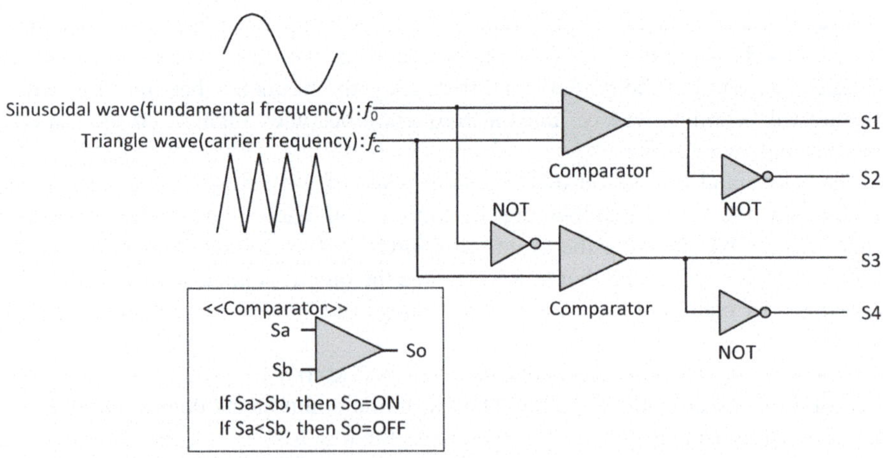

Fig. 8 Decision method of ON/OFF switching timing of four power semiconductors

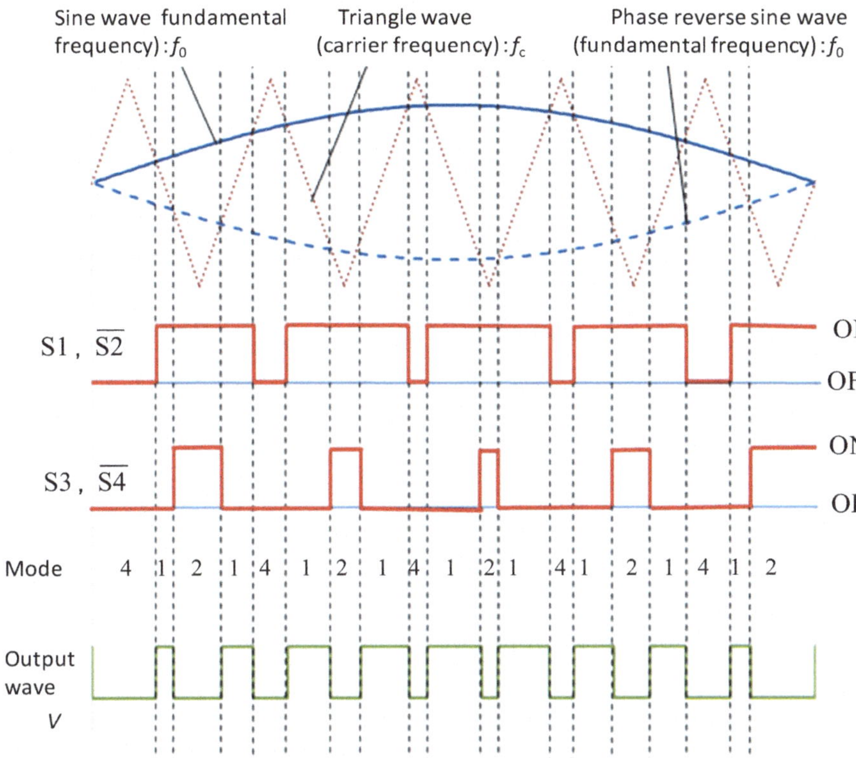

Fig. 9 Time-series of sine wave, triangular wave, switching patterns of four semiconductors, mode patterns and output voltage of inverter

S_1–S_4 are shown in middle part of Fig. 9. Each mode is inferred from the ON/OFF patterns of four power semiconductors, and finally, the output voltage waveform of the inverter is shown in lower part of Fig. 9. An example of time-series waveform of the sine-wave PWM inverter is shown in Fig. 10.

Fig. 10 Example of output voltage waveform of sine-wave PWM inverter

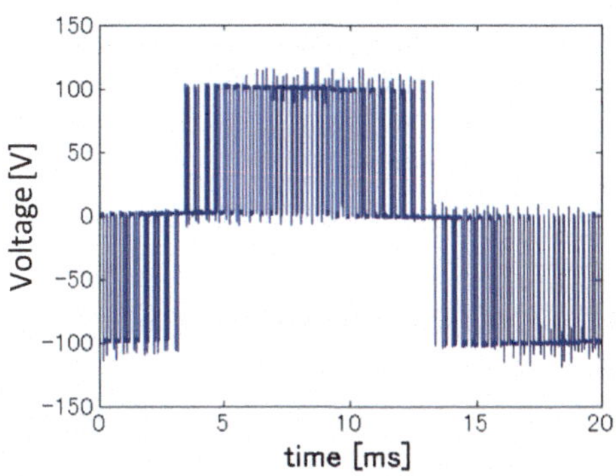

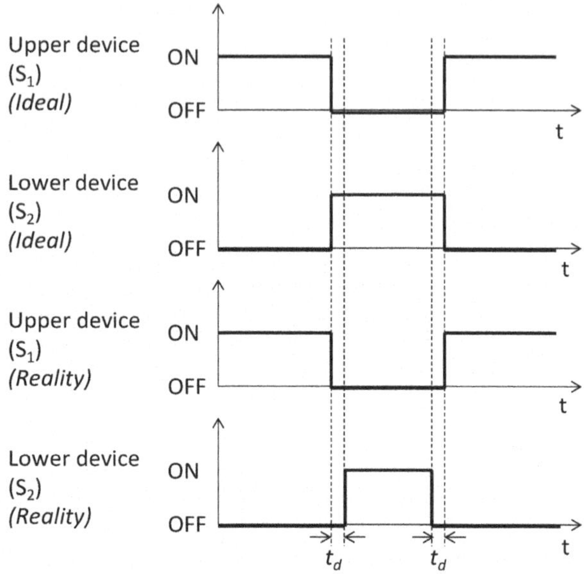

Fig. 11 Dead time in inverter circuit

A change of the ON/OFF switching of the power semiconductor is considered in one leg such as S_1 and S_2 here. When an ON/OFF signal is input to the power semiconductor, an electrical current and a supplied voltage in the power semiconductor are unable to suddenly carry out the ON/OFF operation. A transient state exists that the current and voltage gradually increase and decrease to the ON/OFF state. A rise-up time is gist around 1 μs for a Si device, and it is uneven for each element. When S_1 and S_2 become ON at the same time, a short-circuit electric current will flow and the semiconductors will be destroyed. Therefore, it is necessary to change ON/OFF of S_1 and S_2 having some extent with margin. That is a dead time t_d.

The upper two time-charts in Fig. 11 show the case of an ideal device, where a change of the ON/OFF switching is carried out instantly. However, as each real device has unevenness and rise time, the lower arm lately changes in ON from OFF for a dead time, when the upper arm changes in OFF from ON like lower two time-charts in Fig. 11.

When the carrier frequency f_c is low enough and the responsiveness of the semiconductor is fast enough, the dead time influence is not seen. However, when f_c becomes high and near to the responsiveness of the semiconductor, the influence of t_d begins to appear.

4 Significance of Power Electronics [4–6]

The social use of the electric energy starts to come true by the spread of the electric power network for lighting and motor application in a fan or a blower more than

100 years ago. The frequency and the phase are fixed there in commercial frequency (50/60 Hz) which did not include time-harmonic components, as transformer is used to convert its voltage economically and dynamically. Due to the fixed frequency as 50/60 Hz, the basic theory of the electric energy technique has been constructed assuming the commercial frequency as a result of many efforts of ancient people, and the electromagnetic materials research and development have been also carried out.

However, an appearance of power electronics technology and the full-scale use in the advanced society will greatly break the premise of the commercial frequency use. It is expected that a new operating points and plural frequencies are assumed as follows:

1. The diversification of the operating frequency, particularly a high-frequency operation.
2. Appearance of harmonics components by the switching operation of power semiconductor.

The past electrical engineering was based on electrical mathematics and electromagnetics expressed by complex number assuming a single frequency and linear material constants mainly. A circuit, an electrical apparatus, an electrical power system and its application, measurement method, and materials were discussed and designed assuming a constant frequency. However, in considering a new coming situation that the use of the power electronics with harmonic components and high-frequency operation will reach 80% of the electrical energy in almost about 10–20 years later, a premise of constant frequency may collapse. For example, new electromagnetic materials will be strongly expected in the premise of harmonic components and high-frequency operation. The new magnetic material is particularly required because it accounts for a big ratio in the weight, the price and the loss of the electrical energy equipment.

In considering the spread of power electronics and its influence to give, it is considered to be that the time has already come to construct a new academic theory system for the new electrical engineering based on a power electronics technology assuming the harmonic component and the high-frequency operation.

References

1. Erickson, R.W., Maksimovic, D.: Fundamental of Power Electronics, 2nd edn. Kluwer Academic Publications (2001)
2. Hase, Y.: Handbook of Power Systems Engineering with Power Electronics Applications, 2nd edn. Wiley (2013)
3. Fujisaki, K.: Necessity and problem of magnetic material for electromagnetic actuator system. In: The 2015 Annual Meeting of the Institute of Electrical Engineering of Japan, 5-S22-1, Mar 2015, pp. S22(1)–S22(4) (in Japanese)
4. Fujisaki, K.: Future trend of electrical motor drive system. In: The 39th Annual Conference on MAGNETICS in Japan, Symposium, "Energy Magnetics Improving Motor Efficiency", 09aA-3, Nagoya, H27.9.9, pp. 91–92

5. Fujisaki, K.: Required magnetic property for magnetic material in use of future electrical energy. In: The Japan Association of Bonded Magnetic Materials No. 86, JABM Autumn Symposium, Tokyo, Sept 2014 (in Japanese)
6. Fujisaki, K.: Advanced magnetic material requirement for higher efficient electrical motor design. In: The 38th Annual Conference on MAGNETICS in Japan, Symposium "Challenge of Magnetics to Improve Energy Efficiency", 4aB-2, Sept 2014
7. Fujisaki, K.: In: Annual Meeting of IEEJ, 2-s2-4 (2017) (in Japanese)
8. Fujisaki, K.: Magn. Jpn. **11**(1) (2016) (in Japanese)

Fusion Science and Technology of Electromagnetic Field [1]

Keisuke Fujisaki

Abstract Fusion science and technology of electromagnetic field is proposed here to obtain a higher efficient and downsizing electrical motor drive system for example. Motor drive system does not consist of only one technology such as electrical motor, control, power electronics, material, and production. "Electromagnetic energy apparatus," "electromagnetic field application," and "electromagnetic material" are considered to be related by a transfer of "thing." Each component requires multi-time for power electronics, multi-physics, and multi-scale, respectively. They should be considered to be a fusion science and technology such as electrical engineering, material, electromagnetic field, fluid dynamics, or so. So, they are called first-class fusion. The relation among the apparatus, the application, and the material is considered to be a relation of purpose and means. The purpose introduces what to make, and the means introduces how to make. So, they are called second-class fusion. It is usually difficult and necessary for a lot of time to learn plural department at the same time in the conventional study system. So, some special study system may be required for them.

Keywords Multi-scale · Multi-scale · Multi-time · First-class fusion · Second-class fusion · Purpose and means · Fusion science and technology

1 Introduction

The electromagnetic energy apparatus such as a motor drive system is now considered to be going on two new main developments by using the power electronics technology. It can get an arbitrary electric current, voltage, and frequency in high response. The one is an electrical motorization in the transportation sector which occupies 20–30% of the domestic energy consumption [2, 3]. The other is a microwave application technology of the frequency domain of the GHz degree. Three fields applications are considered as organic chemistry, inorganic material, and metal industry which occupy 30–40% of the domestic energy consumption [4–6].

K. Fujisaki (✉)
Toyota Technological Institute, Nagoya, Japan
e-mail: fujisaki@toyota-ti.ac.jp

K. Fujisaki (ed.), *Magnetic Material for Motor Drive Systems*,
Engineering Materials, https://doi.org/10.1007/978-981-32-9906-1_6

The former is a change from the drive of the internal combustion engine to the drive of the electric motor including the hybrid type, and it presents an aspect of a kind of moving revolution. The latter is accompanied by the new added value that before does not have such as a high speed and low-temperature reaction, reduction, and nitride reaction, and then space-saving process, when microwave is irradiated to organic, inorganic, and metal oxide material, and then a prototypic trial manufacture is partly carried out. Magnetron or so was used for a microwave power supply until now, but the use of the power semiconductor as GaN-FET has also begun by the development of the high-frequency semiconductor material.

Most of the domestic energy consumption presents an aspect used in electric energy in this way, and this trend is thought to be realized by the two technologies. The one is a numerical calculation of electromagnetic field which a physical phenomenon is easy to obtain when an electromagnetic field is irradiated, and the other is a power electronics technology which can give an arbitrary voltage and frequency to our life, society, and industry in high speed, high efficient, and cheaply. Electromagnetic field applications such as electrical motor drive system, microwave application, linear motor application or so should be discussed in the whole system including the related physical phenomena and the related scale and the related equipment to the electromagnetic field, because electromagnetic field has a mutual relation by being applied to the related physical phenomena or so. Therefore, these are placed as an electromagnetic field fusion science and technology like Fig. 1 and decide to consider it in details as follows.

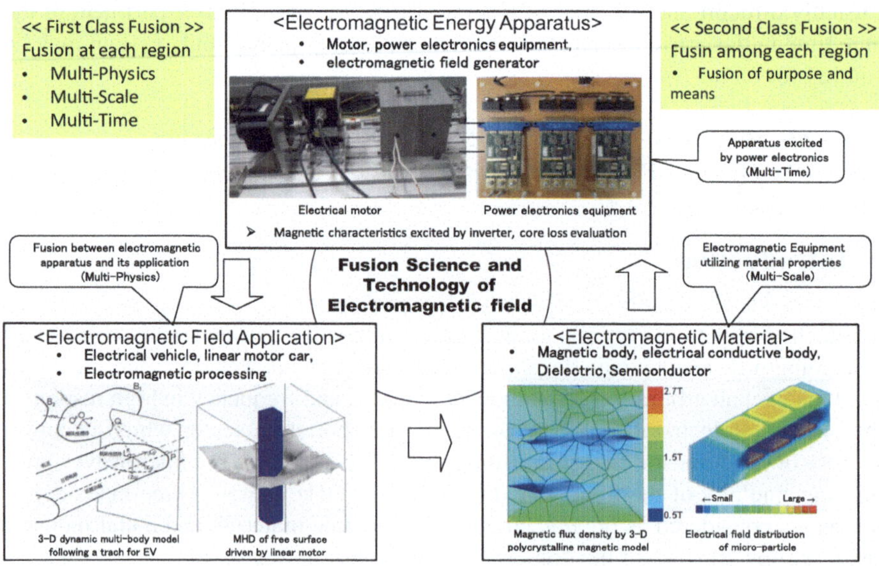

Fig. 1 Electromagnetic field fusion science and technologies [1]

2 Multi-scale, Multi-physics, and Multi-time (Fist Class Fusion) [1]

The conceptual diagram of electromagnetic field fusion science and technology is shown in Fig. 1. "An electromagnetic energy apparatus" and "electromagnetic field application," and "electromagnetic materials" are assumed to be a component and the flow of "the thing" is mainly shown there.

"The electromagnetic energy apparatus" such as the motor drive system and power electronics device is used in "the electromagnetic field application" such as an electric car, a linear motor car, and an electromagnetic process. The electromagnetic process as a part of the application produces the materials such as the electromagnetic steel sheet or semiconductor. A part of the materials manufactured by the process is a raw material of the motor and power electronics device and constitutes "an electromagnetic energy apparatus." In other words, the relations among "an electromagnetic energy apparatus" and "electromagnetic field application" and "electromagnetic materials" are the ones where "thing" itself drifts partly, and it may be said that these three components are the closed relations. The details are shown as follows.

2.1 Electromagnetic Field Application [7–9]

A rotary motor composes of a stator and a rotor and has a small gap between them. It is realized as they are mechanically fixed, and then a high efficiency is realized. However, it needs some mechanical elements such as a gear or a transmission in its application for a linear driving. Therefore, a linear motor drive can realize a downsizing and lighting and high speed due to the direct drive operation.

The direct drive system is a kind of electromagnetic field applied to the object directly to obtain the electromagnetic force. Another meaning of electromagnetic field application is irradiation of electromagnetic field to material process directly to obtain a new material or so. It is thought to be a new development in the electromagnetic field application here.

The electromagnetic process supplies some velocity and heat by giving low-frequency electromagnetic field to material processing and then realizes high quality and high productivity. When microwave magnetic field with high frequency is irradiated to organic matter, an inorganic matter, metal oxide, a new reaction and a new material appear which do not occur before; reduction, nitriding reaction, low-temperature reaction, short-time reaction, and creation of new material.

In the electromagnetic field application, a lot of physical phenomena proceed at the same time such as not only the electromagnetic phenomena but also fluid dynamics, heat transfer, solidification, chemical reaction, control, many rigid bodies, and vibration. So, it is a multi-physics phenomenon of the electromagnetic field. Figure 2

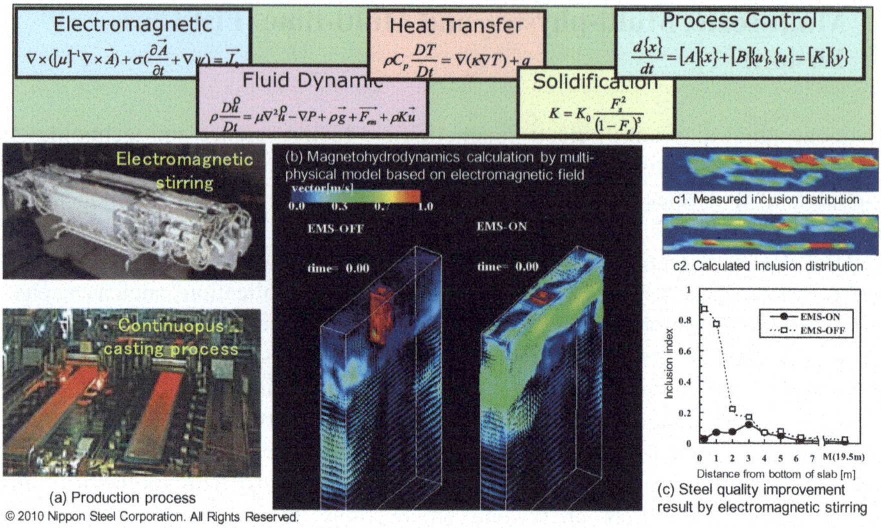

Fig. 2 Electromagnetic field multi-physics for electromagnetic process [7–9] Image courtesy of Nippon Steel & Sumitomo Metal Corporation

is an example of the multi-physics model in the electromagnetic process, which realizes the quality improvement of steel materials by giving electromagnetic force to molten steel (iron which melted) and giving some velocity on the solidification surface. In order to evaluate the solidification thickness and the adhesion situation of nonmetallic inclusion, numerical calculation combined with electromagnetic field, fluid dynamics, heat transfer, solidification, and process control is carried out.

In the multi-physics model, it is a matter which should be considered when the mutual interaction between the different physical phenomena is observed. The constitution of the model varies according to the strength and weakness of the mutual physical interaction. When different physical phenomena make an effect closely, they must be calculated at the same time. When the mutual interaction of the physical phenomena is considered to be a sparse connection, individual calculations of each physical phenomenon can be carried out by putting some assumptions. A way of the combination varies according to the technical problem to be calculated. In other words, the multi-physics analysis changes the combination way of the physical phenomenon and then the analytical technique also changes by a physical phenomenon for the analysis.

A physical phenomenon such as electromagnetic field, heat transfer, mechanical structure, and fluid dynamics is based on a vector analysis of the partial differentiation. So, an analogy is often useful for the other physical model to some extent, if one of the basic concepts and calculation methods is studied deeply. Most of the multiphysical models to be treated are the size of the meter degree.

The electromagnetic processes are one case of "the electromagnetic field application" in Fig. 1, but various kinds of materials are made from a manufacturing process, and "the electromagnetic materials" are one of them.

2.2 *Electromagnetic Materials [10, 11]*

The electromagnetic materials to think about in an electromagnetic field have a magnetic body, a conductive material, a dielectric, and a semiconductor. They are considered to be a multi-scale phenomenon to handle the different physical phenomenon of the scale including atomic structure, crystal, bulk, and composite material. The magnetic body in particular complicates a phenomenon more due to the magnetic domain structure of the microscale between a crystal grain and an atom level. Figure 3 shows the multi-scale model of the magnetic body. The definition and the usage of the micro–macro phenomena vary according to an academic domain or subject which are treated, but decided to think in this way here.

The electric car is a size of the meter (m) degree and is driven by an electrical motor of the macroscale of the centimeter (cm) degree. The motor core is made of electrical steel sheets of the polycrystalline. The size of the crystal grain is the mesoscale of the millimeter (mm) degree, and each crystal grain has some magnetic domain structures of the microscale of the micrometer (μm) degree in the inside. The magnetic domain structure has the saturated magnetization decided with the electronic spin of each atom of the size of the nanometer (nm) degree.

The governed equations are different on each scale. If different scale models can connect each other well, the electric car may be developed and designed at an electronic or atomic level. For example, if there is a magnetic material of the high saturated magnetization that before does not have on a nanoscale, downsizing and high output power motor drive system may be realized for the electric car.

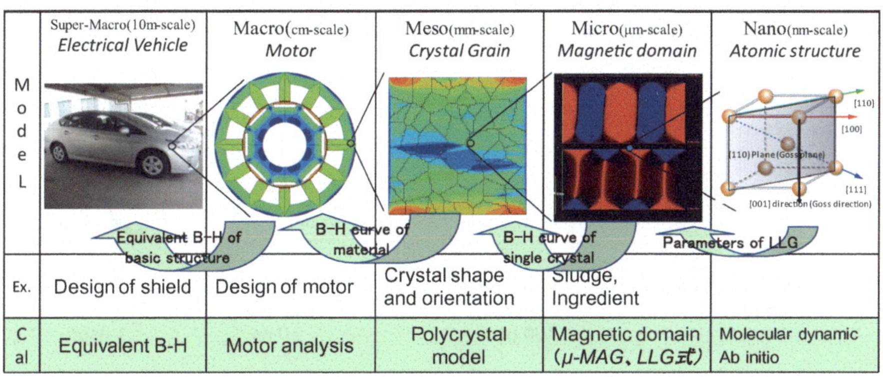

Fig. 3 Multi-scale model for magnetic body [3, 8]

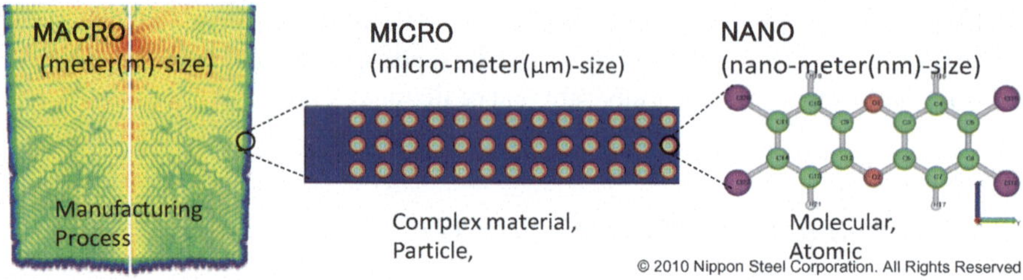

Fig. 4 Multi-scale model for microwave process

Figure 4 is an example of the multi-scale analysis for the microwave. The micro-size structure between the atom and the crystal grain does not exist in the dielectric, though it exists as the magnetic domain structure for magnetic body. However, it might be required when we consider an influence of the composite material such as a shape of some particles.

2.3 Electromagnetic Energy Apparatus

The electromagnetic materials including a magnetic body are used in an electromagnetic energy machine such as the motor drive system. The power electronics technology using the power semiconductor is used as a power supply controlling a motor. As for power electronics technology, a new problem occurs to have to deal with a different frequency and a timescale in the numerical analysis and the measurement system of voltage or current. Figure 5 shows the summarized figure of a multi-time for the power electronics.

Various kinds of frequency domains exist in Fig. 5 such as a fundamental operating frequency f_0, a carrier frequency f_c to reduce harmonic components in voltage or current, a rise time in the switching operation of the power semiconductor t_r, (a rise frequency $f_r = 1/t_r$ which is the reciprocal number of t_r), a sampling frequency f_s to

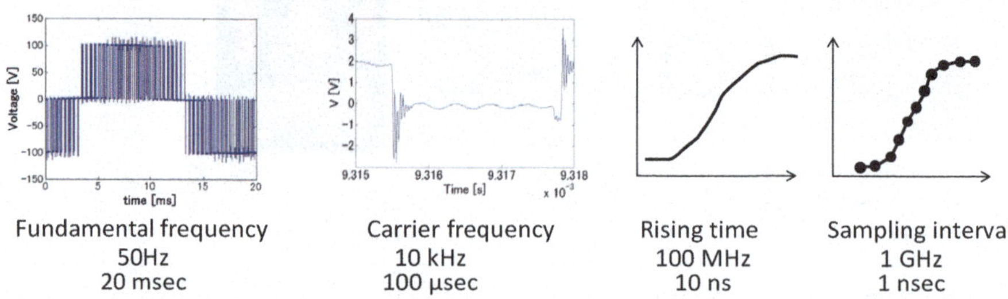

Fig. 5 Multi-time for power electronics

measure the time-series digital data such as voltage or current waveform (how many data are measured by analog digital (AD) conversion in one second).

A serious issue occurs for the electromagnetic field numerical analysis as well as the measurement. Great time for the calculation is required when magnetic nonlinearity or magnetic hysteresis property such as the *B-H* curve of the magnetic material is taken into account.

For example, time-series voltage or current is now considered to be treated for the numerical calculation where $f_0 = 50$ Hz, $f_c = 10$ kHz, $t_r = 10$ ns ($f_r = 100$ MHz) by GaN inverter excitation. When a rising voltage phenomenon is taken into account for the numerical calculation, at least a rise time should be divided into ten or more, and then a division time for nonlinear transient numerical calculation becomes at least $t_s = 1$ ns. Since CPU-Time = 1 day is needed for the calculation time at three-dimensional model in sine wave 50 Hz excitation by a usual personal computer, CPU-Time = 20,000 days = 55 years will be needed for the calculation time at three-dimensional model with GaN inverter excitation. As this is great calculation time, some kind of modeling is necessary.

2.4 *First-Class Fusion*

Fusion science and technology combined with plural technologies such as multi-physics for the electromagnetic application, multi-scale for the electromagnetic material, multi-time for the electromagnetic energy apparatus is introduced as an example. The viewpoint of the fusion science and technology is important and can treat with a further social request such as the downsizing, lightning, high efficiency, and high productivity.

The fusion science and technology is usually hard to insert the conventional study domain. So, some limitations occur as far as the conventional study domain is still used. However, if the different scientific regions can be studied deeply, since the parameters combined with both scientific regions doubles, their combinations will increase exponentially.

If a deep insight can be introduced and it can find out some good combination of better parameters, further development as downsizing, lightning, high efficiency, and high productivity will be realized. For example, a steelmaking process with the further quality improvement may be realized, if the relation between the linear motor parameters as pole number, frequency, structure, and the steel quality of electromagnetic processing can be understood.

To that end, it is insufficient only to remain in a single technique such as electrical engineering and electromagnetism, and then it is necessary to learn more associated technologies and sciences and to think about fusion science and technology including the interaction among them. This concept is placed as first-class fusion here.

The problems to push forward such fusion sciences and technologies are how researchers or engineers with full of specialty deepen the other specialty of the different field. It may be said to learn the other specialty back to a department level at least in order to learn the basic concept of a different study domain such as electrical engineering or materials engineering. However, it is almost impossible for the researcher who was already full of specialty to learn plural lectures one of which usually needs 15 lectures in which each lecture has 90 min like a general student.

Therefore, it is thought to be a solution for a specialist who was already full of specialty but would like to learn a different field to get ready a lecture course with only a few lectures. It may be thought to be enough because he already acquired the process of understanding of the study system, how to be applied to, and how to proceed with the research. It may be said that the different field lecture for such an expert is a clue presenting such fusion sciences and technologies in future.

3 Second-Class Fusion (Fusion of Purpose and Means)

The relation among three components as electric energy apparatus, electromagnetism application, and electromagnetic materials in Fig. 1 is now considered from a point of view of "purpose" and "means."

The purpose and application of the electromagnetic energy apparatus is electromagnetic field application such as a linear motor car or an electromagnetic process. The configurations or the design is different depending on its purpose. For example, a special linear motor is commercially used in on-line process as far as it realizes a high-quality process such as electromagnetic stirring equipment, even if its efficiency is less than 0.1%, though high efficiency such as 90% or more efficiency is usually required for most of the motor. We cannot but wait for the development of the electromagnetic materials which are the component in order to make an improved apparatus in higher efficiency. The material is the means of the apparatus improvement.

What kind of constitution an electromagnetic field application is depends on what kind of product it wants to make. For example, when electromagnetic materials are made with an electromagnetic process, the evaluation is decided whether it can make the materials with a predetermined function. When it is going to be made better, the improvement of itself is possible, but the drastic improvement needs the introduction of the new electric energy apparatus such as the electromagnetic stirrer.

For the question why electromagnetic material wants to be made, the answer should be considered from a point of view of the user side of the material, e.g., it is for the characteristics improvement of "the electromagnetic energy apparatus." When it is required to improve the characteristic of electromagnetic materials, it is necessary to be involved in the manufacturing process.

Here, the electromagnetic field fusion science and technology is required like Fig. 1.

The index such as sensitivity, society, transportation, and structure application does not exist here, and there is a thing except this loop, too, and it is not always the closed loop. However, as far as the contents are considered only in this figure, three technologies as apparatus, application, and material are in a relation of the means and the purpose of each other.

For all the technologies, the clarification of "the purpose" and "the means" is an important matter. "The purpose" shows the directionality as what to make, and "the means" shows the realization as how to make it. It is often considered to fall into an uncontrollable and unstable situation, as the true purpose and the true means often exist other than the area to be treated.

Therefore, if one can grasp the three components in itself, the one can perform "how to make" for "what to make" by oneself. This is a significance of the second-class fusion science and technology, that is, the fusion of "purpose" and "means." The first-class fusion science and technology is a fusion within each technological area such as apparatus, application, and material, and the second-class fusion sciences and technologies are a fusion combined with each technological area.

The motor drive system is said to be one of the examples. Both fusions are a serious problem to be realized, but it may be said to suggest the directionality of future study and education for a long life carrier. So I decided to show it here.

References

1. Fujisaki, K.: Fusion technology of electromagnetic field and motor drive system. In: The 2016 Annual Meeting of the Institute of Electrical Engineering o Japan, pp. 5–025. Fukuoka, March 2016. (in Japanese)
2. Investigating R&D Committee of Magnetic Material for Electromagnetic Actuator: Magnetic Material and Its Evaluation Technology for Electromagnetic Actuator. Technical report of IEEJ, No. 1397,2017.11 (in Japanese)
3. Fujisaki, K.: Necessity and problem of Magnetic Material for Electromagnetic Actuator System. In: The 2015 Annual Meeting of the Institute of Electrical Engineering o Japan, S22(1)–S22(4), 5-S22-1, March, 2015. (in Japanese)
4. Fujisaki, K.: Microwave processing. Handbook of Electrical Engineering, 7th edn. Chapter 41, Section 4.4, pp. 2106–2107 (2013). (in Japanese)
5. Fujisaki, K.: Energy significance of microwave processing. In: The 12th JEMEA (Japan Society of Electromagnetic Wave Energy Applications) Symposium (Sympo2015), O-26, pp. 70–71, 2015.11.20. (in Japanese)
6. Microwave Chemical Co., Ltd.: http://mwcc.jp/. Accessed 9 Jan 2018
7. Fujisaki, K.: Application of electromagnetic field multi-physical model to steel making plant. In: The 2009 Annual Meeting of the Institute of Electrical Engineering of Japan, No. 5–209 (2009). (in Japanese)
8. Fujisaki, K., Hirayama, R., Wajima, K.: Multi physical model application to process solution in electromagnetic field technique. Nippon. Steel Tech. Rep. **379**, 54–58 (2003). (in Japanese)
9. Fujisaki, K.: Electromagnetic field multi-scale by equivalent material constants. Trans. Jpn. Soc. Mech. Eng. (JSME), **119**(1176), 20–23 (2016). (in Japanese)
10. Fujisaki, K., Hirayama, R., Nemoto, Y.: Electromagnetic steel solution in electromagnetic field technique. Nippon. Steel Tech. Rep. **379**, 70–74 (2003). (in Japanese)
11. Fujisaki, K.: Magnetic multi-scale for small and high efficiency motor. In: The 2012 Annual Meeting of the Institute of Electrical Engineering o Japan, No. 5–008 (2012). (in Japanese)

Magnetic Material Excited by Power Electronics Equipment

Magnetic Property and Measurement Excited by PWM Inverter

Keisuke Fujisaki

Abstract A motor drive system is used as a means of transportation such as electric cars, and an electrical steel sheet used as a core of the motor is excited by an inverter for variable velocity operation. Because the electrical steel sheet is usually evaluated under an excitation of a linear amplifier which does not include time-harmonic components, it has a different magnetic property from the one which is excited by the inverter. So, the magnetic property of the electrical steel excited by the inverter is shown in detail. In the inverter excitation, magnetic characteristics of ring core have a lot of minor loops as closed loop and open loop caused by carrier frequency of the inverter and the ON voltage of the power semiconductors. So, the iron loss increases due to the inverter excitation. When the power semiconductor with small ON voltage is used as a switching device in the inverter, the minor loops become small and the iron loss decreases. Higher sampling frequency for AD converter is required to express the pulse-shape voltage of the inverter. The required magnetic characteristics in inverter excitation depend on how to use the magnetic material on the inverter.

Keywords Inverter excitation · Minor loop · Closed loop · Open loop · ON voltage

1 Measurement Equipment of Magnetic Property Excited by Inverter

The electrical steel sheet used for the core of electrical motor has superior magnetic characteristic as high magnetic flux density and low iron loss and economical material.

The magnetic characteristic of the magnetic material such as electrical steel sheet or so is usually decided to be measured by the sine-wave excitation with the linear amplifier which does not include time-harmonics components due to JIS or IEC standard [1, 2]. However, when the electrical steel sheet is used with a core material

K. Fujisaki (✉)
Toyota Technological Institute, Nagoya, Japan
e-mail: fujisaki@toyota-ti.ac.jp

K. Fujisaki (ed.), *Magnetic Material for Motor Drive Systems*,
Engineering Materials, https://doi.org/10.1007/978-981-32-9906-1_7

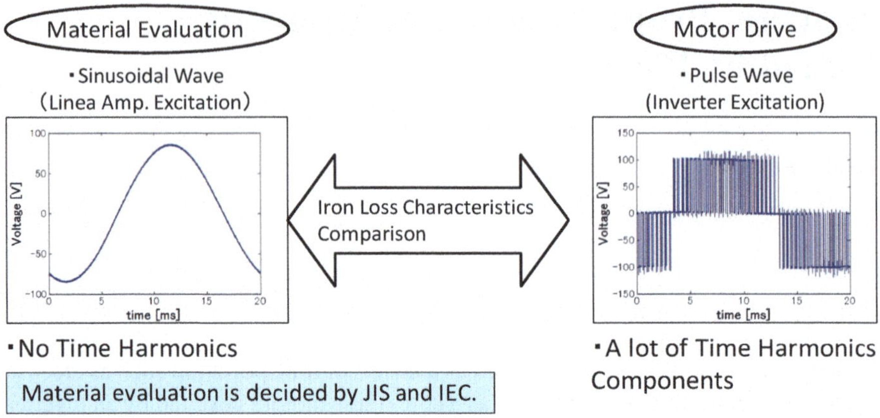

Fig. 1 Sine-wave excitation (standardization by JIS and IEC, left) and inverter excitation of magnetic material (excitation means in motor drive, right) [1, 2, 4–6] JIS C 2552, 2014, P12–13 [7]

of the motor drive for variable velocity operation, it is excited by the inverter due to variable voltage and variable frequency operation like Fig. 1 [3].

Because time-harmonic components are caused by the switching operation of the power semiconductor in the inverter excitation essentially, it is necessary to clarify how it affects the magnetic characteristic. Here, the magnetic characteristic of the magnetic material is clarified in the inverter excitation by comparing it with the magnetic characteristic of the sine-wave excitation.

To make a ring sample, an electromagnetic steel sheet is cut by an electric discharge machine into a ring form, and is laminated, and is entered into the acrylic case, and is coiled with copper coil in the circumference. Two kinds of copper coil are wounded as n_1 turns for the primary coil and n_2 turns for the secondary coil. After the terminal of the primary coil is connected to the inverter and then the voltage is supplied, the current I_1 flows in the primary coil. Magnetic path length l_1 is assumed to be a centerline length of the ring sample, and then the strength of magnetic field H is introduced by the next equation.

$$H = \frac{n_1 i_1}{l_1} \tag{1}$$

The terminal of the secondary coil is connected to a voltage probe and then an induced electromotive force e_2 is measured. Then, the magnetic flux density B is introduced by time integration of e_2 in one period T_0 as the next equation.

$$B = \frac{1}{n_2 S} \int_{t}^{t+T_0} e_2 \mathrm{d}t \tag{2}$$

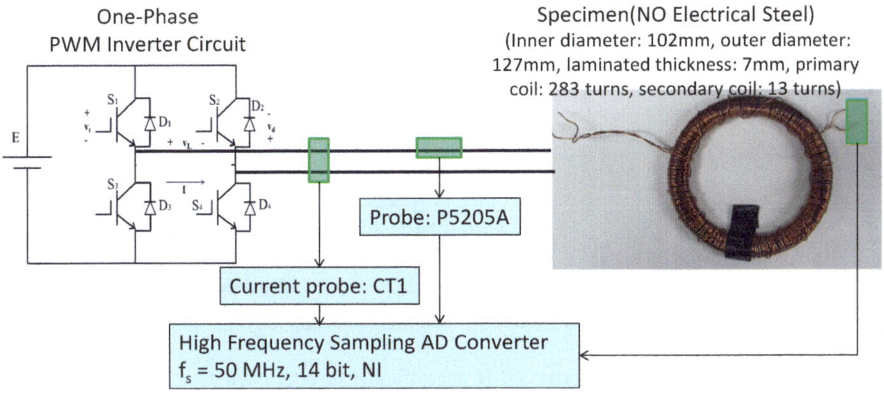

Fig. 2 Magnetic measurement system of ring sample in inverter excitation [3]. **a** Sine-wave excitation ($W_{fe} = 1.22$ W/kg). **b** Inverter excitation (m = 0.4) ($W_{fe} = 1.65$ W/kg)

Here, S is a cross-section of the ring sample. The magnetic characteristic of the B-H curve is introduced, and then iron loss W_{fe} is derived from the next equation

$$W_{fe} = \frac{f_0}{\rho} \int_{t}^{t+T_0} H \mathrm{d}B \tag{3}$$

The ring sample is also connected to the linear amplifier which does not include time-harmonic components, and then the magnetic property in the sine-wave excitation is measured for a comparison.

A sine-wave PWM inverter modulation method is used here in order to reduce the time-harmonics components as much as possible. The ON/OFF time of the power semiconductor in the inverter circuit is introduced by the cross-point with the sine-wave signal having fundamental frequency (f_0) and the chopping wave signal having carrier frequency (f_c). Therefore, the output voltage wave pattern of inverter becomes square waves of small width with a carrier frequency like Fig. 1 left and decreases the lower level of the time-harmonic components' content and lowers the iron loss of the load. Figure 2 is the magnetic measurement system of the ring sample in inverter excitation.

2 Minor Loop Generation in Inverter Excitation

The measured magnetism properties (B-H curve) are shown in Fig. 3. Figure 3a is the one of the electrical steel sheet excited by the sine wave not to include time-harmonics components. It obeys the standardization as JIS or IEC and can get clean B-H curve [1, 2]. In contrast, Fig. 3b is the one excited by the sine-wave PWM

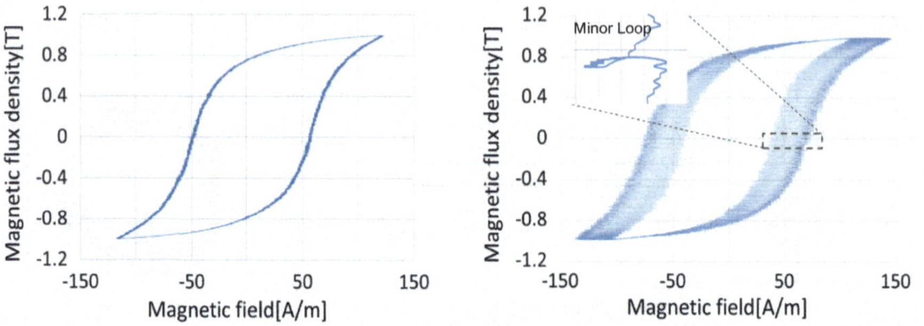

(a) Sine –wave excitation (W_{fe} = 1.22 W/kg) (b) Inverter excitation (m = 0.4) (W_{fe} = 1.65 W/kg)

Fig. 3 Measured magnetism properties (*B*-*H* curve) under inverter excitation [3, 5, 6] (0.35 mm thickness silicon steel, $f_0 = 50$ Hz, $f_c = 10$ kHz, $B_{max} = 1$ T)

inverter. There are a lot of small minor loops depending on the carrier frequency of the inverter in *B*-*H* curve [3]. Both cases are arranged so as that the maximum magnetic flux density in one period becomes 1 *T*.

A minor loop is observed when the ripple component is partially enlarged. As a part of closed loop becomes the double integral in the iron loss calculation, its iron loss increases. In this case when the iron loss with sine-wave excitation and the inverter excitation is compared, the iron loss of inverter excitation is approximately 35% larger than that of sine-wave excitation. In this way, it can be said that the iron loss of the electrical steel sheet increases by a minor loop which occurs by the inverter excitation.

In order to evaluate an influence of the iron loss increase by the inverter excitation, two kinds of electrical steel are selected. They have approximately the same iron loss under the conventional sine-wave excitation and have the same characteristics under the conventional JIS and IES evaluation. Modulation index characteristics of the inverter excitation are measured. A modulation index is a ratio of the sine-wave height level of the fundamental frequency to the chopping wave height level of the carrier frequency whose cross-point decides the ON/OFF signal of the power semiconductor in the inverter. A result is shown in Table 1.

Table 1 Iron loss increase ratio under inverter excitation [3] (The number in the parenthesis is a ratio when iron loss of sine-wave excitation is assumed to be 100)

Excitation method			Thickness of steel	
			Iron loss (W/kg)	
			0.35 mm	0.50 mm
PWM inverter excitation	Modulation index	0.4	1.65 (135)	1.78 (150)
		0.6	1.45 (118)	1.53 (129)
		0.8	1.38 (113)	1.40 (118)
Sine-wave excitation			1.22 (100)	1.19 (100)

Table 2 Grade of electrical steel and iron loss difference in 1 grade up [3] ($B = 1.5$ T, $f_0 = 50$ Hz)

0.35 mm thickness	Iron loss difference ratio (%)	0.50 mm thickness	Iron loss difference ratio (%)
35A230	8.7	50A270	7.4
35A250	20.0	50A290	6.9
35A300	20.0	50A310	12.9
35A360	22.2	50A350	14.3
35A440	–	50A400	17.5

Electrical steels whose thickness is 0.35 mm and 0.5 mm have the same iron loss under the sine-wave excitation, but the iron loss of 0.50 mm thickness is about 5% to 15% larger than one of the 0.35 mm thickness under the same inverter excitation condition. When the thickness of the steel sheet becomes large, it may be said that the eddy current of the harmonics components by the carrier frequency grows.

When the modulation index increases, the iron loss decreases for both steels. The increase of the modulation index causes the decrease of DC voltage of the inverter input voltage, which also causes the decrease of ripple components of the current and then the minor loop of Fig. 3b becomes small.

In other words, it can be said that electrical steels often show different iron loss characteristics under the inverter excitation even if they have the same iron loss under the conventional sine-wave excitation.

The iron loss characteristics in Table 1 have 10–50% increase under the inverter excitation. So, the significance of this iron loss increase ratio is considered here based on the standard of the electrical steel sheet established in JIS, IEC.

Table 2 shows the grade of the electrical steel sheet of 0.35 mm thickness and the 0.5 mm thickness and their iron loss from high grade materials sequentially and also shows how iron loss increased when a steel grade goes up 1. This table shows that the iron loss increases 7–22% degree when a steel grade goes up 1. In other words, it can be said that the iron loss increase ration by the inverter excitation means the 1–2 grades difference of electrical steel. The iron loss increase rate is relatively large.

3 Carrier Frequency Characteristics Under Inverter Excitation

The carrier frequency property, that is, one of the parameters of the inverter control is considered here in order to evaluate an influence on iron loss by the control method as shown in Fig. 4. Both steel sheets of the 0.35 mm thickness and the 0.50 mm thickness have the same iron loss under the sine-wave excitation. However, it is found out that the iron loss of the large thickness steel sheet of 0.50 mm is larger than the small one of 0.35 mm under the inverter excitation. It is thought that the

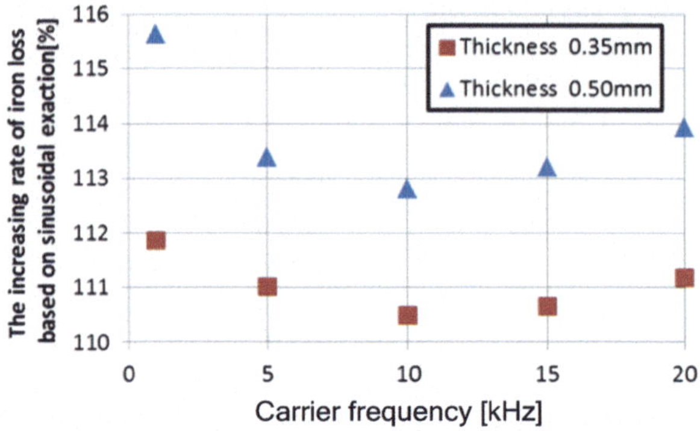

Fig. 4 Carrier frequency and iron loss increase ratio [3]

eddy current flowing in the large thickness steel excited by the harmonic components of the carrier frequency is larger than the one in the small thickness steel.

The iron loss decreases to around 10 kHz when the carrier frequency increases, but it increases at the high carrier frequency such as 20 kHz adversely. When the carrier frequency increases, the eddy current increases, but it may be said that the influence by the harmonics decreases and then the iron loss decreases because the harmonic component by the carrier frequency decreases further and the area where the eddy current flows becomes shallow by the skin effect which occurs in high frequency. In contrast, in higher carrier frequency such as 20 kHz, a pulse-width of inverter output voltage becomes small and then an influence of the dead time in the inverter to the pulse-width comes out, and the amount of pulse-width becomes small. When a maximum magnetic flux density in one period keeps constant as $B_{max} = 1.3\ T$ in carrier frequency properties, since pulse-width becomes small in high carrier frequency, DC voltage which is an input voltage of the inverter and a pulse height of the inverter output must increase. Higher DC voltage increases the ripple components of the current and then the iron loss increases.

In this way, if it raises a carrier frequency, the iron loss decreases, but it can be said that iron loss tends to go up under the influence of dead time adversely when it is going to be given more.

4 Minor Loop Outbreak Due to On-Resistance of Power Semiconductor [4]

As an outbreak of the minor loop under the inverter excitation can become a double integral of the iron loss from the iron loss calculation as Eq. (3), it causes the increase of the iron loss. So, the detailed analysis of its mechanism is tried to examine.

When the minor loop is researched in the *B-H* curve of Fig. 5, in detail, two kinds of minor loop are found out to be observed as a closed loop of Fig. 5a and an open loop of Fig. 5b [8]. The outbreak domain is different in a plus and minus of magnetic

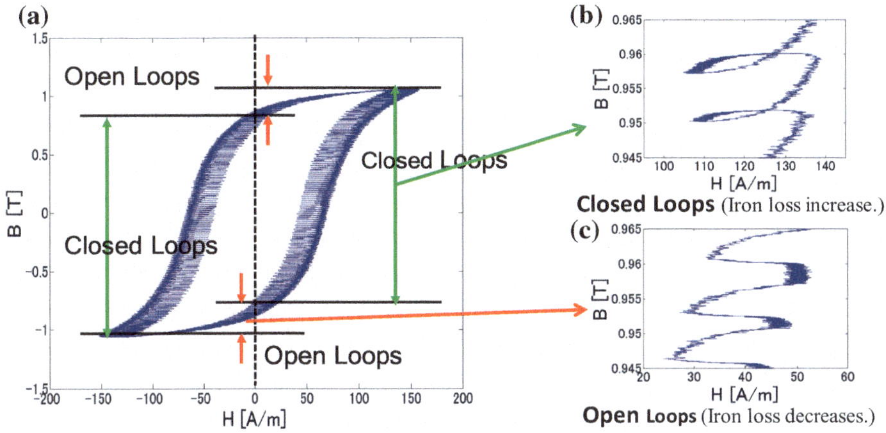

Fig. 5 Closed loop and open loop of minor loop in *B-H* curve under inverter excitation [8]. Reproduced from [8] with the permission of AIP Publishing

field *H* as well as increase and decrease of magnetic flux density *B* like Fig. 5. The closed loop causes the iron loss increase because the closed loop area in the *B-H* curve is double-counted in the iron loss calculation as Eq. (3), but the open loop does not cause the iron loss increase because of no double-counted area.

The closed loop occurs, when *H* is positive and *B* increases or *H* is negative and *B* decreases in *B-H* curve. Conversely, the open loop occurs, when *H* is positive and *B* decreases or *H* is negative and *B* increases in *B-H* curve.

The detailed phenomenon of the closed loop is considered here based on Fig. 6 where the enlarged closed loop and time series of magnetic flux density, voltage, current and inverter circuit are shown. One of the closed loops in Fig. 6a is enlarged

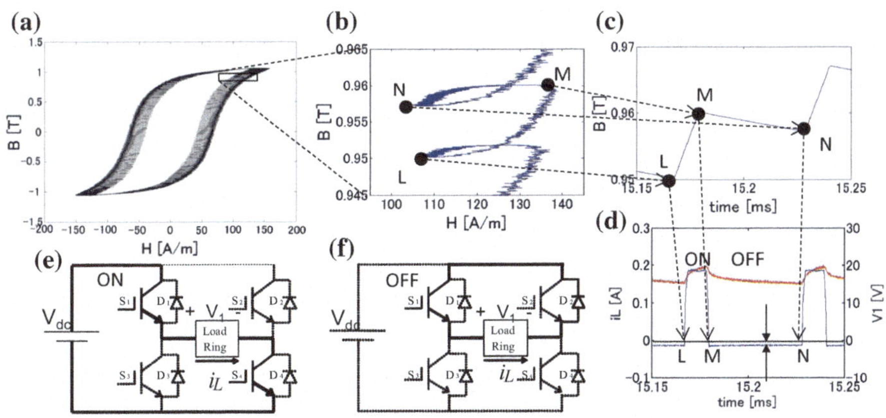

Fig. 6 Closed loop and inverter circuit [8]. Reproduced from [8] with the permission of AIP Publishing

as Fig. 6b, and points L, M, N on it are considered. In the time LM, magnetic flux density and magnetic field increase, and in the time MN, they decrease. These trends are shown in time series of magnetic flux density and magnetic field as Fig. 6c and d. As the magnetic flux density is provided by the time integral of electromotive force of the secondary coil in Fig. 2 and the electromotive force is almost proportion to the voltage of the ring coil, the voltage is plus in the time LM when the magnetic flux density increases, and the voltage is minus in the time MN when the magnetic flux density decreases as shown in Fig. 6d.

The inverter circuit becomes Fig. 6e in the time LM, and it becomes Fig. 6f in the time MN. In the ON state as Fig. 6e, when DC voltage V_{dc}, ON voltage V_{S1} of semiconductor S1, voltage V_l of the load, ON voltage V_{S4} of semiconductor S4 are considered as a closed electrical circuit in the inverter, the next equation is introduced.

$$V_{dc} = V_{S1} + V_l + V_{S4} \tag{4}$$

As the ON voltage of the semiconductor is smaller enough than DC voltage V_{dc}, the load voltage V_l is approximately the same as V_{dc}. In contrast, in the OFF state as Fig. 6f, when ON voltage V_{S1} of semiconductor S1, voltage V_l of the load, ON voltage V_{D4} of diode D4 are considered as a closed electrical circuit in the inverter, the next equation is introduced.

$$0 = V_{S1} + V_l + V_{D3} \tag{5}$$

In other words, the load voltage V_l has a minus value whose strength is a summation of ON voltage V_{D3} of diode D3 and ON voltage V_{S1} of semiconductor S1. It may be said that the ON voltage of the power semiconductor causes a decrease of magnetic flux density and then a closed loop occurs.

Next, the detailed phenomenon of the open loop is considered here based on Fig. 7 where the enlarged open loop and time series of magnetic flux density, voltage, current and inverter circuit are shown. One of the open loops in Fig. 7a is enlarged as Fig. 7b, and points O, P, Q on it are considered. In the time OP magnetic flux density and magnetic field decrease, and in the time PQ, they also decrease. These trends are shown in time series of magnetic flux density and magnetic field as Figs. 7c and d. As the magnetic flux density is provided by the time integral of electromotive force of the secondary coil in Fig. 2 and the electromotive force is almost proportion to the voltage of the ring coil, the voltage is minus in the time OP when the magnetic flux density decreases, and the voltage is minus in the time PQ when the magnetic flux density decreases as shown in Fig. 7d.

The inverter circuit becomes Fig. 7e in the time OP, and it becomes Fig. 7f in the time PQ. In the ON state as Fig. 7e, when DC voltage V_{dc}, ON voltage V_{S3} of semiconductor S3, voltage V_l of the load, ON voltage V_{S2} of semiconductor S2 are considered as a closed electrical circuit in the inverter, the next equation is introduced.

$$V_{dc} = V_{S3} - V_l + V_{S2} \tag{6}$$

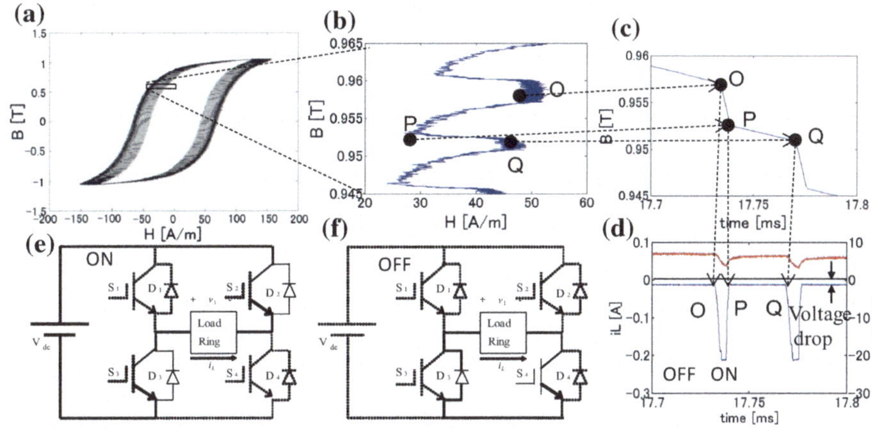

Fig. 7 Open loop and inverter circuit [8]

Here, the load voltage is minus due to its definition.

As the ON voltage of the semiconductor is smaller enough than DC voltage V_{dc}, the load voltage V_1 is approximately the same as $-V_{dc}$ in the ON state. In contrast, in the OFF state as Fig. 7f, when ON voltage V_{S4} of semiconductor S4, voltage V_1 of the load, ON voltage V_{D2} of diode D2 are considered as a closed electrical circuit in the inverter, the next equation is introduced.

$$0 = V_{S4} + V_1 + V_{D2} \tag{7}$$

In other words, the load voltage V_1 has a minus value whose strength is a summation of ON voltage V_{D2} of diode D2 and ON voltage V_{S4} of semiconductor S4. As the load voltage is minus and then magnetic flux density decreases in the ON state as well as OFF state, an open loop occurs.

The above discussion shows that ON voltage properties of the power semiconductor in the inverter circuit are related to a factor of the iron loss increase under the inverter excitation.

5 Power Semiconductor Property and Iron Loss [9]

In the previous section, it was recognized that ON voltage properties of the power semiconductor affected the iron loss increase under the inverter excitation. Therefore, some power semiconductors which vary in the ON voltage are chosen, and an inverter circuit experimentally is produced using it, and the magnetism characteristics under the inverter excitation are evaluated [9].

At first, a bipolar transistor and a MOSFET whose rating is small and ON voltage is small, and an IGBT whose rating is large and ON voltage is large are selected as a switching device. These devices are combined with a diode with small ON voltage and a device with large ON voltage, respectively, and then an inverter circuit is produced experimentally. The detail specification of three kinds of inverter with three different devices is shown in Table 3. These different inverters excite the same ring sample with the same electrical steel as 35H300, and magnetic properties as *B-H* curve are measured.

The results are shown in Fig. 8. All the three *B-H* curves excited by three different inverters seem to have the same magnetism properties overall at a glance like Fig. 8(a-1), (b-1), (c-1), but their minor loops have the different size like Fig. 8(a-2), (b-2), (c-2) when they are enlarged partially. Figures (a-2), (b-2) where the ON voltages are small are a small minor loop, but Figure (c-2) where the ON voltage is large is a large minor loop. As a result, the iron loss excited by the inverters with bipolar transistor and MOSFET which have the small ON voltage is more than 10% smaller than the iron loss excited by the inverter with IGBT which has the large ON voltage.

Next, magnetic property excited by inverters with two different semiconductor materials is measured. The first material is SiC which is expected as a next-generation material of power semiconductor and its device is SiC-FET which is used in inverter and called a SiC inverter here. The second material is Si which is usually used and its devise is IBGT which is used in inverter and called a Si-IGBT inverter here. Table 4 shows the electrical characteristics of the power semiconductors used in the inverters. A new power semiconductor device of SiC-FET has more than one-tenth smaller ON voltage than the conventional Si device.

Magnetic characteristics (*B-H* curve) of the same ring sample excited by the SiC inverter and the Si-IGBT inverter are shown in Fig. 9 [10]. Both magnetic characteristics have almost the same *B-H* curve at a glance as Fig. 9a, but partial enlargement of the minor loop shows a different size as Fig. 9b. In other words, the minor loop size excited by the SiC inverter which has a small ON voltage is smaller than the one excited by the Si-IGBT which has a large ON voltage.

The iron loss properties changing the carrier frequency are shown in Fig. 9, when the ring sample with same electrical steel is excited by both inverters. Since the size of the closed loop is smaller under the SiC inverter excitation, the iron loss excited by the SiC inverter is a little smaller by the one excited by the Si-IGBT inverter.

Figure 10 shows the measured iron loss characteristics excited by two inverters with the different power semiconductors, and it is found out that the iron loss of the magnetic material is different when the excited inverter has different power semiconductor characteristics [11].

Table 3 Power semiconductor characteristics of inverters using three different power semiconductors 2013 IEEE Reprinted, with permission, from [9]

	Switching semiconductor device			Fly wheel diode		
	Type	Maximum rating	Measured ON voltage	Type	Maximum rating	Measured ON voltage
Bipolar transistor–INV	Bipolar transistor 2SC3422 (TOSHIBA)	40 V 3 A	39 mV	Schottky diode 2GWJ42 (TOSHIBA)	40 V 2 A	330 mV
MOSFET–INV	MOSFET 2SK2382 (TOSHIBA)	100 V 12 A	40 mV	Schottky diode 2GWJ42 (TOSHIBA)	40 V 2 A	330 mV
IGBT–INV	IGBT PM75RSA060 (MITSUBISHI)	600 V 75 A	700 mV	Si-diode PM75RSA060 (MITSUBISHI)	600 V 30 A	800 mV

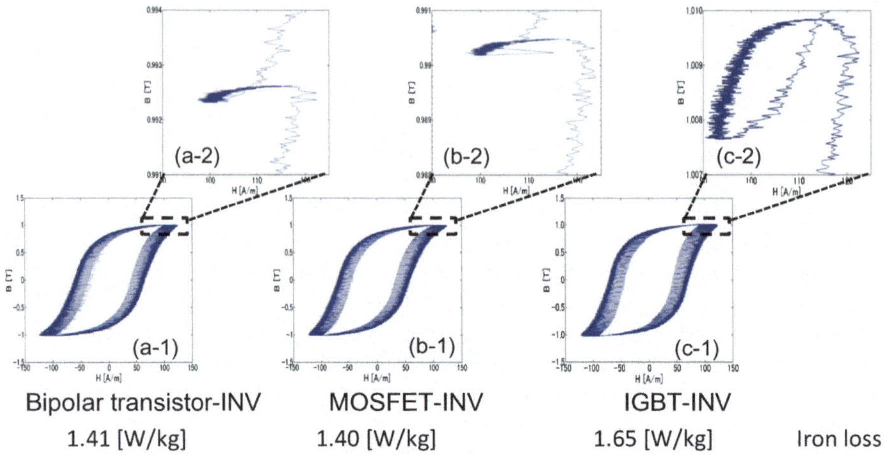

Fig. 8 *B-H* curves ((a-1)–(c-1)) and minor loop shape ((a-2)–(c-2)) and iron loss value excited by inverters with three kinds of power semiconductor IEEE, Reprinted, with permission, from [9]

6 Measurement Technology in Inverter Excitation [12]

Magnetic measurement data excited by the inverter usually memorize digitalized data through analog/digital (AD) converter. So, it is useful to grasp the fundamental characteristics of measured data such as whether the digitalized measured data express a square wave of output voltage of the inverter, minor loop shape in *B-H* curve, how much crror the measured iron loss has. Here, a sampling period (its reciprocal is a number of sampling points per one second) and bit number of the measured data in AD converter are examined.

Figure 11 shows a value of the iron loss when the sampling period is changed from 500 kS (in other words, 500,000 data per second) to 50 MS. Small sampling number is useful to measure the data. The data are measured in 50 MS and then roughened gradually to 500 kS. The measurement data from 50 to 15 MS have almost the same iron loss, but it is a gradually small value after it. Figure 12 showed the shape of the minor loop at each sampling period. The minor loop shape is all the same up to 15 MS, but it begins to change from 5 MS, and when the sampling number is less than 1 MS, the shape greatly changes. Because a rise time of the inverter with Si device is usually about 1 μs, that is, about 1 MHz, a sampling number of 5–15 MS is at least required to express time series digitalized data in the rise time.

Next, a problem of bit number is considered here. Figure 13 shows the *B-H* curve and its partial enlargement in measuring with resolution of 16 bit and resolution of 12 bit. The difference of both minor loop shapes is small enough, and it may be said that measuring with 12 bit resolution is not so enough problem.

Table 4 Power semiconductor characteristics of inverters using SiC material device and Si-IGBT [10]

	Switching semiconductor device			Fly wheel diode		
	Type	Maximum rating	Measured ON voltage	Type	Maximum rating	Measured ON voltage
(a) SiC-INV	SiC-MOSFET SCT2080KE (ROHM)	1200 V 35 A	30 (20)	Schottky diode 1GWJ43 (TOSHIBA)	40 V 1 A	35 (200)
(b) IGBT-INV	IGBT FGW30N60VD (Fuji E.)	600 V 30 A	610 (700)	Si-diode 6A4 (Rectron)	400 V 6 A	720 (650)

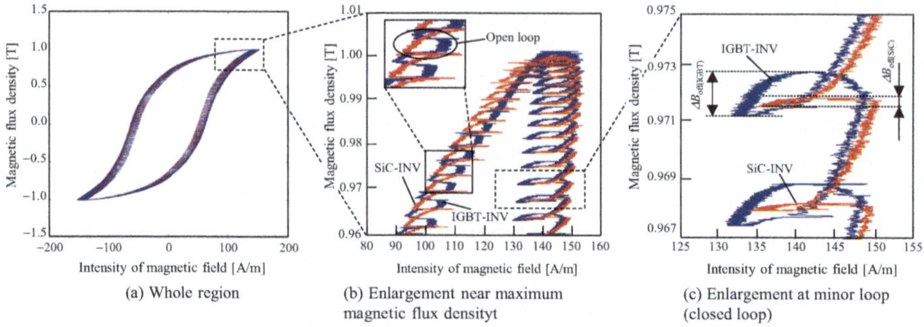

Fig. 9 *B-H* curve and minor loop shape excited by inverter with SiC material device and Si-IGBT ($m = 0.6, f_o = 50$ Hz, $f_c = 10$ kHz) [10]

Fig. 10 Iron loss characteristics excited by inverter with SiC material device and Si-IGBT (modulation index constant as $m = 0.6$) [10]

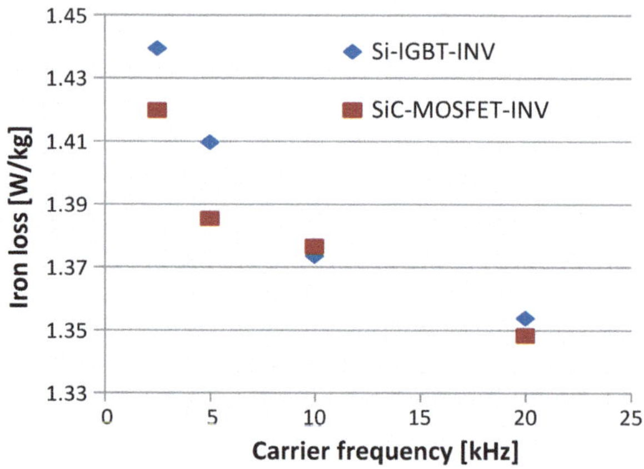

Fig. 11 Characteristics of sampling number per one second in AD converter and iron loss [12]

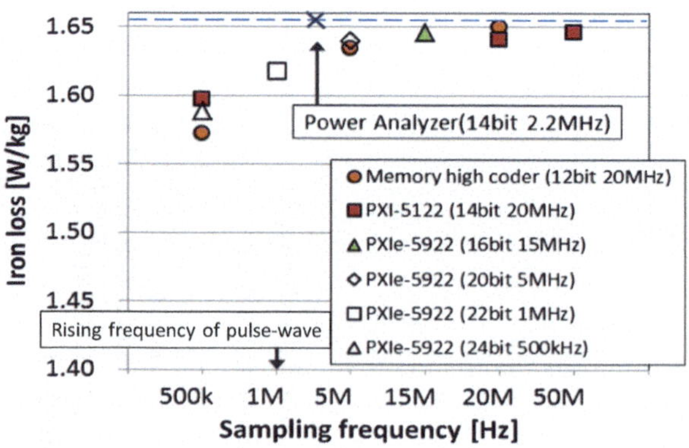

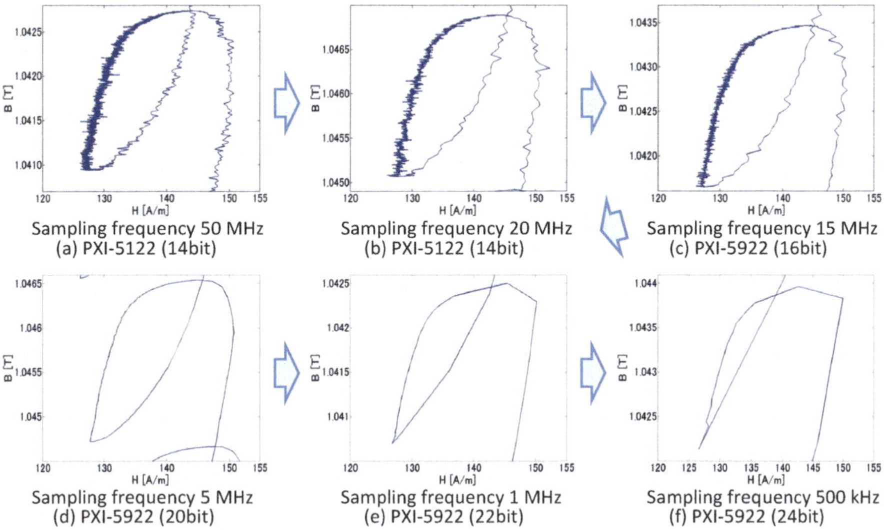

Fig. 12 Minor loop shape in changing sampling period (number per one second) in AD converter [12]

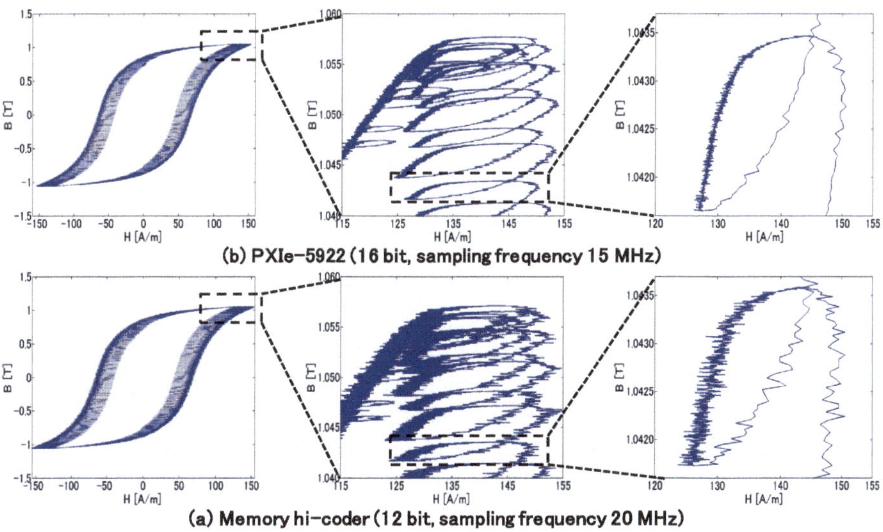

Fig. 13 *B-H* curve and its minor loop in changing bit resolution in AD converter [12]

7 Magnetic Characteristics Required for Magnetic Material [9, 10]

Power electronics excitation including inverter introduces a great change of magnetic material, magnetic property, and magnetic measurement technique as mentioned above. This change is caused by an influence of harmonics components which are derived from a switching operation of power semiconductor in power electronics technology as well as by an operating frequency change which is derived from an arbitrary frequency operation that power electronics excitation makes possible.

Therefore, under the power electronics excitation, it may be said that the magnetic characteristic such as the *B-H* curve required for magnetic material greatly changes depending on how to use it.

As for the magnetic material for electrical apparatus in power system application, soft magnetic material is mainly used in a motor, a transformer, and a reactive coil, and hard magnetic material is mainly used in field magnet of the motor as discussed in the previous sections. Therefore, the required magnetic property based on *B-H* curve depends on what electrical apparatus it is used in and how to use it as shown in Fig. 14.

Figure 14a shows a *B-H* curve standardized by JIS or IEC, which is excited at a commercial frequency of 50 Hz by linear amplifier with no harmonics components. As an electrical motor is excited by an inverter with carrier frequency components, the *B-H* curve includes a lot of minor loops as shown in Fig. 14b. When a transformer is operated in high frequency for the downsizing, the required magnetic property is like Fig. 14c. When a reactor is used in a boost chopper circuit, the operating point has DC bias in *B-H* curve and high-frequency magnetic property such as 20 kHz or so like Fig. 14d.

When permanent magnet is applied to a field magnet of synchronous motor, as angular velocity of moving magnetic field by three-phase alternate current in a stator is the same as the mechanical angular velocity of rotating rotor, the permanent magnet is not excited by a fundamental component of magnetic field. However, there are slot harmonic components derived by the slot shape of the stator and time-harmonics components by a switching operation which depends on the carrier frequency of the inverter in the air gap between the stator and the rotor. So, the permanent magnet is excited by the harmonics components of magnetic field as shown in Fig. 14e.

In this way, the magnetic properties required for magnetic material under power electronics excitation are varying according to how to use it.

Finally, fundamental specifications required for magnetic material are considered here with the progress of the power electronics technology. Figure 15 shows that the specification of electrical energy apparatus used now and in the future is plotted where a horizontal axis is frequency and a vertical axis is electrical power. In the low-frequency domain of the commercial frequency such as 50/60 Hz, large power equipment applied to electrical power network is now used. In the high-frequency domain of MHz to GHz, communication and information appliance is mainly used on information technology (IT) relations in small electrical power as mW. A smartphone,

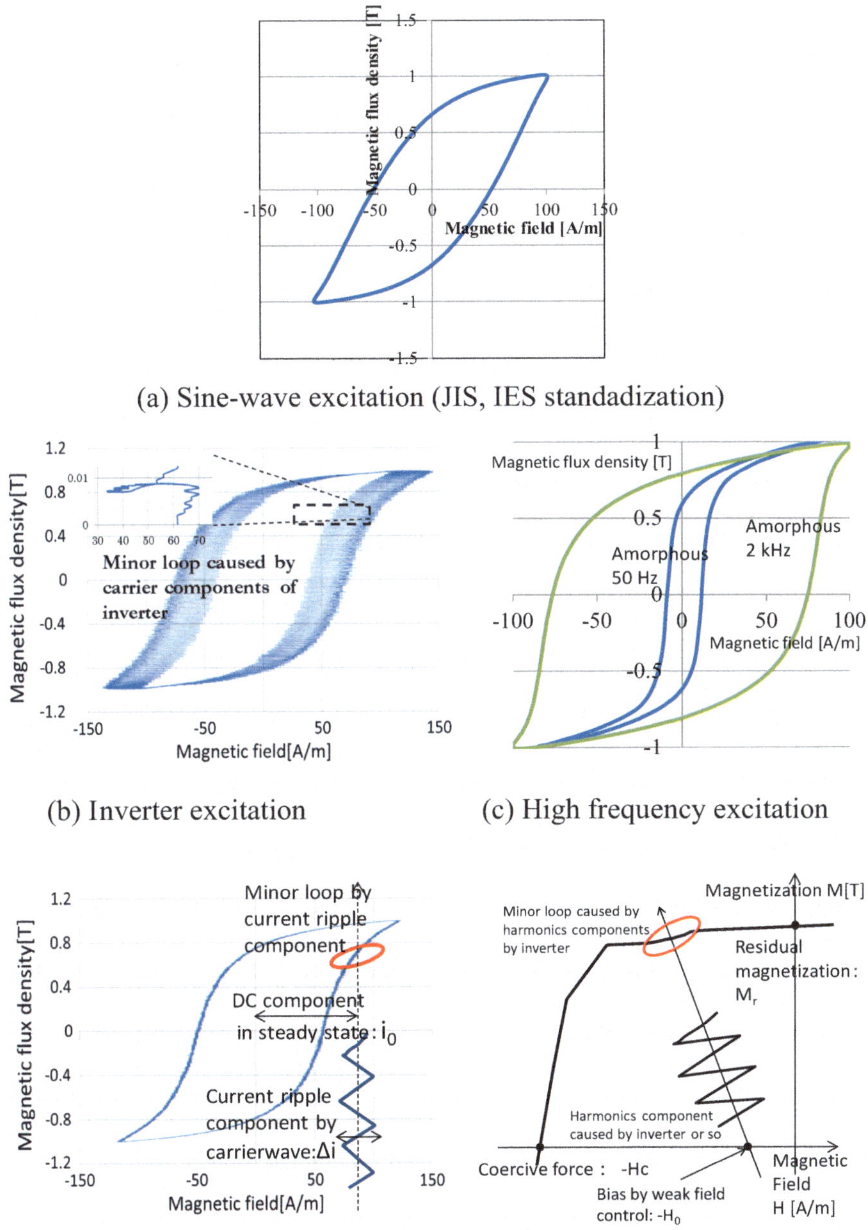

(a) Sine-wave excitation (JIS, IES standadization)

(b) Inverter excitation

(c) High frequency excitation

(d) Local loop inreactor excitation

(e) Permanet magnet excitation in motor

Fig. 14 Magnetic property of magnetic material used in different applications [4, 6, 13–17]. Adapted from [14, 16, 17] with permission ©2015 The Magnetics Society of Japan.

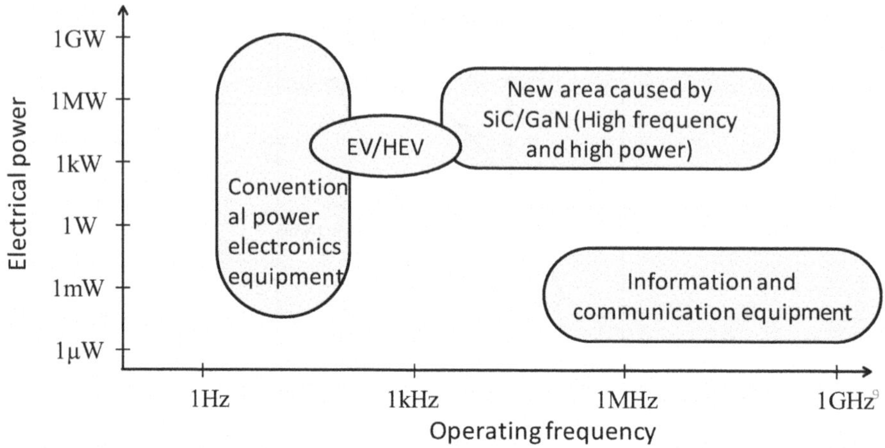

Fig. 15 High frequency and high power of electrical apparatus in the progress of power electronics technology and requirement of magnetic material [4–6, 13, 15, 16, 18–21]. Adapted from [16, 17] with permission ©2015 The Magnetics Society of Japan

a cell-phone, and a note PC are in this level. Electrical vehicle including hybrid electrical vehicle (HEV) driven by motor drive system exists between the large power equipment and the communication and information appliance. Electrical motor of kW degree is excited by power electronics apparatus of kHz degree switching operation.

A problem is that a new semiconductor material of the high frequency and high voltage application such as GaN and SiC is researched and developed in this domain and aims at the practical use. Specific operating point is considered to be in the domain that the frequency is from kHz to MHz degree and the electric power is from kW to MW degree or more. This domain has been not researched, not developed, and not used until now. The magnetic material for the high frequency and the high electrical power are also required for it as well. When a mass production technology of the semiconductor is now researched and developed, a mass production technology of magnetic material is also required as well.

This high frequency and high electric power domain caused by new semiconductor material as SiC or GaN is said to be an untrodden domain. Therefore, though some conventional sciences and technologies are enough applicable to this domain, the others are considered to be not enough. The following technical problems are considered as an insufficient issue to be solved specifically in near future.

At first, the next issues are considered to be a common basic technological problem.

(1) New electromagnetism:

1. Characteristic elucidation of displacement current and eddy current phenomenon taking into account magnetic saturation and magnetic hysteresis phenomenon.

2. Macro- and micro-relation in multi-scale problem taking into account of magnetic saturation and magnetic hysteresis phenomenon.
3. Multi-physics, multi-scale, multi-time modeling for electromagnetism.
4. Numerical analysis technology of electromagnetic field in consideration of the above.

(2) Measurement evaluation technique:

1. Current detector in the high frequency, particularly phase detection.
2. Magnetic characteristic measurement technique in the high-frequency excitation.
3. Measurement and visualization in high frequency and high electric power.

Then, the following is considered as an applied technological problem.

(1) Electromagnetic material in high frequency and high electrical power:

1. Magnetic material for low iron loss in high frequency and of high magnetization.
2. Dielectric and conductive material for low loss in high frequency.
3. Mass production technology of the electromagnetic material.

(2) Electromagnetic wave interference:

1. EMC problem (spatial).
2. Characteristic elucidation of surge voltage and current in electrical circuit.

(3) Influence on human body in high power and high-frequency electromagnetic field and its shielding technology.

In this way, the technical trend of high frequency and high electric power caused by the progress of the power electronics technology needs a solution for new technical problems for its realization. The research and development of the magnetic material are said to be an urgent problem to be solved in particular.

References

1. Japanese Industrial Standard: C 2550–1, 2011,Japanese Industrial Standard, C 2556 (2015)
2. International Electrotechnical Commission: 60404–3, 2nd edn. (1992)
3. Fujisaki, K. Yamada, R., Kusakabe, T.: Difference in iron loss and magnetic characteristics for magnetic excitation by PWM inverter and linear amplifier. IEEJ (The institute of electrical engineering of Japan) Trans. Ind. Appl. **133**(1), 69–76, 2013. (in Japanese)
4. Fujisaki, K.: Technical report of IEEJ, MAG-13-149, 2013. (in Japanese)
5. Fujisaki, K.: Power electronics, **42**, 3–6, JIPE-42-01 (2016). (in Japanese)
6. Fujisaki, K.: Annual meeting of IEEJ, 2-s2-4 (2017) (in Japanese)
7. JIS C 2552, 2014, P12-13
8. Fujisaki, K., Liu, S.: Magnetic hysteresis curve influenced by power-semiconductor characteristics in PWM inverter. J. Appl. Phys. **115**, 17A321 (2014)

9. Kayamori, D., Fujisaki, K.: Influence of power semiconductor on-voltage on iron loss of inverter-fed. In: The 10th IEEE International Conference on Power Electronics and Drive Systems, 22–25 April 201, Kitakyushu, JAPAN, PEDS B3P-Q01-9034, pp. 840–845 (2013.4)
10. Odawara, S., Kayamori, D., Fujisaki, K.: Iron loss characteristics of electrical steel sheet under inverter excitation by using power semiconductor with extremely low on-voltage property. IEEJ (The institute of electrical engineering of Japan) Trans. Ind. Appl. **134**(7), 649–655 (2014). (in Japanese)
11. Odawara, S., Kayamori, D., Fujisaki, K.: Influence of sampling frequency on magnetic characteristic evaluation under inverter excitation. IEEJ (The institute of electrical engineering of Japan) Trans. Fundam. **135**(7), 385–390 (in Japanese)
12. Kayamori, D., Fujisaki, K.: The iron loss compared with the difference from sampling frequency and resolution by excitation of an inverter. IEEJ Tech. Meet. Magn., MAG-12-052, Yokohama (2012.6). (in Japanese)
13. Investigating R&D Committee of Magnetic Material for Electromagnetic Actuator: Magnetic material and its evaluation technology for electromagnetic actuator. Technical report of IEEJ, **1397**,(2017.11). (in Japanese)
14. Fujisaki, K.: Advanced magnetic material requirement for higher efficient electrical motor design. In: The 38th Annual Conference on MAGNTICS in Japan, Symposium "Challenge of Magnetics to Improve Energy Efficiency, pp. 4aB-2 (2014.9)
15. Fujisaki, K.: The 2015 annual meeting of the institute of electrical engineering of Japan, 5-S22-1, March, 2015. (in Japanese)
16. Fujisaki, K.: Magnetics society of Japan, Technical report of 202 (2015). (in Japanese)
17. Fujisaki, K.: Magnetics Jpn., **11**(1), 34–41 (2016). (in Japanese)
18. Fujisaki, K.: Magnetics society of Japan. Technical report of 213 (2017). (in Japanese)
19. Fujisaki, K.: Monthly EMC **349**, 61–64 (2017). (in Japanese)
20. Fujisaki, K.: 22pB-6, Symposium of Magnetics Society of Japan (2017). (in Japanese)
21. Fujisaki, K.: Required magnetic characteristics for future electrical energy magnetic material. In: The 86th Technical Meeting Program of The Japan Association of Bonded Magnetic Materials (2014.9)

Iron Loss Measurement of Interior Permanent Magnet Synchronous Motor

Nicolas Denis

Abstract The International Electrotechnical Commission (IEC) as well as the Japanese Industrial Standards (JIS) proposes standard methods for the measurement of soft magnetic material iron loss density. Under these measurements, the magnetic field applied to the material is strictly unidirectional and purely sinusoidal. However, in an electrical motor, the magnetic path is not straightforward, and the magnetic field can be largely rotating. Moreover, when the motor is driven by an inverter using power electronic devices switching at high frequency, the magnetic field can have large high order harmonic content thus harmonic distortion. Because of these effects, simply using the iron loss density data obtained by standard measurements, in order to predict the motor total iron losses, can be inaccurate. Accordingly, ways to measure and separate the effect of different parameters on the iron losses of an interior permanent magnet synchronous motor (IPMSM) are described in this chapter. In the following sections, experimental methodology and numerical calculation methods are first explained. Then, the effects of the inverter switching frequency, PWM modulation index, and dead-time, along with the effect of the motor load on the total iron losses, are investigated.

Keywords PWM inverter · Permanent magnet synchronous motor · Iron loss · Magnetic material

1 Motor Characteristic and Experimental Methodology

1.1 Interior Permanent Magnet Motor Characteristics

The electrical motor considered in this chapter is an interior permanent magnet synchronous motor (IPMSM) with concentrated windings. It has 8 poles and 12 stator slots. The quarter cross-section view and the related dimensions are illustrated

N. Denis (✉)
Challenergy Inc., Tokyo, Japan
e-mail: nicolas.denis@challenergy.com

K. Fujisaki (ed.), *Magnetic Material for Motor Drive Systems*,
Engineering Materials, https://doi.org/10.1007/978-981-32-9906-1_8

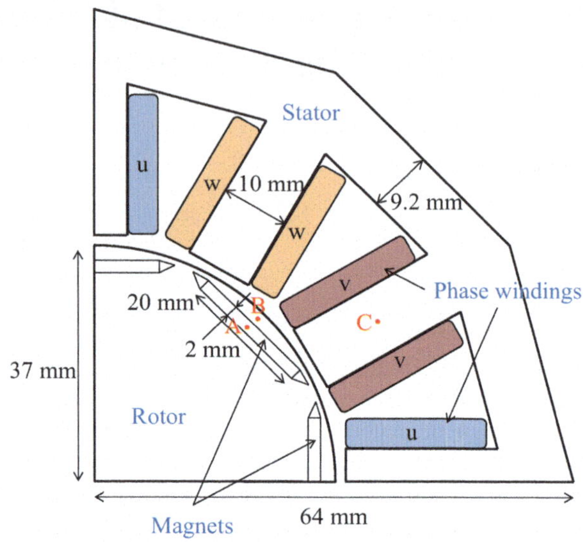

Fig. 1 Quarter cross-section view of the IPMSM. © [2018] IEEE. Reprinted, with permission, from [3] © [2019] IEEJ. Reprinted, with permission, from [12]

Table 1 [2] Electrical characteristics of the IPMSM

Phase resistance R_s	0.498 Ω
d-axis inductance L_d	2.44 mH
q-axis inductance L_q	3.70 mH
Peak line to neutral back EMF at 750 rpm	20.1 V

© [2018] IEEE. Reprinted, with permission, from [3]
© [2018] IEEJ. Reprinted, with permission, from [11]

in Fig. 1. The magnets are sintered NdFeB, and the stator and rotor soft magnetic material is non-oriented silicon steel 35H300.

The electrical characteristics of the motor are presented in Table 1. In this table, the phase resistance is measured under DC condition.

1.2 Measurement Methodology

In this section, three experimental setups used to evaluate the total IPMSM iron losses under three different conditions are presented. These three different conditions are called "no-current," "no-load," and "load", respectively. The total iron losses are considered as the sum of the hysteresis and eddy current losses in the magnets, rotor and stator.

The experimental setup of the no-current test is presented in Fig. 2. During this test, the IPMSM is rotated by an auxiliary brushless DC (BLDC) motor. The tested IPMSM armature windings are in open circuit so that no current flows. A torque

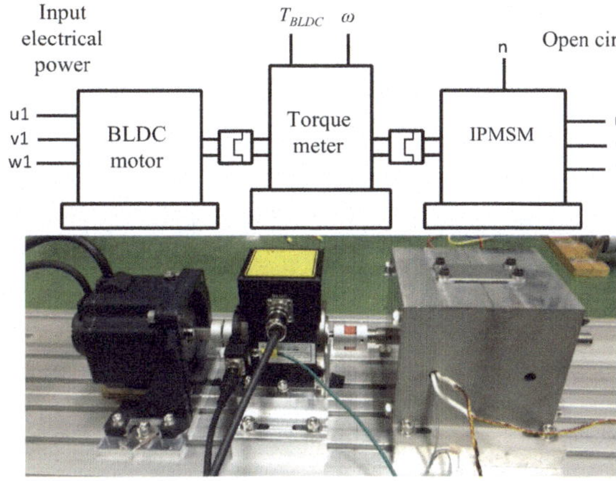

Fig. 2 Experimental setup for the iron loss measurement of IPMSM under no-current condition. © [2018] IEEE. Reprinted, with permission, from [3]. © [2019] IEEJ. Reprinted, with permission, from [12]

meter is connected between the two motors in order to measure the mechanical output power of the BLDC motor P_{BLDC}, which is considered to be the sum of the IPMSM iron losses in no-current condition $P_{i,\mathrm{noc}}$ and its mechanical friction losses P_f. It gives the following equation.

$$P_{\mathrm{BLDC}} = \omega T_{\mathrm{BLDC}} = P_{i,\mathrm{noc}} + P_f \tag{1}$$

where ω is the measured speed in rad/s, and T_{BLDC} is the measured torque in N.m. It is assumed

that, under these conditions, the total iron losses are produced only by the rotating magnets. The torque meter is the TH-3104H from Onosokki. The accuracy of the torque meter is equal to ±0.002 N.m., and its maximum measurable torque is 1 N.m.

Under no-load condition, the IPMSM is driven by a three-phase voltage source pulse-width modulation (PWM) inverter using IGBT. The PWM carrier signal is a triangular fixed frequency wave. A power analyzer (Yokogawa PX8000) is used to measure the IPMSM input active electrical power $P_{3\Phi}$, phase rms currents I_u, I_v, I_w, and the phase voltages. The power analyzer internal shunt resistances, which have a measurement bandwidth of 10 MHz, are used for the current measurement. The voltage measurement modules of the power analyzer have a measurement bandwidth of 20 MHz. The measurement sampling frequency is set to 100 MHz. An encoder is connected to the IPMSM rotor shaft for the speed feedback. Conventional vector control is used for the simultaneous speed and current closed-loop control. The setup is illustrated in Fig. 3. Under these conditions, the IPMSM total iron losses are calculated using the following equation

$$P_{i,\mathrm{nol}} = P_{3\phi} - R_s\left(I_u^2 + I_v^2 + I_w^2\right) - P_f - P_{\mathrm{enc}} \tag{2}$$

where P_{enc} is the encoder mechanical losses, and R_s is the IPMSM phase resistance.

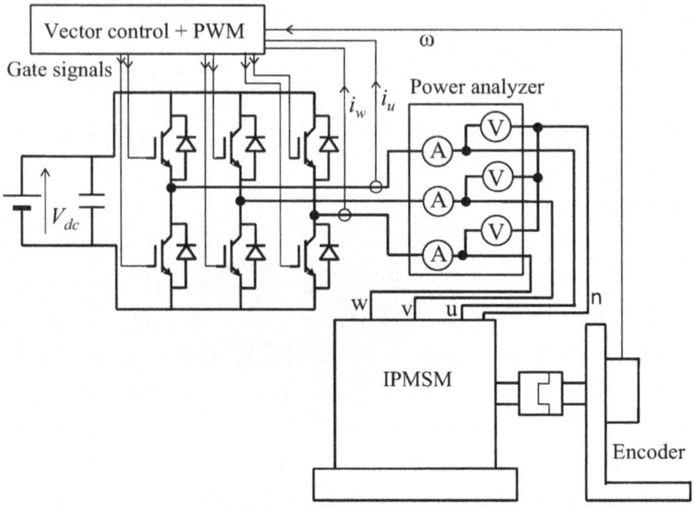

Fig. 3 Experimental setup for the iron loss measurement of IPMSM under no-load condition. © [2018] IEEE. Reprinted, with permission, from [3]. © [2018] IEEJ. Reprinted, with permission, from [11]. © [2019] IEEJ. Reprinted, with permission, from [12]

Under load condition, the IPMSM drives a brushless DC (BLDC) motor whose stator is connected to a resistance network. A torque meter is placed between the IPMSM and the BLDC motor in order to measure the IPMSM output torque T_{out} and the shaft rotational speed ω. The torque meter is the SS-020 from Onosokki. It can measure a maximum torque of 2 N.m., and its accuracy is equal to ±0.004 N.m. The experimental setup is illustrated in Fig. 4. The IPMSM total iron losses under load condition are calculated using the following equation.

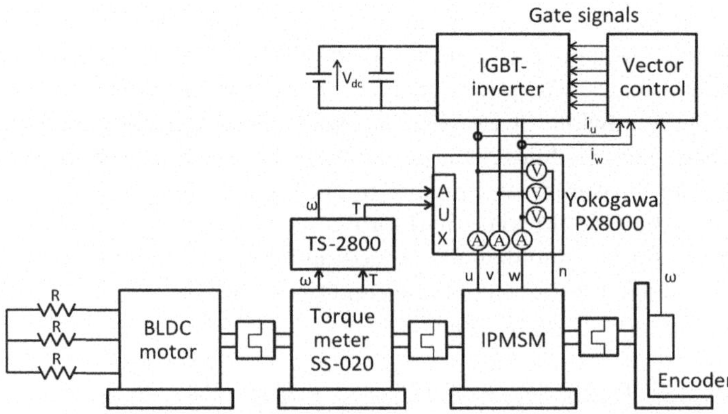

Fig. 4 Experimental setup for the iron loss measurement of IPMSM under load condition. © [2018] IEEE. Reprinted, with permission, from [3]. © [2019] IEEJ. Reprinted, with permission, from [12]

$$P_{i,\mathrm{loa}} = P_{3\phi} - R_s\left(I_u^2 + I_v^2 + I_w^2\right) - P_f - P_{\mathrm{enc}} - \omega T_{\mathrm{out}} \tag{3}$$

Due to imperfect equipment accuracy and perturbations, the measurements of time-average input power, time-average output power, rms voltage, and rms current can variate from one measurement to another. In order to limit the impact of this adverse effect, a wide measurement window of 500 ms is chosen. Moreover, 15 successive measurements are done, and the results illustrated in the following sections present both the average of the 15 measurements and its standard deviation.

2 Calculation of the IPMSM Iron Losses by Finite Element Analysis

A three-dimensional (3D) mesh model built with JMAG software is used for the finite element analysis (FEA) and is presented in Fig. 5. Only one quarter of the IPMSM axial cross-section is considered, and the axial length is equal to the steel sheet thickness (350 μm). A time-stepping magnetic field analysis is performed using the 3D vector potential equation

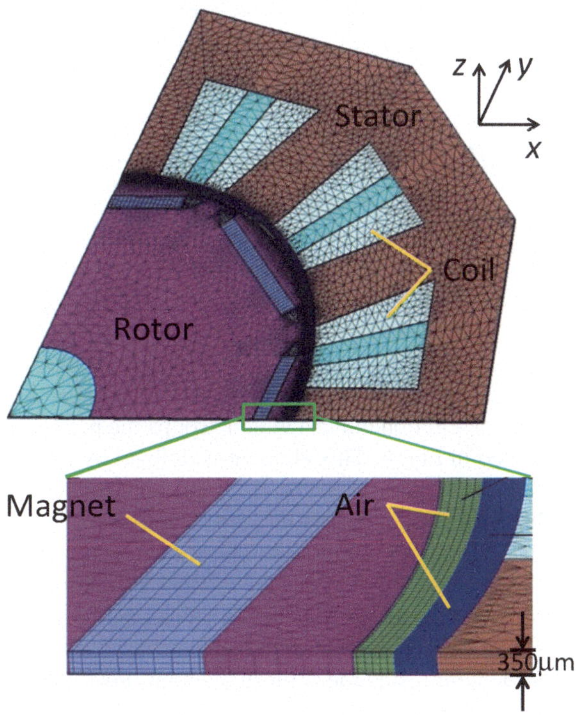

Fig. 5 3D finite element model. © [2018] IEEE. Reprinted, with permission, from [3]. © [2019] IEEJ. Reprinted, with permission, from [12]

$$\nabla \times \left(\frac{1}{\mu}\nabla \times \vec{A}\right) = \overrightarrow{J_0} - \sigma\left(\frac{\partial \vec{A}}{\partial t} + \nabla\vec{\Phi}\right) + \frac{1}{\mu_0}\nabla \times \vec{M} \tag{4}$$

$$\nabla \cdot \left(\sigma\left(\frac{\partial \vec{A}}{\partial t} + \nabla\vec{\Phi}\right)\right) = 0 \tag{5}$$

where $\vec{A}$ is the magnetic vector potential,$\vec{\Phi}$ is the electric scalar potential, $\overrightarrow{J_0}$ is the stator phase current density, σ is the electrical conductivity of the material, μ is the magnetic permeability, and $\vec{M}$ is the PM magnetization. σ and μ depend on the considered material (electrical steel or magnet), and the magnetic permeability of the magnet is considered equal to that of the air μ_0.

The stator phase current density is the input of the model. While it is equal to zero in no-current condition, it is not zero in no-load and load conditions. In the later cases, the current density obtained by measurement is used as input to the model.

Since there are reflection symmetries, periodic boundary conditions are applied at the edges of the model. Moreover, the Dirichlet boundary condition is applied at the interlamination (air) surface boundaries, which means that the vector potential and the axial component of the magnetic flux density B_z are considered locally null in the vicinity of these surfaces. The assumption on the interlamination boundary condition might lead to inaccuracies but allows considering the lamination sheet individually. This has the advantage of considerably decreasing the computation time compared to a case where the full axial length of the motor is considered.

2.1 Calculation of the Hysteresis Losses

The hysteresis losses are calculated using the Steinmetz approximation. Since the Steinmetz equations are usually used for two-dimensional (2D) models only, the following calculation procedure is used to adapt the equations. The magnetic flux density is first averaged over the axial direction. As an illustration, let α be the mesh element which composes the actual mesh of the 3D model and is defined by its radial, tangential, and axial coordinates in the 3D space. Now, let β be the element of a fictional mesh which groups the mesh elements in the axial direction as illustrated in Fig. 6. This element is defined by its radial and tangential coordinates only and does not depend on the axial coordinate. Equations (4) and (5) are solved to obtain the radial, tangential, and axial components of the magnetic flux density in the mesh elements α. They are, respectively, noted $B_{r,}\alpha$, $B_{t,}\alpha$, and $B_{z,}\alpha$. Let (i, j, k) be the coordinate index of a given mesh element α, in which i represents the radial coordinate, j the tangential coordinate, and k the axial coordinate. The magnetic flux density of the β mesh element is the average of the magnetic flux density in each α mesh element inside the β mesh element. For example, the radial component of the magnetic flux density in the mesh element β is

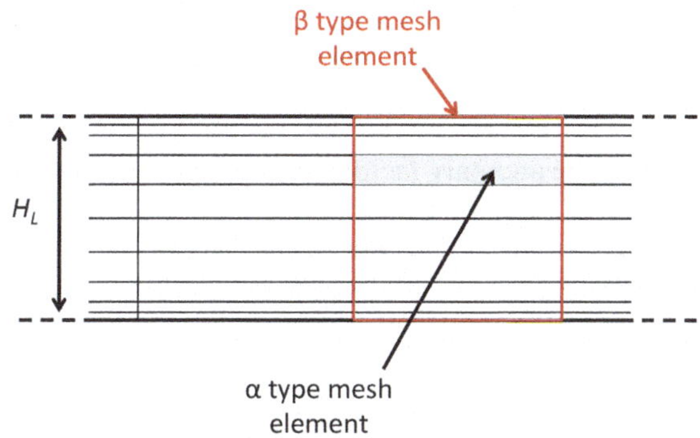

Fig. 6 Fictional β mesh element. © [2018] IEEE. Reprinted, with permission, from [3]. © [2019] IEEJ. Reprinted, with permission, from [12]

$$B_{r,\beta}(i, j) = \sum_{k=1}^{N_z} \frac{B_{r,\alpha}(i, j, k)h_k}{H_L} \tag{6}$$

where N_z is the mesh division in the axial direction ($N_z = 10$), h_k is the dimension of the mesh element α in the axial direction, and H_L is the lamination thickness. The same calculation applies for the tangential component $B_{t,\beta}$, but the axial component $B_{z,\beta}$ is supposed negligible and is not considered in the hysteresis loss calculation for matter of consistency with the 2D nature of the Steinmetz equations. After the above calculation, a fast Fourier transform (FFT) is then performed on the time-varying waveforms of the magnetic flux densities $B_{r,\beta}$ and $B_{t,\beta}$ in order to obtain the magnitude of the harmonic components $B_{r,\beta,i}$ and $B_{t,\beta,i}$. The hysteresis loss density in a given β mesh element is calculated by

$$W_{\text{hys}} = \sum_{i=1}^{N_h} K_{\text{hys},i}\left(B_{t,\beta,i}^2 + B_{r,\beta,i}^2\right) f_i \tag{7}$$

where i is the harmonic order, N_h is the maximum harmonic order, f_i is the frequency of the ith harmonic, and $K_{hys,i}$ is the hysteresis loss coefficient derived from the material specific iron loss data, which depend on the frequency, as denoted by the subscript i.

2.2 Calculation of the Eddy Current Losses

In each mesh element α and at each time-step of the simulation, the magnet, stator and rotor eddy current density vector $\overrightarrow{J_{e,\alpha}}$ is calculated based on the electrical field. This field is itself derived from the electric scalar potential $\vec{\Phi}$. The calculation of the eddy current loss density W_{ed} in a given α element is based on the integration of the eddy current density through time, considering the electrical conductivity σ.

$$W_{\text{ed}} = \frac{\kappa}{T_e} \sum_{t=1}^{N_t} \frac{\left|\overrightarrow{J_{e,\alpha}(t)}\right|^2}{\sigma} \text{d}t \quad (8)$$

where κ is the anomaly factor used as approximation to take the anomalous loss into account, and its value is two for the electrical steel used in the rotor and stator, and one for the magnets. T_e is the electrical period, N_t is the number of time-steps of the analysis, and dt is the time-step duration.

3 Effect of the PWM Carrier Frequency on the IPMSM Iron Losses

The carrier frequency here is referred as the frequency of the triangular PWM carrier signal. In this section, the carrier frequency varies from 1 to 20 kHz, and all the other driving conditions are set constant. The DC bus voltage is set to 180 V, the inverter dead-time is set to 3.5 μs, and the *d*-axis current reference is set to 0 A in the vector control program. The IPMSM is rotated under load condition with a rotational speed of 750 rpm and a constant torque of 1 N.m. The results are presented in Fig. 7, in which the bars represent the standard deviation of the 15 measurements.

The International Electrotechnical Commission (IEC) and the Japanese Industrial Standards (JIS) propose a standard way to measure soft magnetic material iron losses by supplying purely sinusoidal magnetic flux on wound material samples. However, the magnetic flux flowing through the different parts of an inverter-fed IPMSM has low-frequency components caused by the rotor rotation and also high-frequency components at the PWM carrier frequency order. The magnetic flux density is then not purely sinusoidal, and the PWM high-frequency components contribute to increase the iron losses. In order to evaluate this increase experimentally, the iron loss measurement results of Fig. 7 are compared to the iron losses of wound ring cores. Two tests on wound ring core are carried out:

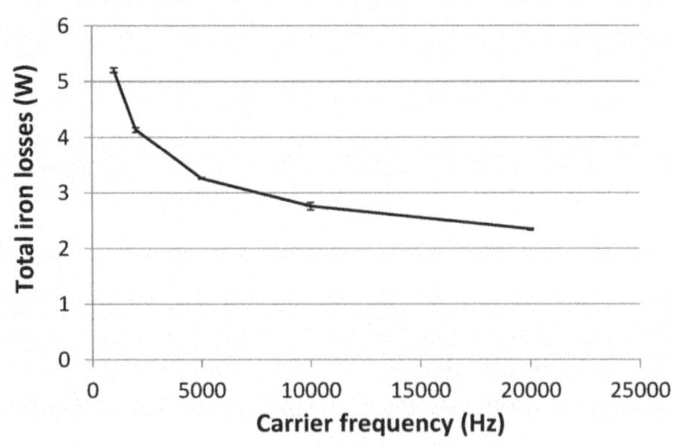

Fig. 7 Effect of carrier frequency on the IPMSM iron losses [9]

(a) Wound ring core supplied with sinusoidal single-phase voltage.
(b) Wound ring core supplied with single-phase inverter output voltage.

By calculating the ratio of the IPMSM iron losses of Fig. 7 to the ring core iron losses following the procedure in [1], it is possible to evaluate the building factor (BF) of the IPMSM stator core. The results are presented in Fig. 8, in which "BF (sine-fed)" comes from the comparison between the IPMSM iron losses and the ring core iron losses under test condition (a), and "BF (inverter-fed)" comes from the comparison between the IPMSM iron losses and the ring core iron losses under test condition (b).

In order to understand the meaning of the calculated BF here, it is important to define the relationships between the iron losses measured in the wound ring under the two test conditions and the iron losses in the IPMSM stator.

Firstly, it is supposed here that the iron loss density measurement through the wound ring core test under sinusoidal voltage supply provides a result close to what would be obtained using IEC or JIS standard measurements. In other words, the sinusoidal-fed ring core test provides the material intrinsic iron loss density for a given frequency and magnetic flux density amplitude. Secondly, when the ring is excited by a PWM voltage waveform, the high-frequency harmonic due the PWM increases the iron loss density compared to the case when the ring is excited by a sinusoidal voltage. Finally, in the inverter-fed IPMSM, some additional phenomena increase the iron losses in the stator core compared to the case of inverter-fed ring core. These additional phenomena are

(1) The fact that the magnetic flux is not only alternating but also rotating [4, 5].
(2) The effect of the magnet shape that distorts the magnetic flux density in the stator and adds low-frequency components [6].
(3) The geometry of the rotor + magnets + stator assembly that creates a particular path for the magnetic flux from silicon steel to magnets to air. More precisely, the air gap is responsible for fringing flux that locally increases the eddy currents in the parts of the core near the air gap [7]. Interestingly, it has been proven for

Fig. 8 Calculated IPMSM stator building factor at different carrier frequency values © [2018] IEEE. Reprinted, with permission, from [1]

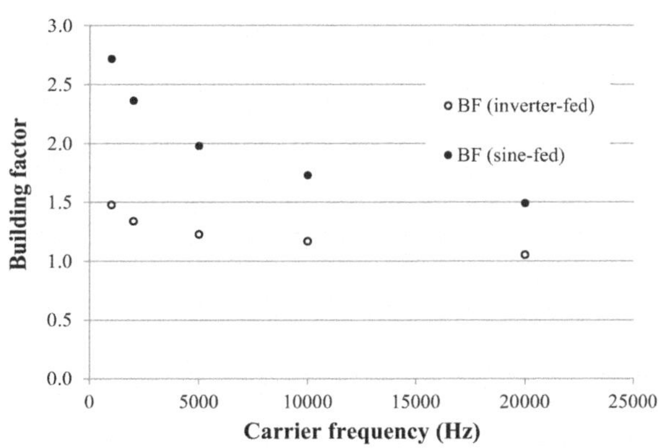

reactors with air gap that this effect decreases with the frequency [8], which could explain the decreasing trend observed on the BF.

Given these considerations, it is thought that the "BF (sine-fed)" of Fig. 8 reflects the stator iron loss increase which is due to the PWM carrier frequency plus the additional phenomena described above, while the "BF (inverter-fed)" reflects the stator iron loss increase which is due to the additional phenomena only.

4 Effect of the PWM Modulation Index on the IPMSM Iron Losses

The modulation index is referred here as the amplitude of the PWM modulation signal. In three-phase PWM control, the rms fundamental of the IPMSM phase voltage $V_{f,\mathrm{rms}}$ is linked to both the modulation index m and the DC bus voltage V_{dc}. This relation can be approximated by the equation

$$V_{f,\mathrm{rms}} = m\frac{V_{\mathrm{dc}}}{2\sqrt{2}} \tag{9}$$

For given constant torque and speed, $V_{f,\mathrm{rms}}$ stays unchanged, which means that an increase of the DC bus voltage automatically leads to a decrease of the modulation index. In order to measure the impact of the modulation index on the iron losses, the rotational speed is first set to 750 rpm with a constant torque of 1 N.m, and then, V_{dc} is changed from 50 to 200 V in order for the modulation index to vary. The iron loss characteristic is measured for 1, 10, and 20 kHz carrier frequencies, the inverter dead-time is set to 3.5 μs, and the d-axis current reference is set to 0 A in the vector control program. The relation between the DC bus voltage and the measured modulation index is illustrated in Fig. 9. Due to the dead-time effect, it also appears that the modulation index increases with the carrier frequency at a given DC bus

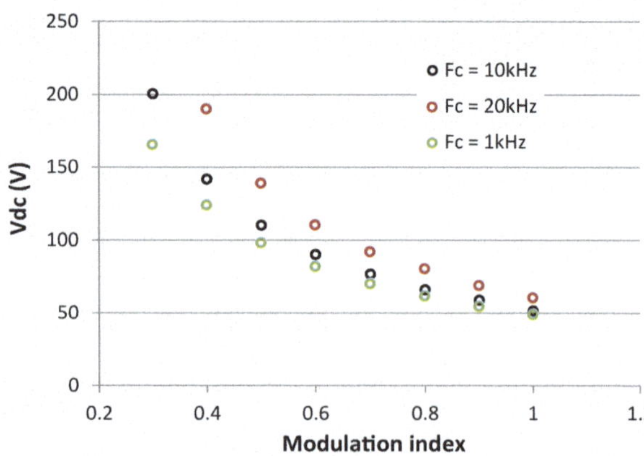

Fig. 9 Relation between V_{dc} and m [9]

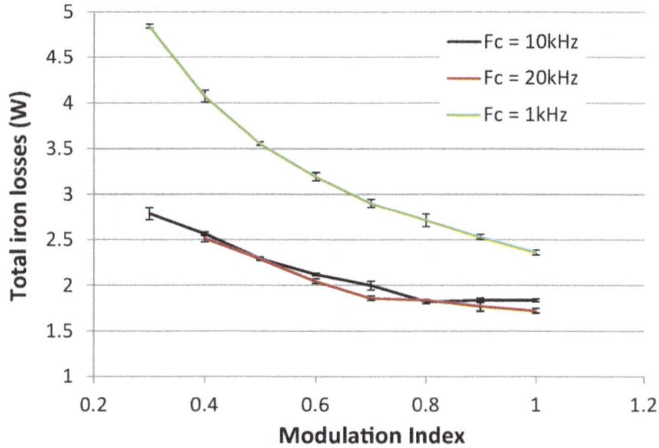

Fig. 10 Effect of modulation index on the IPMSM iron losses [9]

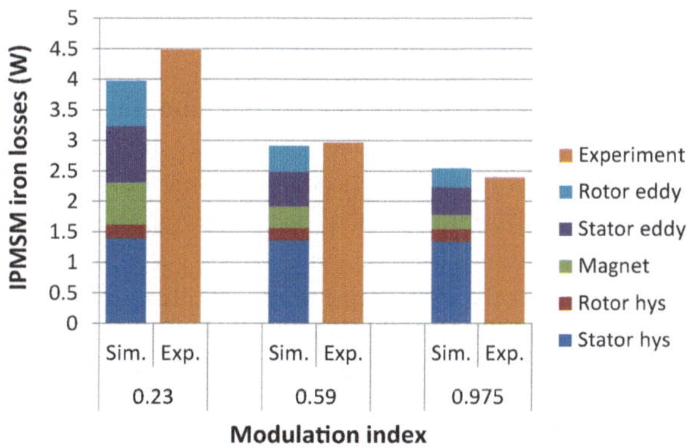

Fig. 11 3D FEA for the calculation of IPMSM iron losses at different modulation index values [10]

voltage value. The iron loss measurement at different modulation index values is presented in Fig. 10.

It appears that higher modulation index is favorable to decrease the IPMSM iron losses. In order to better understand this trend, numerical simulation results are presented in Fig. 11. In Fig. 11, the case of a carrier frequency at 1 kHz under no-load condition is considered. Experimental data and numerical calculation results have the same trend. Moreover, FEA allows us to understand that the modulation index does almost not affect the hysteresis losses. The eddy current losses are, on the contrary, highly affected, the highest loss increase rate occurring in the magnet.

4.1 Analysis of the Phase Voltage and Current

The phase voltage waveforms under no-load condition, carrier frequency of 1 kHz, and for high and low modulation index are presented in Fig. 12. The corresponding

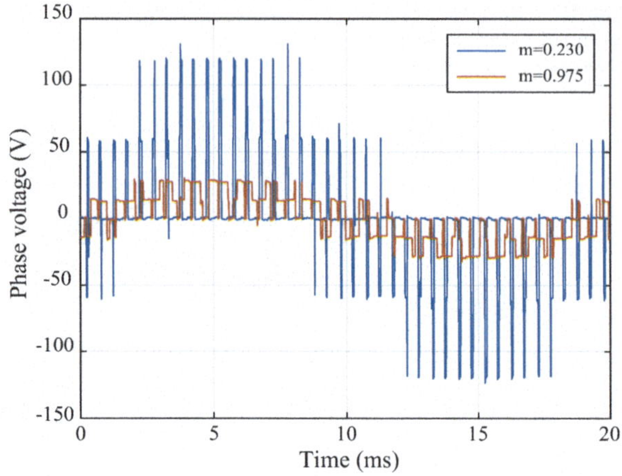

Fig. 12 Phase voltage waveform at low and high modulation index

discrete Fourier transformation (DFT) of the voltage is illustrated in Fig. 13. In accordance with the previous explanation, the fundamental voltage is the same for both modulation index values. However, the variation of modulation index deeply changes the voltage high order harmonic content. This has a direct impact on the high order harmonics of the phase currents, as can be seen in Fig. 14. The high order harmonic content of the current is stronger when a low modulation index is applied. This also directly affects the magnetic flux density in the magnet, stator and rotor, as will be seen in the next section.

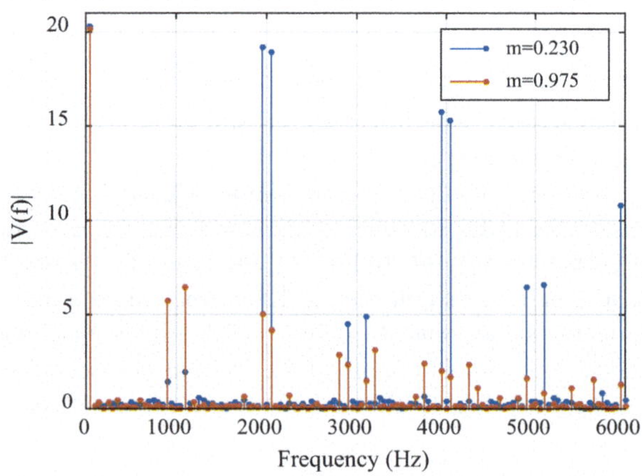

Fig. 13 DFT of phase voltage [10]

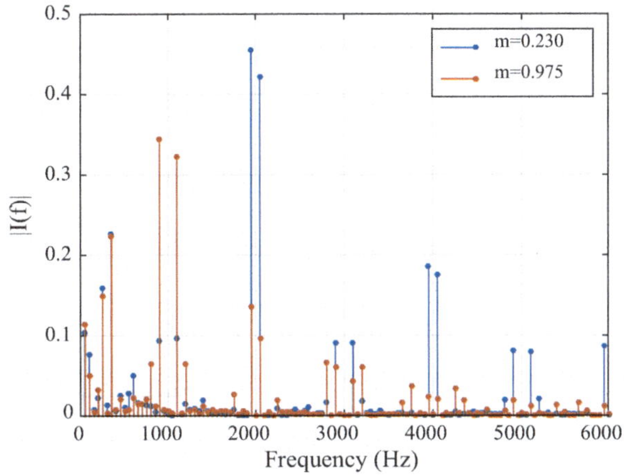

Fig. 14 DFT of phase current [10]

4.2 *Analysis of the Magnetic Flux Density*

The magnetic flux density waveform at a given point of the motor can be observed using FEA. The radial magnetic flux density on the surface of the magnet (point A in Fig. 1) is illustrated in Fig. 15 for high and low modulation indexes, and the corresponding DFTs are shown in Fig. 16. As for the current, the low modulation index decreases the harmonics orders around 1 kHz, while the higher orders from 2 kHz are increased. Since the eddy current losses are roughly proportional to the square of the frequency, the stronger high order harmonics in the magnetic flux density increase the eddy current losses.

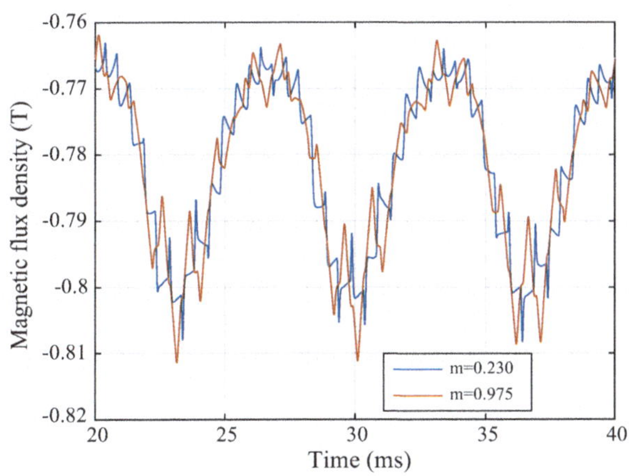

Fig. 15 Radial magnetic flux density in the magnet [10]

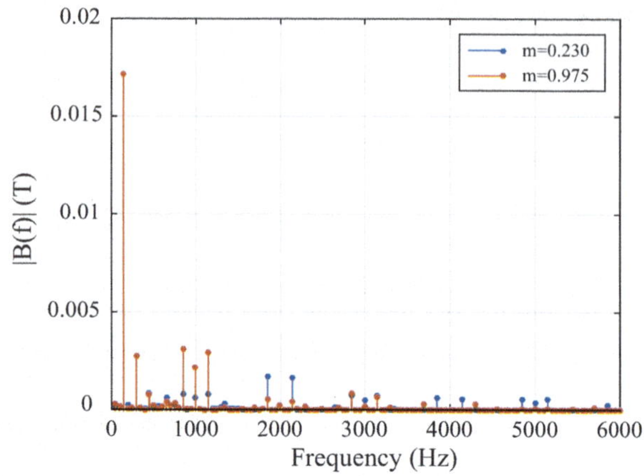

Fig. 16 DFT of the magnetic flux density [10]

5 Effect of the Dead-Time on the IPMSM Iron Losses

In a PWM inverter, when the upper switch of a phase leg turns on, the bottom switch turns off. Inversely, when the upper switch turns off, the bottom switch turns on. Since switching is not instantaneous, upper and bottom switching commands are not made at the same time in order to avoid simultaneous conduction, which would cause a short circuit of the DC bus. Consequently, the conducting switch is first turned off, and the opposite switch is turned on after a set time called dead-time. Changing the dead-time automatically modifies the switching times and then the phase voltage waveform. In this section, the impact of the dead-time on the iron losses is investigated. In order to measure it, the iron losses are measured at the dead-time values 3.5, 4, 4.5, and 5 μs. The rotational speed is set to 750 rpm under no-load condition. The iron loss characteristic is measured for 1, 10, and 20 kHz carrier frequencies, the DC bus voltage is set to 180 V, and the *d*-axis current reference is set to 0 A in the vector control program. The obtained characteristic is shown in Fig. 17.

It appears that the dead-time has almost no effect on the iron losses and that there is no clear increasing or decreasing trend. For every carrier frequency value, the variation of iron loss with the dead-time stays under 0.7%. It is decided here to measure the iron loss characteristics under no-load condition rather than load condition. Measurement under load condition is possible but less accurate than under no-load condition, making it difficult to measure the small impact of the dead-time on the iron losses. For a better understanding, Table 2 provides information about the accuracy in both no-load and load condition.

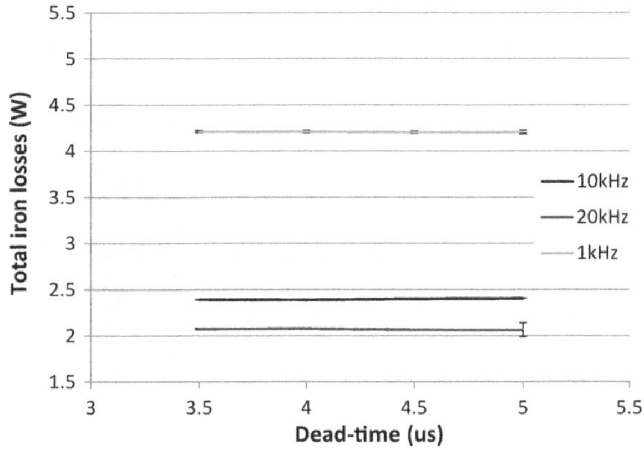

Fig. 17 Effect of dead-time on the IPMSM iron losses [9]

Table 2 Measurement accuracy under load and no-load conditions

	Input power measurement accuracy (Power analyzer)	Output power measurement accuracy (Torque meter)
No-load	±0.005 W (range 150 V/2 A)	–
Load	±0.05 W (range 150 V/5 A)	±0.31 W (at 750 rpm)

6 Effect of the Load on the IPMSM Iron Losses

In order to evaluate the impact of the load on the IPMSM total iron losses, the iron losses are measured under no-load and load conditions. The rotational speed is controlled at a constant value of 750 rpm. The DC bus voltage is constant and equal to 180 V, the dead-time is 3.5 μs, the *d*-axis current reference is equal to 0 A. Both tests are made a 1, 10, and 20 kHz carrier frequency values. The results are illustrated in Fig. 18. It can be seen that at every career frequency, the iron losses are higher under load condition than under no-load condition.

In Fig. 19, the iron losses under no-load and load conditions at a carrier frequency of 1 kHz are compared to the iron losses obtained by no-current test at 750 rpm. 3D FEA results are also presented in Fig. 19. Both numerical calculation and experiments show the largest iron losses for the load condition, closely followed by the no-load condition, while the lowest iron losses are obtained in no-current condition. Observing the FEA results, it can be seen that the hysteresis losses in the stator and rotor are almost equal in every working condition, but there is a significant difference in the eddy current losses.

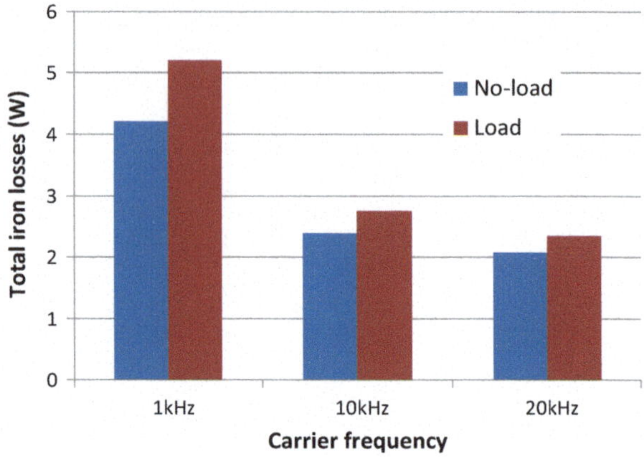

Fig. 18 Effect of load on the IPMSM iron losses [9]

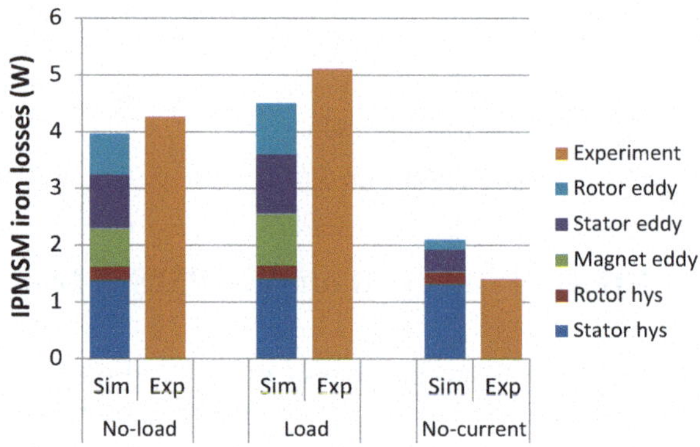

Fig. 19 Comparison of no-load, load, and no-current tests. © [2018] IEEE. Reprinted, with permission, from [3]. © [2019] IEEJ. Reprinted, with permission, from [12]

6.1 Analysis of the Phase Voltage

The *u* phase to neutral voltage measured for the no-current, no-load (carrier frequency 1 kHz), and load (carrier frequency 1 kHz) conditions is illustrated in Fig. 20. In no-current condition, the rotation of the magnet is responsible for a varying magnetic flux in the stator which induces a voltage (black curve) in the phase windings. Contrarily to the no-current condition, high frequencies appear in the phase voltage measured under no-load and load conditions, due to the inverter PWM switching operation. In order to observe the high frequencies, a DFT is performed, and the magnitude of the voltage harmonics is presented in Fig. 21. Only fundamental, 5th and 7th harmonics appear in no-current condition. High order harmonics due to PWM appear only in no-load and load conditions. Moreover, it can be noticed that both the PWM high order harmonics and the fundamental are higher in load condition compared to no-load condition. This can be explained by the fact that during the load test, the IPMSM fundamental current is obviously higher than in no-load condition in order to output a higher torque. Because of the motor phase impedance, higher fundamental

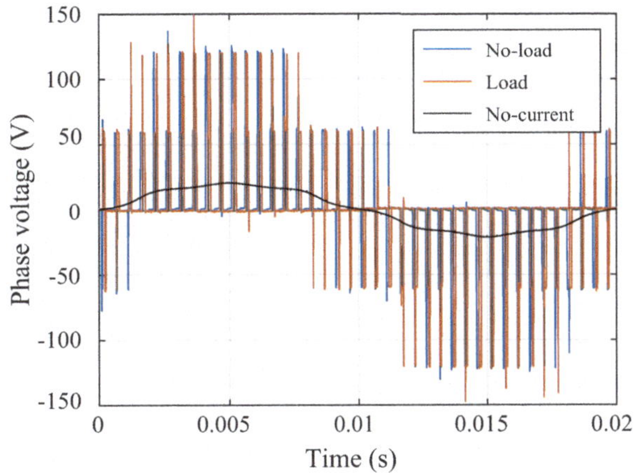

Fig. 20 Comparison of phase voltage waveforms. © [2019] IEEJ. Reprinted, with permission, from [12]

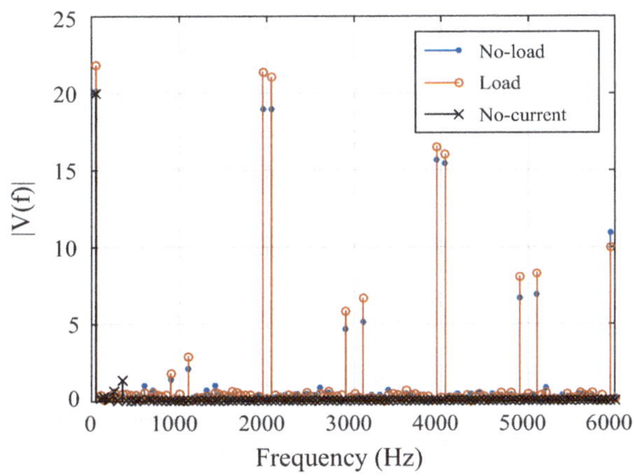

Fig. 21 DFT of phase voltage. © [2019] IEEJ. Reprinted, with permission, from [12]

current needs higher fundamental voltage. Looking at (9) and since V_{dc} is constant, the modulation index has to be increased in order to output a higher fundamental voltage. Accordingly, m is equal to 0.241 in no-load condition and 0.266 in load condition. The increase of modulation index effectively increases the fundamental voltage but also has the adverse effect of increasing the amplitude of the PWM high order harmonics.

6.2 Analysis of the Phase Current

The phase current measured under no-load and load conditions (carrier frequency 1 kHz) is illustrated in Fig. 22. The amplitude of the current harmonics obtained by DFT is illustrated in Fig. 23. The operation in load condition obviously increases

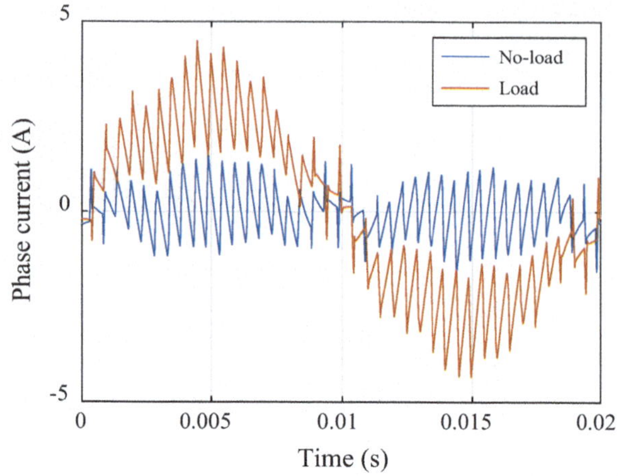

Fig. 22 Comparison of phase current waveforms [10]. © [2019] IEEJ. Reprinted, with permission, from [12]

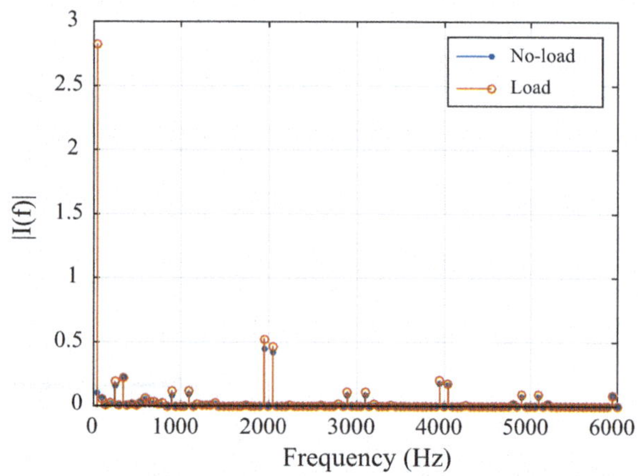

Fig. 23 DFT of phase current [10]. © [2019] IEEJ. Reprinted, with permission, from [12]

the fundamental current, but the high order harmonics due to the PWM also have larger amplitudes. This is the direct consequence of the larger voltage high order harmonics in load condition observed in Fig. 21. The phase current is responsible for magnetomotive forces originating from the stator windings and having an effect on the magnetic flux density in the stator, rotor, and magnets. The next section proposes an analysis of the magnetic flux density in the three operating conditions by use of the 3D FEA numerical calculation results.

6.3 Analysis of the Magnetic Flux Density by Finite Element Analysis

Using FEA, the magnetic flux density waveform can be observed at any point of the motor model. The radial magnetic flux density waveform has been obtained from FEA at three points located at the surface of the model. The three points are located on the magnet (point A), the rotor edge (point B), and the stator tooth (point C), as illustrated in Fig. 1. The magnetic flux density waveform of point A is illustrated in Fig. 24. The fundamental frequency of the waveform is three times the electrical frequency. This is due to the slot effect (the IPMSM has three slots per pole pair). Moreover, the magnetic flux density waveforms under no-load and load conditions (carrier frequency 1 kHz) have more ripple than under no-current condition. This is also observed in the DFT illustrated in Fig. 25 in which high order harmonics

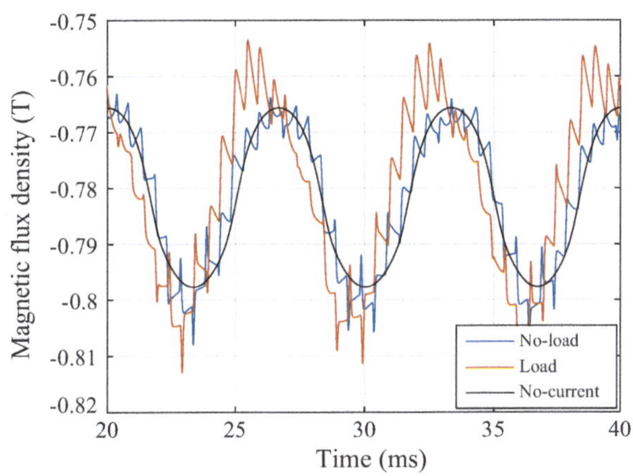

Fig. 24 Radial magnetic flux density in the magnet [10]. © [2019] IEEJ. Reprinted, with permission, from [12]

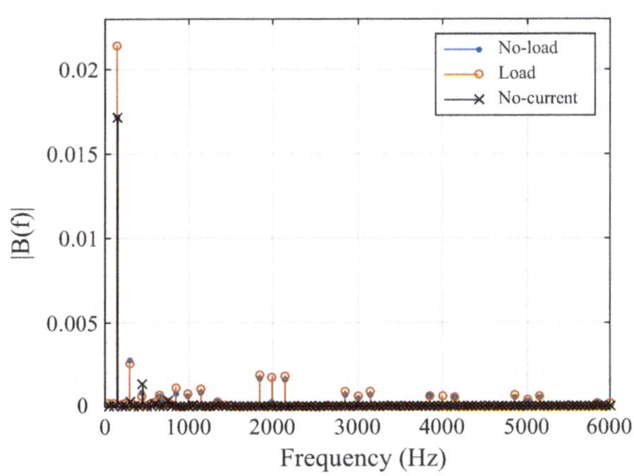

Fig. 25 DFT of radial magnetic flux density in the magnet [10]. © [2019] IEEJ. Reprinted, with permission, from [12]

Table 3 Comparison of fundamental and harmonic content of radial magnetic flux density in no-load and load conditions

		No-load	Load	Increase rate (%)
Magnet (point A)	Fundamental (150 Hz)	0.0172 T	0.0215 T	24.82
	Harmonic content	0.0046 T	0.0052 T	14.31
Rotor (point B)	Fundamental (150 Hz)	0.1802 T	0.1844 T	2.31
	Harmonic content	0.0400 T	0.0440 T	10.06
Stator (point C)	Fundamental (50 Hz)	1.0727 T	1.0785 T	0.54
	Harmonic content	0.0973 T	0.1050 T	7.94

appear only for no-load and load conditions. This is the effect of the PWM high order harmonics in the phase current observed back in Figs. 22 and 23. It can also be seen that both the fundamental and the high order harmonics of the magnetic flux density are larger under load condition than under no-load condition. This is thought to be due to the increase in the current fundamental and high order harmonics under load condition compared to no-load condition, observed back in Fig. 23.

Waveforms at points B and C are not presented for a matter of space, but the same observations, at different scale, can be made. In order to more clearly quantify the effect of the load on the magnetic flux density, the harmonic content of the waveforms is calculated. The harmonic content of the magnetic flux density in a given order range $[N_1; N_2]$ is given by

$$B_{r,\beta,\mathrm{harm}} = \sqrt{\sum_{i=N_1}^{N_2} B_{r,\beta,i}^2} \quad (10)$$

Table 3 gives the amplitude of fundamental and the complete harmonic content excluding the fundamental ($N_1 = 2$, $N_2 = N_h$) of the radial magnetic flux density at points A, B, and C for both no-load and load conditions. Both the harmonic content and the fundamental amplitude are higher in load condition. However, the increase in fundamental is very low in the rotor and stator. Since eddy current losses basically increase with the square of the frequency, it is thought that the increase of high order harmonic content is the main cause of the stronger eddy current losses observed in the magnets, rotor and stator (Fig. 19). However, the hysteresis losses see almost no change, which means that the increase reported in Table 3 is too small to affect them. Moreover, it can be noticed that the strongest increase rate, both for fundamental and high order harmonics, occurs in the magnets. The large increase rate, combined with the fact that the magnet is not segmented and then offers a large radial area, can explain the fact that the eddy current losses increase the most in the magnets.

References

1. Denis, N., Odawara, S., Fujisaki, K.: Attempt to evaluate the building factor of a stator core in inverter-fed permanent magnet synchronous motor. IEEE Trans. Ind. Electron. **64**(3), 2424–2432 (2017)
2. Denis, N., Wu, Y., Fujisaki, K.: Impact of the inverter DC bus voltage on the core losses of a permanent magnet synchronous motor at constant speed. In: Proceeding of International Conference Electrical Machines and Systems (ICEMS), Chiba, Nov. 2016, pp. 1–6
3. Denis, N., Wu, Y., Odawara, S., Fujisaki, K.: Study of the effect of load torque on the iron losses of permanent magnet motors by finite element analysis. In: Proceeding of 11th International 11th International Symposium on Linear Drives for Industry Applications, pp. 1–5. Osaka (Sep. 2017)
4. Ma, L., Sanada, M., Morimoto, S., Takeda, Y.: Prediction of iron loss in rotating machines with rotational loss included. IEEE Trans. Magn. **39**(4), 2036–2041 (2003)
5. Fujisaki, K., Satoh, S., Enokizono, M.: Influence of vector magnetic property with rotational magnetic flux, magnetic hysteresis and angle difference on stator core loss. J. Jpn. Soc. Appl. Electromagn. Mech. **20**, 360–365 (2012)
6. Yamazaki, K., Seto, Y.: Iron loss analysis of interior permanent-magnet synchronous motors—variation of main loss factors due to driving condition. IEEE Trans. Ind. Appl., **42**(4), 1045–1052 (2006)
7. Rylko, M.S., Lyons, B.J., Hayes, J.G., Egan, M.G.: Revised magnetics performance factors and experimental comparison of high-flux materials for high-current DC-DC inductors. IEEE Trans. Power Electron. **26**(8), 2112–2126 (2011)
8. Odawara, S., Denis, N., Yamamoto, S., Sawatari, K., Fujisaki, K., Shindo, Y., et al.: Impact of material on the iron losses of a reactor with air gap. IEEE Trans. Magn. **51**(11), Art. ID 8401404, (Nov. 2015)
9. Collaboration research between DENSO company and Electromagnetic Energy System Laboratory, Toyota Technological Institute
10. Wu, Y.: Finite element analysis for the iron loss investigation of PMSM excited by inverter. Master's thesis, Electromagnetic Energy System Laboratory, Toyota Technological Institute (Feb. 2017)
11. Denis, N., Wu, Y., Fujisaki, K.: Impact of the inverter DC bus voltage on the core losses of a permanent magnet synchronous motor at constant speed. IEEJ J. Ind. Appl. **6**(6), 346–352 (2017)
12. Thao, N.G.M., Denis, N., Wu, Y., Odawara, S., Fujisaki, K.: Study of the effect of load torque on the iron losses of permanent magnet motors by finite element analysis. Accepted for publication in IEEJ J. Ind. Appl., (Mar. 2019)

Electrical Motor Applied by Low Iron Loss Magnetic Material

Keisuke Fujisaki

Abstract To reduce core loss of an electrical motor, I select lower iron loss magnetic material than conventional electrical steel as NO (Non-Oriented) steel, and GO (Grain-Oriented) steel, amorphous steel and nanocrystal material. They are applied to the electrical motor of IPM-SM (interior permanent magnet synchronous motor) and manufactured as GO motor, amorphous motor, and nanocrystal motor. In GO motor, due to the strong magnetic anisotropy, GO steel is divided into some pieces, and they are arranged so as that the main magnetic flux follows the easy magnetization direction of GO steel. The motor core loss is measured in the drag force test where stator core is excited by the permanent magnet of the rotor. The core loss reduces in order of NO motor, GO motor, amorphous motor, and nanocrystal motor. The measured core loss is in good agreement with the calculated core loss of finite element method. Since the core loss includes the stator loss as well as rotor loss and permanent magnet loss of sintered NdFeB, the calculation data show that the stator loss of the nanocrystal motor is about one-twentieth smaller than the one of NO motor.

Keywords GO (Grain-Oriented) steel · Amorphous · Nanocrystal material · Motor core loss · Drag force

1 Introduction

Materials mainly used as core material of an electrical motor are non-oriented steel sheet (NO) now. Because the NO is polycrystalline whose crystal size almost is around 10 to 100 μm, and the polycrystalline is collective organization so as that the crystal orientation of iron atom with magnetic anisotropy is arranged at random, and its magnetic anisotropy is relatively small. When an electrical current flows in the electrical motor, the magnetic flux flows on the plane perpendicular to the plane of the electrical steel sheet. As the magnetic flux flows to any direction in the plane of the electrical steel sheet due to the rotation of the rotor with permanent magnet or so

K. Fujisaki (✉)
Toyota Technological Institute, Nagoya, Japan
e-mail: fujisaki@toyota-ti.ac.jp

K. Fujisaki (ed.), *Magnetic Material for Motor Drive Systems*,
Engineering Materials, https://doi.org/10.1007/978-981-32-9906-1_9

and the three-phase alternating current, it is desirable that the electrical steel should have small magnetic anisotropy on the two-dimensional plane.

As an electromotive force occurs in the parallel direction to the alternating current which flows in the electrical motor, an electrically insulating layer of μm degree thickness is applied to the outer layer of the electrical steel sheet to prevent an eddy current in the steel sheet. In addition, in order to prevent the eddy current flow in the steel sheet, around several percent of Si is contained in it, and then, it has a big electrical resistance.

Furthermore, the electrical steel sheet realizes low price because it is produced by a mass production process such as casting and rolling in the production. Because the electromagnetic steel sheet has the superior property as high performance and cost reduction in this way, it is widely used as core materials of the motor.

However, when a motor drive system comes to be applied to all the moving bodies such as a ship, an airplane as well as a car, the driving system must be on the machine from on the ground, where it moves with the moving bodies. As the application objects of the motor drive system increase in this way, it may be said that the demand specifications should be changed. Then, one of the most important demands is considered to be an extreme high performance as low iron loss of electrical motor even if it is some expensive. Therefore, it is decided to work on the study of the low core loss motor using commercially available low iron loss materials.

As commercially available low iron loss materials are, GO materials, amorphous materials, nanocrystal materials, they are taken up for low iron loss motor production, and their characteristics are summarized in Table 1. As for NO materials, 35H300 is taken up here as relatively low iron loss NO material.

The GO material is grain-oriented steel sheet whose crystal grain size is almost around 10 to 100 mm [1, 2]. GO steel has the strong magnetic anisotropic characteristics and then it realizes the low iron loss characteristics so as that the crystal orientation of GO aligns the easy magnetization direction to the rolling direction as much as possible. It is mainly used for a transformer by arranging the cutting GO steel so as that the main magnetic flux in the transformer flows in the easy magnetization direction of GO steel. The iron loss of easy magnetization direction is about the half one of NO materials. Saturated magnetization, steel sheet thickness, and electrical

Table 1 Characteristics comparison of conventional (NO) and low iron loss materials [4–6]

Soft Mag. Mat.	Nanocrystal	Amorphous	GO steel	NO steel	Unit, remark
Iron loss	0.049	0.224	0.632[a] [1]	1.105	50 Hz, 1T, W/kg
Saturated mag. M_s	1.23	1.63	2.03	2.11	T
Thickness	18	25	350	350	μm
Electrical resistance	1.2	1.2	0.46	0.52	μΩ-m
Density	7.3	7.33	7.65	7.65	kg/dm^3
Picked up material	FT-3 M	2605HB1 M	35ZH135	35H300	

[a][1] Iron loss at easy magnetization direction

resistance of GO steel are approximately the same ones as NO materials. When GO steel is applied to an electrical motor core, it is necessary to consider the magnetic anisotropic characteristic of GO steel.

The amorphous material has a structure not to have crystal organization by quenching it at the time of casting. So, it has no obstruction of the domain wall displacement caused by the grain boundary and has thin thickness and large electrical resistance. Steel sheet thickness of the amorphous is around 25 μm which is extremely thinner than the one of NO materials. Therefore, it has around a one-fifth smaller iron loss than NO steel. In addition, it is magnetic isotropic material which does not have magnetic anisotropy, but saturated magnetization is small because it consists of a lot of kinds of elements. It has a problem of cutting the sheets for core production due to high hardness. In addition, it has a high mechanical stress sensitivity for the magnetic characteristic.

The nanocrystal material is produced by reheating amorphous materials and recrystallizing a lot of small size of crystals as around 10 to 100 nm in the amorphous which has single magnetic domain structure [3]. So its magnetization process occurs only magnetization rotation without magnetic wall movement. Therefore, it has approximately one-20th lower iron loss than the one of NO material. The steel sheet thickness and the electrical resistance are the same level as the amorphous material, but the saturated magnetization becomes smaller than the one of the amorphous material because it contains a further number kinds of elements. Because it makes a brittle easily due to high-temperature recrystallization, it has some problems such as integral molding to be as electrical motor core and securing electrical insulation in the layer.

Another nanocrystal material is also studied as a high B_s material which increases saturated magnetization than amorphous materials, but the nanocrystal materials with low iron loss are only paid attention to because the purpose here is to realize a low iron motor.

So an iron core of an electrical motor is produced experimentally using such as low iron loss materials, and its core loss is measured and compared with the numerical calculation results of core loss. The details are shown as follows.

2 Magnetic Anisotropic Motor Using Grain-Oriented Steel

The magnetic and iron loss characteristics of GO steel and NO steel are shown in Fig. 1. The characteristics of the easy magnetization direction and the transverse direction are also shown as for the GO steel. The GO steel shows a big difference in the easy magnetization direction and the transverse direction, and it is found out to have an extremely big magnetic anisotropy. The NO materials are located at the middle.

It is insufficient only to replace the NO steel to the GO steel in an electrical motor, when the GO steel is used as a motor core, because the GO steel has the strong magnetic anisotropy that magnetization and iron loss characteristics vary according

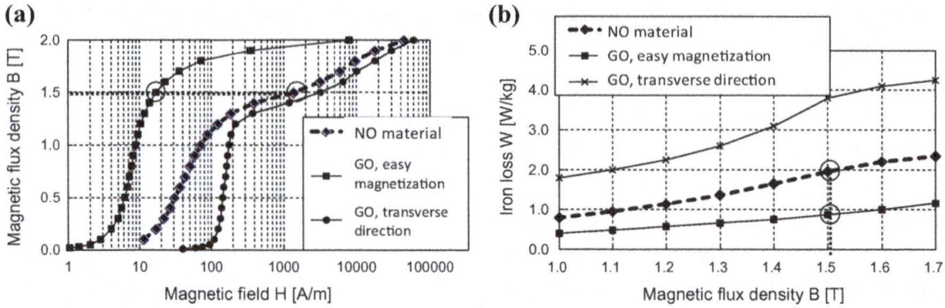

Fig. 1 Magnetic characteristics of GO steel and NO steel: **a** Magnetization characteristics, **b** Iron loss characteristics [8]

to a direction of the magnetic flux density vector. As magnetic flux flows through not only the easy magnetization direction but also the transverse direction of the GO steel with the motor, and this causes an increase of the motor iron loss. When the magnetic flux flow in the motor core is considered, it is found out that the magnetic flux in the teeth part mainly flows in a radial direction, and the one in the yoke part mainly flows in a rotatory direction.

So, the GO steel is decided to be divided into some pieces and reconstructed as a stator core of the electrical motor like Fig. 2 so as that the divided piece has the easy magnetization direction which directs to the radial direction in the teeth part, and it has the easy magnetization direction which directs to the rotational direction in the teeth part [7, 8]. This kind of motor is called "magnetic anisotropic motor."

Because the magnetic anisotropic motor using the divided GO pieces has a separated structure as yoke part and teeth part, it is necessary to pay attention to the external form precision, particularly the machine precision of the inside diameter as the motor core. This is because the rotor is arranged inside the stator core, and it rotates with high rotational speed with constant clearance between the stator and the rotor. Therefore, when the divided GO steel pieces are used to produce the stator core, a special jig is prepared to keep an internal shape of stator core constant for a predetermined size like Fig. 3 [7].

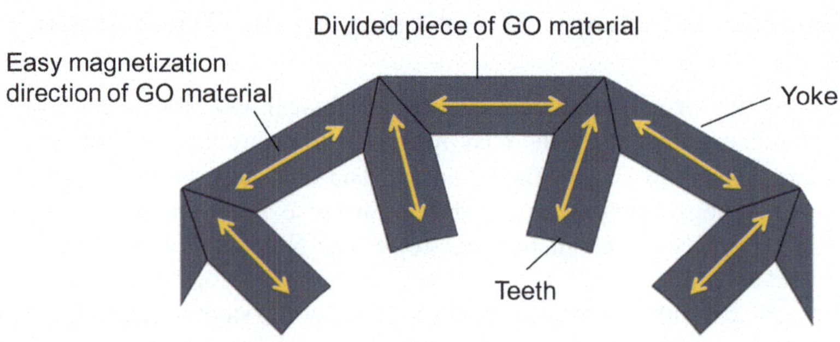

Fig. 2 Magnetic anisotropic motor using divided pieces of GO steel [13]

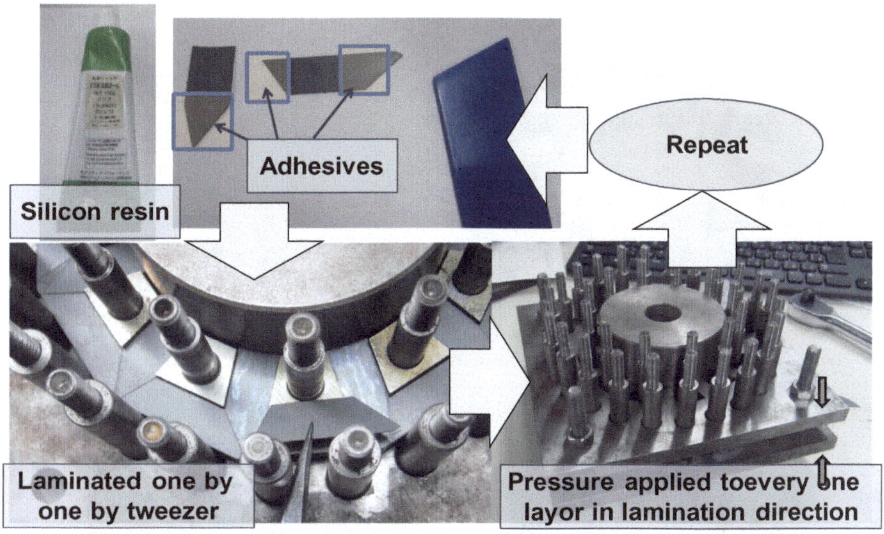

Fig. 3 Manufacturing process of magnetic anisotropic motor [13] 2014 IEEE. Reprinted, with permission, from [7]

Adhesive of silicon resin is applied to the surface of GO divided piece one by one, the GO divided piece is set on a predetermined core with tweezers, and finally, total 12 GO divided pieces with 6 pieces for teeth part and 6 pieces for yoke part are arranged on the core in one layer. A steel plate with 5 mm thickness and stator core shape is put on the 12 pieces, and then, they are tightened with a vise and are kept in 30 min. After that, pieces are applied for the next layer. The divided pieces are piled up on the laminating direction of 47 mm with 134 layers, and then, totally 1608 pieces are used for producing the stator core. It took around two months to produce the stator core in this time due to the first trial manufacture.

A photograph of the stator core of the magnetic anisotropic motor is shown in Fig. 4. The photograph of the NO core with the same shape and the same size is also shown for comparison. The outer diameter of the stator is 128 mm, the inside diameter is 76. 5 mm, and the clearance between the stator and the rotor is 1.25 mm. It is 12 slots and 8 poles (4 pole pairs), a concentration winding motor. The NO motor core laminates NO materials with one punching as a stator core shape. Since the teeth part and the york part are separated and it is difficult to keep the fixed one-body, the anisotropic motor core uses a lap connection in the T-shape region that is a junction of the teeth part and the yoke part. Due to an asymmetry shape of the teeth part from the centerline of the teeth part, the teeth part joins with the york part of the layer under one with surface-bonding, and then a mechanical bond strength with the teeth part and the yoke part is strong.

The drag force test is carried out to measure the core loss characteristic of the magnetic anisotropic motor and the conventional NO motor. In the drag force test, magnetized permanent magnets are installed in a rotor, and the rotor is driven by an

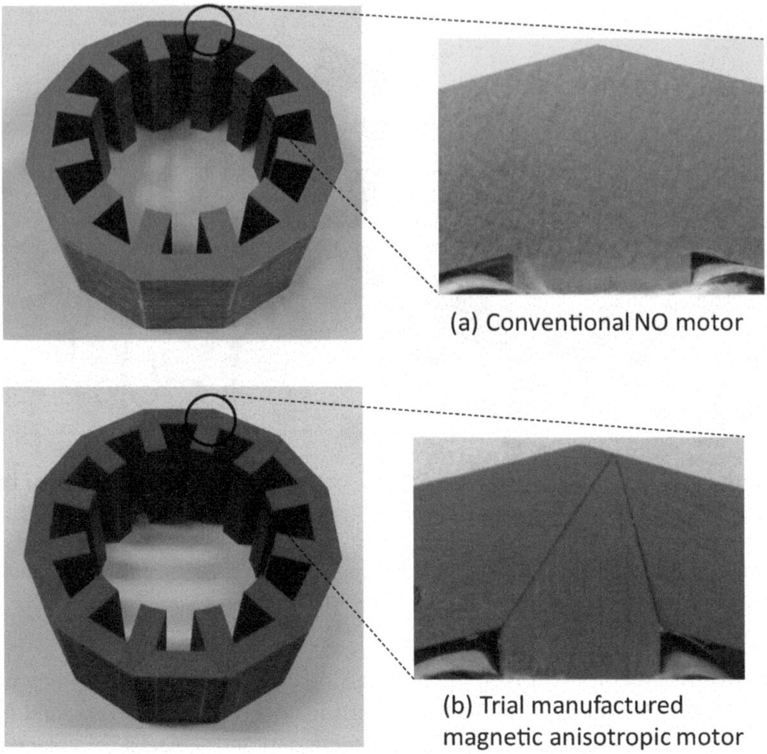

Fig. 4 A photo of **a** conventional NO motor and **b** magnetic anisotropic motor [7, 8, 13]. 2014 IEEE. Reprinted, with permission, from [7]

external motor. The magnetic flux derived from the permanent magnet in the rotor excites the stator core, and an electromagnetic torque generated at that time due to iron loss of the motor is measured by a torque detector. The drag force loss is calculated by pulling a machine loss from a machine input multiplying an angular velocity by the measured electromagnetic torque. The machine loss is introduced by the electromagnetic torque when the rotor with permanent magnets without magnetization is driven by the external motor at the same rotational speed as the drag force test. The core material of the rotor is the conventional NO materials.

At the same time, two-dimensional electromagnetic field non-linear transient numerical calculation is carried out at the same condition as the drag force test. Time series data of magnetic flux density at each divided element are expanded to Fourier series, and then, an iron loss is calculated by applying Steinmetz equation to each frequency component of magnetic flux density. The magnetic anisotropy is expressed by the two-axis anisotropy model where magnetic flux density vector is divided into easy magnetization element and transverse element and then initial non-linear magnetization *B*-*H* curve of each direction is applied to. Iron loss characteristics (*B*-*W* curve) of the easy magnetization direction and the transverse direction

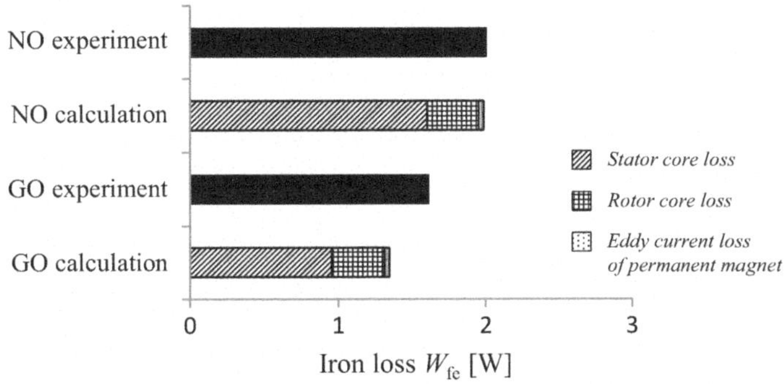

Fig. 5 Measurement and calculation results of drag force loss of magnetic anisotropic motor (GO) and conventional NO motor

are measured at some limited frequencies. The iron loss at no measured frequency is obtained by two-frequency iron loss characteristics at the nearest measured two frequencies.

The magnetic flux derived from the permanent magnet in the rotor flows in the stator core, returns to the rotor again, and then forms a closed magnetic circuit, but the slot shape of the stator causes slot harmonics components in the magnetic flux density of the rotor. Therefore, the measured core loss includes the rotor iron loss and the permanent magnet eddy current loss as well as the stator iron loss. Electrical conductivity is set in the permanent magnet as 1.6×10^{-6} Ωm to express an eddy current in it, and then, the eddy current loss is calculated. Figure 5 shows the measured core loss as well as the numerical calculated core loss which separates the stator iron loss, the rotor iron loss, and the permanent magnet eddy current loss.

The results show that the measured iron losses of the GO divided magnetic anisotropic motor and the conventional NO motor have almost the same values as the numerical calculation. As the stator loss, the rotor loss and the permanent magnet loss cannot be separated in the measured drag force loss, and they are separated in the calculation and analyzed.

Because the permanent magnet and the rotor of the NO motor and the GO motor have the same materials and the same shape, the permanent magnet loss and the rotor loss are almost the same. When the stator loss that changed the materials from the NO to the divided GO is paid attention to, it is found out to become almost the half.

3 Amorphous Motor [9–11]

Amorphous material has no crystal grain boundary, and has very thin thickness, and has a large electrical resistance. It has smaller iron loss than the one of the easy magnetization directions of the GO steel. For further iron loss reduction, the amorphous material is considered to be applied to the stator core material of electrical motor.

Figure 6 shows iron loss characteristics of the amorphous material and the conventional NO material. As the NO material has saturated magnetization up to 2 T, the iron loss characteristic can be shown up to it, but the saturation magnetization of amorphous material is smaller as 1.63 T, the iron loss characteristic is only shown up to it. However, the iron loss of the amorphous is about one-fifth smaller than NO.

As the amorphous material is magnetic isotropy, one punching or cutting process in unity is possible like the NO materials, and the machine strength is high. However, due to the high hardness, it is difficult to cut the amorphous material, and it is ultrathin with 25 μm thickness. The film to electrically insulate is not put on the surface in particular. Though an electrically insulating layer of oxide of around 10 nm thickness is said to exist on the outer surface layer, an electrical insulation in layer is a problem in motor core production. In addition, as stress sensitivity is extremely high, a mechanical residual stress to suffer from molding it as a motor core from material is also a problem to be solved.

Around 1888 sheets of the amorphous are laminated for one core here. A photo of the amorphous core is shown in Fig. 7. Copper coil is inserted in each tooth of the stator core.

Figure 8 shows a drag force loss of the amorphous motor and the conventional NO motor in comparison with measurement and analysis. Analysis model is two-dimensional non-linear transient electromagnetic field analysis, and iron loss is calculated by using a Steinmetz equation derived from time series of magnetic flux density in each element. The measured drag force loss of the NO motor accords with the analytical one well, but the measured loss of amorphous motor is larger than the analytical one. The reason why the core loss increases in measured data may be said to be not enough electrical insulation in laminated direction, mechanical stress supply in motor core production, and short circuit in cutting the steel sheet.

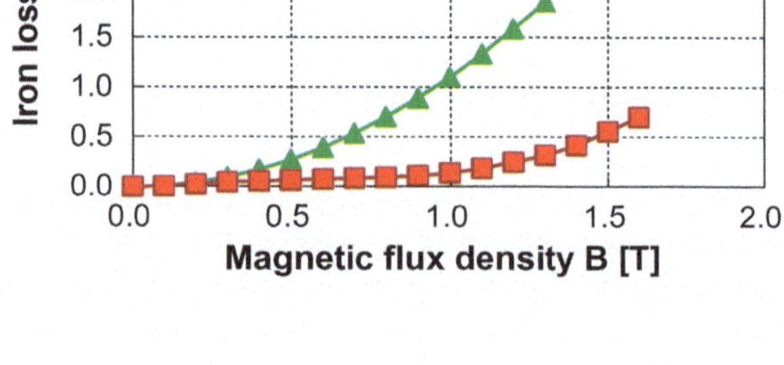

Fig. 6 Iron characteristics of amorphous (Metglass (2605HB1M), Hitachi Metal) and NO steel (35H300, Nippon Steel Sumitomo Metal) (50 Hz) [9]

Fig. 7 A photo of amorphous core

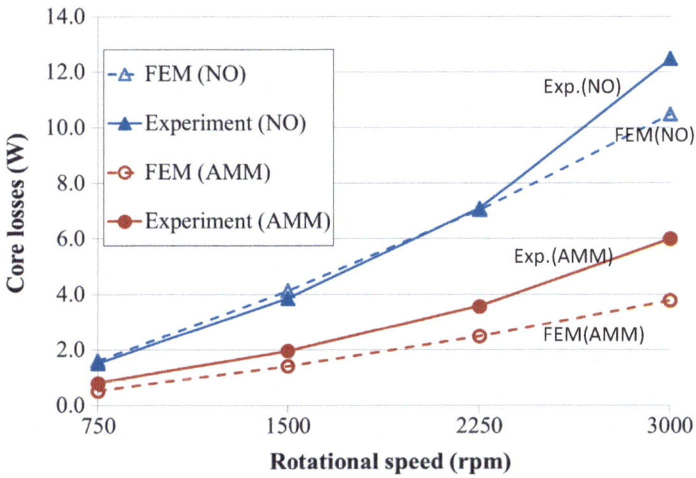

Fig. 8 Drag force loss comparison with amorphous motor and NO motor (comparison with measurement and calculation) 2015 IEEE. Reprinted, with permission, from [11]

A breakdown of the stator loss, the rotor loss, and the permanent magnet loss is shown in Fig. 9 by the analysis result. Since as for the amorphous motor and the NO motor the same material and the same shape are used in the rotor and the permanent magnet, their rotor loss and permanent loss are the same value.

When a comparison is carried out with the core loss of the stator in which the material changes from NO to amorphous, the stator core loss of the amorphous motor is found to be around from one-fifth to one-seventh smaller than the one of the NO motors. The core loss of the NO motor becomes large in high frequency, but the one of the amorphous motors is not so large. So, the core loss difference becomes large in high frequency. High performance of the amorphous motor is considered to be caused by being ultrathin and high electrical resistance.

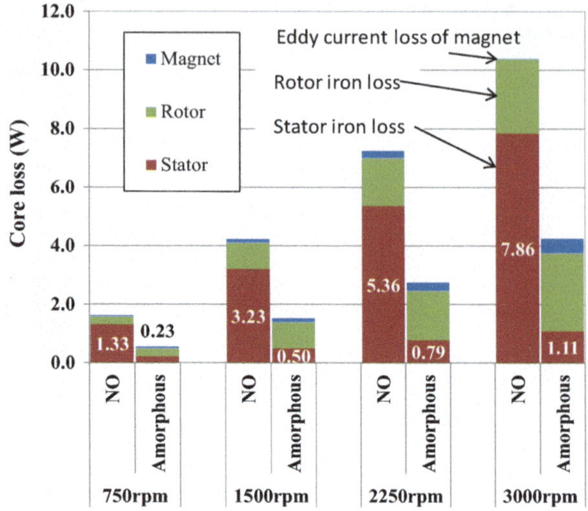

Fig. 9 A breakdown of motor core loss by calculation data 2015 IEEE. Reprinted, with permission, from [11]

4 Nanocrystal Motor [12]

Because the obstruction of domain wall movement at the crystal grain boundary happens when a crystal grain size becomes small, iron loss increases in inverse proportion to crystal grain size such as the electromagnetic steel sheets, but iron loss is known to decrease adversely in proportion to the crystal grain size to the 6th power as it is at the magnetization process by the magnetized rotation of the single magnetic domain when the crystal grain size is in lower than 100 nm degree. The nanocrystal materials are made so as to recrystallize a single crystal whose crystal size is dozens of nano-meter on the amorphous materials, and the machine strength is extremely fragile, but iron loss becomes remarkably small from the amorphous. In this time, nanocrystal material is applied to a stator core of the motor and produces a nanocrystal motor experimentally.

Figure 10 shows a stator core produced by nanocrystal materials. The cut sheet of nanocrystal materials is dipped in a resin row, the resin is impregnated into the nanocrystal materials under high pressure, and it is cut by a wire cut in oil for a stator core shape. A quarter size of the stator core is produced due to the limitation of width of nanocrystal materials, and then, the four quarters are joined for one stator core.

Motor core loss is measured by the drag force test using the nanocrystal motor trial manufactured in this time. For comparison, the motor core loss of NO motor as well as GO motor (magnetic anisotropic motor) and amorphous motor is also shown in Fig. 11. It is understood that core loss decreases in order of the NO motor, the GO motor, the amorphous motor, and the nanocrystal motor.

Since the motor core loss includes not only the stator core loss but also the rotor core loss in the conventional material as NO and the permanent magnet eddy current loss, numerical calculation result is also shown in Fig. 12 for the comparison

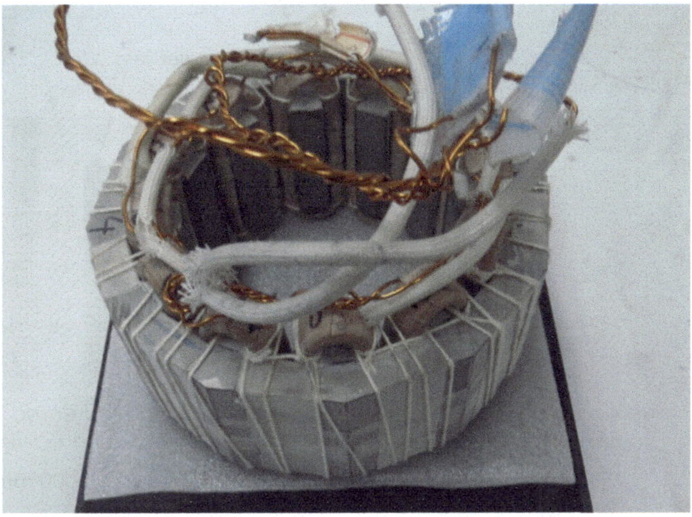

Fig. 10 Photo of stator core made of nanocrystal magnetic material [12] 2017 IEEE. Reprinted, with permission, from [12]

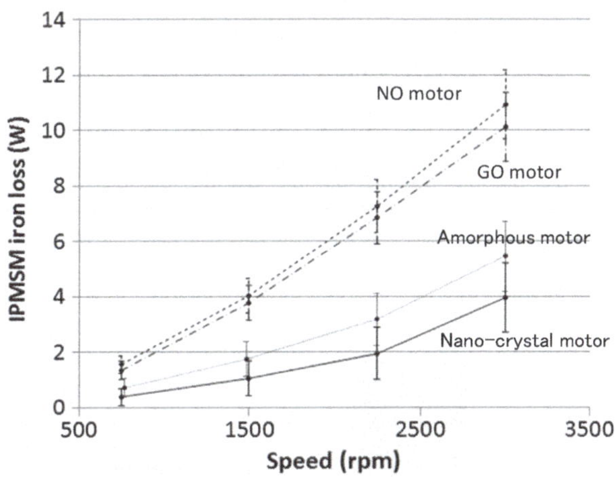

Fig. 11 Measured core loss in drag force test 2017 IEEE. Reprinted, with permission, from [12]

with the experimental data as well as for the breakdown of the motor core loss. The calculation is carried out by two-dimensional non-linear transient electromagnetic field numerical analysis. The calculated core loss expresses well the measured core loss for the NO motor and the nanocrystal motor. The rotor core loss and the permanent magnet eddy current loss of NO motor are found to be the same as the ones of the nanocrystal motor. The calculation results show that the stator core loss of the nanocrystal motor is around one-20th smaller than the one of the NO motors.

The significance of the core loss of the nanocrystal motor which has one-20th smaller core loss than NO motor is considered in comparison with the other motor

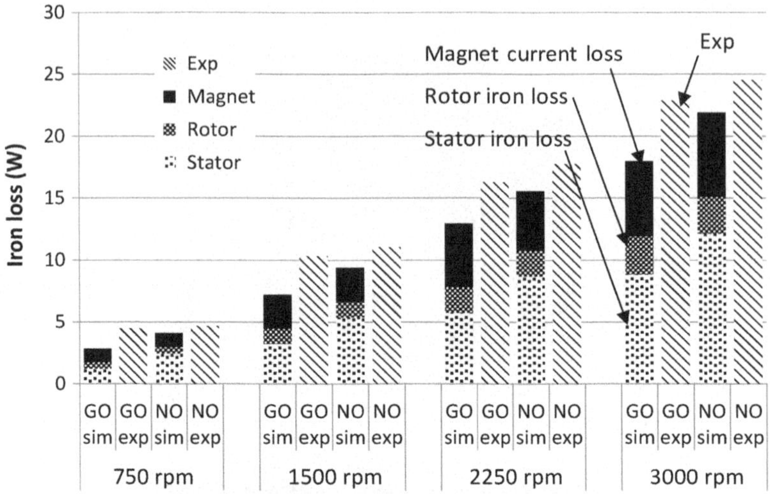

Fig. 12 Comparison of drag force loss with measurement and calculation 2017 IEEE. Reprinted, with permission, from [12]

losses as copper loss and mechanical loss. Figure 13 is shown for it where the core loss is calculated, and the copper loss and mechanical loss are measured in no-load condition. When the stator material changes from the conventional NO to the nanocrystal material, the motor loss decreases largely, but the rotor core loss and the permanent magnet loss still remain. So when the nanocrystal material is applied to the rotor core, and bond NdFeB permanent magnet is applied the permanent magnet, so-called iron loss extremely becomes small. The bond NdFeB magnet has smaller

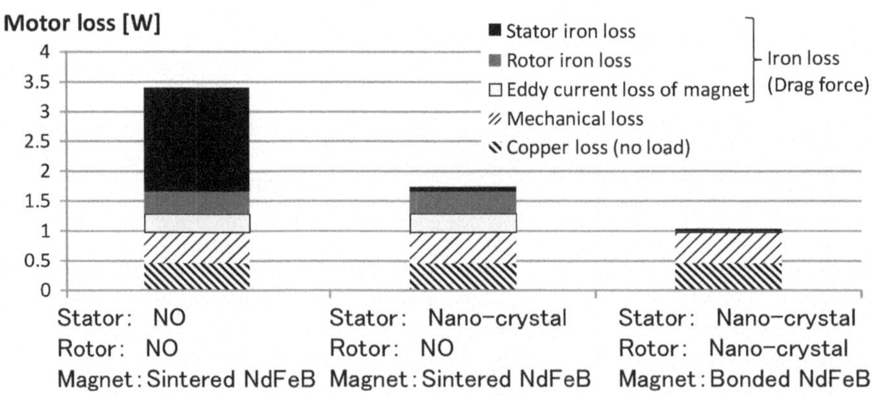

Fig. 13 Contents of motor loss changing the material (750 rpm, 50 Hz, stator core loss and rotor core loss, and permanent magnet eddy current loss are calculation value in drag force test, and mechanical loss and copper loss are measured in no-load condition)

residual magnetic flux density than the conventional sintered NdFeB, but it has large electrical resistance and almost no eddy current flows in it.

In the case of the nanocrystal stator and rotor and the bond NdFeB permanent magnet, the iron loss that added up the stator core loss, the rotor core loss and the permanent magnet eddy loss are extremely smaller than the machine loss or the copper loss. In other words, since the motor core loss of the nanocrystal motor is extremely small enough in comparison with the machine loss and the copper loss, though it has a technical problem such as saturation magnetization being small, it may be said that the nanocrystal motor is an ultimate material for the iron loss reduction.

It may be said that there is a motor which reduced the motor loss to the ultimate if it can be realized that a superconducting coil is applied to power supplied coil for the copper loss reduction, vacuum or almost the same circumstance and a magnetic bearing are used for the machine loss reduction, and the nanocrystal materials are applied to the motor core for the iron loss reduction.

References

1. Goss, N.P.: Electrical sheet and method and apparatus for its manufacture and test. U.S. Patent No.1965559 (1934)
2. Yamamoto, T., Taguchi, S., Sakakura, A., Nozawa, T.: Magnetic properties of grain-oriented silicon steel with high permeability Orientcore HI-B. IEEE Trans. Magn. **8**(3), 677–681 (1972)
3. Yoshizawa, Y., Oguma, S., Yamauchi, K.: New Fe-based soft magnetic alloys composed of ultrafine grain structure. J. A. P. **64**, 6044 (1988)
4. Catalogue of Nippon Steel & Sumitomo Metal. http://www.nssmc.com/product/sheet/list/motors-transformers.html (2017.12.10)
5. Catalogue of Hitachi Metals: Nano-crystal magnetic material (FINEMET). https://www.hitachi-metals.co.jp/products/elec/tel/pdf/hl-fm9-h.pdf#search=%27finemet%27 (2016.4)
6. Catalogue of Hitachi Metals: (Amorphous). http://www.hitachi-metals.co.jp/products/infr/en/pdf/hj-b10-b.pdf (2017.12.10)
7. Takeda, S., Fujitani, K., Odawara, S., Fujisaki, K.: Trial manufacture of magnetic anisotropic motor and evaluation of drag loss characteristics. In: Proceedings of the 2014 International Conference on Electrical Machines (ICEM), Berlin, Germany, pp. 2049–2055 (2014)
8. Fujitani, K., Haruta, N., Fujisaki, K.: Evaluation of stator core iron loss of 'magnetic anisotropic motor'," EV Tec, Yokohama, 20144081 (2014)
9. Inoue, M., Odawara, S., Fujisaki, K.: Investigation of manufacturing processes of amorphous alloy for motor cores. In: IEEJ Technical Meeting of Linear Drives Technical Committee, RM-16-098, LD-16-106 (Sept. 2016)
10. Ieki, M., Kitsunezaki, N., Fujisaki, K.: Iron loss characteristics in manufacturing IPMSM amorphous motor core. In: IEEJ Technical Meeting of Linear Drives Technical Committee, LD-16-022 (Jan. 2016)
11. Okamoto, S., Denis, N., Ieki, M., Fujisaki, K.: Core loss reduction of an interior permanent magnet synchronous motor using amorphous stator core. IEEE Trans. Ind. Appl. **52**(3), 2261–2268 (2016)
12. Denis, N., Inoue, M., Fujisaki, K., Itabashi, H., Yano, T.: Iron loss reduction of permanent magnet synchronous motor by use of stator core made of nanocrystalline magnetic material. IEEE Trans. Magn. **53**(11), 8110006 (2017)
13. Takeda, S., Fujitani, K., Fujisaki, K.: Trial making of magnetic anisotropic motor and investigation of dragging loss characteristic. Technical meeting on magnetics of the institute of electrical engineering o Japan, MAG-14-024, Tokyo, March, 2014. (in Japanese)

Magnetism and Its Modelling

Origin of Magnetism 90 Years of Misunderstanding

Yoshiyuki Kawazoe

Abstract After Slater published his paper on Hund's rule, almost all standard textbooks follow his idea of explaining the origin of magnetism (the reason of stabilization of the magnetic ground state) based on electron exchange energy. However, when his theory was invented, computer was not available to numerically solve many-body electron system, and he omitted the most important contribution from the nucleus–electron interaction (nucleus charge has a sharp peak, but he assumed uniform distribution). We have proved that this omission makes a serious problem of violation of virial theorem, which is a necessary condition for any Coulomb system. We have solved the Schrödinger equation numerically satisfying the virial theorem and proved that the contribution of the nucleus–electron interaction is the fundamental part to reduce the energy for magnetic ground state. Based on the virial theorem satisfaction, we can now distinguish correct and incorrect theories. Most striking fact is that the two-level models assuming the same space wavefuntions such as Hubbard model is incorrect and they should be excluded from the theory of magnetism.

Keywords Virial theorem · Coulomb interaction · Electron correlation · Electron–nucleus interaction · Density functional theory · Quantum Monte Carlo method · Incorrect theory

1 Introduction

Most of the people recall their interest in science during their student days where they would collect iron present in the sand using a horseshoe magnet and observe peculiar patterns called lines of magnetic force. However, they lose their way while

---

Y. Kawazoe (✉)
New Industry Creation Hatchery Center, Tohoku University, Sendai, Japan
e-mail: kawazoe@imr.edu

Physics and Nanotechnology, SRM Institute of Science and Technology, SRM Nagar, Kattankulathur, Kancheepuram District, Tamil Nadu 603 203, India

School of Physics, Suranaree University of Technology, 111 University Avenue Muang, Nakhon Ratchasima 30000, Thailand

K. Fujisaki (ed.), *Magnetic Material for Motor Drive Systems*,
Engineering Materials, https://doi.org/10.1007/978-981-32-9906-1_10

trying to understand magnetism. They were taught that there are N and S poles at the ends of a magnet. When cut it into two pieces, once again N and S poles appear on the ends of each of the pieces. Even if the process of cutting the magnet continues, we still obtain N and S poles at the ends of each piece. In order to understand this phenomenon, several questions of fundamental nature arise in the minds of clever students: To begin with, what are N and S poles? Are there independent N or S poles? What are the lines of magnetic force? The general response from most of the teachers to these questions would be that "magnet is such a material" without any deep understanding. With such an off-handed argument, such clever students immediately understand that science is a subject of categorization and its concepts are required to be memorized in order to obtain good grades. With such a belief, they begin to dislike science and find that it is difficult to understand the real concepts this way.

Another example, which leads to this situation, includes students being taught that the principle of a motor is given by Fleming's left-hand rule, and that, the right-hand rule is for the electric dynamo, in high school. It is important to understand that strange phenomena are governed by the laws of nature, and the principles or rules are developed to obtain a reasonable understanding of them. In this regard, once again students stop utilizing their brains to understand the mechanism of operation at a deeper level. Even though teachers ask their students to think more, or try to encourage them through various science system promotion plans, students are not interested in the rote way science is being taught. This method of teaching is further continued at the university level. University courses teach students the concepts of para-, ferro-, antiferro-, and ferrimagnetisms, i.e., a more detailed classification is provided, but the fundamental understanding of magnetism (i.e., the origin of magnetism) itself is not covered. In standard textbooks on magnetism, this phenomenon arises from aligned electron spins. This is the correct explanation. However, as a suitable quantum mechanical example for an introductory course, the concept of Coulomb, direct, and exchange integrals is introduced to explain the origin of magnetism to be the exchange energy to stabilize the magnetic ground state. This way of describing magnetism started 90 years ago, when Slater applied his perturbation theory to explain Hund's rule [1].

Recently, we proved that his perturbation theory misses the most important contribution, i.e., the nucleus–electron interaction and is therefore completely wrong. Misunderstandings in such textbooks are normally discussed for historical reasons, but there are several such problems in mathematics and sciences. Basically, the newspaper reporters and teachers who have graduated with literature and social science majors do not understand our problems or believe that normal people are not interested in mathematics or the sciences.

Moreover, concerning quantum mechanics, most people give up understanding it and simply believe in textbooks, i.e., they feel that something is wrong with the concepts in the textbooks, but either find it difficult or are not motivated to resolve them. As will be introduced below, this situation does not originate from the difficulty of quantum mechanics, but actually, the concept of the many-body problem. (There are many subjects in classical mechanics, for which a similar level of mathematics is required. For example, the precession of top, for which the polar coordinate system and Langrangian formulation are necessary to solve it. The Hamiltonian formulation was also introduced to separate position and momentum independently, i.e., to change the second-order differential Newton's equation into a set of two first-order differential equations, which is the basis to express uncertainty principle in quantum mechanics.) Due to the difficulty in understanding the many-body problem, many models have been developed to easily explain physical properties. As a result, these are used frequently. However, the approximations introduced in these models are frequently erroneous fundamentally. Recent significant progresses in supercomputing technology have made it possible to solve exactly the many-body problem, and therefore, finally, to correctly understand the origin of magnetism, which is completely different from the traditional explanation expressed in most of the standard textbooks. This paradigm shift in understanding will certainly enable beginners to attain clarity in the concept of magnetism. It is, therefore, imperative to rewrite textbooks and redesign teaching curricula accordingly. It is of paramount importance that the basics of magnetism are clarified presently, so that industrial applications arising from it are possible in the near future.

This chapter begins with the introduction of the difficulty in the many-body problem with an example of the trajectory of the moon around the sun and then proceeds to describe the correct method to understand the electronic states in atoms, molecules, and materials in general from a very fundamental viewpoint. In the traditional university courses, it is taught in an incorrect manner and is the reason why students cannot understand the basics. Based on the understanding of the many-electron system, the final aim of this chapter is to explain correctly, the origin of magnetism.

2 Difficulties in the Many-Body Problem

The so-called common sense of the people termed alchemy as a "nonsensical science" in the past, simply because to convert normal metal to valuable noble metal, nuclear conversion energy is necessary which was impossible at the time. In addition, in the twenty-first century nanotechnology has made it possible to realize useful new materials, based on manipulation of atomic and molecular configurations that are environmentally safe and possess fundamentally new functionalities, necessary for a sustainable society. Does quantum mechanics makes understanding of alchemy or nanotechnology difficult? No. As stated earlier, the difficulty arises from the many-body problem.

In elementary school, we were taught that the Moon (M) is a satellite. This "common sense" is already wrong! Several years ago, a major discussion was initiated to determine the number of planets, when nobody thought about the Moon. From Fig. 1, it is obvious that the Moon is not moving around the Earth (E). It moves around the Sun (S) similarly to the Earth. Computing the gravitational forces F between the three objects, F(S-E) is, of course, the strongest. But the second strongest force is F(S-M) and not F(E-M), and F(S-M) $2\times$F(E-M) (easily computed). Therefore, we

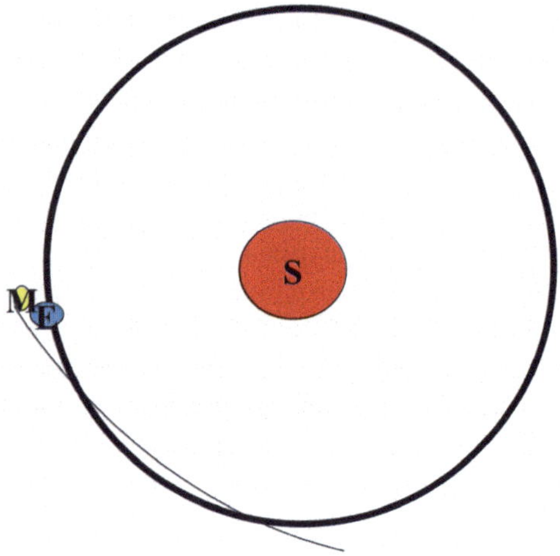

Fig. 1 Trajectory of the Earth and Moon around the Sun. They move together around the sun, nearer or farther to it (the figure is drawn exaggeratingly. In reality the two trajectories cannot be distinguished in this size). Other planets, such as Saturn or Jupiter, have satellites, which are located far away from the Sun and much lighter than the planets, and the interaction between the Sun and satellite can be treated as perturbation. The Moon is closer to the Sun and does not have negligible mass compared to the Earth, and hence, cannot be classified as a satellite according to the definition of perturbation, but based on our selfish misunderstanding that the center is the Earth, we classify the Moon as the satellite of the Earth (which moves around the Sun)

cannot treat M as a satellite of E (i.e., perturbation (neglecting F(S-M))) cannot be applicable). E and M should be categorized as binary stars (planets) [2]. Although E and M move around the S at speeds of 100,000 km/h, we do not feel this extraordinary high speed due to the law of inertia, and therefore, believe that M is moving around E.

Believing in common sense and textbooks in schools, and in the historically incorrect idea that the Moon is moving around the earth, indicates that we are still constrained to the Ptolemaic system. This classic example illustrates that perturbation theories are dangerous and we should, therefore, understand and try to resolve the difficult subject of the three-body (and by extension, many-body) problem. We should not categorize subjects based on common sense, but we should treat them quantitatively to understand them correctly. In the words of Feynman, "What I can't compute, I do not understand."

3 Truth Behind the Formation of Molecules and Crystals from Atoms

About 100 years ago, when experimental studies in physics and chemistry greatly progressed, it was realized that not only bio or non-bio systems, but also all materials are composed of nucleus and electrons, and that all physico-chemical properties are determined by electronic states in them. However, although the governing equations, i.e. Schrödinger (or Dirac) equations treated nucleus and electrons as fundamental compositions and accurately modeled the interaction between them as Coulombic, it was completely impossible at that time to solve the equations to determine the atomic configurations and subsequently, the physico-chemical properties in them through accurate numerical calculations.

We start with the simplest example of the electronic structure model for the hydrogen molecule, which is composed of two hydrogen atoms. In the famous Heitler–London (henceforth referred to as HL) model [3], as described in standard quantum mechanics textbooks, molecular orbital in hydrogen molecule is expressed by the superposition of two 1 s electron orbitals, satisfying Pauli's exclusion principle. The standard explanation is that "electronic clouds from each hydrogen atom overlap and a molecular state is realized." With such a statement, students develop a misconception that the electron is a "cloud," which is the worst starting point for them to (mis)understand nature. As a result, they cannot understand quantum mechanics correctly. Electron is not a cloud by nature, but a particle (one of the elementary particles). Owing to its extremely light weight (10^{-30} kg weight) and small size (comparable to photon, i.e., quantum unit of light), it is not possible to determine its position and momentum simultaneously. This uncertainly principle provides an area within which the electron can exist and this area is called "cloud." To summarize, an electron is a particle that is distributed in a region of space. Therefore, the concept of "electron cloud" comes from the observation problem.

Based on accurate calculations of the electronic states in hydrogen molecule, the results are fundamentally different from HL. In HL, two hydrogen atoms come together and the electron "clouds" overlap with kinetic energy T reducing and potential energy V increasing (this reduces the total energy $E = T + V$) to stabilize the molecule. However, we can prove that this way of stabilization contradicts with what is observed in nature as follows: An electron has negative charge, and hence, overlapping of two electrons gives rise to a repulsive force. This is not only a classical viewpoint, but also observed in quantum mechanics. (If one thinks carefully, they will realize that this is obvious, and that the textbooks for the introductory course on quantum mechanics are incorrect. By mentioning statements such as "quantum mechanics is different from classical mechanics" and "electron is a cloud," students are brainwashed into believing them and lose the ability to think scientifically and objectively. If understood correctly, quantum mechanics is not a difficult subject to explain the nature of physical existence. The non-scientific community may not have any problems, but the problem is for the students in science and engineering departments, who should understand quantum mechanics correctly, since that forms the basis of nanotechnology. As mentioned earlier, the most fundamental problem starts at the elementary school level, where students believe that science is a subject of remembering theory and principles, and they do not emerge from this misunderstanding.)

To be able to reduce the total energy to stabilize the hydrogen molecule, there is simply an attractive Coulombic interaction between the nucleus (proton with positive

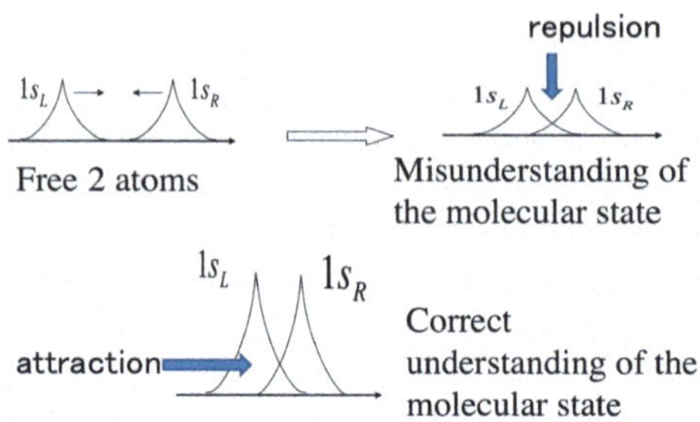

Fig. 2 Correct understanding of the formation of the hydrogen molecule. It is facilitated by the attraction between the nuclei and the electrons rather than the widely accepted explanation of the overlap between electron clouds, as stated in the HL model

charge) and electrons (negative charge). Therefore, to reduce V, the electrons should come closer to the nucleus while forming the molecule. As a result, the electron starts moving faster to avoid being absorbed into nucleus and therefore, T increases. The total energy $E = T + V$ decreases due to this difference between V and T. In Fig. 2, a schematic picture is shown for the change in the electronic charge distribution when the hydrogen molecule is formed from two hydrogen atoms.

In Sect. 5, we explain that T and V are not independent and in the equilibrium state $2T + V = 0$ (virial theorem). This is a necessary condition for any many-body system interacting via Coulomb force. The increment in T is just half the decrement in V, and it is easily proved that $-T = V/2$ is the energy of stabilization. The numerical example is shown in Fig. 3 for hydrogen molecule, where we can understand the importance of the virial theorem satisfaction. Most of the recent theories only aware of the total energy, but we should check the ratio of V and T. Then, we can distinguish correct and incorrect theories. Basically, the two-level models such as the Hubbard model assuming the same space wavefuntions for the two levels are incorrect, since they violate the virial theorem. We should be aware that in the textbook by Prof. Fujinaga [4], it is indicated that the HL is incorrect and the correct theory concerning the stabilization of molecular systems is introduced. By the virial theorem satisfaction criterion, we now have a new tool to distinguish between correct and incorrect theories.

From Table 1, we can understand the importance of the virial theorem, which affects more than the electron correlation for the stabilization of the magnetic state. In the Hartree–Fock (HF) theory, by definition, there are no electron correlations included, but it explains correctly the origin of stabilization of hydrogen molecule by decrement of the potential energy V. As a matter of course, DMC solves directly the Schrödinger equation and explains the origin of stabilization exactly numerically and the result matches with the experimental observation [5]. Table 1 shows the correlation energy, which is the difference between the exact DMC (satisfies the virial theorem and the cusp condition) value and HF. The correlation energy is small and around 5% of the total energy.

Not only accurate, but polarity (+-) reversed

(eV)	condition	HL	MO	HF	DMC	experiment
ΔE	-	-2.48	-2.88	-3.63	-4.75	-4.75
ΔT	+	-5.01	-4.22	3.63	4.8(1)	
ΔV	-	2.53	1.33	-7.27	-9.6(1)	
$-\Delta V/\Delta T$	2	0.5	0.3	2.0	2.0(1)	

Complete agreement

Traditional explanation for the hydrogen molecule contradicts the virial theorem. This is a fundamental misunderstanding since virial theorem is based simply on the 3 dimensional space.

Fig. 3 Energy change happens when two hydrogen atoms come together to form a hydrogen molecule. Heitler–London model (HL), molecular orbital calculation with only 1s electrons (MO), Hartree–Fock method (HF), diffusion quantum Monte Carlo method (DMC), and experimentally obtained values are shown. Although the total energy values are not so much different, if one computes the kinetic energy (T) and potential energy (V) separately and their changes (ΔT and ΔV), ΔT is negative and ΔV is positive for HL and MO. This numerical result contradicts with nature, since it violates the necessary condition of the virial theorem

This problem is not limited to the hydrogen molecule, but for all materials including molecules, crystals, amorphous, etc., equally. HL–Slater–Pauling (HLSP) method (valence electron connection method), which is an extension of HL, is applied to multi-atom molecules, but also based on the same mistake of overlapping electron clouds (thereby, reducing T) to stabilize the system. This serious problem in standard quantum mechanics textbooks is testimony to the fact that these incorrect models are widely accepted because they can explain experimental observations. We have clearly shown that by using a correct way of explanation based on satisfying the virial theorem, this fundamental problem is solved theoretically.

A standard argument provided by theorists is that numerical results of the many-body problem provide only numerical information, which is not understandable, and models give a clear understanding of experimental observations provided that it is "correct." However, as stated above, the results from the HL and similar models have even positive values changed to negative, which are completely nonsensical. These proposed models explain materials' properties only by electron–electron interactions. They assume that the nucleus is a uniformly distributed positive background charge, and based on this assumption, correlations among multi-electrons are discussed. The nucleus size is around 10 fm and compared to the atomic size of Angstroms, it is nearly five orders smaller. Moreover, it is assumed generally, that the effect of the nucleus only extends to the core electrons (named as "localized electrons") and the valence electrons, which contribute to the bonding between atoms, are not affected by the nucleus and move freely in the system (named as "itinerant electrons"). However,

Table 1 [Source 5] Accurate numerical result for hydrogen molecule by DMC

	E	T	V_{en}	V_{ee}	V_{nn}	V	Virial ratio	R_{eq}
HF	−1.13365(2)	1.1333(1)	−3.6484(1)	0.66113(2)	0.7214	−2.2669(1)	2.0003(2)	1.386
DMC	−1.17447(4)	1.1749(14)	−3.6535(17)	0.5880(4)	0.7137	−2.3518(19)	2.002(3)	1.401
Correl.	−0.04081(5)	0.0416(14)	−0.0041(17)	−0.0731(4)	−0.077	−0.0848(18)	2.040(69)	

The correlation energy values (which is the difference between DMC and HF) is shown for each term. Since the virial theorem is satisfied by both DMC and HF, the correlation energy also satisfies the virial theorem. The electron correlation contributes to the increase in kinetic energy, with the potential energy decreasing to twice the value of the kinetic energy increment. This is the fundamental principle of the electron correlation as understood from the virial theorem from accurate all-electron calculations, it is clearly shown that all the electrons (core and valence) are affected strongly by the nucleus, and therefore, the standard model fails. (Even outermost electrons are attracted to the nucleus, and the core electrons are expanded.) Therefore, before performing quantitative calculations, it is incorrect to categorize electrons as local and itinerant, because this leads to completely different conclusions from those observed in nature.

4 Even to Solve One Atom Is a Many-Body Problem

The reason why the quantum states in hydrogen atom are solved exactly is that it is a binary system composed of a proton and an electron. The next atom in the periodic table, Helium (He), has already become difficult to be solved accurately, because it is composed of three particles, one nucleus and two electrons, and as we know, it is difficult to solve three-body problem (the Sun–Earth–Moon analogy). In standard chemistry textbooks, it is written that the He ground state is composed of two electrons located in 1 s orbitals. However, these two electrons are not only attracted by the nucleus, but also have repulsive Coulomb interaction between them (the strength is known to be about 1/7 compared to the nucleus–electron interaction). Therefore, this is not simple and we should solve the three-body problem to obtain highly accurate electronic states. If one applies the molecular orbital calculation program with full configuration interaction (CI) with over 100 bases in a basis set, a converged state is obtained. However, the calculated electron correlation is only about 85% compared to the experimental value. The rest of the 15% is very difficult to be reached. This is attributed to the method utilized to solve the problem. However, it does not mean the problem is not solvable. Applying diffusion quantum Monte Carlo (DMC) method, we can reach the real ground state easily. A summary of this problem is shown in Fig. 4. It contains the numerical results on hydrogen molecule with two electrons and two protons (four-body problem), which is also difficult to be solved by the standard full CI method.

We have spent several years to solve the problems plaguing the transition metal Hund's rule. For example, we have been partially successful to reproduce the experimental level ordering for all low-lying levels in Ni atom (very difficult to reach the satisfactorily accuracy, and still working to finalize this case for publication) [6]. Is a single atom so difficult to be solved? Yes, it is difficult, since Ni atom is composed of 28 electrons and a nucleus, and we should solve the 29-body quantum mechanical problem accurately. Heavier atoms than transition metal atoms are much more difficult and are not yet solved. In the twenty-first century, it is still a fundamental and important problem for researchers to solve single atom quantum mechanical states accurately. It is imperative to solve the quantum states in not only atoms, but also, molecules, crystals, multi-grain crystals with defects, organic molecules, protein, DNA, etc., in the future. However, the present-day supercomputers do not yet have sufficient power to solve nature itself (in silico = in computer). We, therefore, frequently use density functional theory (DFT), which reduces dramatically the

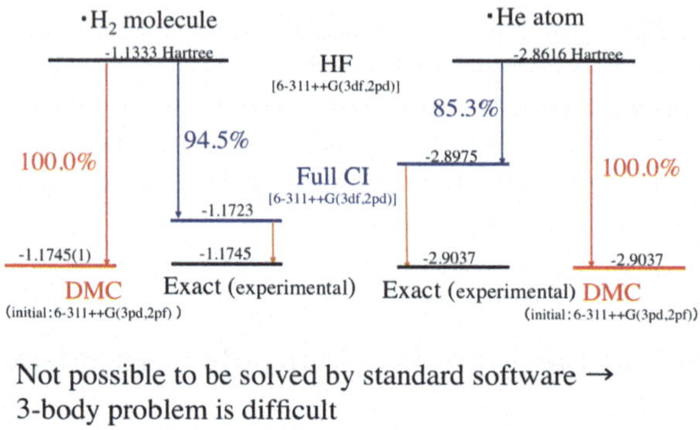

Fig. 4 Accurately solving a 3-(and 4-) body problem. Here, the hydrogen molecule (composed of 4 particles = 2 protons and 2 electrons) and the He atom (3 particles = 1 nucleus and 2 electrons) are considered. The full CI method is compared with DMC. As can be seen, in case of He atom, the full CI is only 85% accurate, while DMC results match with experiments in 5 digits

amount of computation (from 3 N dimensional space (N = number of electrons) to 3 dimension) to exactly solve the many-body equation for the non-degenerate ground state. However, in DFT, exchange–correlation functional (E_{xc}) is not given. Immense research has been done using E_{xc} with phenomenological parameters that try to reproduce the experimental observations. In doing so, the quality of the calculation is not improved, but it is possible to solve for a larger number of atoms in the system. This way of applying DFT is only phenomenology and cannot be said to be truly *ab initio* (or the first principles). The recently used Van der Waals (vdW) exchange-correlation functionals to obtain the layer distance in two-dimensional materials are good example of this kind of misunderstanding, since vdW interaction arises from the dipole–dipole excitation, and is not related to exchange-correlation at all. In response to this severe misunderstanding, we have proposed a new method, which computes dipole excitation of both sides (molecules) by time-dependent DFT (TDDFT) and estimate the C_6 coefficient [7]. We should now make a fundamental paradigm shift "from explanation to prediction" of new materials with confidence and reliability using real *ab initio* calculations. Selecting an E_{xc} which reproduces experimental results and parallelizing the corresponding software for the present-day high-performance computers is not a reliable way to perform research in the area of computational materials science. For the first time, we have a good chance to overcome this low level of computer simulations.

5 Virial Theorem

In recent textbooks on solid-state physics, we rarely see the virial theorem. In most of the classical famous textbooks, however, the virial theorem is introduced as follows "In the material systems interacting via Coulomb forces at the equilibrium phase

T and V are not independent and $-V/T = 2$, which is called the virial ratio." This theorem is a conclusion from our observation that our universe is three-dimensional; therefore, it is one of the necessary conditions for the system to satisfy. The Coulomb force is proportional to $1/r^2$ and this is deduced from the universal condition that the total effect of a point charge is the same at any distances. (The surface area at distance r from the point charge is $4\pi r^2$, and therefore Coulomb force decreases inversely as $1/r^2$.) The value $G = 2T + V = 3P\Omega$(P is pressure and Ω is volume)is called virial.

Virial theorem is known very well in classical mechanics and is proved as follows [8]. Virial G is defined as a sum of products of coordinate r_i and momentum p_i of all elements in the system as

$$G = \sum_i r_i \cdot p_i. \tag{1}$$

Differentiating G with respect to time and using the definitions from classical mechanics such as $T = \frac{1}{2}\sum_I p_i v_i$ and $\frac{\mathrm{d}p_i}{\mathrm{d}t} = F_i$ from Newton's equation of motion, where F_i is the force on particle i

$$\frac{\mathrm{d}G}{\mathrm{d}t} = 2T + \sum_i r_i \cdot F_i. \tag{2}$$

Assuming the area of particle motion is finite, the time average value of G (obtained by integrating both sides from 0 to t, dividing by t and setting t to infinity) should be zero, i.e.,

$$0 = 2\langle T\rangle + \left\langle \sum_i r_i \cdot F_i \right\rangle. \tag{3}$$

The force on each particle is defined by the differential of the potential energy V (conservative force)

$$\sum_i r_i \cdot F_i = -\sum_i r_i \cdot \nabla_{r_i} V. \tag{4}$$

If V is the Coulomb force, with coefficient a (in MKS unit, with charge q_i, $q_i \times q_j/4\pi\varepsilon_0$, ε_0 is the dielectric constant of the vacuum)

$$V = a/r, \tag{5}$$

The average becomes

$$\left\langle \sum_i r_i \cdot \nabla_{r_i} V \right\rangle = -\langle V\rangle. \tag{6}$$

Finally, the virial theorem for equilibrium state is obtained as

$$0 = 2\langle T\rangle + \langle V\rangle. \tag{7}$$

In quantum system, applying the scaling law to the system Hamiltonian H, the virial theorem is obtained by the variational method [4]. H is the sum of kinetic energy operator T and potential operator V as

$$H = T + V. \tag{8}$$

T is the second derivative of the space coordinate and V is inversely proportional to the space coordinate. By introducing a scale factor λ and setting $T(\lambda)$, $V(\lambda)$ with coordinate multiplied by λ, then relationships are

$$T = \lambda^2 T(\lambda),\ V = \lambda V(\lambda), \tag{9}$$

and

$$E(\lambda) = \lambda^2 T(\lambda) + \lambda V(\lambda). \tag{10}$$

For the equilibrium state, total energy E should have the minimum, and

$$\delta E/\delta\lambda_{\lambda=1} = 2T + V = 0. \tag{11}$$

Therefore, we can obtain for a quantum mechanical system interacting via Coulomb forces at the equilibrium state, the virial value $G = 2T + V = 0$. This theorem can be applied to all the states in equilibrium phases, such as isolated two hydrogen molecules or the hydrogen molecule. Therefore, the total energy E should be

$$E = T + V = -T = V/2. \tag{12}$$

This ratio $V/T = -2$ is called the virial ratio and is a good measure to evaluate the reliability in electronic structure calculations at the equilibrium state using the variational principle [5].

This ratio holds in each equilibrium state, and therefore, the energy difference between before and after the formation of a molecule is

$$\Delta E = \Delta T + \Delta V = -\Delta T = \Delta V/2. \tag{13}$$

From Eq. (13), it is now clear that in the previous section, in the process of hydrogen molecule formation, potential energy V decreases and the kinetic energy T increases by just half of its value. Checking with virial ratio is an important measure to obtain "quality assurance" in any *ab initio* calculations. A number of models can be developed which can explain physical phenomena and reproduce experimental

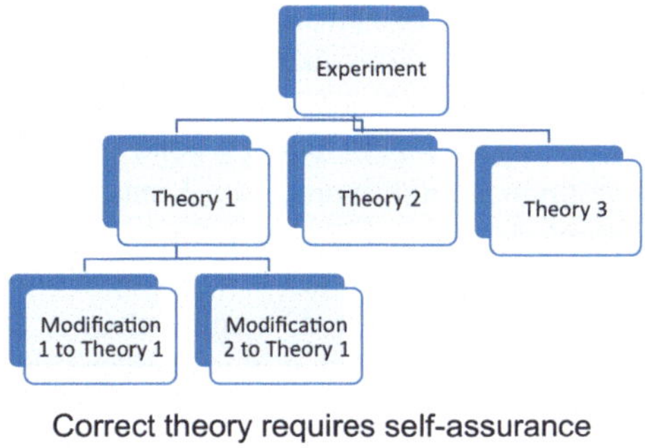

Fig. 5 This figure shows how out of an experiment, several theories arise to explain the phenomena observed in it. However, the theories must undergo rigorous modification, in the form of inclusion of additional parameters, in order to better describe the experimental observations

values by simply varying some parameters. However, this does not ensure the theory is correct. Especially, for models concerning materials interacting via Coulomb forces, many of the existing theories violate the virial theorem (not rare cases, such as widely used Hubbard model). These need to be classified as completely incorrect models, since they do not correspond to nature. Figure 5 shows a schematic of the same. As seen here, if there exists some discrepancy in the explanation, more parameters are added to a model to better explain the experiments. Our proposal to evaluate the virial ratio is a new scheme of quality assurance in computational materials science. Henceforth, we strongly believe that all models proposed in the future should be constructed by satisfying the virial theorem.

6 Heisenberg's Exchange Interaction Cannot Explain the Origin of Magnetism-Explanation of Hund's Rule for Atoms–

One of the important properties in materials is magnetism, which has been explained by the Heisenberg model [9]. This model was created when magnetism was known to be related to the electronic states, and as the governing equation quantum mechanics was established. In 1920 s, scientists assumed that all electrons were the same and that exchanging two electrons would lead to only phase change, (which was a new concept at that time) unlike in classical mechanics, which is based on individual/non-identical particles. Therefore, magnetism was misunderstood to be directly related to exchange energy. As was shown in the previous section, a single transition metal atom is still difficult to be solved accurately, and as a matter of fact, models are still needed. Only recently using supercomputers, this problem was solved accurately, and the numerical results by HF and accurate enough theories, which satisfy the virial theorem, showed that not the electron correlation, but nucleus–electron interaction provides the main energy gain to realize the magnetic ground state and it is this interaction that correctly describes Hund's rules.

Hund found two phenomenological rules on electronic configuration in the ground state in 1925, before quantum mechanics was formulated [10]. (1) Hund's first rule: In the *LS* terms (degeneracy of$(2S + 1) \times (2L + 1)$), the largest total spin *S* state is the most stable, (2) Hund's second rule: If the spin states are also degenerated, the maximum angular momentum *L* state is the ground state. These rules are applicable not only to atoms, but also molecules.

Magnetism originates from many electrons aligning their spins in the field created by the nucleus and "coming close to the nucleus to reduce the total energy." It is not possible to explain this phenomenon by considering only the electron–electron interaction as in most of the traditional theories. This new explanation from Davidson [11], (later, the works of Katriel–Pauncz [12], Boyd [13] etc., all indicate that Slater's perturbation theory is incorrect) is proved as follows: Firstly, if we compute the energy satisfying the virial theorem, electron–electron repulsive energy even increases for more stable states. Importantly, ignoring the difference in the nucleus–electron interaction at each level has led to the old models contradicting to the nature for over 90 years. Figure 6 shows this situation.

From the less screening mechanism by Boyd [13], we can explain Hund's first rule as follows: Aligned spins repel each other at short distances from Pauli's exclusion principle and nucleus screening by the Hartree term is weakened by the exchange term at short distances. Therefore, the electrons effectively feel a more attractive field created by the nucleus. In this way, the outermost (valence) electrons "shrink" to the nucleus and stabilize the system. The difficulty in understanding the origin of magnetism is very carefully outlined in the famous textbook by Kubo [14], in the last chapter "difficulty in the fundamental discussion in magnetism." Unfortunately, most of the standard textbooks after his one have never appreciated his deep concerns.

Figure 7 shows the energy expression for the spin state in the standard textbooks on magnetism. It is an easily understandable equation that twice the exchange energy *K* corresponds to the energy difference between spin and non-spin states. Since it is

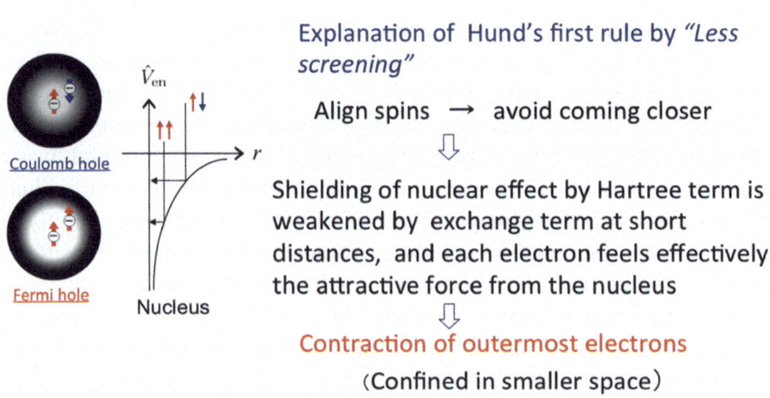

Fig. 6 Conceptual viewgraph of correct interpretation of Hunt's rule based on the less screening mechanism. This way is fundamental to understanding the origin of magnetism (the reason why magnetic state is the ground state)

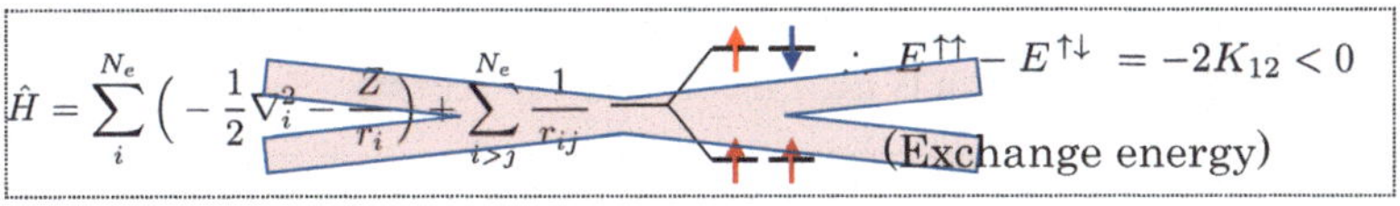

Fig. 7 Incorrect magnetic energy expression given in standard textbooks

a simple and beautiful equation, it is frequently used as an example in introductory quantum mechanics on magnetism. Unfortunately, we have proved that this equation is fundamentally wrong and should be deleted from all the textbooks on magnetism.

We have analyzed this problem in detail, and the results were astonishing [15–18]. Overall, we have been successful to prove that Slater's perturbation theory, which has been believed to be correct for over 90 years, is in fact, incorrect. In the standard textbooks traditional explanation of Hund's rule is explained by the energy difference in the magnetic ground state occurs by the gain in the exchange energy. This explanation is completely wrong, and the energy difference comes from mainly the nucleus–electron attractive energy, some from electron kinetic energy, and electron–electron repulsive energy in each state.

Figure 8 shows quantitatively the fundamental misunderstanding still written in standard textbooks on magnetism concerning the origin of magnetism with an example of carbon atom. This example shows the results for only the singlet and triplet states in carbon atom, and it is easy to conclude the incorrect explanation provided by Slater's perturbation theory [19].

In Fig. 9, the accurate numerical results of the energy levels in the carbon atom by including Hund's second rule [20 (the first theoretical paper on the second rule)]. The upper columns show the numerical results based on the standard model which assumes the same special wavefunctions for all the three low-lying (degenerate) ^3P, ^1D, ^1S levels. The total energies are in the order of experiment (which is the traditional understanding). However, if we solve for all the levels, the T and V values independently, by satisfying the virial theorem, these values are different for each level as shown in the lower columns. Importantly, the decrement in nucleus–electron interaction is far larger than the increment in electron–electron repulsive energy. Even the trend in the electron–electron interaction is reversed compared to the traditional results. This result indicates that the traditional explanation of Hund's rule is based on the misunderstanding of the system.

DMC	E_{total}	V_{ee}	V_{en}	V_{total}	*Virial ratio*
Triplet	-37.8280(7)	12.595(6)	**-88.253(27)**	-75.658(27)	1.99996
Singlet	-37.8107(5)	12.430(4)	-88.053(18)	-75.622(18)	1.99997

Fig. 8 [Source 19] Numerical results obtained by very accurate diffusion quantum Monte Carlo method (DMC) on Hunt's rule for carbon atom. Energy unit is Hartree. By this high accuracy DMC results, it is clearly understood that the traditional way of explanation of Hund's rule is incorrect. Magnetic ground state (triplet state) becomes the most stable, because of the decrement in nucleus–electron interaction V_{en}, not by V_{ee} (only which is encountered in the traditional explanation. Even this value is higher in the Triplet state)

(I) Traditional explanation of Hund' s rule: Orbitals optimized to ^3P term is used for ^1D and ^1S terms

state	E	V_{en}	V_{ee}	T	$-V/T$
C (^3P)	-37.6886	-88.1369	12.7596	37.6886	2.0000
C (^1D)	-37.6302		12.8180		1.9985
C (^1S)	-37.5426		12.9056		1.9961

(II) Self-consistent HF calculation: All orbitals optimized for each ^3P, ^1D, ^1S

State	E	V_{en}	V_{ee}	T	$-V/T$
C (^3P)	-37.6886	-88.1369	12.7596	37.6886	2.000000004
C (^1D)	-37.6313	-87.9910	12.7283	37.6313	2.000000004
C (^1S)	-37.5496	-87.7662	12.6670	37.5496	2.000000004

Fig. 9 [Source 20] Numerical results including Hund's second rule. In (I), the same special wavefunction is assumed for the three levels of ^3P, ^1D, ^1S. In (II) special wavefunctions are independently optimized for the three levels of ^3P, ^1D, ^1S. Similar to Fig. 8, the results of (II) satisfy the virial theorem and the main origin of Hund's first and second rules is the decrement in the attractive force between nucleus and electrons. The order of repulsive forces between electrons is reversed in (I) and (II). Note that both of (I) and (II) do not include electron correlation effects. Energy unit is Hartree

7 High Accuracy Calculation on Magnetism in Molecules

For a long time in chemistry, magnetism has rarely been studied. After the discovery of ferromagnetism in a type of charge-transfer salt by Miller in 1987 [21], many researches have actively pursued this area. However, these salts show very low Curie temperatures of 10–20 K and are far away from realistic applications. One-dimensional ferromagnetism in the direction of double bridge was observed in chain structure molecules [22], which is more transparent than metallic magnetic materials, and is expected to be useful in high-end devices such as magnetic imaging. Some of the recent molecules exhibit ferromagnetism at room temperature and they show interesting composite physical properties than traditional metallic magnetic materials.

We have selected Hund's rule consideration in methylene molecule CH_2 as you understand the reason shown below. Darvesh et al. have published a theoretical paper on methylene [23]. Their results support the "traditional explanation" of energy gain obtained by decrement in electron–electron repulsive part and they categorized this methylene case as an exception of the newly stated explanation provided by Davidson [11]. However, when the value of the virial ratio was checked in their calculation, it was completely different from −2, and their calculation was not reliable even qualitatively. The reason is that at that time the computer power was not good enough and they did not variationally optimize the atomic structures of the methylene molecule for each level, and were constrained to use the experimental value. Unfortunately, the experimental accuracy was also not sufficiently good and the atomic configuration was away from the equilibrium even at the ground state. We performed a high accuracy electron correlation calculation with multireference configuration interaction (MRCI) method for the high and low spin states, variationally relaxed the atomic configurations for each level, and successfully computed electronic structures of two levels while satisfying the virial theorem. Figure 10 shows each atomic configuration, and Fig. 11 indicates the energy components for all terms and the virial ratio. Accordingly, we have successfully shown that the molecule methylene does not follow the

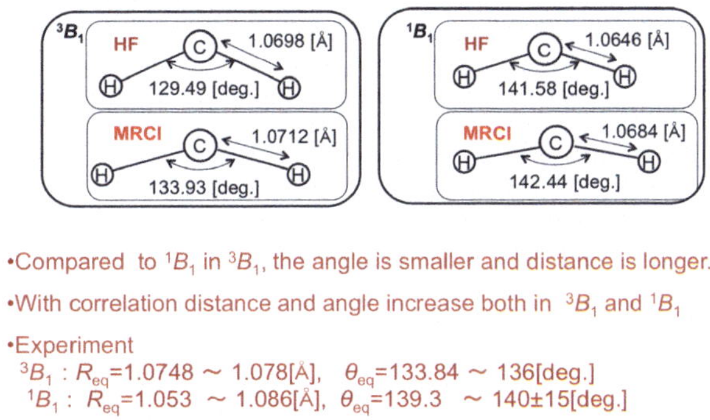

Fig. 10 [Source 24] Atomic configurations in methylene molecule determined theoretically. To understand correctly the molecular systems with many electrons, it is important to use the atomic configurations, which are optimized for each energy level

State	E	T	V	$-V/T$	V_{ee}	V_{en}	V_{nn}
$^3B_1$	-39.1259	39.1229	-78.2488	2.0001	18.5764	-103.0217	6.1965
$^1B_1$	-39.0733	39.0703	-78.1436	2.0001	18.5281	-102.8767	6.2051

$^3B_1$-$^1B_1$ spliting	ΔE	ΔT	ΔV	$-\Delta V/\Delta T$	ΔV_{ee}	ΔV_{en}	ΔV_{nn}
Present study	-0.0526	0.0526	-0.1052	2.0009	0.0483	-0.1449	-0.0086
Darvesh *et al.* [8]	-0.0625	0.0150	-0.0775	5.1508	-0.0799	0.1425	-0.1401

Present results : $-\Delta V/\Delta T = 2.0009$ ➡ $\Delta V_{ee} > 0,\ \Delta V_{en} < 0,\ \Delta V_{nn} < 0,$

Highly accurate and satisfies virial theorem

$$|\Delta V_{en} + \Delta V_{nn}| > |\Delta V_{ee}|$$

$$\Delta E = \frac{1}{2}(\Delta V_{ee} + \Delta V_{en} + \Delta V_{nn}) < 0$$

Correlation does not change HF results qualitatively

Darvesh *et al.* : $-\Delta V/\Delta T = 5.1508$ ➡ $\Delta V_{ee} < 0,\ \Delta V_{en} > 0$

Not optimized the atomic configuration

$$\Delta V_{nn}(\exp.) < 0$$

[8] K. V. Darvesh and R. J. Boyd, *J. Chem. Phys.* **87**, 5329 (1987).

Fig. 11 [Source 24] Comparison of each energy term in methylene molecule for the numerical results by Darvesh et al. which use the experimentally determined ground state atomic configuration, and our MRCI calculation optimized for each level. The results by Darvesh et al. do not satisfy the virial theorem, and their conclusion to support traditional interpretation of Hund's rule is completely wrong traditional explanation, but it follows the new explanation by Davidson [24]. This study indicates the importance of virial theorem to understand the laws of nature.

The misunderstanding in system stabilization based on the gain in exchange energy has been taught in universities without any fundamental changes in more than 90 years. Now, based on our accurate calculations, it is clearly known that the traditional explanation of origin of magnetism is incorrect. Physical properties in different spin states are caused by the difference in electron exchange–correlation interaction. They do not simply lead to the difference in the electron–electron repulsive force, and more importantly, they affect the electronic charge distribution, difference in atomic configuration (Hellmann–Feynman theorem). This is the nature. Therefore, in each spin state, the valence part in each term (electron–electron, electron–nucleus, and nucleus–nucleus) varies significantly. In most cases, contradictory to traditional explanations, the electron–electron repulsion increases at high spin states. The origin of magnetism (the reason of magnetic ground state) is based on the larger energy gain by the nucleus–electron attractive interaction. Nature is more skillful in terms of system stabilization with highly self-consistent variational way than human beings. This is the conclusion taught by the virial theorem.

8 Cr@Si_n Clusters

For carbon systems starting from fullerene, many new nano-structured materials have been found experimentally and theoretically. However, for silicon systems, new nanostructures are yet to be found. The first successful example in this direction is the experimental observation and theoretical explanation of Cr@Si_{12} cluster by a group from Tsukuba University [25]. We have extended their finding to more general silicon clusters theoretically and found successfully that a number of metal inclusion silicon fullerenes are stable. We applied the density functional theory (DFT) to analyze their properties, but the results depend strongly on the selected E_{xc}. Figure 12 shows the numerical results of Cr@Si_n clusters [26]. This figure indicates a large difference in the binding energies computed by GGA and B3PW91. For example, for the 12-body system, GGA gives 1 eV and B3PW91 gives 2 eV.(estimated for $n = 12$ cluster compared to $n = 11,13$ systems.)Most of the present-day DFT works select E_{xc} which provides a good match to experimental observations. However, this way is simply phenomenological and does not prove the validity of the theory as a first principles method. To overcome this contradiction, we have applied diffusion Monte Carlo method (DMC), which gives results similar to those of B3PW91 [27]. However, this result does not prove that B3PW91 is generally good for unknown materials (parameters fit mainly to organic molecules). Reliable theory should not contain uncertain parameters, such as E_{xc}, which is one of the fundamental problems in DFT.

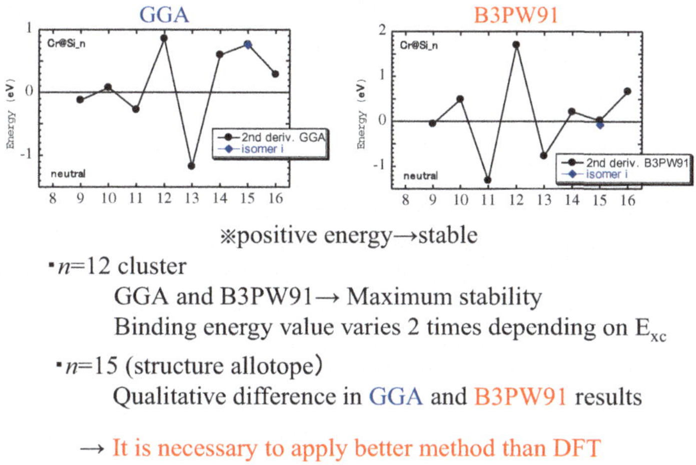

Fig. 12 [Source 26] Stability in Cr@Si_n clusters. From the comparison of the results using GGA and B3PW91 for the electron exchange–correlation functionals, there is a large difference in binding energy values, which indicates that a better theoretical method than DFT is necessary. We have applied DMC method for a cluster with $n = 12$ and concluded chemical stability with high accuracy without experimental supports

9 Conclusions—Seeking for a Satisfactory Condition

Based on the progress in supercomputer power, it is possible to apply high accuracy theories to analyze the many-body Coulombic systems satisfying the necessary condition of the virial theorem. Below, we discuss the evaluation of existing theories based on the virial theorem satisfaction. HF theory does not include electron correlations (by definition), but completely satisfies the virial theorem because of the self-consistent variational theory. Based on the HF, the theoretical hierarchy treating electron correlation, i.e., multi-electron wavefunction method, Green's function method, DFT, all of them potentially satisfy the virial theorem, since they are all based on the variational principles. However, based on the implementation of these methods (lack of knowledge), a very small number of existing software gives the results satisfying the virial theorem to high precision. For example, a typical method in electron correlation configuration interaction method (CI) has difficulty in satisfying the virial theorem to high accuracy, since it does not optimize the orbitals nonlinearly. In such a case, multi-configuration self-consistent field method (MCSCF) should be used. Although the mostly used Gaussian software [28] in computational chemistry outputs the virial ratio ($-V/T$), T and V in DFT are not exactly the kinetic and potential energies in the system and they use E_{xc} with phenomenological parameters, and should therefore, be carefully checked. To fit the HOMO–LUMO gap value to experimental value, modification of E_{xc} (such as recently frequently used hybrid functional, LDA+U, etc.) is becoming widely applied, but these are simply phenomenological concepts and should not be labeled as the first principles calculations (band gap value is fit to experimental value, but other properties are much damaged and not correspond to the experimental observations).

On the other hand, models such as the Heitler–London model, which basically reduces the kinetic energy (extension of the electron clouds) to stabilize the system, are to be categorized as completely incorrect, since they do not follow the nature. This category also includes the Hubbard model, which is used often to explain high-T_c superconductor, assuming that spacial wavefunctions for the two levels are the same. However, most important contribution to stabilize the ground state, as we have shown above, comes not from the exchange energy part but from the difference in the spacial wavefunctions. Accordingly, it is now clear that the necessary condition of virial theorem satisfaction is the criterion of correctness of multi-electron quantum mechanical theories. Finally, we consider what is the sufficient condition when we solve the many-body system consisting of Coulombic interactions. Although E_{xc} is very small compared to the total energy of the system, it does affect strongly the physico-chemical properties in the system. The only way to model a system close accurately to the nature is to compute E_{xc} as accurately as possible.

To treat theoretically the periodic systems, such as molecular crystals and so-called strongly correlated systems, DFT has a serious difficulty in satisfying the virial theorem to high accuracy, since no reliable E_{xc} functionals have been developed. On the contrary, quantum Monte Carlo methods are extremely costly and can be applied for a very small unit cell, but they can form an important part in E_{xc} (not for long range part) with careful treatment of small unit cell effect numerically [29]. Therefore, for our future studies, electron correlation and finite size correction will form the most important subjects in *ab initio* simulations. Many people have the opinion that important basic theories were established prior to the twenty-first century and that we are expected to work only on the applications aspect of materials. Also, it is said that scientific and engineering methods are composed of theory, experiment, and simulation. Unfortunately, these statements are not applicable at the moment in the scenario of the quantum many-body system as shown in Fig. 13 (in another heavy user of the present-day supercomputer and hydrodynamics simulation, this situation is already realized after the history of 200 years). It is necessary to study this subject

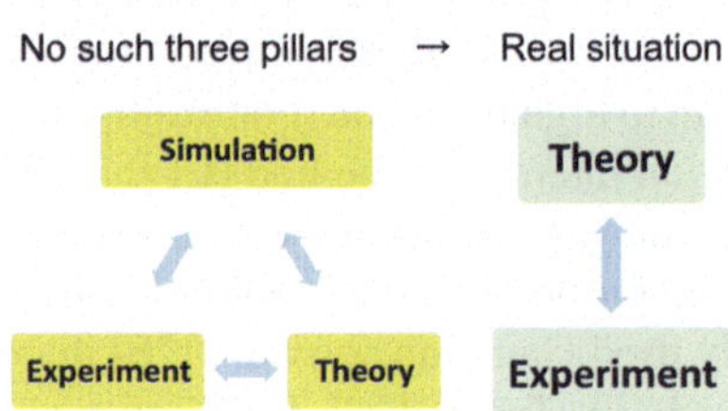

Fig. 13 It is impossible to express important physical and chemical properties by simple models. The present so-called state-of-the-art simulation method as the third rung in materials science actually does not exist. Therefore, we should restart from the very basic theory to overcome the traditional misunderstanding and reconstruct a new paradigm, where we can predict new useful materials with confidence without any experimental help

still from the basics and propose a correct way of solving such complicated systems that satisfy a necessary condition of the virial theorem. With this approach in mind, we will be able to predict new materials with confidence prior to experiment, and if experimental and theoretical results do not match each other, we can say that the experimental results have errors.

Acknowledgements This chapter is based on the important suggestion by Professor Emeritus Hiroshi Yasuhara of Tohoku University, who passed away 4 years ago. He proposed to follow the virial theorem, which is a necessary condition for any of the many-electron systems. It is important to note that all of the future theoretical papers on solid-state physics, quantum chemistry, and computational materials science should adopt this important message. This chapter indicates a fundamental misunderstanding concerning the origin of magnetism, and we have carefully checked the contents before publication. All the results have been numerically obtained by my former students, Kenta Hongo, Takayuki Oyamada, and Yohie Maruyama. The author is thankful to Dr. Aaditya Manjanath for correcting his English. This manuscript is an English translation from the issue in Kinzoku from AGNE Gijutsu Center ("Origin of Magnetism," Vol. 84, No. 9 (2014) pp. 1–13).

References

1. Slater, J.C.: The theory of complex spectra. Phys. Rev. **34**, 1293–1322 (1929)
2. Logergist: About the movement of moon. In: Butsuri no Sanpomichi No.5, Iwanami Shoten, Publishers, pp. 154–170(1972)(in Japanese)
3. Heitler, W., London, F.: Wechselwirkung neutraler Atome und homöopolare Bindung nach der Quantenmechanik. Z. Phys. **45**, 455–472 (1927)
4. Fujinaga, S.: Molecular Orbital Method. Iwanami Shoten, Publishers (1980), 「Introduction to Molecular Orbital Method", (1990), Kodansha Scientific, Publishers (in Japanese)
5. Hongo, Kenta, Kawazoe, Yoshiyuki, Yasuhara, Hiroshi: Diffusion monte carlo study of correlation in the hydrogen molecule. Int. J. Quantum Chem. **107**(6), 1459–1467 (2007)
6. T. Oyamada, unpublished
7. Belosludov, R.V., Mizuseki, H., Sahara, R., Kawazoe, Y., Subbotin, O.S., Zhdanov, R.K., Belosludov, V.R.: Computational materials science and computer-aided materials design and processing. In: Handbook of Sustainable Engineering, Springer, pp. 1215–1247 (2013)
8. Marc, G., McMillan, W.G.: The virial therorem. Adv. Chem. Phys. **58**, 209–361 (1985)
9. Heisenberg, W.: On the theory of ferromagnetism. Z. Phys. **49**, 619–636 (1928)
10. Hund, F.: Concerning the interpretation of complex spectra, especially the elements scandium to nickel. Z. Phys. **33**, 345–371 (1925); Hund, F.: Interpretation of the complicated spectra II", Z. Phys. **34**, 296–308 (1925); Hund, F.: Linienspektren und Periodisches System der Elemente. (Springer-Verlag, Berlin, 1927)
11. Davidson, E.R.: Single-Configuration Calculations on Excited States of Herium. J. Chem. Phys. **41**, 656–658 (1964); Single-configuration calculations on excited states of herium. II. J. Chem. Phys. **42**, 4199–4200 (1965)
12. Katriel, J., Pauncz, R.: Generalized branching diagrams. Adv. Quantum Chem. **10**, 143–151 (1977)
13. Boyd, R.J.: A quantum mechanical explanation for Hund's multiplicity rule Nature, **310**, 480–480 (1984)
14. Kubo, Ryogo: Magnetism. Publishers, Iwanami Shoten (1956). (in Japanese)
15. Hongo, Kenta, Oyamada, Takayuki, Maruyama, Youhei, Kawazoe, Yoshiyuki, Yasuhara, Hiroshi: Correct Interpretation of Hund's Rule and Chemical Bonding Based on the Virial Theorem. Mater. Trans. **48**(4), 662–665 (2007)

16. Hongo, K., Oyamada, T., Maruyama, Y., Kawazoe, Y., Yasuhara, H.: Correct interpretation of hund's multiplicity rule for atoms and molecules. J. Magn. Magn, Mater. **310**(2), e560–e562 (2007)
17. Hongo, Kenta, Kawazoe, Yoshiyuki, Yasuhara, Hiroshi: Diffusion monte carlo study of atomic systems from Li to Ne. Mater. Trans. **47**(11), 2612–2616 (2006)
18. Oyamada, T., Hongo, K., Yoshiyuki Kawazoe and Hiroshi Yasuhara, "The Influence of Correlation on the Interpretation of Hund's Multiplicity Rule: A Quantum Monte Carlo Study", J. Chem. Phys., **125**[1], pp. 014101-(1–9) (2006)
19. Hongo, K., Maezono, R., Kawazoe, Y., Yasuhara, H., Towler, M.D., Needs, R.J.: Interpretation of hund's multiplicity rule for the carbon atom. J. Chem. Phys., **121**(15), 7144–7147 (2004)
20. Oyamada, T., Hongo, K., Kawazoe, Y., Yasuhara, H.: Unified interpretation of hund's first and second rules for 2*p* and 3*p* Atoms. J. Chem. Phys. **133**(16), 164113–(1–19) (2010)
21. Miller, J.S., Calabrese, J.C., Rommelmann, H., Chittipeddi, S.R., Zhang, J.H., Reiff, W.M., Epstein, A.J.: Ferromagnetic behavior of [Fe(C5Me5)2]+.bul. [TCNE]-.bul.. Structural and magnetic characterization of decamethylferrocenium tetracyanoethenide, [Fe(C5Me5)2]+.bul. [TCNE]-.bul..cntdot.MeCN and decamethylferrocenium pentacyanopropenide, [Fe(C5Me5)2]+.bul. [C3(CN)5]-. J. Am. Chem. Soc., 109:769–781 (1987)
22. Verdauer, M., Gleizes, A., Rendard, J.P., Seiden, J.: Susceptibility and Magnetization of CUMN(S2C2O2)2.7.5H2O—1st experimental and theoretical characterization of a quasi-one-dimensional ferrimagnetic chain. Phys. Rev. B **29**, 5144–5155 (1984)
23. Darvesh, K.V., Boyd, R.J.: Hund rule and singlet-triplet energy differences for molecular-systems. J. Chem. Phys. **87**, 5329–5332 (1987)
24. Maruyama, Youhei, Hongo, Kenta, Tachikawa, Masanori, Kawazoe, Yoshiyuki, Yasuhara, Hiroshi: *Ab Initio* Interpretation of hund's rule for the methylene molecule: variational optimization of its molecular geometries and energy component analysis. Int. J. Quantum Chem. **108**(4), 731–743 (2008)
25. Hiura, Hidefumi, Miyazaki, Takehide, Kanayama, Toshihiko: Formation of Metal-Encapsulating Si Cage Clusters. Phys. Rev. Lett. **86**, 1733–1736 (2001)
26. Kawamura, Hiroaki, Kumar, Vijay, Kawazoe, Yoshiyuki: Growth, magic, behavior, and electronic and vibrational properties of Cr-doped Si Clusters. Phys. Rev. B **70**(245433), 1–10 (2004)
27. Hongo, Kenta, Kumar, Vijay, Kawazoe, Yoshiyuki, Yasuhara, Hiroshi: Quantum monte carlo study of electron correlation in chromium-doped silicon cluster $Cr@Si_{12}$. Mater. Trans. **47**(11), 2617–2619 (2006)
28. Gaussian、Gaussian Inc
29. Averill, F.W., Painter, G.S.: Phys. Rev. B **24**, 6795 (1981); Levy, M., Perdew, J.P.: Phys. Rev. A **32**, 2010 (1985)

Magnetic Domain Structures and Techniques in Micromagnetics Simulation

Fumiko Akagi

Abstract Ferromagnetic materials exhibit spontaneous magnetizations even without an applied magnetic field. Their internal structure contains regions called magnetic domains in which magnetic moments are aligned in the same direction. Magnetic moments in different domains point in different directions. For example, when the total magnetization in a ferromagnetic material is close to zero (magnetically neutral state or demagnetized state), the directions of the magnetic moments are determined so as to minimize magnetic energy of the ferromagnetic material. To understand the magnetic domain structure, the behaviors of magnetic moments and the magnetic energy of the ferromagnetic material need to be analyzed. Techniques in micromagnetics simulation are extremely effective for analyzing them. In this chapter, the basic terminologies concerning magnetic domains of ferromagnetic materials are introduced, techniques in micromagnetics simulation are explained, and calculation examples of the magnetic domain structures for the ferromagnetic materials are presented.

Keywords Ferromagnetic material · Magnetic domain · LLG simulation

1 Magnetic Structure of Grain-Oriented Electrical Steel (GOES)

Magnetic domain structures of grain-oriented electrical steel (GOES) observed by type-II Lorentz scanning electron microscope (SEM) are shown in Fig. 1 [1]. Black and white regions indicate the left and right directions of magnetization, respectively. As shown in the SEM images, magnetizations in the left and right directions in the GOES are lined up as stripes, the structure of which is called a "stripe magnetic domain." Also, as shown in Fig. 1a, fine linear grooves in the vertical direction are formed on the surface of the GOES. By this formation of grooves, the width of the stripe magnetic domains becomes smaller than that in the original "plain material"

F. Akagi (✉)
Kogakuin University, Shinjuku, Japan
e-mail: fumiko.akagi@cc.kogakuin.ac.jp

K. Fujisaki (ed.), *Magnetic Material for Motor Drive Systems*,
Engineering Materials, https://doi.org/10.1007/978-981-32-9906-1_11

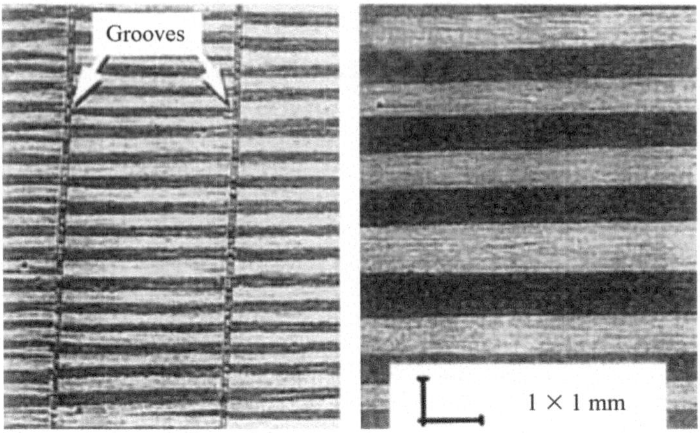

(a) Grooved material (b) Plain material

Fig. 1 Magnetic domain structures observed by type-II Lorentz SEM method (grain-oriented electrical steel, GOES) Reprinted from [1], Copyright 2005, with permission from JFE Steel Corporation

shown in Fig. 1b. The width of a stripe magnetic domain is called "magnetic domain width," which is explained in the next section.

GOES is mainly used for iron cores of transformers, and it is required to have two magnetic characteristics. One is high permeability to obtain high magnetic flux density even if a small magnetic field is applied to it. The other is a low thermal loss (iron loss) on transformation between electrical and magnetic energies. High permeability can be achieved by using soft ferromagnetic materials, which have coercivity close to zero. One means of decreasing iron loss is to form fine linear grooves to decrease the magnetic domain widths, as shown in Fig. 1a. To make the desired GOES, it is important to investigate both magnetic characteristics and structures of ferromagnetic material.

2 Magnetic Domain and Domain Wall

A ferromagnetic material is an aggregate of microscopic magnetos known as "magnetic moments" at an atomic level. As shown in Fig. 2a, each magnetic moment has a strong uniaxial magnetic anisotropy that aligns it in one direction without an applied magnetic field. As a result, a ferromagnetic material has spontaneous magnetization M. However, as shown in Fig. 2b, a magnetic field in the opposite direction of magnetization, called demagnetizing field H_d, is generated by the north (N) pole and south (S) pole at respective ends of the ferromagnetic material. When the magnetic anisotropy is weak, the ferromagnetic material is divided into regions of magnetic moments with different directions, as shown in Fig. 3a, b, because the magnetic moment at each end of the ferromagnetic material is disturbed by the demagnetizing

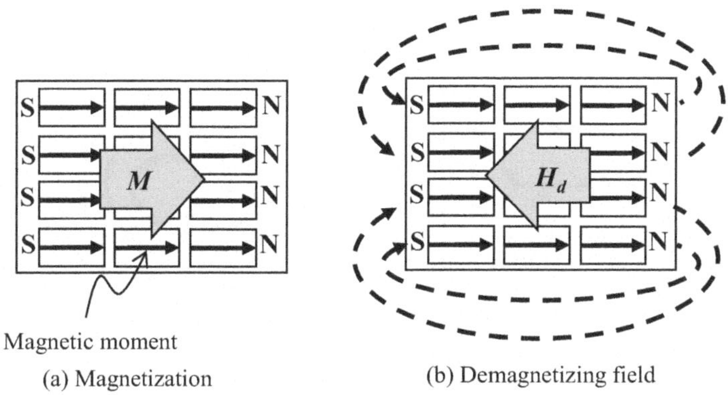

Fig. 2 Magnetization and demagnetization fields of a single-domain structure

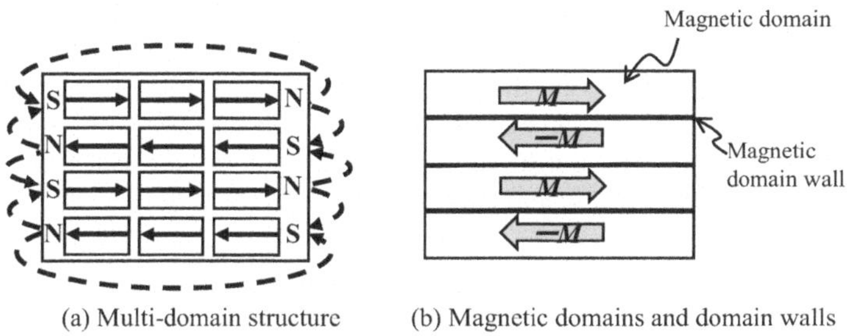

Fig. 3 Magnetization of multi-domain structure, magnetic domain, and domain wall

field. Each region is called a magnetic domain, and the boundary between magnetic domains is called a magnetic domain wall. The magnetic structure with only one magnetic domain shown in Fig. 2 is called a "single-domain structure." In contrast, the magnetic structure shown in Fig. 3 is called a "multi-domain structure." In the case of GOESs, magnetic domain width is several hundred micrometers, and in the case of non-oriented electrical steels, it is several dozen micrometers [2].

The multi-domain structure in ferromagnetic materials reduces and stabilizes the magnetostatic energy introduced by the demagnetization field. As shown in Fig. 3, therefore, the magnetostatic energy decreases as the number of magnetic domains increases; however, it cannot be infinitely small. The domain wall energy, which is determined by magnetic anisotropy and exchange energies, increases as magnetostatic energy decreases. The magnetic domain structure is determined when the sum of the magnetostatic, Zeeman, magnetic anisotropy, and exchange energies becomes a minimum. Each of the energies is explained in Sect. 4. As explained in the previous section, a striped pattern as shown in Fig. 3 is called a "stripe magnetic domain structure." As for electrical steels, magnetic domain width should be as small as possible. This is because large iron losses are generated by eddy currents caused by the change of external fields when the magnetic domain width is large.

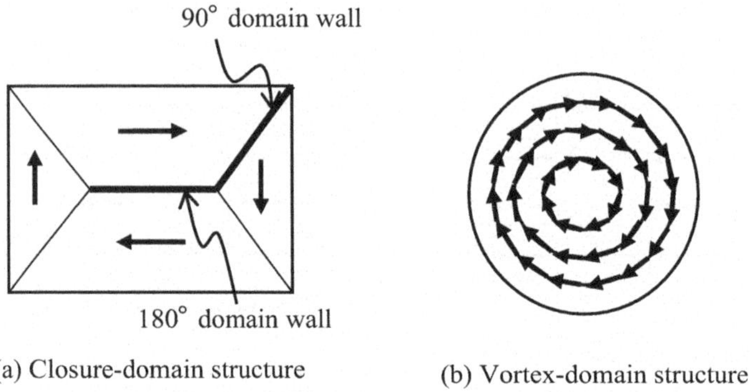

Fig. 4 Closure and vortex domain structures

Other magnetic domain structures, namely a closure domain and a vortex domain, are shown in Fig. 4a, b, respectively. In both structures, no magnetic charges exist at the edge of the magnetic material; therefore, total magnetization in each case is zero. The closure domain structure can be commonly seen in ferromagnetic material. The vortex structure is formed when the radius of the magnetic material is smaller than the characteristic length of magnetostatic field ζ in Eq. (1), giving

$$\zeta = \sqrt{\frac{2\mu_0 A}{N M_s^2}}, \tag{1}$$

where μ_0 is magnetic permeability of a vacuum, A is the exchange stiffness constant, N is the demagnetizing factor, and M_s is saturation magnetization [3].

A magnetic domain wall is a transition region between adjacent magnetic domains, where the directions of the magnetic moments gradually change. As shown in Fig. 4a, a magnetic domain wall is known as a 180° domain wall when the directions of adjacent magnetizations are opposite, and the magnetic domain wall is known as a 90° domain wall when the directions of adjacent magnetizations are orthogonal. When the directions of two adjacent magnetizations are different, the exchange energy increases the wall thickness to make the magnetic moments parallel, whereas uniaxial anisotropy energy narrows the wall. As a result, the width of the magnetic domain wall is determined by minimizing the sum of exchange energy and uniaxial anisotropy energy. Moreover, the directions of the magnetic moments in the domain wall are determined by the demagnetizing field, which depends on the thickness of the magnetic material. A Bloch wall, of which the magnetic moments rotate in the thickness direction of the magnetic material when it is thick, is shown in Fig. 5a. A Neel wall, of which the magnetic moments rotate in-plane of the magnetic material when it is thin, is shown in Fig. 5b. In the case of the Bloch wall, the demagnetizing field in the thickness direction is weak, but in the case of the Neel wall, it is strong.

Bloch wall width, d_B, and magnetic wall energy, σ_B, are given by

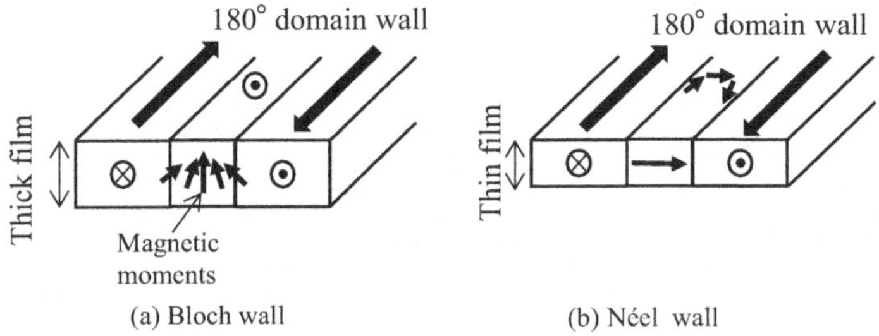

Fig. 5 Bloch and Néel walls

$$d_B = \pi\sqrt{\frac{A}{K}}, \tag{2}$$

$$\sigma_B = 4\sqrt{AK}, \tag{3}$$

where K is the anisotropy constant [3, 4].

If the magnetic material is iron, with A of 1×10^{-11} J/m and K of 2×10^{-4} J/m^3, Bloch wall width and its energy are about 70 nm and 18×10^{-8} J/m^2, respectively. Wall widths in soft and hard magnetic materials are generally from 100 to 1000 nm and 2 to 3 nm, respectively [5].

3 Magnetization Processes

Ferromagnetic material is divided into domains, as mentioned in the previous section. Although the magnetic domains have spontaneous magnetizations, the directions of the magnetizations differ from those of adjacent domains. Since the magnetizations in the domains cancel each other out, the magnetic material is in a magnetic neutral state or a demagnetized state. An initial magnetization curve of a ferromagnetic material with the closure domain structure in magnetically neutral state is shown in Fig. 6. The magnetic domain structure when no magnetic field is applied to the ferromagnetic material (in the demagnetized state) is shown in Fig. 6a. When a magnetic field is applied diagonally to the right, the magnetic domain in which the magnetic moment points to the right gradually expands, as shown in Fig. 6b; namely, domain wall motion of a 180° domain wall occurs. However, the speed of the domain wall motion is low; that is, the magnetization increases slowly because of lattice defects or impurities in the magnetic material. If the applied magnetic field is weak, the magnetization returns to the magnetically neutral state after the applied magnetic field is removed. The range of this applied magnetic field is called the "reversible magnetic domain wall motion range." As the applied magnetic field is strengthened, the domain wall motion is accelerated, and the magnetization never returns to the neutral state after the applied magnetic field is removed. The range of

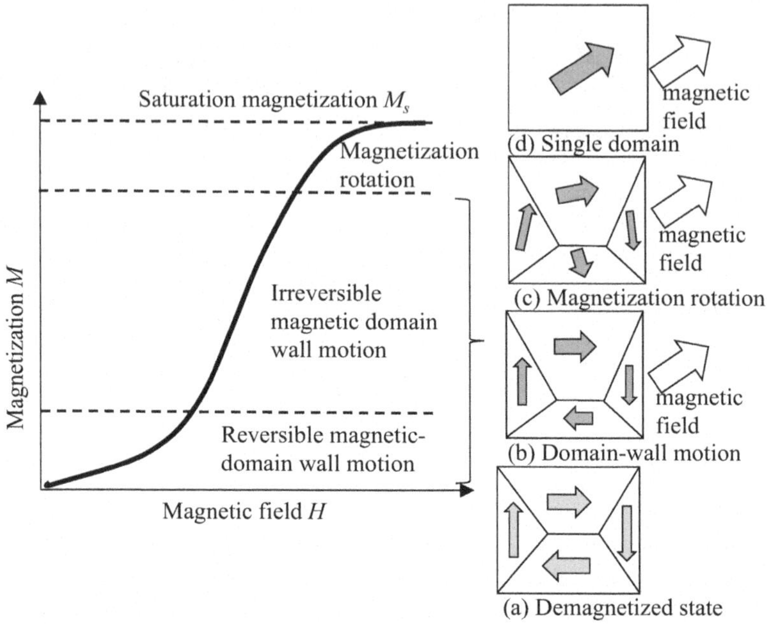

Fig. 6 Initial magnetization curve and domain structures

this magnetic field is called the "irreversible magnetic domain wall motion range." If the applied magnetic field is much stronger, as shown in Fig. 6c, the magnetization gradually rotates. Finally, as shown in Fig. 6d, the ferromagnetic material becomes a single magnetic domain, and the magnetization becomes saturated and is called the "saturation magnetization."

Regarding soft magnetic materials, when the applied magnetic field is rapidly changed, namely a high-frequency field, an eddy current is generated by the fast change of the magnetization due to the fast magnetic domain wall motion, which is governed by Faraday's electromagnetic induction law [Eq. (4) below]. The eddy current causes large iron loss and, as a result, induces a large amount of heat [Eqs. (5) and (6)]. This heat generation is a major problem for soft magnetic materials. The eddy current losses in electrical steel can be decreased by increasing the number of magnetic domains and decreasing their widths, as shown in Fig. 1. That is, motions of magnetic domain walls are restricted, and their speed is reduced.

$$\mathrm{rot}\vec{E} = -\frac{\partial \vec{B}}{\partial t}, \tag{4}$$

$$\vec{J} = \sigma \vec{E}\,[A/m^2], \tag{5}$$

$$Q = \frac{1}{\sigma} \vec{J}^2 \, [J], \tag{6}$$

where $\vec{E}$ is the electrical vector, $\vec{J}$ is the eddy current vector,σ is the conductivity and Q is Joule heat.

4 Micromagnetics Simulation

As mentioned in the previous section, magnetic domain wall motions and magnetic rotations are caused by the behaviors of microscopic magnetic moments. Micromagnetics is a method for analyzing magnetic moments in magnetic materials, clarifying magnetic phenomenon, and obtaining new knowledge about them. In recent years, micromagnetics simulations of magnetic materials, which apply models with a huge number of magnetic moments, can be performed in a short time by using high-performance computers (with CPU clock frequency higher than 3.0 GHz) or massively parallel processing using multicore, multi-CPU, or general-purpose GPU (GPGPU). Micromagnetics is therefore used for simulating magnetic recording systems and industrial magnetic devices with small size, high performance, and high frequency, and medical equipment utilizing biomagnetism. Representative equations used for micromagnetics simulation are the Landau–Lifshitz equation (LL equation) or the Landau–Lifshitz–Gilbert equation (LLG).

4.1 LL and LLG Equations

The origin of a magnetic moment is electron spins in an atom. Electrons (having a negative charge) rotate outside a nucleus (having a positive charge) and also rotate themselves. When the electrons rotate, currents are generated. Therefore, the atom is equivalent to a magnetic dipole having a magnetic moment. The interaction between magnetic spin $\vec{S}$ and magnetic field $\vec{H}$ is represented by the Bloch equation, given as [6]

$$\frac{d\vec{S}}{dt} = -\left(\vec{S} \times \vec{H}\right). \tag{7}$$

This equation indicates that a torque in the direction orthogonal to the magnetic spin and magnetic field is generated, and the magnetic spin precesses around the magnetic field vector, namely a phenomenon called "precession."

When $\vec{M}$ and μ_s denote the magnetic moment vector generated by the spin and its strength, respectively, $\vec{M} = \mu_s \vec{S}$. Therefore, when $\vec{M}$ is in magnetic field $\vec{H}$, the magnetic moment is processed by the torque $-\left(\vec{M} \times \vec{H}\right)$ around the magnetic field vector, as shown in the following equation,

$$\frac{\mathrm{d}\vec{M}}{\mathrm{d}t} = -\left(\vec{M} \times \vec{H}\right). \tag{8}$$

This equation indicates that the precession permanently continues, keeping a steady angle between the magnetic moment and field vectors. However, the magnetic moment is actually oriented parallel to the direction of the magnetic field vector, and the precession stops. In the LL Equation (9), the first term is the precession term, and the second term is added as a damping term, in which the magnetization points in the direction of the magnetic field vector, and the precession stops.

$$\frac{\mathrm{d}\vec{M}}{\mathrm{d}t} = -\gamma\left(\vec{M} \times \vec{H}_{\mathrm{eff}}\right) - \lambda\vec{M} \times \left(\vec{M} \times \vec{H}_{\mathrm{eff}}\right), \tag{9}$$

where γ is a gyromagnetic ratio (i.e., the ratio of magnetic moment to angular momentum in units of $\mathrm{s}^{-1} \cdot \mathrm{T}^{-1}$) and λ is a phenomenological damping parameter, given as

$$\lambda = \frac{\alpha\gamma}{M_s}, \tag{10}$$

where α is the Gilbert damping constant and M_s is saturation magnetization. For a ferromagnetic material, α is less than 0.1 [7, 8]. $\vec{H}_{\mathrm{eff}}$ is the sum of the magnetic fields applied to the magnetic material, which is called "effective magnetic field" (which is explained in the following subsection).

In the LLG equation, the second term of the LL equation is replaced with the time derivative of magnetization, as shown in Eq. (11). The LLG equation equals the LL equation in terms of mathematical expression.

$$\frac{\mathrm{d}\vec{M}}{\mathrm{d}t} = -\gamma\left(\vec{M} \times \vec{H}_{\mathrm{eff}}\right) + \frac{\alpha}{M_s}\left(\vec{M} \times \frac{\mathrm{d}\vec{M}}{\mathrm{d}t}\right) \tag{11}$$

The relationship between the effective magnetic field and magnetization represented by the LL or LLG equation is illustrated in Fig. 7.

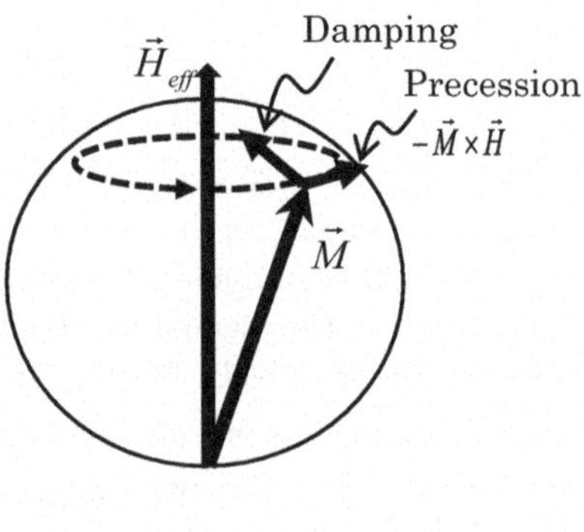

Fig. 7 Precessional motion of the magnetic moment

4.2 Effective Magnetic Field

The effective magnetic field applied to magnetic moments of a magnetic material is the sum of the externally applied (H_a), static (H_s), anisotropy (H_k), and exchange (H_e) magnetic fields. If the energies of these fields are, respectively, denoted as E_a, E_d, E_k, and E_e, each magnetic field can be calculated by partial differentiation of energy with respect to $\vec{M} = (M_x, M_y, M_z)$, where the magnetic fields are given as

$$\vec{H}_a = (H_{a,x}, H_{a,y}, H_{a,z}) = -\frac{\partial E_a}{\partial \vec{M}}, \tag{12}$$

$$\vec{H}_d = (H_{d,x}, H_{d,y}, H_{d,z}) = -\frac{\partial E_d}{\partial \vec{M}} = \sum_{\text{cell}} \begin{bmatrix} S_{xx} & S_{xy} & S_{xz} \\ S_{yx} & S_{yy} & S_{yz} \\ S_{zx} & S_{zy} & S_{zz} \end{bmatrix} \begin{bmatrix} M_x \\ M_y \\ M_z \end{bmatrix}, \tag{13}$$

$$\begin{aligned} \vec{H}_k &= (H_{k,x}, H_{k,y}, H_{k,z}) = -\frac{\partial E_k}{\partial \vec{M}} \\ &= -\frac{2K_u}{M_s^2}\left(\vec{k} \cdot \vec{M}\right)\vec{k} = -\frac{H_k}{M_s}\left(\vec{k} \cdot \vec{M}\right)\vec{k}, \end{aligned} \tag{14}$$

$$\vec{H}_e = (H_{e,x}, H_{e,y}, H_{e,z}) = -\frac{\partial E_e}{\partial \vec{M}} = \sum_{\text{neighbor cell}} \frac{2A}{M_s^2} \begin{bmatrix} \nabla^2 M_x \\ \nabla^2 M_y \\ \nabla^2 M_z \end{bmatrix}. \tag{15}$$

The tensor for calculating the static magnetic field in Eq. (13) is explained by using the left figure in Fig. 8. It is assumed that the static magnetic field is applied to cell n from cell m, where a cell represents a calculation model of a crystal or a particle (which is explained in detail in the next section) https://ejje.weblio.jp/content/discussed. The static magnetic field is the sum of the fields applied to the gravity of cell n from the magnetic poles of the surfaces of cell m determined by the direction of magnetization of it. The components of the tensor are determined

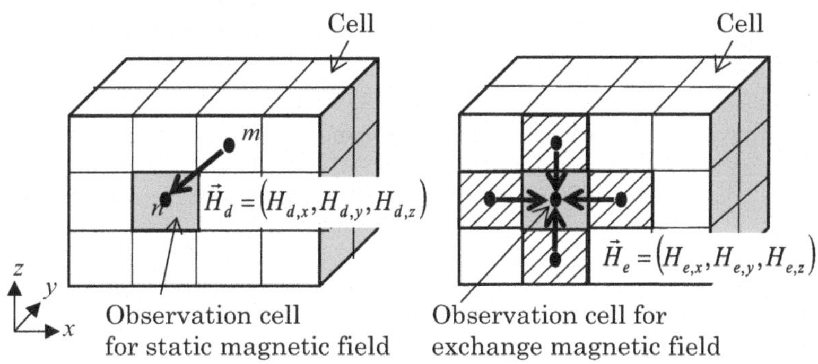

Fig. 8 Relation between magnetic cells and static and exchange fields

by the positional relationship between cells and the shape of the cells. If the shape of the cells is a rectangular, the components of the tensor are given by Eq. (16). (See detailed calculation methods in Refs. [9, 10]).

$$S_{u,v} = -\frac{1}{4\pi\mu_0}\sum_{\beta}\int \mathrm{d}s_\beta \frac{(r_{m,u}-r_{n,u})}{|\vec{r}_m-\vec{r}_n|^3} n_{\beta,v}, \tag{16}$$

where u and v are any one of the components of x, y, and z, μ_0 is vacuum permeability, β is any one of the six surfaces of rectangular cell m, $\vec{r}_m$ and $\vec{r}_n$ are position vectors of the gravities of cells m and n, respectively, $r_{m,u}$ and $r_{n,u}$ are u components of the position vectors, and $n_{\beta,v}$ is the v-component of an outward normal vector of surface β. The total static magnetic field applied to cell n is the sum of the static fields from all cells including the cell n in the magnetic model.

K_u (unit: J/m^3) is the anisotropy constant, $\overrightarrow{k}$ is a unit vector of the easy axis direction in Eq. (14), and A (unit: J/m) is the exchange stiffness constant in Eq. (15). The exchange field is the sum of the exchange fields between the observation and nearest-neighbor cells, as shown in Fig. 8.

5 Numerical Methods for First-Order Initial Value Problem

Numerical methods for solving the LLG equation are explained as follows. The LLG equation is an ordinary differential equation including a first-order derivative. In addition, solving the LLG equation is a first-order initial value problem, where the initial conditions are given by magnetic structures at time zero. Numerical methods include the finite element method (FEM), finite difference method (FDM), boundary element method (BEM). FDM is discussed in the following. FDM solves differential equations by finite difference approximation, where "finite difference" is the time for solving the LLG equation. Finite difference can be categorized as forward (explicit), backward (implicit), and central. Concerning forward difference, the Euler method, and the Runge–Kutta methods are described below. For the other two, refer to the other technical books.

The Euler method is explained as follows. Time (variable) (t) and first-order ordinary differential equations of function $y(t)$ and the initial condition are given as

$$\frac{\mathrm{d}y}{\mathrm{d}t} = y'(t) = f(t, y(t)) \quad (t_0 \le t \le t_N), \tag{17}$$

where initial value

$$y(t_0) = y_0. \tag{18}$$

If solution $y(t)$ of Eqs. (17) and (18) is obtained, $y(t)$ is obtained by Taylor series expansion around point t in interval $t_0 \leqq t \leqq t_N$ according to

$$y(t + \Delta t) = y(t) + \Delta t y'(t) + \frac{(\Delta t)^2}{2!} y''(t) + \frac{(\Delta t)^n}{n!} y^{(n)}(t) + \cdots, \tag{19}$$

where Δt in the above equation is time increment (sampling width), and the ith time $t_i (i = 0, 1, 2, \ldots)$ is described as $t_i = t_0 + i \cdot \Delta t$. It is thus assumed that Δt is sufficiently small, and the second and subsequent terms of Eq. (19) are zero. When t, $t + \Delta t$, and $y'(t)$ are replaced by t_i, t_{i+1}, and $f(t_i, y(t_i))$, respectively,

$$y(t_{i+1}) = y(t_i) + \Delta t f(t_i, y(t_i)). \tag{20}$$

This equation indicates that $y(t_{i+1})$ at $(i + 1)$-th time is calculated by $y(t_i)$ at the ith time; thus, $y(t)$ is sequentially calculated by using this equation with the Euler method.

As for the Runge–Kutta method, the terms other than the second term in Eq. (9) are taken into consideration. The nth-order Runge–Kutta method takes the nth differential equation into consideration. Therefore, the Euler method is a first-order Runge–Kutta method.

The second-order Runge–Kutta method (also called the Heun method) is explained below. The equation that takes the second-order differential in Eq. (19) into consideration is given by

$$y(t_{i+1}) = y(t_i) + \Delta t f(t_i, y(t_i)) + \frac{(\Delta t)^2}{2} f'(t_i, y(t_i)), \tag{21}$$

where

$$f'(t_i, y(t_i)) = \left.\frac{\partial f(t, y(t))}{\partial t}\right|_{t=t_i} + \left.\frac{\partial f(t, y(t))}{\partial y}\right|_{t=t_i} \cdot f(t_i, y(t_i)). \tag{22}$$

Moreover, by rearranging Eq. (21), the following equation is obtained.

$$y(t_{i+1}) = y(t_i) + \frac{\Delta t}{2}(k_1 + k_2), \tag{23}$$

where

$$k_1 = f(t_i, y(t_i)), \tag{24}$$

$$k_2 = f(t_i + \Delta t, y(t_i) + \Delta t f(t_i, y(t_i))) = f(t_i + \Delta t, y(t_i) + \Delta t k_1). \tag{25}$$

See details in Refs. [11, 12].

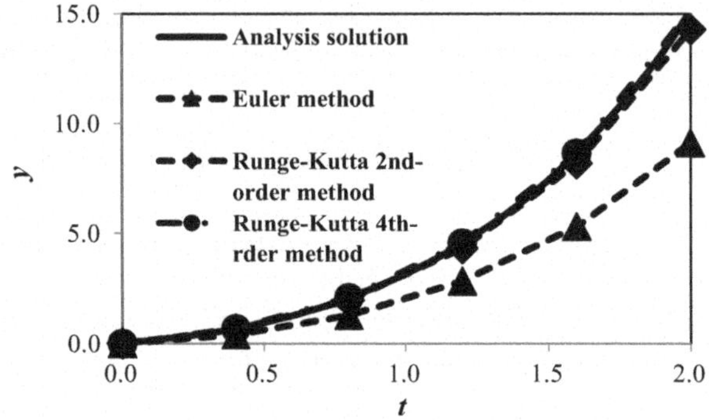

Fig. 9 Comparison between analysis and numerical solutions for ordinary differential equation $y' = 2t + y + 1$ with initial condition $y(0) = 0\,(\Delta t = 0.4)$

Comparing Eqs. (21)–(22) and (23)–(25) reveals that the former indicates that $f(t, y)$ and partial differentiation of $f(t, y)$ must be calculated once and twice, respectively, whereas the latter represents that $f(t, y)$ has only to be calculated twice.

Runge–Kutta methods of third or higher order are given in the same way. Numerical solutions calculated by the Euler, second-order Runge–Kutta and fourth-order Runge–Kutta methods with Eq. (26) are compared with the analytical solution (where Δt is 0.4) in Fig. 9.

$$\frac{\mathrm{d}y}{\mathrm{d}t} = y' = 2t + y + 1 \qquad (0 \leq t \leq 2), \tag{26}$$

initial condition:

$$y(0) = 0. \tag{27}$$

From the figure, the solution obtained by the fourth-order Runge–Kutta method is the closest to the analytical solution. Thus, as the order of Runge–Kutta method increases, the solution gets closer to the analytical solution. However, as for the high-order Runge–Kutta method, care needs to be taken in regard to calculation times in the case of using computers because the calculation is time-consuming. Realistically, the fourth-order Runge–Kutta method is widely used.

The error in the case of the nth order Runge–Kutta method is given by $O(\Delta t^{\,n})$ (where O is Landau's symbol or Big O notation). That is, when sampling time Δt decreases by half, the error decreases by about $(1/2)^n$. The solution for Δt of 0.2 is plotted in Fig. 10. In contrast to the solution shown in Fig. 9, it is clear that the results are given by the Euler and second-order Runge–Kutta methods are close to the analytical solution. As for the second-order Runge–Kutta method, the error at Δt of 0.2 decreases 0.5 times less compared to that at Δt of 0.4, where $0 \leqq t \leqq 50$. However, the error depends on t.

To solve the LLG equation by using FDM, Δt is ordinary 1×10^{-12}–1×10^{-14} s. The value of Δt depends on the cell size of the calculation model and magnetic parameters. In addition, when Δt is small, or magnetization is difficult to converge,

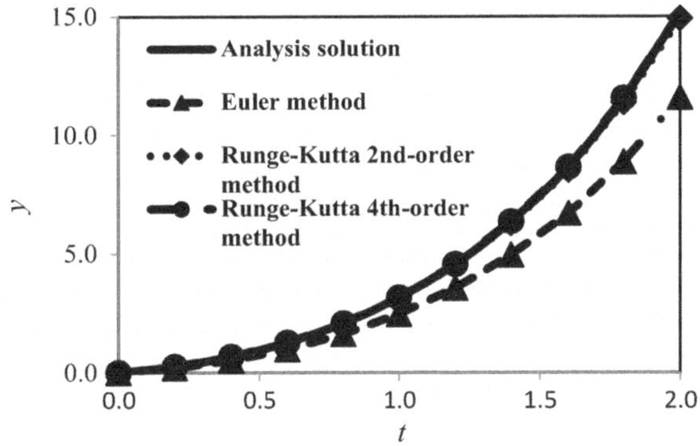

Fig. 10 Comparison between analysis and numerical solutions for an ordinary differential equation $y' = 2t + y + 1$ with initial condition $y(0) = 0\,(\Delta t = 0.2)$

the solutions for time intervals of seconds, minutes, and hours are significantly time-consuming.

6 Magnetic Domain Structures Calculated by Using the LLG Equation

Examples of magnetic domain structures calculated by using the LLG equation, which was solved by the fourth-order Runge–Kutta method, are described as follows. Two calculation models of magnetic materials, with sizes of 1280 nm × 640 nm × 10 nm and 1280 nm × 640 nm × 100 nm, are shown in Fig. 11. The calculation models are divided into small cells as mentioned in Sect. 4.2. The size of a cell is, for example, equal to size of a crystal, a particle, or an atom. In these calculations, the shape of the cell is a hexagonal prism, and the diameter of the hexagonal surface (x-y surface) is 10 nm, as shown in Fig. 11. The cell is not divided in the z-direction

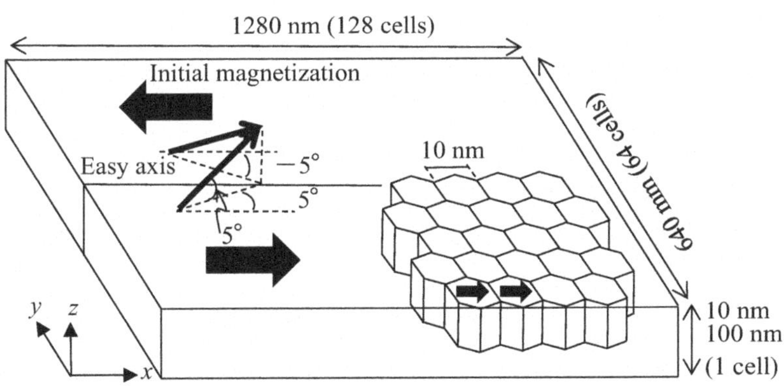

Fig. 11 Model of magnetic film for LLG simulation

(i.e., cell thickness); namely, the number of cells is one, even if the thicknesses of the magnetic material are 10 nm and 100 nm. The diameter of the *x*-*y* surface is determined as a value less than the exchange length of a soft magnetic film. The exchange length is given by $\sqrt{A/K_u} = \sqrt{0.5 \times 10^{-15}} = 22\,\text{nm}$ where exchange stiffness $A = 1.0 \times 10^{-11}$ J/m and anisotropy constant $K_u = 2 \times 10^4$ J/m^3. Saturation magnetization M_s is 2.0 T. The anisotropy field is given by $H_k = 2K_u/M_s = 4000$ A/m [13, 14]. As for the easy axis, the angle between the *x*- and easy axes is 5°; namely, the upper and lower parts of the model are −5° and +5°, respectively. The elevation angle in the z-direction is 5°. The initial magnetization vectors of the upper and lower parts of the model are assumed to point in the −x- and x-directions, respectively. The magnetizations were relaxed by the applied field at zero for 51 ns with Δt of 1.0 $\times\ 10^{-13}$ s; namely, the time step was 510,000.

The calculated magnetization vectors are shown in Fig. 12. Arrows in the figures indicate the magnetization vectors of the cells of the calculation model. The thicknesses of the soft magnetic films in Fig. 12a, b are 10 and 100 nm, respectively. The figures indicate the magnetization distribution on the *x*-*y* surface, and the enlarged views on the right indicate the region near the center of the magnetic materials, where the number of the arrows is thinned out in the left figures. As shown in Fig. 12a, when the film is thin (10 nm), a Neel wall is observed, and the magnetization vectors rotate in the *x*-*y* plane of the domain wall. On the other hand, as shown in Fig. 12b, when the film is thick, a Bloch wall is observed, and the magnetization vectors rotate out

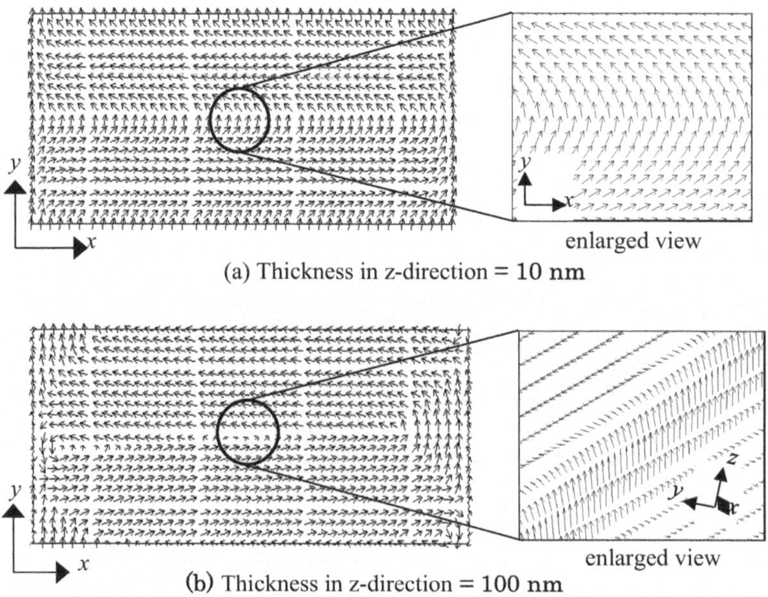

Fig. 12 Comparison of magnetizations of soft magnetic films with two different thicknesses simulated by using the LLG equation

of the plane of the domain wall. As a result, it was confirmed by the LLG simulation that when the film is thicker, the magnetization wall changes from a Neel wall to a Bloch wall (as explained in Sect. 2).

References

1. Sadahiro, K., Gotoh, S., Uenosono, S.: Soft Magnetic Materials of JFE Steel Group,JFE technical report, No. 8, 1–6 (2005)
2. Kubota, T.: Denji-kohan Jisei-zairyo to shiteno "tetsu" no tokusei wo saidaigen ni hikidasu (last volume). Nippon Steel Monthly **17**, 9–12 (2005)
3. Hubert, A., Schäfer, R.: Magnetic Domains. Springer, Berlin (1998)
4. Aharoni, A.: Introduction to the Theory of Ferromagnetism. Oxford Science Publication (1996)
5. Stanislas ROHART: Basic Concepts on Magnetization Reversal (1) Static Properties: Coherent Reversal and Beyond. http://magnetism.eu/esm/2011/slides/rohart-slides1.pdf (8/1/2017)
6. Bloch, F.: Nuclear Induction. Phys. Rev. **70**, 460–473 (1946)
7. Iihama, S., Mizukami, S., Inami, N., Hiratsuka, T., Kim, G., Naganuma, H., Oogane, M., Miyazaki, T., Ando, Y.: Observation of precessional magnetization dynamics in L10-FePt thin films with different L10 order parameter values. Jpn. J. Appl. Phys. **52**, 073002 (2013)
8. Fujita, N., Inaba, N., Kirino, F., Igarashi, S., Koike, K., Kato, H.: Damping constant of Co/Pt multilayer thin-film media. J. Magn. Magn. Mater. **320**, 3019–3022 (2008)
9. Hayashi, N., Uesaka, Y., Nakatani, Y., Fukushima, H.: Micromagnetics simulation II. Magnetics Jpn. **3**(7) (2008)
10. Yoshida, K., Kanai, Y., Greaves, S., Takagishi, M., Akagi, F.: Application of micromagnetics to magnetic recording I. Magnetics Jpn. **4**(4) (2009)
11. Hasegawa, T., Yoshida, T., Hosoda, Y.: Kougaku no tame no suuchi-keisan. Suurikougaku-Sha (2012)
12. Yanagida, E., Nakaki, T., Mimura, M.: Numerical computation. Shokabo Tokyo (2014)
13. Sagawa, M.: Neojimu Jisyaku no Subete. AGNE Gijutsu center (2011)
14. Mitsumata, C.: Methodology of Analysis on Magnetics. Kyoritsu Shuppan (2013)

Polycrystalline Magnetic Calculation

Keisuke Fujisaki

Abstract To express the polycrystalline characteristics of magnetic material in numerical calculation, three-dimensional polycrystalline magnetic field analysis is proposed here. LLG calculation for magnetic domain structure should be used, but it usually has a mesh explosion issue for a lot of crystal grains magnetic material. A local coordinate system is set in each crystal grain to express the crystal magnetic anisotropy and a whole coordinate system is set in the polycrystalline of magnetic material to express magnetic flux continuity. Coordinate transform is used to express each physical parameter defined in the different for coordinate system. Each crystal is assumed to have the same magnetic characteristics as *B–H* curve of a single crystal. Finite element method is used for its numerical calculation, because magnetic flux density distribution is obtained in the minimum electromagnetic energy condition. Two cases of GO steel sheet of 80 mm square are picked up as 2 crystal grains and 56 crystal grains. Magnetic flux density distribution of the GO steel is measured by needle method for magnetic flux density and H-coil for magnetic field. Thc proposed calculation result of magnetic flux density distribution expresses well the measured result.

Keywords Polycrystalline · Coordinate transform · Finite element method · *B–H* curve of single crystal

1 Introduction

Magnetic materials such as electrical steel sheets are used in an electric apparatus including a motor or a transformer, and most of them have a polycrystalline structure that many crystals are collectively organized. It may be said that the magnetism properties with polycrystalline depend on a size and shape of each crystal, its crystal orientation, and the magnetism characteristic of single crystal. As the magnetic characteristic of such a polycrystalline makes a serious influence on the design or

K. Fujisaki (✉)
Toyota Technological Institute, Nagoya, Japan
e-mail: fujisaki@toyota-ti.ac.jp

K. Fujisaki (ed.), *Magnetic Material for Motor Drive Systems*,
Engineering Materials, https://doi.org/10.1007/978-981-32-9906-1_12

the development of the electrical motor or the transformer, it may be said that it is an important technical issue to be solved for a high efficiency and downsizing of the electrical apparatus.

As the magnetic material has a magnetic domain structure, applications of the numerical analysis of the magnetic domain structure that accords with the orientation and shape of polycrystalline are usually effective for such a magnetic characteristic grasp. For that reason, the numerical calculation based on LLG (Landau–Lifshitz–Gilbert) equation [1] should be carried out for the polycrystalline by dividing the space to be calculated into a lot of meshes. However, as it introduces a mesh explosion to divide all the crystal grain of the polycrystalline into detailed meshes and to carry out the numerical calculation, it may be said that it is impossible to calculate it even if we use a cutting edge super computer now.

Therefore, it is usually required for such a polycrystalline magnetic calculation to develop a simplified calculation method using some kinds of numerical calculation model. Here, I describe a numerical modeling method which uses a conventional static magnetic field calculation and a coordinate transformation technology in a case of grain-oriented steel (GO material) [2] with high degree of integration of the crystal orientation.

2 Polycrystalline Magnetic Analysis Model [3]

The polycrystalline such as an electrical steel sheet is collective organizations having a lot of crystal grains whose borders are grain boundary. Each crystal grain has an independent crystallographic orientation, and has a magnetic characteristic influenced with the magnetic domain structure so as that magnetic energy is minimized.

Therefore, following items are considered on modeling a magnetic phenomenon of such a polycrystalline

1. A magnetism property of each crystal grains is assumed to have the same magnetic property as a single crystal. The magnetic property is assumed to be expressed by a *B–H* curve that takes into consideration of nonlinearity and magnetic anisotropy for the magnetic characteristic [4].
2. A local coordinate system based on the crystal orientation is set in each crystal grain, and a whole coordinate system is set in the whole polycrystalline. The magnetic anisotropy and the nonlinearity are treated in the local coordinate system, and the continuity of the magnetic flux is treated in the whole coordinate system. Coordinate transformation of physical electromagnetic parameters is used for a transformation between the local coordinate system and the whole coordinate system.
3. The mesh division is carried out along the crystal grain shape, and different crystal orientation is given every crystal grain.

4. The magnetic characteristic behavior among the crystal grains is introduced so as that magnetic energy of the whole polycrystalline is minimized. In other words, a finite element method of the static magnetic field is used here.

Magnetic energy decides magnetic domain structure usually, and it consists of four clauses like the next equation, when an influence of the mechanical stress is ignored [5].

$$\chi_{\mu-\mathrm{MAG}} = \iiint \left(E_{\mathrm{ex}} + E_{\mathrm{qns}} + E_{\mathrm{app}} + E_{d}\right)\mathrm{d}x\mathrm{d}y\mathrm{d}z \tag{1}$$

Here, E_{ex} is an exchange energy, E_{qns} is a magnetic anisotropic energy, E_{app} is a Zeeman energy, and E_d is a static magnetic energy. In contrast, in the three-dimensional polycrystalline magnetic field analysis model proposed here, each clause is considered and modeled as follows.

The exchange energy is a proximity effect at the atom level and affects a magnetic wall movement, but the magnetic domain structure itself such as the domain wall movement is not thought here, and the behavior is assumed to be expressed by a *B*–*H* curve. The magnetic anisotropic energy is assumed to be considered as a *B*–*H* curve in consideration of magnetic anisotropy. The Zeeman energy is considered to be a quantum energy difference by an external magnetic field, and is assumed to be expressed by a magnetization process of a *B*–*H* curve.

The static magnetic energy is considered to be influenced by demagnetization. So it is assumed to be expressed by a grain shape of each crystal grain. From the above considerations and assumptions, the polycrystalline magnetic field analysis proposed here is considered to have the same magnetic characteristics as the magnetic domain structure analysis. Magnetic modeling differences between the normal magnetic domain structural calculation and the polycrystalline magnetic field analysis proposed here are summarized and shown in Fig. 1 [3].

Each crystal grain in the polycrystalline is thought to have an inherent local coordinate system based on the crystal orientation, but as the magnetic flux flowing on the whole steel with polycrystalline must be continuous and close, magnetic flux density should be considered on the whole coordinate system. Therefore, a local coordinate system (U_i, V_i, W_i) $(i = 1, 2, \ldots, N$, U_i is an easy magnetization direction, V_i is a transverse direction on the steel plane, W_i is a perpendicular direction to the $U_i - V_i$ plane, N is a number of crystal grains) are set up on each crystal grain of the polycrystalline [6] as shown in Fig. 2.

A whole coordinate system (*X*, *Y*, *Z*), (*X* is a rolling direction, *Y* is a vertical direction on the steel plane, and *Z* is a perpendicular direction to the steel plane) is set in the whole polycrystalline (steel sheet) at the same time. In other words, when magnetic flux density flows on each crystal, the magnetism characteristic such as nonlinearity or anisotropy is expressed by local coordinate system. When the whole property as magnetic flux continuity is thought about, magnetic flux density or so is treated in whole coordinate system as shown in Fig. 2.

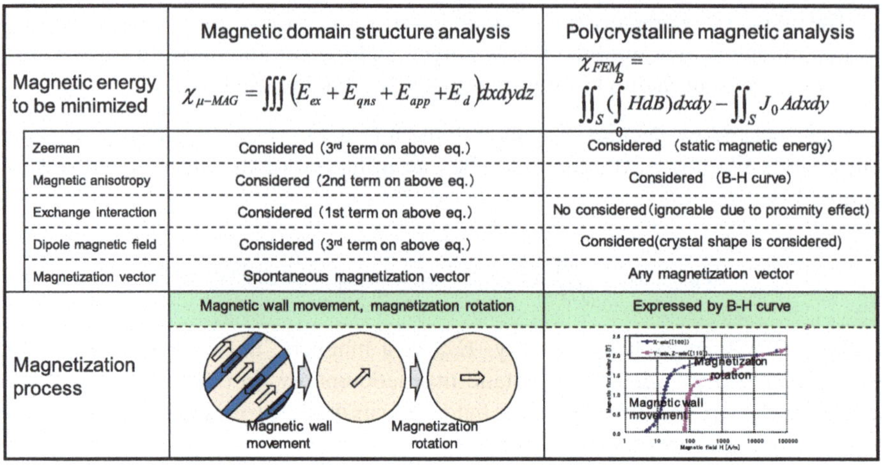

Fig. 1 Comparison with magnetic domain structure analysis and polycrystalline magnetic field analysis 2009 IEEE. Reprinted, with permission, from [3]. Image courtesy of Nippon Steel & Sumitomo Metal Corporation

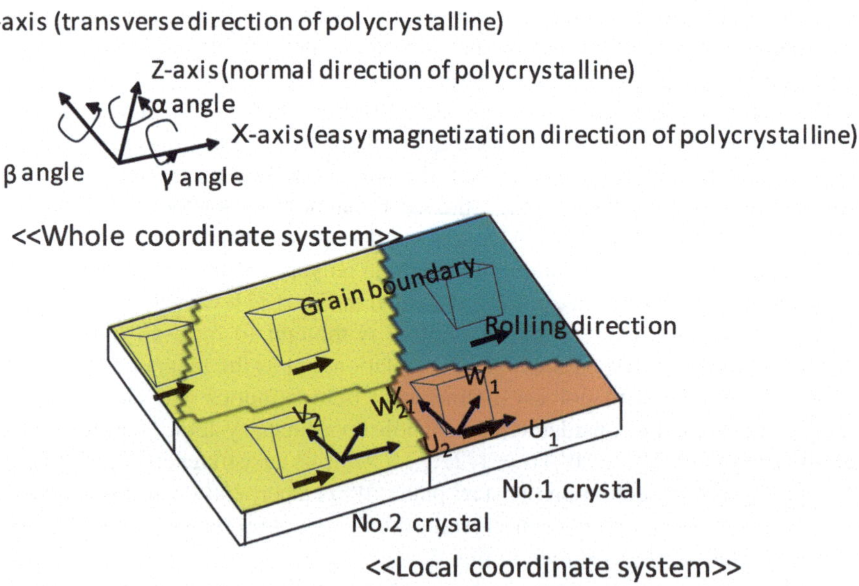

Fig. 2 Local coordinate system and whole coordinate system for polycrystalline magnetic field analysis [16]. Adapted from [16] with permission ©2016 The Magnetics Society of Japan

Observing from a point of view of a whole coordinate system, crystal orientation of each crystal grain is defined so as that a rolling angle around *Z*-axis is α, a rolling angle around *Y*-axis is β, and a rolling angle around *X*-axis is γ. If each axial rotary angle is small enough, a coordinate transformation from a whole coordinate system to a local coordinate system is expressed as the next equation [7].

$$[T] = \left[R_z(\alpha)\right]\left[R_y(\beta)\right]\left[R_x(\gamma)\right] \tag{2}$$

In this way, when a certain magnetic flux density vector $\vec{B}$ is $\left\{ B_u \ B_v \ B_w \right\}^{\mathrm{T}}$ (*T* with the top is a transpose matrix) when it is expressed by a local coordinate system, and it is $\left\{ B_x \ B_y \ B_z \right\}^{\mathrm{T}}$ when it is expressed by a whole coordinate system; the coordinate transformation between them is related as follows.

$$\begin{Bmatrix} B_u \\ B_v \\ B_w \end{Bmatrix} = [T] \begin{Bmatrix} B_x \\ B_y \\ B_z \end{Bmatrix} \tag{3}$$

The magnetism properties defined with each crystal grain can be treated in a whole coordinate system.

Each crystal grain in the polycrystalline is assumed to have single-crystal magnetism properties like Fig. 3. In the local coordinate system (U_i, V_i, W_i), U_i is an easy magnetization direction, and V_i and W_i shall have a magnetic characteristic of transverse direction from Goss direction [8].

An overall calculation flow of the polycrystalline magnetic field analysis is shown in Fig. 4. In this time, the static magnetic field is carried out where an eddy current does not flow in a steel sheet, and magnetic flux density is calculated. The input data of this calculation is the crystal orientation and the crystal shape of each crystal in the polycrystalline and the single-crystal magnetic characteristics. The crystal shape and the crystal orientation are measured by using X-rays analysis [9]. In this way, polycrystalline static magnetic field calculation is carried out by finite element method (FEM). FEM calculation introduces magnetic flux density distribution so as that electromagnetic energy becomes minimized.

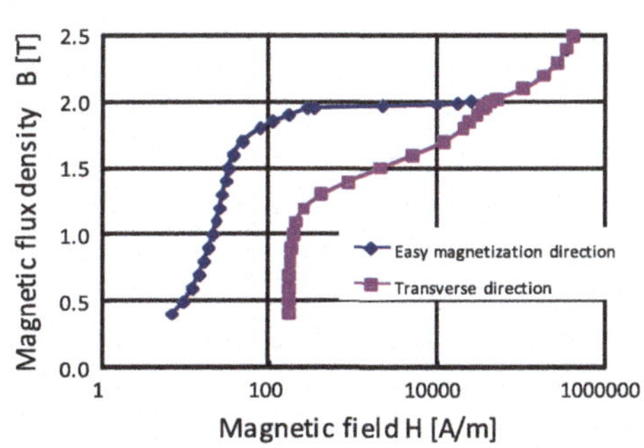

Fig. 3 Magnetic property of single crystal [12–14] 2009 IEEE. Reprinted, with permission, from [3]. Image courtesy of Nippon Steel & Sumitomo Metal Corporation

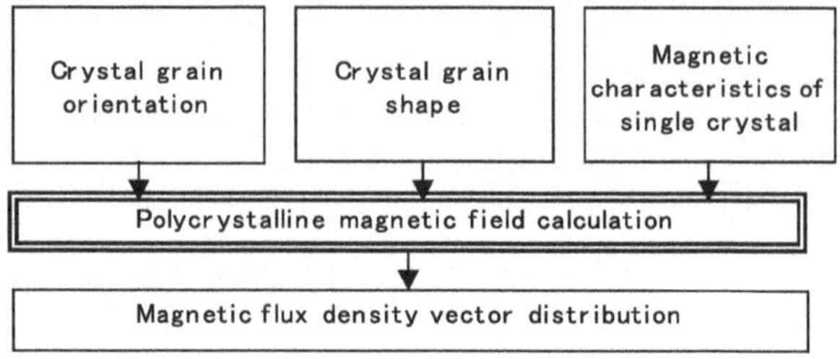

Fig. 4 Flow chart of polycrystalline magnetic field analysis [12] 2009 IEEE. Reprinted, with permission, from [3] Image courtesy of Nippon Steel & Sumitomo Metal Corporation

3 Model Validity Verification

At first, to verify the validity of the polycrystalline magnetic field calculation model, I consider a case as shown in Fig. 5, where a GO steel sheet of 80 mm square has 2 crystal grains of different crystal orientations. An external magnetic field is excited horizontally in the steel sheet and is adjusted so as that the maximum magnetic flux density becomes 1 T in one period of 50 Hz AC on the center line of the steel on average.

To obtain the measured magnetic characteristics of the GO steel sheet, a needle method is used for magnetic flux density distribution [10], and a H-coil method is used for magnetic field distribution [11]. The induced voltage between the needle terminals is measured and assumed to be that induced area where magnetic flux density flows is half of the steel thickness and the needle distance. The needle method and the H-coil can be measured *X*-direction and *Y*-direction and be moved on an arbitrary point on the steel sheet. The size of needle distance and H-coil is 15 mm and they move in 5 mm pitch. So their distribution is obtained. Figure 5a is a measurement result, and Fig. 5b is a calculation result of magnetic flux density distribution [12]. It is revealed that the situation that the magnetic flux concentrates on the upper side of the polycrystalline steel is observed on the measurement data as well as the calculation data due to the crystal orientations.

The next specimen is thought as shown in Fig. 6, where a GO polycrystalline material has 56 crystal grains in the steel sheet of the 80 mm square. Each crystal grain of the polycrystalline has different crystal orientation of α angle, β angle, γ angle, and each angle distribution of the GO steel sheet is shown in Fig. 6a–c, respectively, in contour expression.

The same measurement method as the 2 polycrystalline is applied to the 56 polycrystalline. An angle difference φ between the easy magnetization direction of the steel sheet and the measured magnetic flux density vector can be obtained. Distribution of the angle difference φ is shown in Fig. 7. As the magnetic flux flow is considered to follow the magnetization easy axial direction of each crystal grain, this angle difference φ should be compared with α angle of each crystal grain. In comparison with α angle distribution of Fig. 6a, the same distribution is observed.

Distribution of the size of magnetic flux density vector of the polycrystalline magnetic field analysis result is shown in Fig. 8a. To compare it with the measured data, the calculated magnetic flux density is shown to be averaged in 15 mm square

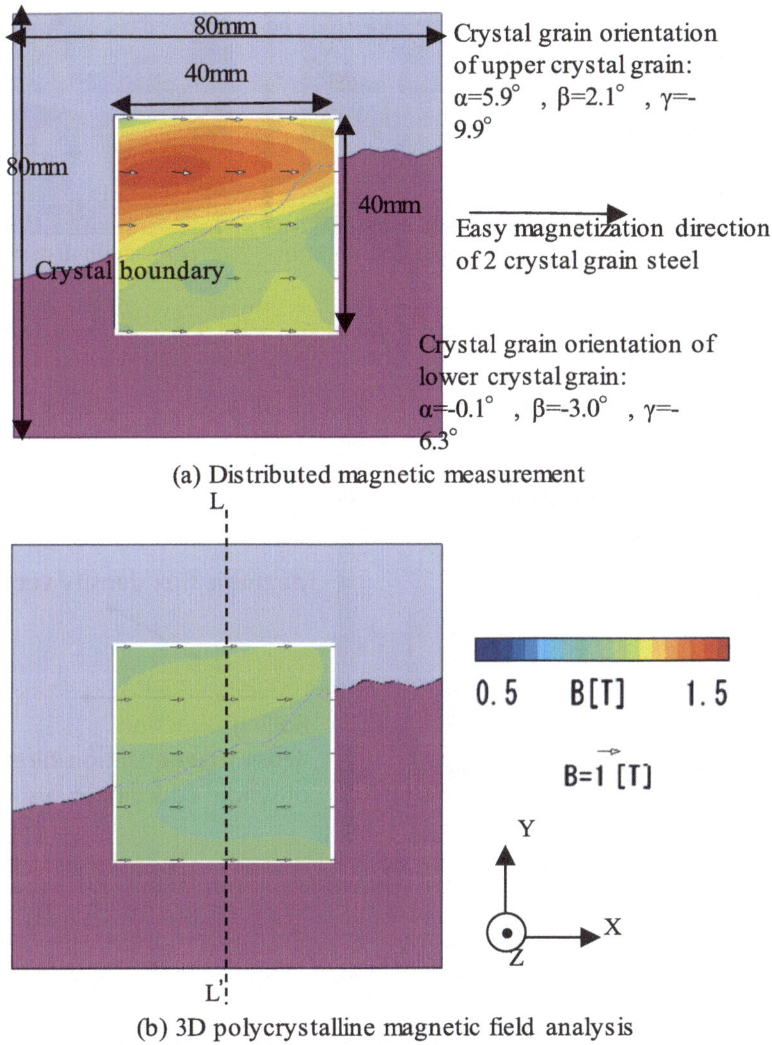

Fig. 5 Magnetic flux density distribution of 2 polycrystalline in comparison with calculation data by the polycrystalline magnetic field analysis and the measured data [3.5%Si–Fe, 0.19 mm thickness,averaged magnetic flux density on the centerline of the steel is 1 T in maximum of one period (easy magnetization direction of the steel sheet)] [12]. Image courtesy of Nippon Steel & Sumitomo Metal Corporation. **a** α-angle distribution, **b** β-angle distribution, **c** γ-angle distribution

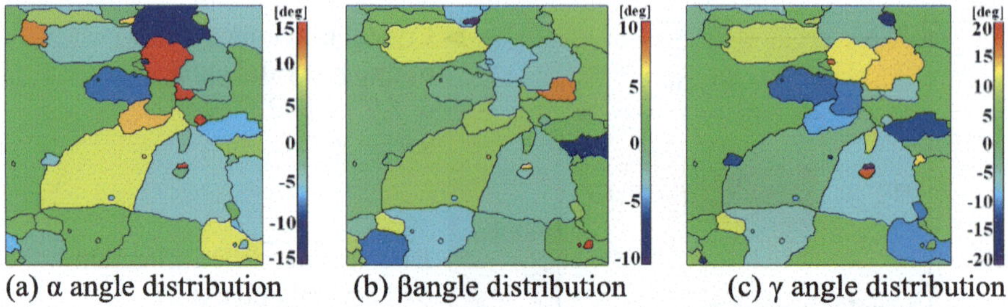

Fig. 6 Crystal orientation of GO polycrystalline steel sheet having 56 crystal grains in 80 mm square steel [12, 15] 2009 IEEE. Reprinted, with permission, from [3]. Image courtesy of Nippon Steel & Sumitomo Metal Corporation

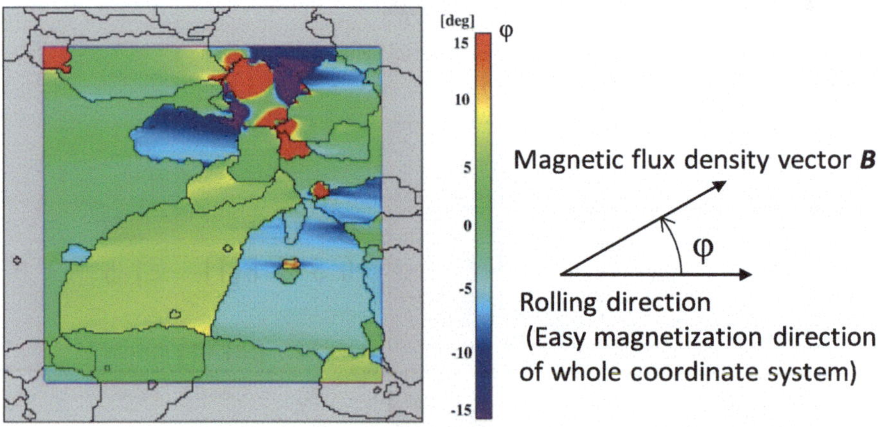

Fig. 7 Calculated angle difference distribution between magnetic flux density vector and easy magnetization direction of the steel for each element [12, 16] 2009 IEEE. Reprinted, with permission, from [3]. Image courtesy of Nippon Steel & Sumitomo Metal Corporation. Adapted from [16] with permission ©2016 The Magnetics Society of Japan

and its operation is carried out in 5 mm pitch in order that the calculated condition of magnetic flux density distribution coincides with the measured one. The measured distribution is also shown in Fig. 8b [12].

The measured magnetic flux density distribution as Fig. 8b shows that the magnetic flux density concentrates on the center part of the steel and it is small on its upper and lower side. The same situation is found to be also expressed by the calculation data as Fig. 8a. By using this technique, the study of the crystal grain shape is carried out [13, 14].

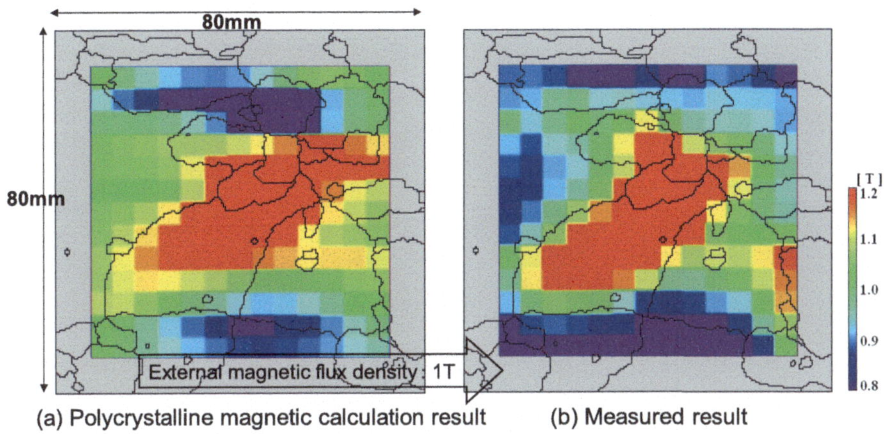

Fig. 8 Magnetic flux density distribution of GO polycrystalline with 56 crystal grains in comparison with calculated data by polycrystalline magnetic field analysis and measured data [12, 15, 16]. Image courtesy of Nippon Steel & Sumitomo Metal Corporation. Adapted from [16] with permission ©2016 The Magnetics Society of Japan

References

1. Brown, W.F.: Micromagnetics. Wiley, New York (1963)
2. Taguchi, S., Yamamoto, T., Sakakura, A.: New grain-oriented silicon steel with high permeability "Orientcore HI-B". IEEE Trans. Magn. **10**(2), 123–127 (1974)
3. Fujisaki, K., Tamaki, T.: Three-dimensional polycrystal magnetic field analysis of thin steel. IEEE Trans. Magn. **45**(2), 687–693 (2009)
4. Chikazumi, S.: Physics of Ferromagnetism. Shoukabou, Tokyo (1978). (in Japanese)
5. Hubert, A., Schafer, R.: Magnetic Domain. Springer, Berlin (1998)
6. Bunge, H.J.: Texture Analysis in Materials Science. Butterworth and Co, London (1982)
7. Crouch, T.: Matrix Method Applied to Engineering Rigid Body Mechanics, Pergamon (1981)
8. Goss, N.P.: Electrical Sheet and Method and Apparatus for its Manufacture and Test, U.S. Patent No. 1965559 (1934)
9. Cullity, B.D.: Elements of X-ray diffraction. Addison-Wesley Publication Company Inc, Reading, Massachusetts (1977)
10. Pfützner, H., Krismanic, G.: The needle method for induction tests: sources of error. IEEE Trans. Magn. **40**(3), 1610–1616 (2004)
11. Enokizono, M., Tanabe, I., Kubota, T.: Localized distribution of two-dimensional magnetic properties and magnetic domain observation. J. Magn. Magn. Mater. **196–197**, 338–340 (1999)
12. Fujisaki, K., Tamaki, T., Yasuhiro, S.: Comparison of 3D polycrystal magnetic field analysis and distributed magnetic measurement. IEEJ (The Institute of Electrical Engineers of Japan) Trans. Fundam. Mater. **129**(11). 821–826 (2009) (in Japanese)
13. Fujisaki, K.: Crystal grain shape aspect of grain oriented steel by three dimensional polycrystalline magnetic field analysis. J. Jpn. Soc. Appl. Electromagn. Mech. **21**(2), 129–134 (2013). (in Japanese)
14. Fujisaki, K.: IEEJ Trans. Fundam. Mater. **133**(11), 585–590 (2013) (in Japanese)
15. Investigating R&D Committee of Magnetic Material for Electromagnetic Actuator, "Magnetic Material and Its Evaluation Technology for Electromagnetic Actuator," Technical report of IEEJ, No. 1397, 11 (2017) (in Japanese)
16. Fujisaki, K.: Polycrystalline magnetic field analysis of electrical steel for magnetic multi-scale. In: The 42nd Annual Conference on MAGNTICS in Japan, 13pC-2, 2018.9

Magnetic Hysteresis Represented by Play Model

Tetsuji Matsuo

Abstract The play model is a simple and efficient hysteresis model that is mathematically equivalent to the classical Preisach model. The play model requires only one-dimensional integral, whereas the Preisach model is described by a two-dimensional integral basically. The relation between the Preisach distribution function and the shape function of the play model is explained. Based on the Everett integral for the Preisach model, a method for identification of the play model is presented, where the shape function is determined from given symmetric hysteresis loops. For the application to the magnetic field analysis using the magnetic vector potential, B-input play model is also explained. Isotropic and anisotropic vector play models are presented, which are required for 2D and 3D magnetic field analyses.

Keywords Identification · Play model · Preisach model · Silicon steel · Vector hysteresis

1 Introduction

In order to represent hysteretic properties of magnetic materials, the Preisach model [1, 2], the Jiles–Atherton (JA) model [3], and the play model [4, 5] are often used. The Preisach model has higher representation capability than the JA model and, however, requires a higher computational cost. The play model requires smaller computational cost than the Preisach model because of its simple mathematical description and has an equivalent capability to the Preisach model. Since the vectorization of the play model is also simple, it can be applied to the finite element magnetic field analysis.

T. Matsuo (✉)
Kyoto University, Kyoto, Japan
e-mail: matsuo.tetsuji.5u@kyoto-u.ac.jp

K. Fujisaki (ed.), *Magnetic Material for Motor Drive Systems*,
Engineering Materials, https://doi.org/10.1007/978-981-32-9906-1_13

2 Play Model

Figure 13.1a illustrates the property of play hysteron. The branch $\Gamma_R : p_\zeta = X - \zeta$ is used only when the input X is increasing, whereas the branch $\Gamma_L : p_\zeta = X + \zeta$ is used only when X is decreasing. Therein, ζ is a nonnegative parameter giving the width of play hysteron. When X turns from increasing/decreasing to decreasing/increasing, the point (X, p_ζ) moves horizontally between Γ_R and Γ_L. The point can move both right and left on the horizontal links. The property of play hysteron is represented as

$$p_\zeta(X) = \max(\min(p_\zeta^0, X + \zeta), X - \zeta)), \tag{13.1}$$

where p_ζ^0 is the value p_ζ at the previous timepoint.

The play model describes the hysteretic relation between the input X and the output Y by summing up play hysterons with various ζ $(= \zeta_1, \zeta_2, \ldots)$ (Fig. 13.1b) as

$$Y = P(X) = \sum_{n=1}^{Np} f(\zeta_n, p_{\zeta n}(X)), \tag{13.2}$$

where N_p is the number of play hysterons, $p_{\zeta_n}(n = 1, \ldots, N_p)$ are play hysterons having width ζ_n, and f is a single-valued function of ζ and p_ζ; f is called shape function.

As illustrated in Fig. 13.1a, the play hysteron yields a counterclockwise hysteresis loop. Accordingly, it is natural for the play model to represent the hysteretic property from the input of the magnetic field H to the output of the magnetic flux density B or the magnetization M. When the shape functions f are continuous with respect to p_ζ, the function $P(X)$ is continuous. It is easy to differentiate Eq. (13.2) with respect to X, which implies that the play model (13.2) can be combined with the Newton method to be applied to magnetic field analyses.

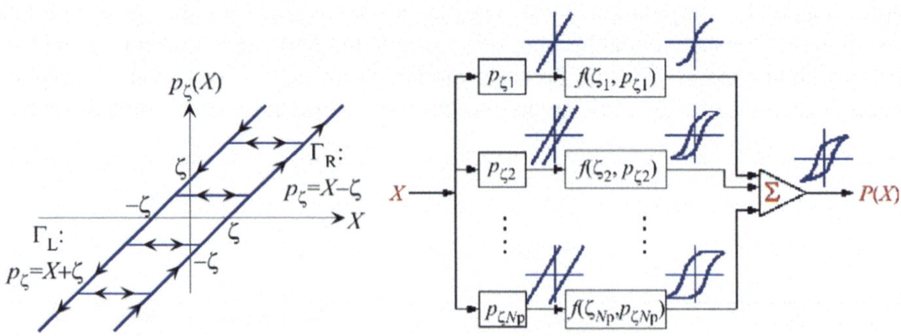

(a) play hysteron (b) superposition of play hysterons

Fig. 13.1 Play model

Suppose that $P(X)$ becomes a single-valued function when $|X| \geq X_S$ because of the saturation, where X_S is a positive constant. Then $f(\zeta_n, p_{\zeta n})$ with $\zeta_n \neq 0$ needs to saturate and be constant for $|p_{\zeta n}| \geq X_S - \zeta_n$ or $p_{\zeta n}$ needs to saturate and be constant as $p_{\zeta n} = \pm(X_S - \zeta_n)$. In this case, the domain for $f(\zeta_n, p_{\zeta n})$ with $0 < \zeta_n < X_S$ is given as

$$0 < \zeta_n < X_S, -X_S + \zeta_n \leq p_{\zeta_n} \leq X_S - \zeta_n. \tag{13.3}$$

Hysterons for $\zeta_n \geq X_S$ are unnecessary because $p_{\zeta n}$ can be set to 0. When $\zeta_n = 0$, $p_{\zeta n}(X) = X$ has no hysteretic property, which means shape function f for $\zeta_n = 0$ represents the reversible component with the domain $-\infty < p_{\zeta n} < \infty$. The reversible component generally has a different property from the irreversible component, which results in the discontinuity of $f(\zeta, p)$ at $\zeta = 0$.

3 Comparison with the Preisach Model

It is known that the play model is mathematically equivalent to the classical Preisach model. Consequently, it is possible to transform the Preisach distribution function into the shape functions and vice versa.

The Preisach model is represented by the superposition of dipoles whose property is shown in Fig. 13.2. If the dipole has the positive direction, it has an output of 1/2, whereas its output is −1/2 if its direction is negative. When the input X increases and reaches u, the dipole turns to the positive direction and its output jumps to 1/2 from −1/2. When X decreases to v, the output jumps to −1/2 from 1/2 by the reversal to the negative direction. The distribution of dipoles is defined by a function $K(u, v)$ called Preisach distribution function. The number of dipoles having positive and negative reversal points in the intervals $[u, u + du]$ and $[v, v + dv]$ is given by $K(u, v)dudv$. Using $K(u, v)$, the output of Preisach model is given as

$$Y(X) = \iint_{D_K} K(u, v) D_{uv}(X) dudv$$

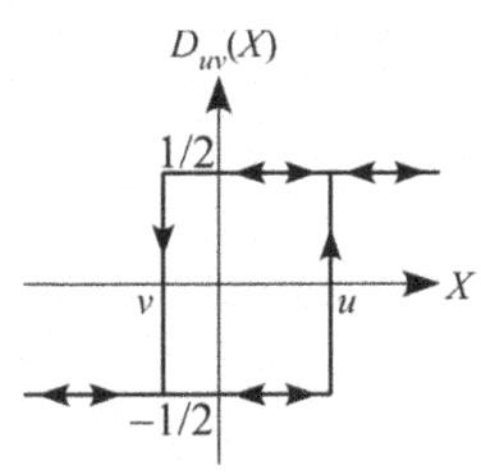

Fig. 13.2 Preisach operator

$$= \iint_{D_K} K(u, v) D'_{uv}(X) \mathrm{d}u \mathrm{d}v + Y(-X_S) \tag{13.4}$$

$$D'_{uv}(X) = D_{uv}(X) + 1/2 \tag{13.5}$$

$$Y(-X_S) = -\frac{1}{2} \iint_{D_K} K(u, v) \mathrm{d}u \mathrm{d}v \tag{13.6}$$

$$D_K : \ -X_S \le v \le u \le X_S \tag{13.7}$$

where D_K is the domain of K and the status of dipoles is represented by function D'_{uv} given by

$$D'_{uv}(X) = \begin{cases} 1 & (u \le X) \\ 0 & (X \le v) \\ \text{unchanged} & (v < X < u) \end{cases} . \tag{13.8}$$

For comparison, the play model (13.2) is rewritten in a continuous form with respect to ζ as

$$P(X) = \int_0^{X_S} f(\zeta, p_\zeta(X)) \, \mathrm{d}\zeta. \tag{13.9}$$

Using a partial differentiation of $f(\zeta, p)$ with respect to p as

$$\mu(\zeta, p) = \partial f(\zeta, p)/\partial p \tag{13.10}$$

in the domain (13.3), Eq. (13.9) is rewritten as

$$P(X) = \int_0^{X_S} \int_{-X_S+\zeta}^{p_\zeta(X)} \mu(\zeta, p) \, \mathrm{d}p \mathrm{d}\zeta + P(-X_S). \tag{13.11}$$

By setting

$$K(u, v) = \mu(\zeta, p)/2 \tag{13.12}$$

with the variable transformation

$$u = p + \zeta, v = p - \zeta \tag{13.13}$$

Equation (13.11) becomes equivalent to the Preisach model (13.4) having the distribution function (13.12) [4, 6].

The play model (13.9) is discretized with ζ by setting

$$\zeta_n = (n-1)X_{\mathrm{S}}/N_{\mathrm{p}}\big(n = 1, 2, \ldots, N_{\mathrm{p}}\big). \tag{13.14}$$

The shape functions are given in the form of piecewise linear functions as

$$f(\zeta_n, p) = f_{n,j-1} + \mu'(n, j)(p - p_{n,j-1}) \tag{13.15}$$

within $p_{n,j-1} \leq p \leq p_{n,j}$, where

$$p_{n,j} = 2jX_{\mathrm{S}}/N_{\mathrm{p}} - X_{\mathrm{S}} + \zeta_n = j\Delta p - X_{\mathrm{S}} + \zeta_n$$

$$(j = 0, \ldots, N_{\mathrm{p}} - n + 1),\ \Delta p = 2X_{\mathrm{S}}/N_{\mathrm{p}} \tag{13.16}$$

and

$$f_{n,j} = f(\zeta_n, p_{n,j}) \tag{13.17}$$

$$\mu'(n, j) = (f_{n,j} - f_{n,j-1})/\Delta p. \tag{13.18}$$

By defining a distribution function table,

$$K'(i, j) = \int\limits_{X_j-1}^{X_j} \int\limits_{X_i-1}^{X_i} K(u, v)\mathrm{d}u\mathrm{d}v\ \big(1 \leq j \leq i \leq N_{\mathrm{p}}\big) \tag{13.19}$$

$$X_i = 2iX_{\mathrm{S}}/N_{\mathrm{p}} - X_{\mathrm{S}}\big(i = 0, \ldots, N_{\mathrm{p}}\big) \tag{13.20}$$

a discretized Preisach model is given as

$$Y(X) = \sum_{i=1}^{N_{\mathrm{p}}} \sum_{j=1}^{i} K'(i, j)D'(i, j) + Y_{\mathrm{min}} \tag{13.21}$$

$$D'(i, j) = \begin{cases} 1 & (X_i \leq X) \\ 0 & (X \leq X_{j-1}) \\ D^0(i, j) & (X_j < X < X_{i-1}) \\ \Delta D(X, i) & (X_{i-1} \leq X \leq X_i, i = j) \\ \max(\Delta D(X, i), D^0(i, j)) & (X_{i-1} \leq X \leq X_i, j < i) \\ \min(\Delta D(X, j), D^0(i, j)) & (X_{j-1} \leq X \leq X_j, j < i) \end{cases} \tag{13.22}$$

Fig. 13.3 Relation between distribution function table K' and slopes μ' of shape functions

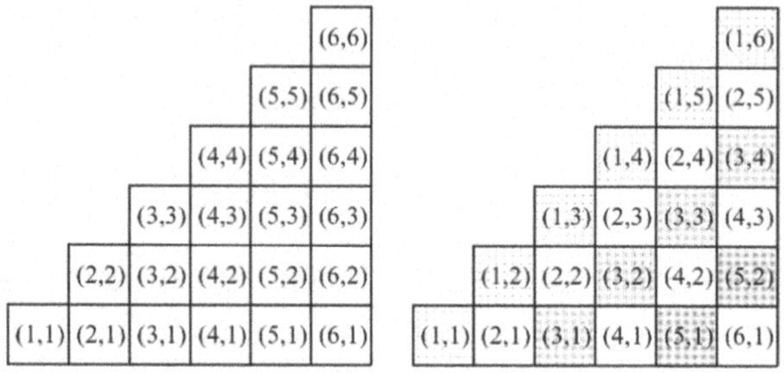

(a) $K'(i,j)$ (b) $\mu'(n,j)$

$$\Delta D(X,i) = \frac{X - X_{i-1}}{X_i - X_{i-1}}, Y_{\min} = -\frac{1}{2}\sum_{i=1}^{Np}\sum_{j=1}^{i} K'(i,j) \tag{13.23}$$

where $D^0(i,j)$ is the value of $D'(i,j)$ at the previous timepoint.

The relation between the slope of shape function of play model and the Preisach distribution function table is given as

$$\mu'(n,j)\Delta p = K'(n+j-1,j). \tag{13.24}$$

For example, when $N_p = 6$, $K'(i,j)$ in Fig. 13.3a corresponds to $\mu'(n,j)$ in Fig. 13.3b. Figure 13.3b shows that the play model stores the state of Preisach dipoles in the oblique direction, where the boundary between positive and negative states is represented by $p_{\zeta n}(X)$.

It is necessary for the Preisach model to store the state of diploes in a two-dimensional distribution $D'(i,j)$. In contrast, the play model only needs to store a one-dimensional array of $p_{\zeta n}$. Consequently, the play model has advantages in the memory cost and the simplicity of the algorithm. The simplicity of play model is also understood from the fact that the implementation of Eqs. (13.1), (13.2) is easy.

4 Identification of the Play Model

Since the scalar play model is equivalent to the Preisach model, the identification method for the Preisach model is applicable to the play model. First-order reversal curves (FORCs) [1] are often used to obtain the Everett integral for the identification of Preisach model. Symmetric loops are also used for the identification [5], which is useful to determine the shape functions of play model as follows.

Suppose that the number of play hysterons, N_p, is an even number. The play model is identified from $N_p/2$ symmetric loops having amplitudes of X as

$$A_m = 2mX_S/N_p \left(m = 1, \ldots, N_p/2\right). \tag{13.25}$$

The ascending and descending curves of the symmetric loop with amplitude A_m are denoted by (Y=) $h^+(A_m, X)$ and $h^-(A_m, X)$, respectively. Using the symmetric loops, the Everett integral $E(v, u)$ is defined in domain $-X_S \le v \le u \le X_S$ as

$$E(v,u) = \int_v^u \int_v^U K(U,V)\mathrm{d}V\mathrm{d}U = \begin{cases} h^-(u,u) - h^-(u,v) & (u+v \ge 0) \\ h^+(|v|,u) - h^+(|v|,v) & (u+v \le 0) \end{cases}. \tag{13.26}$$

Corresponding to the second-order differential of Everett integral, $\mu'(n,j)$ is given as

$$\begin{aligned} \mu'(n,j)\Delta p &= K'(i,j) \\ &= E(X_{j-1}, X_i) - E(X_{j-1}, X_{i-1})\ i = n + j - 1 \\ &\quad - E\left(X_j, X_i\right) + E(X_j, X_{i-1}) \end{aligned} \tag{13.27}$$

where X_i is given by Eq. (13.20) and $E(X_i, X_{i-1}) = 0$.

5 B-Input Play Model

As mentioned before, it is natural for the play model to represent a hysteretic function having input H and output B or M because the play model has a phase delay of its output to input. However, the magnetic field analysis often requires a function giving output H from input B. This means that an inverse function of the play model is required. The inverse distribution function method [7] for the Preisach model enables the play model to provide output H from input B directly, which fortunately improves the representation accuracy of the magnetic property of silicon steel.

The inverse distribution function method for the Preisach model simply replaces the input–output pair of (H, B) by (B, H) and obtains the distribution function. As a result, the distribution function K and function table K' become negative except for the reversible component. Accordingly, the shape functions of the play model have negative slopes μ and μ'. The reversible components $\mu(0, p)$ and $\mu'(1, j)$ are positive and cancel the negative slopes in the irreversible components.

Figure 13.4 shows the simulated hysteretic property of non-oriented silicon steel sheet JIS: 50A290 given by the B-input play model. The play model with 40 hysterons is identified from 20 symmetric BH loops. BH loops are accurately represented.

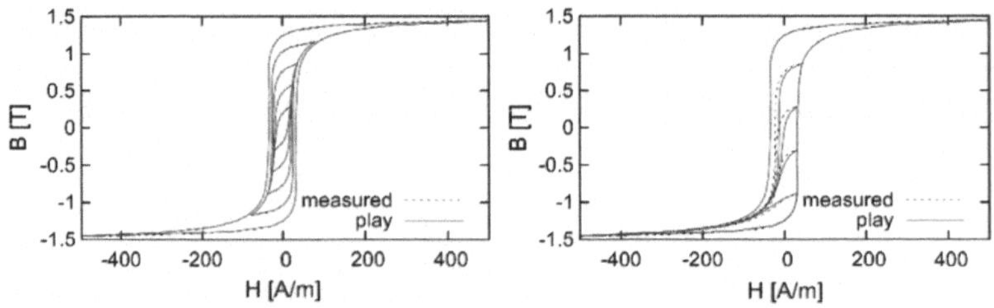

(a) symmetric loops (b) descending curves

Fig. 13.4 BH loops given by the play model

6 Vector Play Model

There are two vector versions of the play model: One is a superposition of scalar models [8], whereas the other is a geometric extension of the scalar model.

6.1 Superposition of Scalar Models

A basic version of vector play model using superposition is constructed as

$$\boldsymbol{Y} = \boldsymbol{P}(\boldsymbol{X}) = \int_{-\pi/2}^{\pi/2} \boldsymbol{e}_{\varphi} P_2(\boldsymbol{X} \cdot \boldsymbol{e}_{\varphi}) \mathrm{d}\varphi \tag{13.28}$$

where $\boldsymbol{e}_{\varphi}$ is the unit vector along the φ-direction, P_2 is a scalar play model and φ is the angle of input vector $\boldsymbol{X}$. This version is equivalent to the vector Preisach model proposed by Mayergoyz [1], whose mathematical properties are known including the identification method of P_2. This model has a weakness to be a magnetic hysteresis model because its rotational hysteresis loss saturates and becomes constant for large $\boldsymbol{X}$. The rotational hysteresis loss of real magnetic material decreases with the increase in the amplitude of the magnetic field near the saturation. This is because the single-domain state does not arise domain wall pinning. Several generalized models have been proposed to overcome this drawback. However, the generalized models still have weakness; they are not accurate enough to represent silicon steel [8] even though the computational cost for the superposition is high.

6.2 Geometric Extension of the Scalar Model

The play hysteron is vectorized as

$$\boldsymbol{p}_\zeta(\boldsymbol{X}) = \boldsymbol{X} - \frac{\zeta(\boldsymbol{X} - \boldsymbol{p}_\zeta^0)}{\max(\zeta, |\boldsymbol{X} - \boldsymbol{p}_\zeta^0|)} \tag{13.29}$$

where ζ is a nonnegative constant giving the radius of vector play hysteron and $\boldsymbol{p}_\zeta^0$ is the value of $\boldsymbol{p}_\zeta$ at previous timepoint. The geometrical meaning of the vector play hysteron is illustrated in Fig. 13.5. A point X positioned by vector $\boldsymbol{X}$ is located in a circle of which the radius and center are ζ and P positioned by vector $\boldsymbol{p}_\zeta$, respectively. When X moves inside the circle, P does not move (Fig. 13.5a). When X reaches the circumference of the circle, the circle and its center P move together with X (Fig. 13.5b).

Using the vector play hysteron above, a vector play model [8, 9] is constructed as

$$\boldsymbol{Y} = \boldsymbol{P}(\boldsymbol{X}) = \int_0^{X\mathrm{s}} \boldsymbol{f}(\zeta, \boldsymbol{p}_\zeta(\boldsymbol{X}))\mathrm{d}\zeta \tag{13.30}$$

$$\boldsymbol{f}(\zeta, \boldsymbol{p}) = f(\zeta, |\boldsymbol{p}|)\frac{\boldsymbol{p}}{|\boldsymbol{p}|} \tag{13.31}$$

where $f(\zeta, p)$ is the shape function. For unidirectional inputs, the vector model (13.30) is reduced to the scalar model. This means that f can be determined only from unidirectional measured data.

The dashed line in Fig. 13.6a shows the simulated rotational hysteresis loss per cycle of non-oriented silicon steel sheet, JIS: 35A230, where the input and output of vector play model are B and H vectors. The shape function f was identified from the

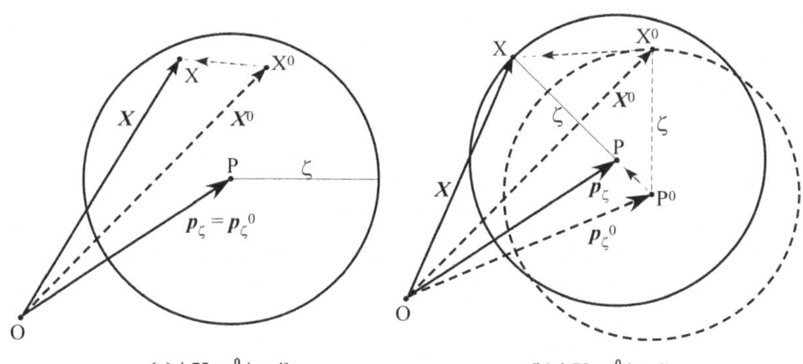

(a) $|\boldsymbol{X}-\boldsymbol{p}_\zeta^0| < \zeta$ (b) $|\boldsymbol{X}-\boldsymbol{p}_\zeta^0| \geq \zeta$

Fig. 13.5 Vector play hysteron

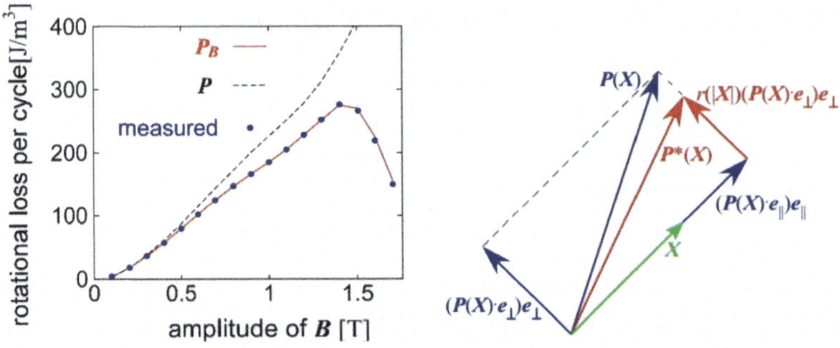

(a) rotational hysteresis loss (b) correction of rotational hysteresis loss

Fig. 13.6 Rotational hysteresis losses per cycle given by isotropic vector play models

averaged unidirectional properties. The simulated loss is larger than the measured value because the vector model is identified only from the alternating property.

To correct the rotational hysteresis loss, the vector play model is modified as

$$P^*(X) = (P(X) \cdot e_{||})e_{||} + r(|X|)(P(X) \cdot e\perp)e\perp \tag{13.32}$$

where $e_{||}$ and $e_{\perp}$, respectively, denote parallel and perpendicular unit vectors to X and $r(X)$ is given as

$$r(|X|) = \frac{L_{\text{rot}}^{\text{mea}}(|X|)}{L_{\text{rot}}^{\text{sim}}(|X|)} \tag{13.33}$$

where $L_{\text{rot}}^{\text{sim}}$ and $L_{\text{rot}}^{\text{mea}}$ are the simulated rotational loss given by P with amplitude $|X|$ and the measured loss. The relation between P and P^* is presented in Fig. 13.6b. This modification does not affect the unidirectional properties of P because $e_{\perp}$ becomes zero.

The solid line in Fig. 13.6a shows the rotational hysteresis loss given by P^*, where the modified play models obtain accurate rotational loss.

6.3 Anisotropic Vector Play Model

By multiplying a matrix representing anisotropy by the isotropic model (13.32), a simple anisotropic version [10, 11] is given as

$$P_X(X) = W_X(X)P^*(X) \tag{13.34}$$

where W_X is an anisotropy matrix of which components are single-valued functions of X. $W_X(X)$ is determined to reconstruct the anisotropic alternating property approximately from the averaged alternating property as

$$W_X(X) = \text{diag}\left(\frac{Y_{\text{alt}x}(X_a, \varphi_X)}{P_{\text{ave}}(X_a)\cos\varphi_X}, \frac{Y_{\text{alt}y}(X_a, \varphi_X)}{P_{\text{ave}}(X_a)\sin\varphi_X} \right) \tag{13.35}$$

(see Fig. 13.7). Therein $(Y_{\text{alt}x}(X_a, \varphi_X), Y_{\text{alt}y}(X_a, \varphi_X))$ is the measured vector $\boldsymbol{Y}$ for amplitude X_a with fixed angle φ_X of $\boldsymbol{X}$. $P_{\text{ave}}(X)$ is the azimuthally averaged alternating property that is represented by the isotropic vector play model (13.32). This model assumes that the anisotropy depends on $\boldsymbol{X}$, which can be extended to have dependence on both $\boldsymbol{X}$ and $\boldsymbol{P}$.

Figure 13.8 portrays the loci of $\boldsymbol{H}$ of non-oriented silicon steel sheet, JIS: 35A230, which is stimulated by the anisotropic vector play model for counterclockwise rotational inputs of $\boldsymbol{B}$. The anisotropy matrix was determined from the unidirectional

Fig. 13.7 Construction of anisotropy matrix

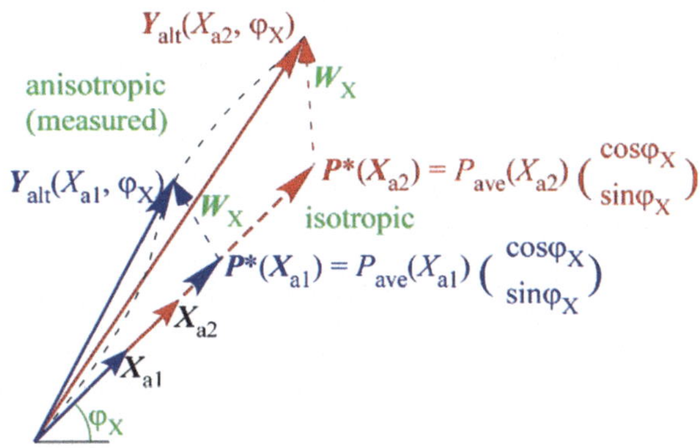

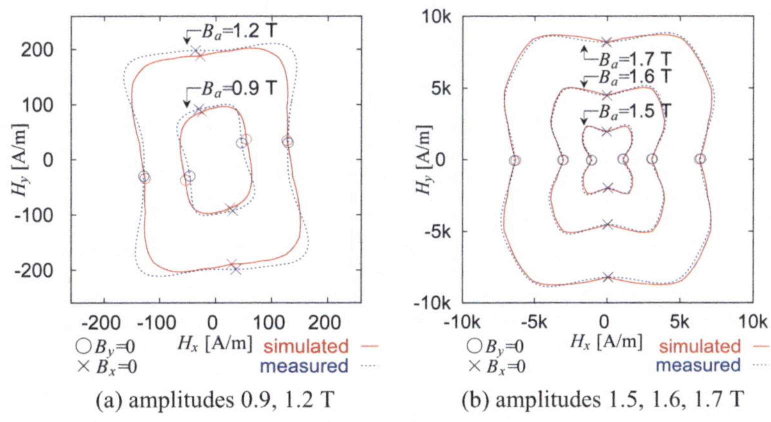

Fig. 13.8 Simulated loci of $\boldsymbol{H}$ for rotational inputs of $\boldsymbol{B}$, where $B_x = 0$ or $B_y = 0$ at "o" (simulated) and "×," respectively

properties. The anisotropy model (13.34) approximately agrees with the measured loci.

The representation (13.34) using the anisotropy matrix is a simple approximation. By using an angular-dependent shape function [12] in (13.30), a more accurate vector play model can be constituted.

This chapter presents static hysteresis models that do not depend on the time variation of input. For example, the silicon steel presents dynamic hysteretic property under AC excitation because of the eddy current. Dynamic hysteresis models [13] can be constructed by combining the method for modeling eddy current field, presented in Chap. 15.

References

1. Okumura, K., Kishima, A.: Digital simulation of hysteretic characteristics and its application. Trans. IEE Japan (B) **103-B**(7), 451–458 (1983). (in Japanese), English translation is available in Electrical Engineering in Japan, **103**(4), 32–39 (1983)
2. Mayergoyz, I.: Mathematical Models of Hysteresis and their Applications. Elsevier, Academic Press (2003)
3. Jiles, D.C., Atherton, D.L.: Theory of ferromagnetic hysteresis. J. Magn. Magn. Mater. **61**, 48–60 (1986)
4. Bobbio, S., Miano, G., Serpico, C., Visone, C.: Models of magnetic hysteresis based on play and stop hysterons. IEEE Trans. Magn. **33**(6), 4417–4426 (1997)
5. Matsuo, T., Shimasaki, M.: An identification method of play model with input-dependent shape function. IEEE Trans. Magn. **41**(10), 3112–3114 (2005)
6. Brokate, M.: Some mathematical properties of the Preisach model for hysteresis. IEEE Trans. Magn. **25**(4), 2922–2924 (1989)
7. Takahashi, N., Miyabara, S., Fujiwara, K.: Problems in practical finite element analysis using Preisach hysteresis model. IEEE Trans. Magn. **35**(3), 1243–1246 (1999)
8. Matsuo, T., Shimasaki, M.: Two types of isotropic vector play models and their rotational hysteresis losses. IEEE Trans. Magn. **44**(6), 898–901 (2008)
9. Bergqvist, A.: Magnetic vector hysteresis model with dry friction-like pinning. Phys. B **233**, 342–347 (1997)
10. Matsuo, T.: Anisotropic vector hysteresis model using an isotropic vector play model. IEEE Trans. Magn. **46**(8), 3041–3044 (2010)
11. Matsuo, T., Miyamoto, M.: Dynamic and anisotropic vector hysteresis model based on isotropic vector play model for non-oriented silicon steel sheet. IEEE Trans. Magn. **48**(2), 215–218 (2012)
12. Fujinaga, T., Mifune, T., Matsuo, T.: Anisotropic vector play model incorporating decomposed shape functions. IEEE Trans. Magn. **52**(3), 7300604 (2016)
13. Shindo, Y., Miyazaki, T., Matsuo, T.: Cauer circuit representation of the homogenized eddy-current field based on the Legendre expansion for a magnetic sheet. IEEE Trans. Magn. **52**(3), 6300504 (2016)

From a Thermodynamic Model to a Magnetic Hysteresis Model

Fumiaki Ikeda

Abstract The magnetic characteristics of magnetic materials are determined by various factors such as the type and arrangement of atoms comprising the microstructure, that is, the crystalline structure. This model uses a coarse-grained modeling approach to capture the relationship of this microscale phenomenon as it relates to the macroscale behavior of magnetic materials. One of the methods of coarse-grained modeling is defined according to state variables having no correlation with the history of the material such as temperature and pressure, taking into account thermodynamic state variables such as energy and entropy. However, the phenomenon of hysteresis that appears in magnetism is clearly dependent on the history of the material; therefore, there are various difficulties involved in this sort of thermodynamic approach. At the heart of the phenomenon of hysteresis exists a state variable called free energy, which can be described according to thermodynamics, but it is obfuscated by a phenomenon resembling macroscale friction. Therefore, here we introduce a model to describe the resulting historically dependent magnctic properties.

Keywords Thermodynamics of magnetic materials · Free energy · Hysteresis

1 The Thermodynamics of Magnetic Materials

The magnetic behavior of magnetic materials is complex, due to it involving the relationship of two vector quantities, the magnetic field and the magnetic flux density. Further, these magnetic properties are generally affected by hysteresis, which is dependent on the history of the material. Because the consideration of these magnetic properties is a necessary part of magnetic field analysis, this is currently a topic of much study.

First, assuming that the magnetic material is isotropic, the magnetic field **H** and the magnetic flux density **B** have the same orientation and are written as follows:

$$\mathbf{B} = \mu \mathbf{H} \tag{1}$$

F. Ikeda (✉)
PHOTON Co., Ltd, Kyoto, Japan
e-mail: ikeda@photon-cae.co.jp

K. Fujisaki (ed.), *Magnetic Material for Motor Drive Systems*,
Engineering Materials, https://doi.org/10.1007/978-981-32-9906-1_14

Here, the constant of proportionality μ represents the magnetic permeability and varies generally according to the strength of the magnetic field. In this case, because the magnetic field and the magnetic flux density are always oriented in the same direction, as long as the magnitudes of the magnetic permeability, magnetic field, magnetic flux density, etc., are known, this relationship is relatively manageable.

Next, when handling an anisotropic magnetic material, Eq. (1) is expanded as follows to express the components of the magnetic field and magnetic flux density:

$$\begin{bmatrix} B_x \\ B_y \\ B_z \end{bmatrix} = \begin{bmatrix} \mu_{xx} & \mu_{xy} & \mu_{xz} \\ \mu_{yx} & \mu_{yy} & \mu_{yz} \\ \mu_{zx} & \mu_{zy} & \mu_{zz} \end{bmatrix} \begin{bmatrix} H_x \\ H_y \\ H_z \end{bmatrix} \tag{2}$$

In this expression, unlike in Eq. (1), the magnetic permeability is a tensor quantity represented by a matrix of three columns and three rows.

Generally, because the magnetic permeability tensor components vary according to the strength of the magnetic field, considering this number of components in the relationship is necessary for an anisotropic magnetic material. In addition, in the case of hysteresis, each component is dependent on its own respective history, and expressing the magnetic characteristics of the material becomes very difficult.

However, it is not necessarily the case that the components of magnetic permeability are all truly independent variables. For example, if the magnetic permeability tensor is isotropic, the independent variables reduce from nine to six, and according to the coordinate axis that is chosen, it is possible to reduce the number of independent variables further to three. Consequently, under these conditions, it becomes fairly simple to describe an anisotropic magnetic material.

According to the laws of thermodynamics, when the magnetic field is relatively small such that the component of magnetic permeability is constant, the magnetic permeability tensor can be generally represented as isotropic. Accordingly, we make use of the thermodynamics of magnetic materials to explore here what can generally be said regarding magnetic behavior.

When electric current is run through a coil-shaped magnetic material, assuming there is no loss of energy from the source of the electricity, magnetic energy is accumulated in the material and in the surrounding space. If the magnetic flux density is varied by changing the electric current, the magnetic energy varies in proportion to that change. Representing the internal energy contained within a unit volume of the magnetic material as U, this change is expressed according to the variation in magnetic flux density as follows:

$$dU = \mathbf{H} \cdot d\mathbf{B} \tag{3}$$

As this expression is derived generally using Maxwell's equations in matter, it is independent of the specific magnetic properties of the given material. Because there is typically an exchange of heat between a magnetic body and its surroundings, it is necessary to introduce the influx quantity of heat $d'Q$ on the right side of the equation.

$$dU = d'Q + \mathbf{H} \cdot d\mathbf{B} \tag{4}$$

If the change is adiabatic, the energy of the magnetic body can be expressed as follows by integrating Eq. (3):

$$U(\mathbf{B}_2) - U(\mathbf{B}_1) = \int_{\mathbf{B}_1}^{\mathbf{B}_2} \mathbf{H} \cdot d\mathbf{B} \tag{5}$$

Because energy is a state quantity, it is dependent only on the initial magnetic flux density $\mathbf{B}_1$ and the final magnetic flux density $\mathbf{B}_2$, not on the integral path shown on the right side of the expression. That is, Eq. (3) is an exact differential equation and can be written as follows:

$$\begin{aligned} dU &= \frac{\partial U}{\partial B_x} dB_x + \frac{\partial U}{\partial B_y} dB_y + \frac{\partial U}{\partial B_z} dB_z \\ dU &= \frac{\partial U}{\partial \mathbf{B}} \cdot d\mathbf{B} \end{aligned} \tag{6}$$

This relationship can be expressed with the following notation.

$$\frac{\partial U}{\partial \mathbf{B}} = \left(\frac{\partial U}{\partial B_x}, \frac{\partial U}{\partial B_y}, \frac{\partial U}{\partial B_z} \right) \tag{7}$$

Therefore, in an adiabatic process, the following relationship holds between energy and the magnetic field.

$$\mathbf{H} = \frac{\partial U}{\partial \mathbf{B}} \tag{8}$$

This equation expresses the highly critical fact that if the energy of a magnetic material is given as a function of the magnetic flux density, then the magnetic field component can be obtained by differentiating that function by the magnetic flux density component. That is, the relationship between the two vector quantities magnetic field and magnetic flux density can be represented by one scalar quantity, energy.

Therefore, the scalar quantity of the energy of the magnetic material can be adequately described as a function of the magnetic flux density, regardless of the relationship between the two vector quantities. However, this conclusion is only valid for a magnetic material held in an adiabatic state.

Returning to the general expression Eq. (4), the first term on the right side of this equation can be written as follows for a quasi-static change:

$$d'Q = TdS \tag{9}$$

where T is the absolute temperature, and S is the entropy per unit volume of the magnetic material. Substituting this into Eq. (4), the following expression is obtained.

$$dU(S, \mathbf{B}) = TdS + \mathbf{H} \cdot d\mathbf{B} \tag{10}$$

Here, the left side shows that the energy of the magnetic body is a function of entropy and magnetic flux density. Here, the free energy of the magnetic material is introduced, as defined by the following equation.

$$F(T, \mathbf{B}) = U(S, \mathbf{B}) - TS \tag{11}$$

Free energy is a function of temperature and magnetic flux density, wherein internal energy is a function of entropy and magnetic flux density.
Differentiating this and employing Eq. (10) yields the following equation.

$$dF(T, \mathbf{B}) = -SdT + \mathbf{H} \cdot d\mathbf{B} \tag{12}$$

In the case of isothermal change, the first term on the right side disappears, and Eq. (8) is modified as follows:

$$\mathbf{H} = \frac{\partial}{\partial \mathbf{B}} F(T, \mathbf{B}) \tag{13}$$

However, in this differentiation, temperature must be constant. Further, if the magnetic flux density is taken to be constant and is differentiated according to temperature, the following equation is obtained.

$$S = -\frac{\partial}{\partial \mathbf{B}} F(T, \mathbf{B}) \tag{14}$$

Here, denoting the vector components with the subscripts i and j, we obtain the following expression from Eq. (13).

$$\frac{\partial H_j}{\partial B_i} = \frac{\partial}{\partial B_i} \frac{\partial F}{\partial B_j} = \frac{\partial}{\partial B_j} \frac{\partial F}{\partial B_i} = \frac{\partial H_i}{\partial B_j} \tag{15}$$

As a special case, if every component of the magnetic permeability tensor shown in Eq. (2) is constant, then this relationship shows that the magnetic permeability tensor is isotropic.

$$\mu_{ij} = \mu_{ji} \tag{16}$$

Thus, by introducing the free energy, which is a function of the temperature and magnetic flux density, while considering the thermodynamics of the magnetic material, it is possible to treat the magnetic properties as a scalar during isothermal change.

2 Free Energy and Hysteresis

Our perspective on the properties of magnetic materials is significantly improved by using free energy as the means of analysis rather than considering the relationship between vector quantities. Here, we examine in further detail how free energy relates to magnetic behavior.

Firstly, in the case that there is no magnetic field, the following relationship holds true.

$$\frac{\partial}{\partial \mathbf{B}} F(T, \mathbf{B}) = 0 \tag{17}$$

This relationship shows that if the temperature is constant, the free energy assumes an extreme value as the magnetic flux density varies. Consequently, if the free energy is taken to be a function of the magnetic flux density and is expressed as shown below,

$$F(\mathbf{B}) = aB^2 = a\left(B_x^2 + B_y^2 + B_z^2\right) \tag{18}$$

then we arrive at the following relationship in place of Eq. (17), and it is known that the free energy takes an extreme value when the magnetic flux density is zero.

$$\frac{\partial F}{\partial B_x} = 2B_x = 0 \qquad \frac{\partial F}{\partial B_y} = 2B_y = 0 \qquad \frac{\partial F}{\partial B_z} = 2B_z = 0 \tag{19}$$

This result is valid for a magnetic material having no hysteresis. However, if free energy is a function of the absolute value of the magnetic flux density **B** and is expressed as below,

$$F(\mathbf{B}) = aB^2\left(B^2 - 2b^2\right) \qquad (a > 0) \tag{20}$$

then the case where the magnetic field is zero is expressed as follows:

$$\frac{\partial F}{\partial B} = 4aB\left(B^2 - b^2\right) = 0 \tag{21}$$

From this expression, it is apparent that the extrema are found where the magnetic flux density is zero and at a spherical surface of radius b within the magnetic flux density space. Differentiating again, the function becomes negative where the magnetic flux density is zero, and this therefore constitutes the maximum; the minimum is found at the spherical surface of radius *b* because there the differentiated function becomes positive. That is, when the magnetic flux density is zero, the system is in an unstable state; whereas, it is stable at the spherical surface. This conclusion indicates that even if the value of the magnetic field is zero, the magnetic flux density is a nonzero, finite value.

In the case of a ferromagnetic material, even if a magnetic field bearing residual magnetic flux density is zero, the magnetic flux density is not zero. The minimum value of the free energy of a magnetic material in this state is, as depicted in this example, not at the origin of the magnetic flux density space, but rather is defined by the magnitude of the residual magnetic flux density. When a magnetic material exists in such a state, it is said to exhibit spontaneous magnetization. Since it is possible to describe spontaneous magnetization by considering the free energy of a magnetic material in this manner, we posit that the phenomenon of hysteresis can be understood in a similar fashion.

It is known that a ferromagnetic material is composed of regions called magnetic domains within which magnetization is aligned in one direction, and at the border of each of these magnetic domains exists on a region boundary called the magnetic domain wall that separates it from adjacent domains bearing differently oriented magnetization.

In the absence of magnetism, the average of the magnetization of these magnetic domains is zero; thus, the material as a whole is not magnetized. When an external magnetic field is applied, the magnetic domains composing the ferromagnetic material that are oriented similarly to the external magnetic field grow larger, and the magnetic domains opposing the orientation of the external magnetic field become smaller. These changes occur due to the movement of the magnetic domain walls, and as a result, the magnetization of the material as a whole approaches alignment with the external magnetic field. Conversely, when the external magnetic field is reduced to zero from this state, the reverse of the aforementioned process occurs, and the material returns to its initial non-magnetized state as the average magnetization of the magnetic material as a whole becomes zero. According to this line of reasoning, a ferromagnetic material does not exhibit hysteresis.

However, in reality, ferromagnetic materials do exhibit hysteresis; therefore, there is a flaw somewhere in this reasoning. Indeed, the above discussion assumes that the ferromagnetic material is composed of an ideal crystal structure, in which case, the movement of the magnetic domain wall is reversible because the position of the magnetic domain wall is determined by the factors dependent on the crystal structure such as anisotropic energy, exchange energy and magnetostatic energy. In contrast, a real ferromagnetic material does not have a uniform crystal structure but rather is composed of many crystal grains, and the movement of the magnetic domain walls is hindered by lattice defects such as grain boundaries and impurities. Movement against these defects generates thermal energy, which is not a reversible change. Thus, this is the source of hysteresis.

It was initially thought that spontaneous magnetization induced by free energy was the cause of hysteresis, but this is the cause of uniform magnetization within a

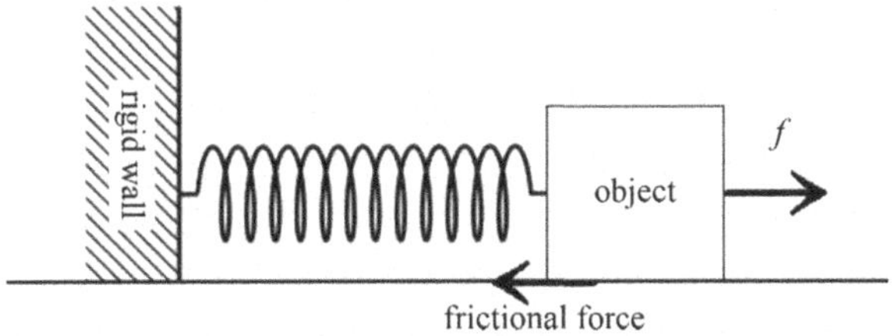

Fig. 1 Spring affixed to a rigid wall

single magnetic domain. The cause of macroscale hysteresis in a magnetic material is an irreversible change occurring across many magnetic domains.

3 A Friction Model of Hysteresis

Given that hysteresis in a ferromagnetic material is due to the obstruction of the movement of magnetic domain walls by crystal grain boundaries and lattice defects, and given that hysteresis does not occur for an ideal crystal structure, hysteresis represents a phenomenon similar to friction.

Consider an object connected to a spring as an example of friction (Fig. 1).

The other end of the spring is affixed to a rigid wall, and when force is applied to the object, the object is displaced up to the point at which that force balances the opposing force of the spring. Taking the displacement of the object as x and the spring constant as k, the force f applied to the object is as follows:

$$f = kx \tag{22}$$

Here, the elastic energy of the spring is expressed as below,

$$E = \frac{1}{2}kx^2 \tag{23}$$

and Eq. (22) can be rewritten as follows:

$$f = \frac{\partial E}{\partial x} \tag{24}$$

This equation is comparable to equations such as (8) and (13) describing the magnetic field and magnetic flux density: Force corresponds to the magnetic field and displacement corresponds to magnetic flux density in an analogous relationship to energy. Parenthetically, because energy and free energy are equivalent when the spring constant is independent of temperature, this expression represents the total energy.

When the relationship between force and displacement is linear in this manner, there is naturally no historical dependence. However, when the force of friction f_C

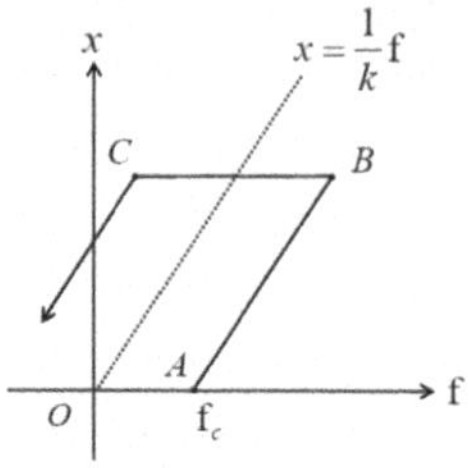

Fig. 2 Historically dependent relationship between force and displacement

is nonzero, the displacement of an initially static object remains zero as long as the force applied to the object does not exceed the frictional force. When the force applied exceeds the frictional force, the displacement increases as follows:

$$f - f_C = kx \tag{25}$$

When that applied force is then decreased, the displacement does not change until the system is as described below:

$$f + f_C = kx \tag{26}$$

Thus, the relationship between force and displacement becomes historically dependent, as depicted in Fig. 2.

Consequently, Eq. (24) can be modified as follows for the case of friction:

$$f = \frac{\partial F(x)}{\partial x} + f_S \tag{27}$$

for the following conditions.

$$\begin{aligned} f_S &= f_C & f &> f_C \\ f_S &= f & |f| &\le |f_C| \\ f_S &= -f_C & f &< -f_C \end{aligned} \tag{28}$$

We utilize this spring model of friction as a model for hysteresis in a magnetic material. Thus, in light of Eqs. (26), (13) is rewritten as follows:

$$\mathbf{H} = \frac{\partial F(\mathbf{B})}{\partial \mathbf{B}} + \mathbf{H}_S \tag{29}$$

where the following conditions apply, and where $\mathbf{H}_S$ represents the magnetic field analog of frictional force, which we call the hysteresis magnetic field.

$$\begin{aligned} \mathbf{H}_S &= H_C \frac{\mathbf{H}}{|\mathbf{H}|} && |\mathbf{H}| > H_C \\ \mathbf{H}_S &= \mathbf{H} && |\mathbf{H}| \leq H_C \end{aligned} \tag{30}$$

Upon solving Eq. (29) via the finite element method, rotating both sides, and slightly rearranging the terms, the expression is written as below:

$$rot\left\{\frac{\partial F(\mathbf{B})}{\partial \mathbf{B}} + \mathbf{H}_S\right\} = rot\mathbf{H} = \mathbf{J} \tag{31}$$

where although the magnetic field $\mathbf{H}$ is a controlled variable, it cannot be directly determined but is controlled according to the forced current $\mathbf{J}$. Therefore, Eq. (30) is solved through iteration as follows.

Iterating n times, the following expression is obtained for the magnetic flux density $\mathbf{B}^{(n)}$ and the hysteresis magnetic field $\mathbf{H}_S^{(n-1)}$.

$$\mathbf{H}^{(n)} = \frac{\partial F}{\partial \mathbf{B}^{(n)}} + \mathbf{H}_S^{(n-1)} \tag{32}$$

Solving until it converges yields the new hysteresis magnetic field.

$$\begin{aligned} \mathbf{H}_S^{(n)} &= H_C \frac{\mathbf{H}^{(n)}}{\left|\mathbf{H}^{(n)}\right|} && \left|\mathbf{H}^{(n)}\right| > H_C \\ \mathbf{H}_S^{(n)} &= \mathbf{H}^{(n)} && \left|\mathbf{H}^{(n)}\right| \leq H_C \end{aligned} \tag{33}$$

Generally, the magnetic field H_C introduced here varies according to the magnetic flux density.

$$H_C = H_C(\mathbf{B}) \tag{34}$$

When the magnetic flux density is zero, this value coincides with the coercivity; therefore, this is referred to as the coercivity function. Since this depends on the orientation of the magnetic flux density, it is necessary to consider the proper function of magnetic flux density, rather than the absolute value function. An example of the variation of the magnetic field is depicted in Fig. 3.

In Fig. 3, the lines of the initial magnetization curve and minor hysteresis loop are parallel to the axis representing the magnetic field, which is unrepresentative of a real magnetization curve. This is due to the fact that a ferromagnetic material is composed of many magnetic domains, and the coercivity corresponding to the frictional forces at each respective magnetic domain wall differs in value. It is thought that the macroscopic coercivity is the average of these individual coercivity values. That is, even if a magnetic field which is smaller than the macroscopic coercivity is applied, at the magnetic domain scale, magnetization proceeds to those magnetic

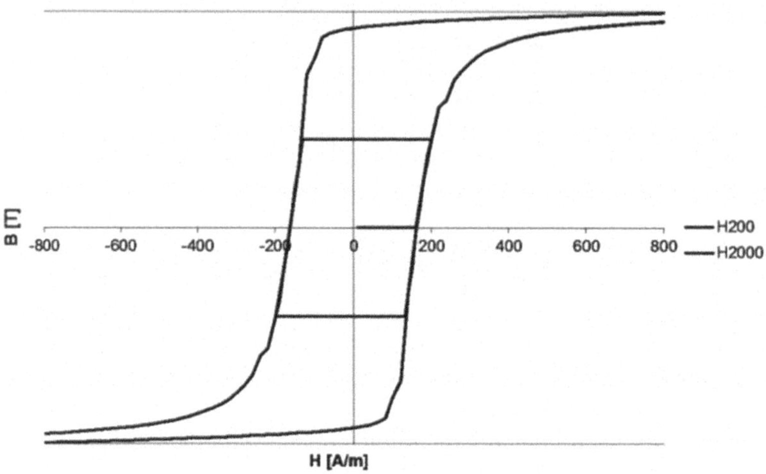

Fig. 3 Magnetic field and magnetic flux density

domain walls having comparatively smaller coercivity. Therefore, the magnetic flux density is not held at zero but rather increases along the initial magnetization curve.

In the finite element method, magnetization is evaluated on an element-by-element basis, but the scale of magnetization is considerably larger than a typical magnetic domain and is effectively macroscale, as is the coercivity. Thus, considering a magnetic material as an aggregate of magnetic bodies bearing disparate coercivities, where each of those respective magnetic bodies responds independently to the magnetic field, the magnetic flux density is an average thereof.

To illustrate via the spring example, it is as if many springs are each connected to objects with differing frictional forces, and the same external force is applied to all. In such a system, the objects with the smallest frictional forces are displaced first, and as the applied force increases, the other objects will begin to move as that force exceeds their respective frictional forces. Thus, the relationship between the force applied to the objects and the average of the objects' displacement can be obtained.

Accordingly, Eq. (33) is modified as below upon the introduction of parameter α

$$
\begin{aligned}
\mathbf{H}_S &= H_C(\mathbf{B})\frac{\mathbf{H}}{|\mathbf{H}|} \qquad & |\mathbf{H}| > H_C(\mathbf{B}) \\
\mathbf{H}_S &= \alpha\mathbf{H} \qquad & |\mathbf{H}| \le H_C(\mathbf{B})
\end{aligned}
\tag{35}
$$

for the following condition.

$$0 < \alpha < 1 \tag{36}$$

A magnetic material is considered thusly to be a conglomeration of any number of component magnetic bodies having differing values of parameter α, and the magnetic

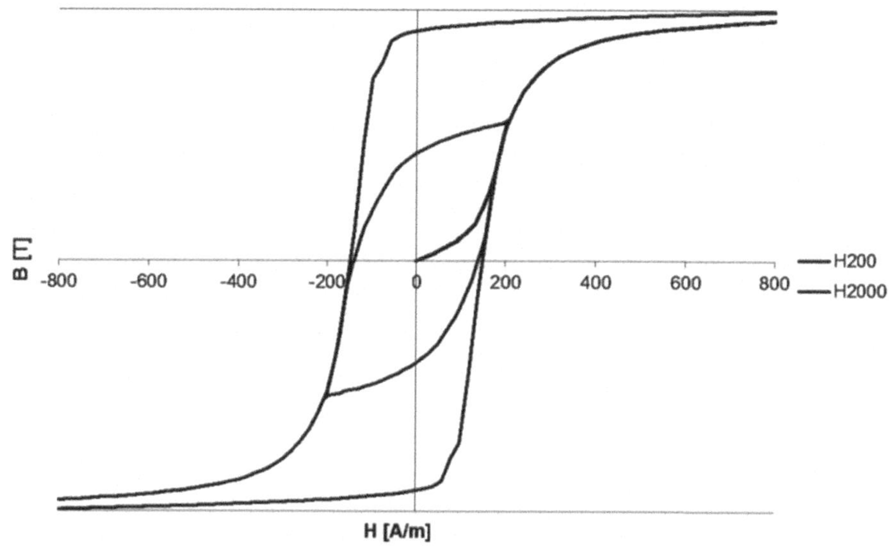

Fig. 4 Effect of the introduction of parameter α

field $\mathbf{H}_S$ is determined as an average thereof. An example of the magnetic field varying in this manner is displayed in Fig. 4.

By introducing parameter α in this manner, we can see that hysteresis as exhibited by a realistic magnetic material can be reproduced.

Finally, the acquisition of the magnetic material data required for use of this model is as follows. First, in order to obtain the free energy of the magnetic material, the measured value of the major hysteresis loop in the direction orthogonal to the easy axis of magnetization is used. Because these curves depend on the hysteresis magnetic field (analogous to frictional force), the magnetization curve is obtained as the average of the ascending and descending magnetic field, and the free energy in this direction can be determined.

Depicting this free energy on a contour plot, it forms an ellipsoid drawn about the easy axis of magnetization in the magnetic flux density space. Further, the coercivity function is obtained from the difference between the ascending and descending magnetization curves. Next, parameter α is determined such that the actual measurement is possible using the initial magnetization curve. In this manner, a major feature of this model is that little data is required of the magnetic material.

Equivalent Circuit of AC Magnetic Fields

Yuji Shindo

Abstract An effective equivalent circuit method suitable for high-frequency usage is explained in this chapter. This equivalent circuit is based on the ladder network originally proposed by Wilhelm Cauer (1900–1945) in the field of network synthesis. This method enables significant reduction of computations and also enables coupled analysis of magnetic field and electric circuit. Models for the magnetic sheet and magnetic wire with cylindrical shape are presented. Moreover, the universal method for modeling of arbitral eddy-current fields is introduced.

Keywords Eddy current · Complex permeability · Continued fraction · Ladder network

1 Eddy Current and Complex Permeability

The transient analysis and frequency analysis of magnetic fields become more important as the power electronics devices become widespread. There, the accurate formulation of the eddy current in magnetic materials is required. As illustrated in Fig. 1, the AC magnetic field applied to the laminated magnetic sheets yield the eddy current in the sheets, and this eddy current opposes the flux change. Then, the changes of flux in the sheets are lagged and weaken. Thus, the magnetic permeability in the AC magnetic field is usually denoted by the complex expression as $\dot{\mu} = \mu' - \mathrm{j}\mu''$. The negative sign for the imaginary part is preferred to express the flux delay. The actual complex permeability of the nanocrystalline soft magnetic material is shown in Fig. 2, as an example.

The eddy-current field is described by Eq. (1). Here, the magnetic field and flux density are denoted by H and **B**, respectively, while **E** and **J** denote the electric field and current density. The symbols μ and σ denote the DC magnetic permeability and electric conductivity defining the relations $\mathbf{B} = \mu \mathrm{H}$ and $\mathbf{J} = \sigma \mathbf{E}$. The equations for eddy-current filed (1) are derived from Maxwell's equations by neglecting the electric flux and charge.

Y. Shindo (✉)
Kawasaki Heavy Industries, Ltd., Akashi, Japan
e-mail: shindo_yuji@khi.co.jp

K. Fujisaki (ed.), *Magnetic Material for Motor Drive Systems*,
Engineering Materials, https://doi.org/10.1007/978-981-32-9906-1_15

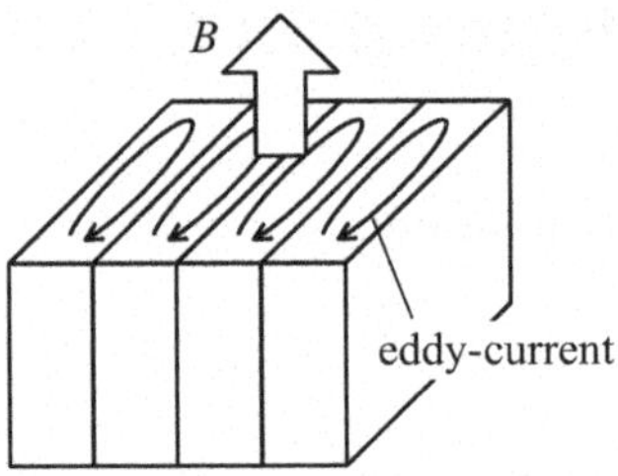

Fig. 1 Laminated magnetic sheets

$$\mathrm{rot}\,\mathbf{H} = \mathbf{J}, \quad \mathrm{rot}\,\mathbf{H} = -\partial \mathbf{B}/\partial t, \quad \mathrm{div}\,\mathbf{B} = 0, \quad \mathrm{div}\,\mathbf{J} = 0 \tag{1}$$

Let us consider the infinite magnetic sheet as Fig. 3. The formulation of this plate can be treated as one-dimensional problem as (2). Here, $\dot{B}$ and $\dot{H}$ have only y-directional component, while $\dot{E}$ and $\dot{J}$ have only z-directional component. The dot on the each variable implies that the variable is treated as a phasor. The time differentiation operator is denoted by $j\omega$. The AC uniform magnetic field $\dot{H}_0$ whose angular frequency is ω is applied externally as a boundary condition.

$$\frac{\partial \dot{E}}{\partial x} = j\omega\mu\dot{H}, \qquad \frac{\partial \dot{H}}{\partial x} = \sigma\dot{E} \tag{2}$$

Differentiating the both sides of the second equation by x and substituting the differential of the electric field with the first equation give the following.

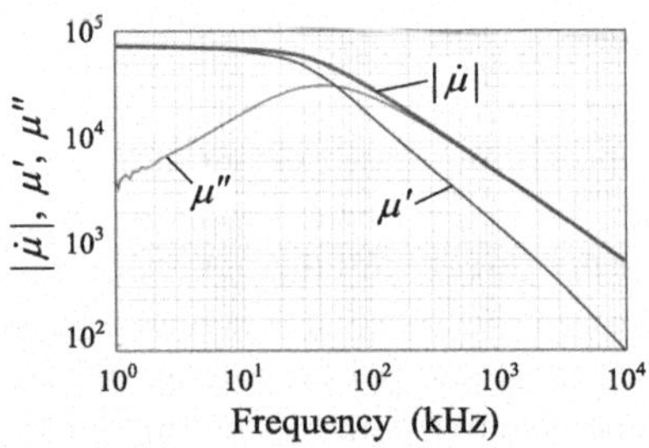

Fig. 2 Complex permeability of FineMet®, by Hitachi Metals [1]

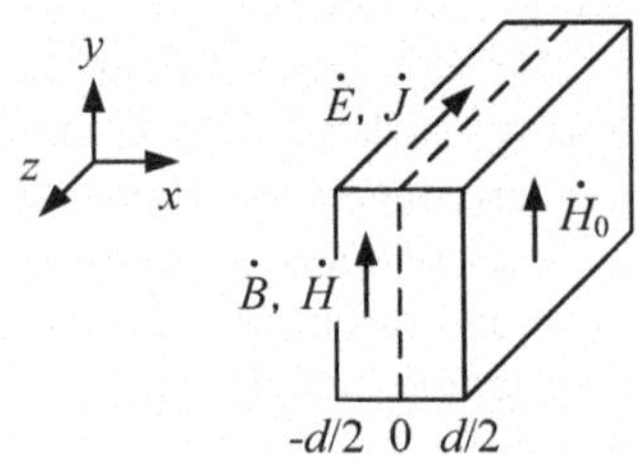

Fig. 3 Magnetic sheet [3]

$$\frac{\partial^2 \dot{H}(x)}{\partial x^2} - j\omega\sigma\mu\dot{H}(x) = 0 \tag{3}$$

Solving this with the boundary condition $\dot{H}(\pm d/2) = \dot{H}_0$ brings

$$\dot{H}(x) = \frac{\cos kx}{\cos(kd/2)}\dot{H}_0. \tag{4}$$

Here, the variable k is complex and denoted by

$$k = \sqrt{-j\omega\sigma\mu}. \tag{5}$$

The total flux $\dot{\Phi}$ in the magnetic sheet is obtained by integrating the flux density over the sheet as (6). The equivalent permeability of the sheet is given by complex permeability as (7) because the mean value of the flux density is the flux per the thickness. Note that the tangent appeared here is supposed to be a complex function.

$$\dot{\Phi} = \int_{-d/2}^{d/2} \mu\dot{H}(x)\mathrm{d}x = \frac{2\mu}{k}\tan(kd/2)\dot{H}_0 \tag{6}$$

$$\dot{\mu} = \frac{\dot{\Phi}}{\mathrm{d}\dot{H}_0} = \mu\frac{2}{kd}\tan\left(\frac{kd}{2}\right) \tag{7}$$

The flux distribution of a magnetic sheet obtained by (4) is shown in Fig. 4, and the complex permeability is shown in Fig. 5. Note that the amplitude of the permeability in high-frequency region is inversely proportional to the square root of the frequency.

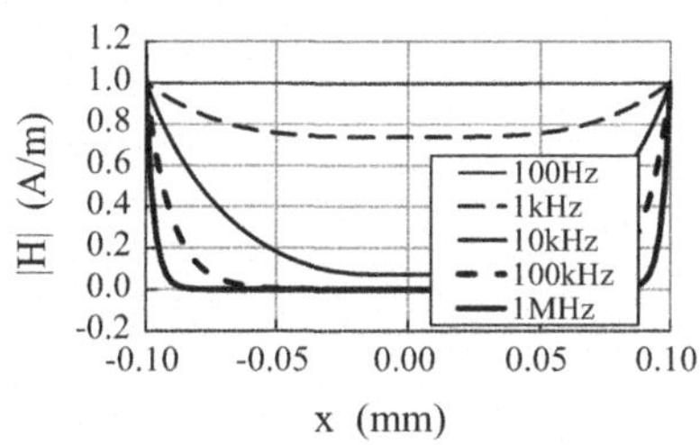

Fig. 4 Magnetic field in a sheet [3]

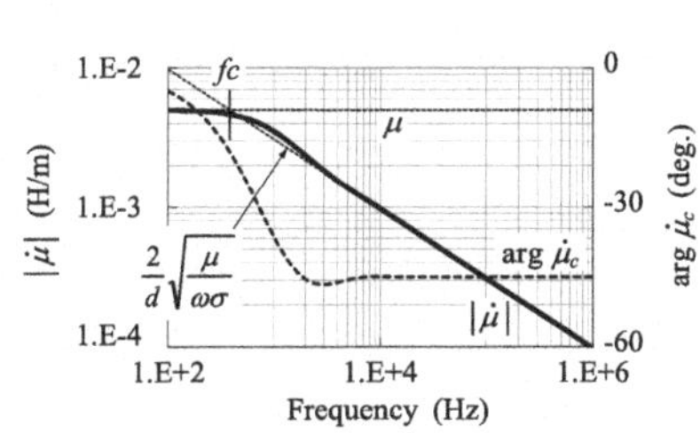

Fig. 5 Complex permeability of a sheet [2]

It is known that the corner frequency f_c of the permeability in Fig. 5 is derived from (7) and given by (8) [2].

$$f_c = \frac{2}{\pi d^2 \sigma \mu} \tag{8}$$

2 Equivalent Circuit of Complex Permeability

The equivalent circuit for the magnetic core is derived here. Figure 6 shows a reactor with laminated magnetic sheets. Let N denotes the coil turns, and the S and l denote the cross section of the core and the core length, respectively. The terminal voltage and current are denoted by $\dot{E}$ and $\dot{I}$. Then, the magnetomotive force of the winding is denoted by $N\dot{I}$. The magnetic reluctance is denoted by $\dot{\mu}S/l$, and the flux in the core is denoted by $\dot{\Phi} = (\dot{\mu}S/l)N\dot{I}$. Thus, the winding voltage is given by

$$\dot{E} = N\frac{\mathrm{d}\dot{\Phi}}{\mathrm{d}t} = \frac{SN^2}{l}j\omega\dot{\mu}\dot{I}. \tag{9}$$

As the equivalent circuit represents the relation of the voltage and the current, it is convenient that the quantity $j\omega\dot{\mu} = j\omega\dot{B}/\dot{H}$ in right hand of (9) is represented by the equivalent circuit. Here, the coefficient SN^2/l is considered later.

Firstly, the Foster equivalent circuit is introduced [3]. The following partial fraction expansions hold.

$$\frac{1}{z}\tan z = -2\sum_{n=1}^{\infty}\frac{1}{z^2 - \left[(2n-1)\pi/2\right]^2} \tag{10}$$

$$z\cot z = 1 + 2z^2\sum_{n=1}^{\infty}\frac{1}{z^2 - (n\pi)^2} \tag{11}$$

Applying these formulae to (7) with the substitution $z = kd/2$ yields

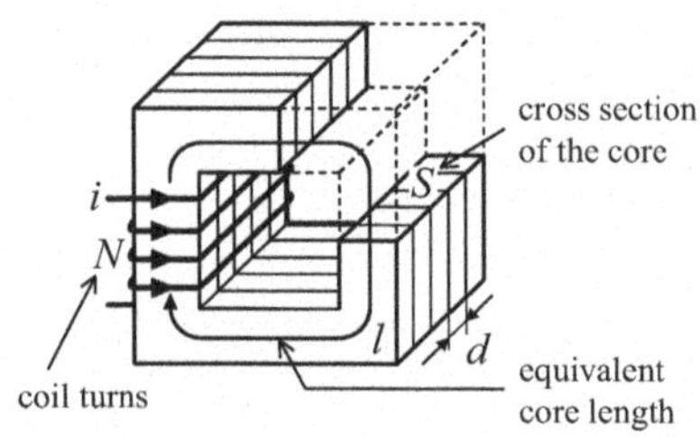

Fig. 6 Reactor [3]

$$j\omega\dot{\mu} = \sum_{n=1}^{\infty} \frac{1}{(1/2R) + (1/j\omega L_n)} \tag{12}$$

and

$$\frac{1}{j\omega\dot{\mu}} = \frac{1}{j\omega L} + \sum_{n=1}^{\infty} \frac{1}{(j\omega L/2) + R_n}. \tag{13}$$

Here, the constants are set as $L = \mu$, $R = 4/\sigma d^2$, and

$$L_n = \frac{2}{(n-1/2)^2\pi^2}L, \quad R_n = \frac{n^2\pi^2}{2}R. \tag{14}$$

Equation (12) presents the series circuit of parallel of resistors and inductors, while (13) presents the parallel circuit of series of resistors and inductors. Each circuit is shown in Fig. 7. Here, the variables $j\omega\dot{B}$ and $\dot{H}$ are regarded as the terminal voltage and the current flowing to the terminal. Then, the network impedance corresponds to $\mathrm{j}\omega\dot{\mu} = \mathrm{j}\omega\dot{B}/\dot{H}$. The network in 7(a) and (b) is known as the Foster realizations in I and II forms, respectively. Note that these networks are the representation for the complex permeability, and the units of resistors and inductors differ from the units in the electric circuits. Finally, multiplying the coefficient SN^2/l in (9) yields the impedance of the reactor winding.

One more realization for the equivalent circuit is called as the Cauer realization, where the elements of continued fraction are used. The following rational expression is derived by using the power series expansions for sine and cosine functions.

$$\frac{1}{z}\tan z = \frac{1}{\left(1 - \frac{z^2}{2!} + \frac{z^4}{4!} - \frac{z^6}{6!} + \cdots\right) \Big/ \left(1 - \frac{z^2}{3!} + \frac{z^4}{5!} - \frac{z^6}{7!} + \cdots\right)} \tag{15}$$

Extracting the quotient at $z = 0$ from the denominator of right side gives

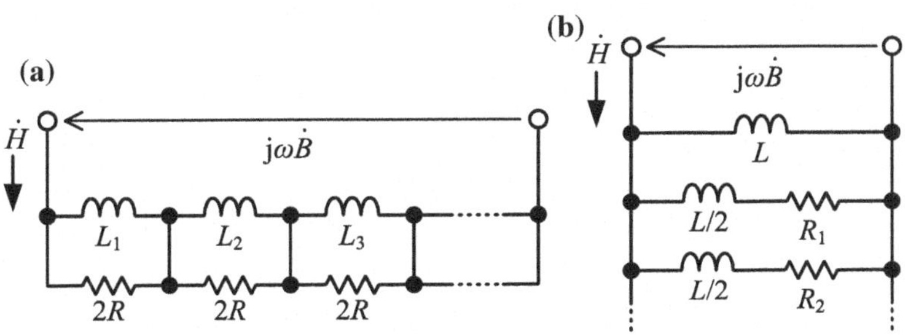

Fig. 7 Foster realizations in I form a, and II form b [3]

$$
\begin{aligned}
\frac{1}{z}\tan z &= \cfrac{1}{1+\cfrac{\left(1-\frac{z^2}{2!}+\frac{z^4}{4!}-\frac{z^6}{6!}+\cdots\right)-\left(1-\frac{z^2}{3!}+\frac{z^4}{5!}-\frac{z^6}{7!}+\cdots\right)}{1-\frac{z^2}{3!}+\frac{z^4}{5!}-\frac{z^6}{7!}+\ldots}} \\
&= \cfrac{1}{1-\cfrac{z^2}{\left(1-\frac{z^2}{2!}+\frac{z^4}{4!}-\frac{z^6}{6!}+\cdots\right)\Big/\left(\frac{1}{1!\cdot 3}-\frac{z^2}{3!\cdot 5}+\frac{z^4}{5!\cdot 7}-\cdots\right)}}
\end{aligned} \quad (16)
$$

Applying this operation repeatedly to the last rational expression in the second raw of (16) gives the fractional expression as the (17). The fractions having such nested structure are called continued fractions. The notation in right side hand is used to avoid the bulks of usual fractional expressions.

$$
\frac{1}{z}\tan z = \cfrac{1}{1-\cfrac{z^2}{3-\cfrac{z^2}{5-\cfrac{z^2}{7-\ddots}}}} = \frac{1}{1-}\,\frac{z^2}{3-}\,\frac{z^2}{5-}\,\frac{z^2}{7-}\cdots \quad (17)
$$

Using this continued fraction to (7) and setting $z = kd/2$ give the following representation for the complex permeability

$$
j\omega\dot{\mu} = \frac{1}{1/\mathrm{j}\omega L+}\,\frac{1}{3R+}\,\frac{1}{5/\mathrm{j}\omega L+}\,\frac{1}{7R+}\cdots \quad (18)
$$

Here, $L = \mu$ and $R = 4/\sigma d^2$. The Eq. (18) is presented by the Cauer ladder network shown in Fig. 8.

Both of the Foster and Cauer equivalent circuits should be truncated to the finite length when they are applied to actual objects. Figure 9 shows the Bode plots when the networks are truncated to finite length, where n is the number of stages that corresponds to the number of inductors. Bode plots for the Foster realizations of the I and II forms are also plotted. The Cauer network is terminated by the inductor, and the subsequent resistor is cut off. The Cauer network in five stages covers about three-decade frequency range. On the other hand, the errors are still observed in Foster realization with sixteen stages. The Cauer ladder network stands out in the convergence property.

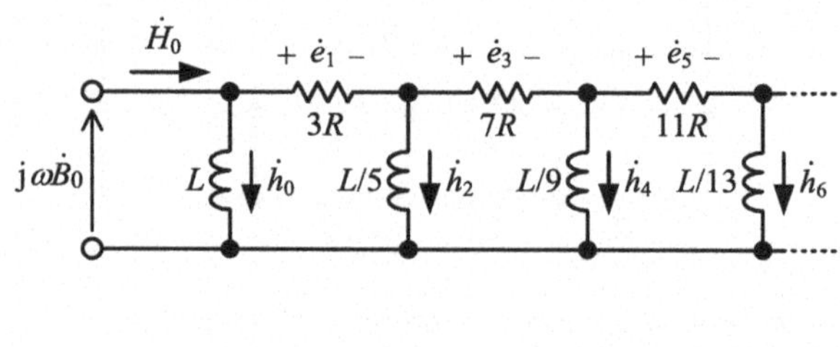

Fig. 8 Cauer ladder network [2]

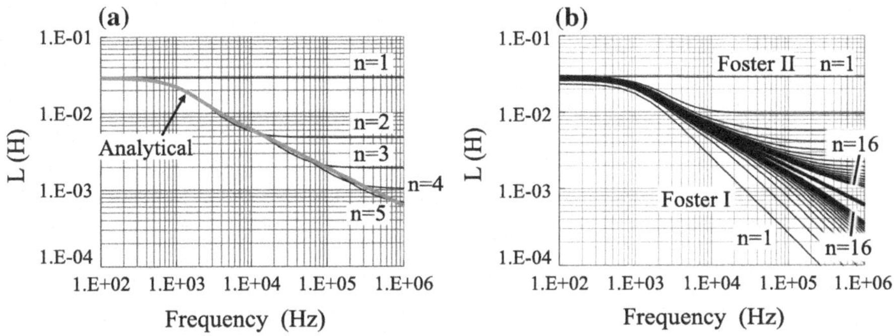

Fig. 9 Bode plots in truncated network of Cauer realization a and Foster realization b [3]

3 Synthesis of Magnetic Fields

As shown in previous sections, the network realization of the complex permeability is obtained from the complex impedance. In this and next section, it is shown that the magnetic field in the magnetic sheet and magnetic wire can be expressed by the orthogonal functions and the Cauer ladder network, although the derivation is discussed in the last section.

The equations for the eddy-current field can be solved by the variable separation method. The magnetic and electric fields can be represented as (19) and (20). Here, the complex variables $\dot{h}_{2m}$ and $\dot{e}_{2m+1}$ are the current of the inductors and voltage of the resistors of the Cauer ladder network shown in Fig. 8, respectively. The functions $P_{2m}(x)$ and $P_{2m+1}(x)$, defined on the interval [-1, 1], represent the spatial distribution.

$$\dot{H}(x) = \sum_{m=0}^{\infty} \dot{h}_{2m} P_{2m}(2x/d) \tag{19}$$

$$\dot{E}(x) = \sum_{m=0}^{\infty} \dot{e}_{2m+1} P_{2m+1}(2x/d) \tag{20}$$

If we apply the Legendre polynomials (21) to the distribution functions, then the magnetic and electric fields (19) and (20) coincide to those of magnetic sheet. The Legendre polynomials are defined on the interval $[-1, 1]$ and have the orthogonality defined by (22).

$$P_n(x) = \frac{1}{2^n n!} \cdot \frac{d^n}{dx^n}(x^2 - 1)^n \tag{21}$$

$$\int_{-1}^{1} P_n(x) P_m(x)\, dx = \begin{cases} 0 & (m \neq n) \\ 2/(2n+1) & (n = m) \end{cases} \tag{22}$$

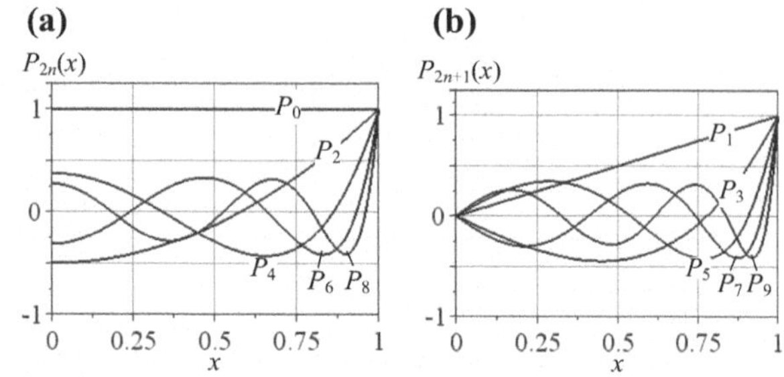

Fig. 10 Legendre polynomials, a even order and b odd order [2]

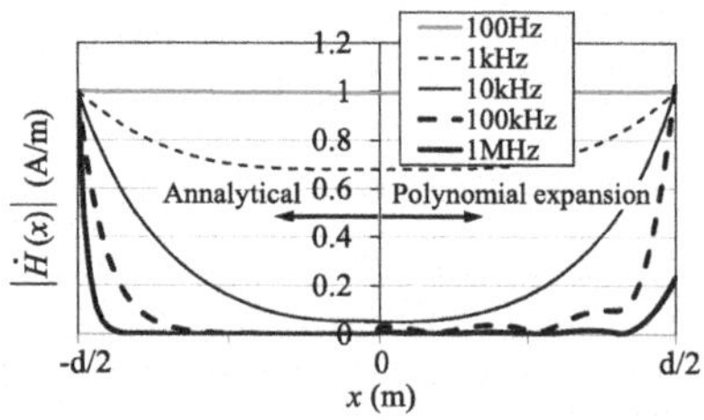

Fig. 11 Approximation of magnetic fields by orthogonal polynomials [2]

Figure 10 shows the graphs of the Legendre polynomials. These polynomials and the ladder network truncated to four stages give the magnetic distributions shown in Fig. 11. Compared with the analytic field shown in Fig. 4, the approximation errors are small even in the high-frequency range. Thus, the ladder network not only gives the frequency response but also provides the distributions of magnetic fields.

4 Modeling for Wire

The high-frequency current in the magnet wire tends to flow on the surface of the wire, what is called the skin effect. The method explained in preceding section is also applicable to the simple cylindrical shape shown in Fig. 12.

Assume that the electric field $\dot{E}_0$ is applied axial direction of the cylinder with the diameter D. The differential equation for the eddy-current field in the wire is represented by (23) in the cylindrical polar coordinates.

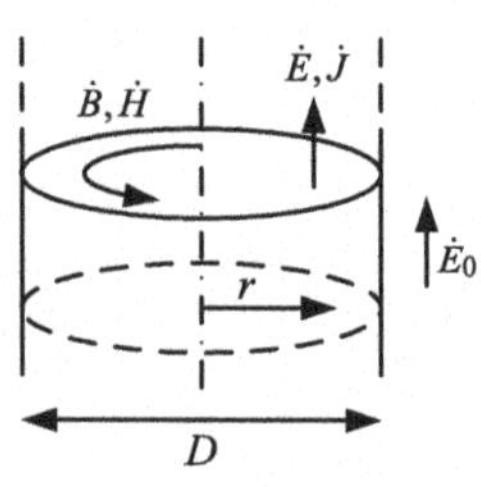

Fig. 12 Cylindrical model [2]

$$\left.\begin{aligned} \frac{\partial \dot{E}}{\partial r} &= j\omega \dot{B} = j\omega\mu \dot{H} \\ \frac{1}{r}\frac{\partial}{\partial r}\left(r\dot{H}\right) &= \dot{J} = \sigma \dot{E} \end{aligned}\right\} \tag{23}$$

Substituting the variable $\dot{H}$ in the second equation with the first gives the following.

$$\frac{\partial^2 \dot{E}(r)}{\partial r^2} + \frac{1}{r}\frac{\partial \dot{E}(r)}{\partial r} - j\omega\sigma\mu \dot{E}(r) = 0 \tag{24}$$

Solving this equation with the boundary condition $\dot{E}(D/2) = \dot{E}_0$ yields the electric field in the wire and resultantly the mean value of the current as

$$\dot{E}(r) = \frac{J_0(kr)}{J_0(kD/2)}\dot{E}_0, \tag{25}$$

and

$$\dot{I}_0 = \int_0^{D/2} 2\pi r\sigma \dot{E}(r)\mathrm{d}r = \frac{\pi D^2\sigma}{2}\left(\frac{2}{kD}\right)\frac{J_1(kD/2)}{J_0(kD/2)}\dot{E}_0. \tag{26}$$

Here, the r denotes the radial distance from the z-axis, and $J_n(z)$ denotes the n-th order Bessel function of the first kind. The variable k is already defined in (5). The following continued fraction expansion holds, similarly to (17).

$$\frac{1}{z}\frac{J_1(z)}{J_0(z)} = \frac{1}{2-}\frac{z^2}{4-}\frac{z^2}{6-}\frac{z^2}{8-}\frac{z^2}{10-\cdots} \tag{27}$$

Then, the effective impedance of the cylinder $\dot{R} = \dot{E}_0/\dot{I}_0$ is given by

$$\frac{1}{\dot{R}} = \frac{1}{R_0 + \dfrac{1}{1/(j\omega L_0/2) + \dfrac{1}{3R_0 + \dfrac{1}{1/(j\omega L_0/4) + \dfrac{1}{5R_0 + \dfrac{1}{1/(j\omega L_0/6) + \cdots}}}}}}, \tag{28}$$

$$R_0 = 4/\pi D^2\sigma, \quad L_0 = \mu/4\pi. \tag{29}$$

The ladder network corresponding to this equation is shown in Fig. 13. Similar to the Legendre polynomial for magnetic sheets, the electric field in the cylinder can be represented by a certain orthogonal polynomials [2]. The analytical current distribution in the cylinder is shown in the left-half plane of Fig. 14, while the

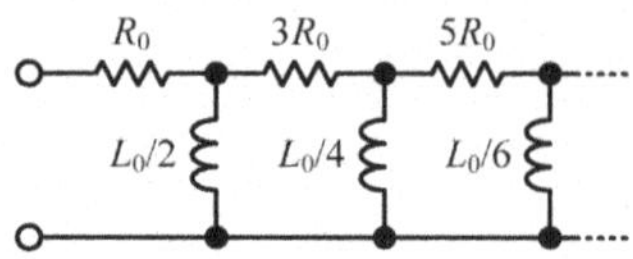

Fig. 13 Equivalent circuit for the cylinder [3]

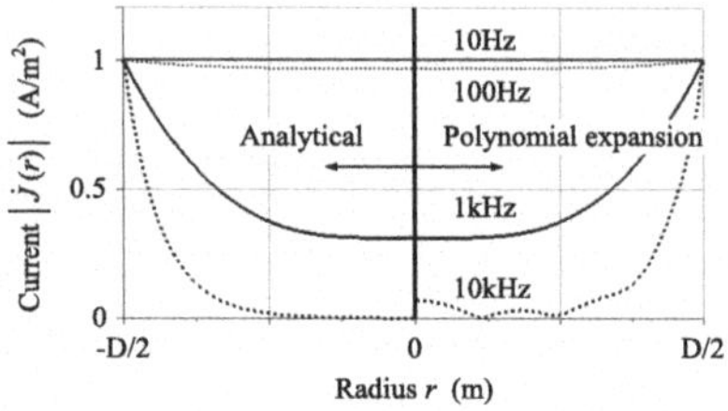

Fig. 14 Current distribution approximated by polynomials [2]

approximated value is shown in the right half where the network is truncated to four stages.

5 Arbitrary Eddy-Current Field

The method stated in the previous sections uses the analytic functions and their continued fraction expansions. However, such approach has the limitation that the method is applicable only in cases that the shape of the magnetic materials is simple such as the plane or the cylinder. Recently, it was found that the ladder network representation is possible for arbitrary eddy-current field as shown in Fig. 15. This method uses numerical methods such as finite element methods to obtain the orthogonal functions and network constants for the eddy-current fields. The orthogonal functions appeared in plane or cylinder can also be obtained by this method. The transient analysis is performed by the circuit simulator by using the ladder network. The magnetic field at any point of time is reconstructed by using the orthogonal functions and the voltages and currents of the network component at that moment. The procedure of this method is constructed as follows.

Decompose the magnetic and electric field as (30) instead of (30) and (20). The values of the resistors and inductors are defined by (30).

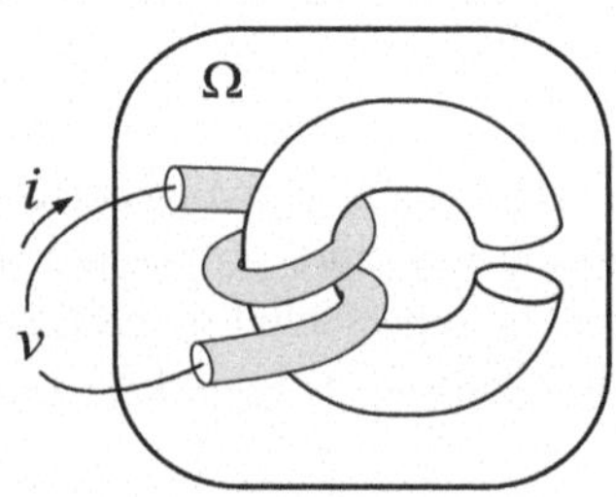

Fig. 15 Eddy-current field

$$\left.\begin{aligned}\mathbf{E} &= \sum_{n=0}^{\infty}\mathbf{E}_{2n} = \sum_{n=0}^{\infty} e_{2n}\bar{\mathbf{E}}_{2n} \\ \mathbf{H} &= \sum_{n=0}^{\infty}\mathbf{H}_{2n+1} = \sum_{n=0}^{\infty} h_{2n+1}\bar{\mathbf{H}}_{2n+1}\end{aligned}\right\} \tag{30}$$

$$\left.\begin{aligned}\lambda_{2n} &= 1/R_{2n} = \int_{\Omega} \sigma\,\bar{\mathbf{E}}_{2n}\cdot\bar{\mathbf{E}}_{2n}\mathrm{d}V \\ \lambda_{n2+1} &= L_{2n+1} = \int_{\Omega} \mu\,\bar{\mathbf{H}}_{2n+1}\cdot\bar{\mathbf{H}}_{2n+1}\mathrm{d}V\end{aligned}\right\} \tag{31}$$

Then, the following steps yield the series of orthogonal functions and the circuit constants of the ladder network.

Step 0: Solve

$$\nabla\times\mathbf{E}_0 = 0 \tag{32}$$

based on the given boundary conditions. Set $\bar{\mathbf{E}}_0 = \mathbf{E}_0/v$ and $\mathbf{H}_{-1} = 0$.

Step 1: Solve

$$\nabla\times\tilde{\mathbf{H}}_{2n-1} = \sigma\,\bar{\mathbf{E}}_{2n-2}/\lambda_{2n-2}, \tag{33}$$

and set $\bar{\mathbf{H}}_{2n-1} = \tilde{\mathbf{H}}_{2n-1} + \bar{\mathbf{H}}_{2n-3}$.

Step 2: Solve

$$\nabla\times\tilde{\mathbf{E}}_{2n} = -\mu\,\bar{\mathbf{H}}_{2n-1}/\lambda_{2n-1}, \tag{34}$$

and set $\bar{\mathbf{E}}_{2n} = \tilde{\mathbf{E}}_{2n} + \bar{\mathbf{E}}_{2n-2}$.

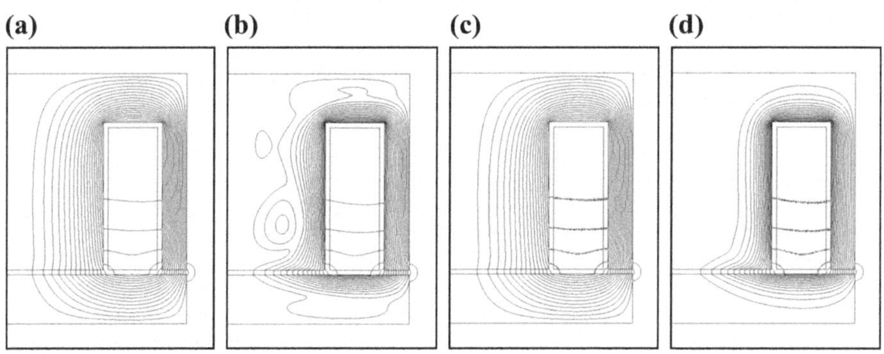

Fig. 16 By ladder network and conventionals [6], a Cauer 1 Hz, b Cauer 10 Hz, c Conventional 1 Hz, d Conventional 10 Hz

Return to Step 1.

By using this method, the magnetic field of an electromagnet with cylindrical shape is constructed as shown in Fig. 16 [6]. The excitation frequencies are chosen to 1 Hz and 10 Hz, while the ladder network is truncated to four stages. The magnetic fields by the finite element complex analysis are also shown for the comparison. When the magnetic flux distribution due to CLN at 10 Hz in Fig. 16b is compared with that by the conventional method in Fig. 16d, some error is observed in the magnetic flux density inside the magnetic material. However, there is no significant difference in the magnetic flux in the air gap that determines the electromagnetic force of the electromagnet.

The method to construct the equivalent circuit for linear eddy-current field is explained up to here. The distribution of the currents in the Cauer ladder network is determined by the frequency of the applied voltage. In case of DC, the current only flows through the inductor of the first stage. By using this property, it is possible to formulate the models for nonlinear hysteresis magnetic materials [2].

References

1. Hitachi Metals: "Nanocrystalline soft magnetic material FINEMET®", https://www.hitachi-metals.co.jp/products/elec/tel/pdf/hl-fm9-h.pdf (2016)
2. Shindo, Y., Kameari, A., Matsuo, T.: High frequency nonlinear modeling of magnetic sheets using polynomial expansions for eddy-current field. IEEJ Trans. Power Energy **137**(3), 162–172 (2017)
3. Shindo, Y., Noro, O.: Simple circuit simulation models for eddy current in magnetic sheets and wires. IEEJ Trans. Fundam. Mater. **134**(4), 173–181 (2014)
4. Kameari, A., Ebrahimi, H., Sugahara, K., Shindo, Y., Matsuo, T.: Cauer ladder network representation of eddy-current fields for model order reduction using finite element method. IEEE Trans. Magn., TMAG.2017.2743224 (2017)
5. Shindo, Y., Kameari, A., Sugahara, K., Matsuo, T.: Dynamical model of an electromagnet using cauer ladder network representation of eddy-current fields. IEEJ J. Ind. Appl. **7**(4), 305–313 (2018)
6. Shindo, Y., Kameari, A., Sugahara K., Matsuo T.: Dynamical model of an electromagnet by Cauer ladder network representation of eddy-current fields. In: Proceedings of 11th International Symposium on Linear Drives for Industry Applications (LDIA) 1–6 (2017)

Coupled Analysis of Semiconductor Characteristics and Magnetic Properties

Shunya Odawara and Nguyen Gia Minh Thao

Abstract This chapter presents an analysis with mutual consideration of the characteristics of semiconductor components used as switching devices in a DC–AC inverter and the characteristics of the magnetic material constituting iron core of electric equipment. The on-voltage characteristics of semiconductor components affect the magnetic hysteresis characteristics of the magnetic material considerably. Besides, we introduce a numerical calculation method based on the play model that can take into account the interaction between these characteristics efficiently. The play model is utilized to consider and express the magnetic hysteresis phenomenon; this model can calculate a magnetic flux density waveform (magnetic field strength waveform) from an arbitrary magnetic field strength waveform (magnetic flux density waveform) by preparing several magnetic hysteresis loops for identification process. Because the conventional play model only can express the magnetic hysteresis in the DC field, the classical eddy current theory and the Cauer equivalent circuit theory are additionally applied in this study for considering the eddy currents in the AC field; in detail, the Cauer theory is used to analyze the reaction magnetic field caused by the carrier frequency in the pulse-width modulation (PWM) inverter excitation, where the skin effect cannot be ignored. To validate effectiveness of the designed calculation method, an example of magnetic characteristics under the PWM inverter excitation using two different types of semiconductor components with large and low on-voltages are conducted; finally, magnetic hysteresis loops obtained by the calculation method and experimental measurements are compared and evaluated.

Keywords Coupled analysis method · On-voltage characteristics of semiconductor · Magnetic hysteresis characteristics · Inverter excitation · Minor loop · Play model

S. Odawara
Kitami Institute of Technology, Hokkaido 090-8507, Japan

N. G. M. Thao (✉)
Electromagnetic Energy System Laboratory, Toyota Technological Institute, Nagoya 468-8511, Japan
e-mail: ngmthao@toyota-ti.ac.jp

K. Fujisaki (ed.), *Magnetic Material for Motor Drive Systems*,
Engineering Materials, https://doi.org/10.1007/978-981-32-9906-1_16

1 Introduction

In recent years, with the advent of semiconductor devices with high switching speed, low on-resistance, high-withstand voltage using Silicon Carbide (SiC) and Gallium Nitride (GaN), the field of power electronics has been improved more and more, as well as power conversion circuits such as inverters and converters can adjust voltage and frequency freely. In particular, variable speed control for motor becomes possible and more efficient by controlling the changeable output voltage and frequency, so it is popular for the electric motor to be driven using a DC–AC inverter. Wherein, the iron core of motor is often composed of a magnetic material such as an electrical steel plate. Therefore, when the motor is driven by an inverter, the magnetic material constituting this motor is inevitably excited by the inverter; however, when the magnetic material is excited by the inverter, it is well known that the iron loss increases as compared to a case of excitation with a pure sinusoidal wave [1, 2]. This means that the characteristics of the power electronic devices used in inverter, motor, and magnetic material are mutually related. That is, until now, although their characteristics have been independently evaluated in each field of electric circuits, semiconductors, electric equipment, and magnetic materials, respectively, further advanced technologies utilized in electric machines being driven as shown above cannot be well developed and implemented without the mutual evaluation on their respective characteristics.

Hence, in this chapter, we describe the interaction between the characteristics of semiconductor components serving as the switching devices in the DC–AC inverter and the characteristics of the magnetic material constituting iron core of electric equipment such as motors, transformers, and reactors. This chapter also presents a calculation method in detail with an illustrated example.

2 Semiconductor Characteristics and Magnetic Properties

2.1 Inverter Excitation and Magnetic Properties

When a magnetic material such as an electromagnetic steel plate based on an amorphous material or a nanocrystalline material is excited by an inverter, the harmonic component inevitably contained in the inverter output causes a minor loop in the magnetic hysteresis loop as described in Fig. 1. Iron loss is calculated in (1) and is proportional to the internal integral of the magnetic hysteresis loop; hence, the iron loss increases by the area of the minor loop.

$$W_{\text{fe}} = \frac{f}{\rho} \int H \mathrm{d}B \tag{1}$$

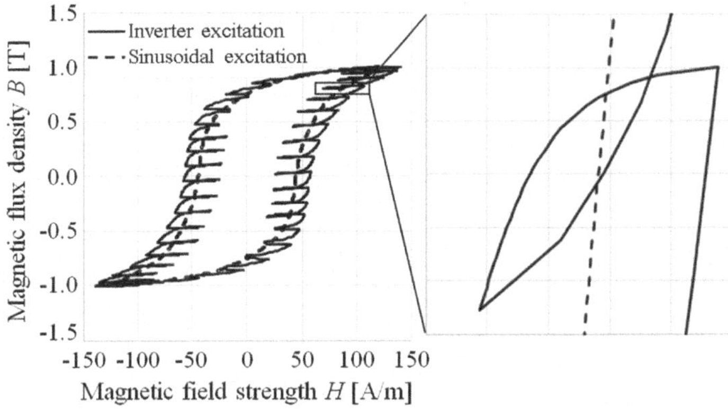

Fig. 1 Magnetic hysteresis loop under inverter excitation measured by ring core (right: enlargement of minor loop)

W_{fe}: iron loss, f: frequency, ρ: density of magnetic material, H: magnetic field strength, B: magnetic flux density.

2.2 *Semiconductor Characteristics and Inverter Output Voltage*

It has also been reported that the manner, in which minor loops are formed, is affected by various characteristics of semiconductor components used in the inverter [3, 4]. This is due to a change in the output voltage waveform significantly affected by characteristics of the semiconductor components used in the inverter; the on-voltage characteristics and the output voltage waveform are presented and considered in this subsection. A general voltage-type single-phase inverter circuit diagram is represented in Fig. 2, and a case where the load (magnetic material) is excited by this inverter is considered in this research. Under inverter excitation, the voltage is applied to the load mainly in four modes by on/off operation of the semiconductor component as shown in Fig. 3.

First, we consider a case where the on-resistance of the semiconductor component is zero (ideal state), it means that the voltage drop does not occur in the semiconductor components when they turn on in this case. The voltage V_1 applied to the load in mode 1 is fully generated by the DC voltage source V_{dc} as follows.

$$V_1 = V_{dc} \tag{2}$$

Since there is no voltage source in mode 2, the voltage V_2 applied to the load is zero. Similarly, the voltage V_3 applied to the load in mode 3 is V_{dc}, and the voltage V_4 applied to the load in mode 4 is zero. Here, the mode (i.e., mode 1 and mode 3),

Fig. 2 Circuit of voltage-type single-phase inverter excitation

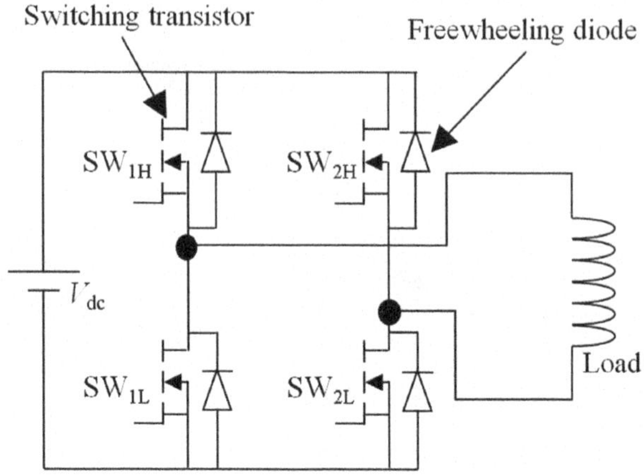

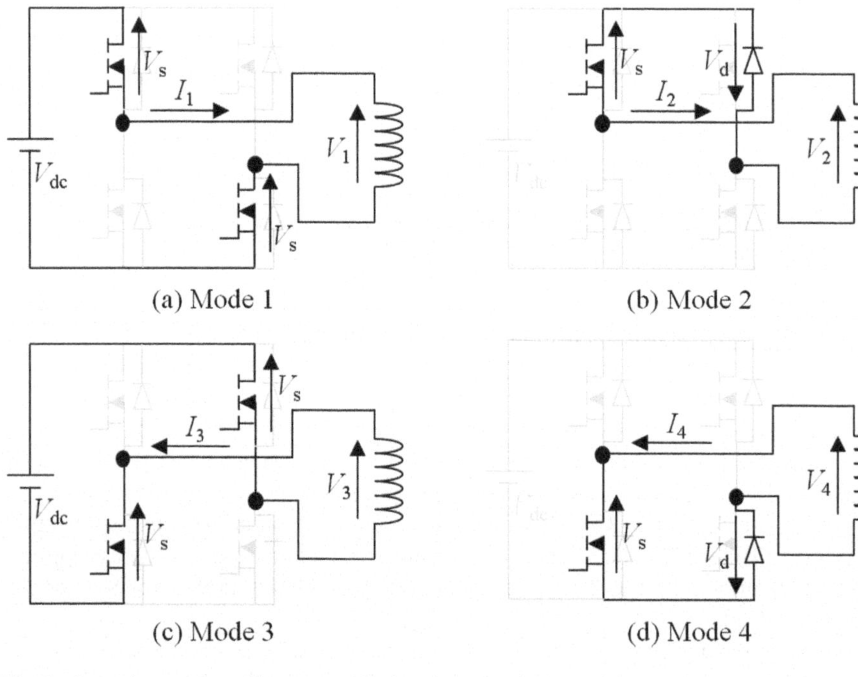

Fig. 3 Operation modes of inverter excitation

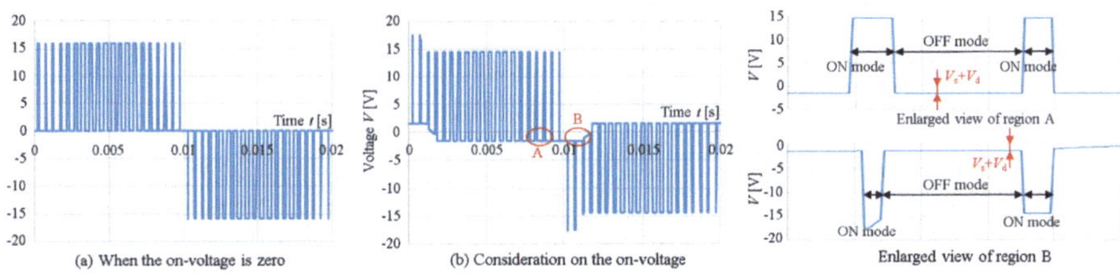

Fig. 4 Voltage waveform applied to the load

in which the DC voltage source V_{dc} is applied to the load, is referred as "ON mode"; and the mode (i.e., mode 2 and mode 4) in which V_{dc} is not applied to the load, is referred as "OFF mode."

Next, we consider the case where the semiconductor component has the on-resistance. In this case, a voltage drop (on-voltage) is generated by this on-resistance. It is assumed that the on-voltage of the switching transistor is V_s and the on-voltage of the freewheeling diode is V_d ; the voltage applied to the load in each mode is from the Kirchhoff's second law as expressed in the following four equations.

$$\text{Mode 1:}\quad V_1 = V_{dc} - V_s - V_s = V_{dc} - 2V_s \tag{3}$$

$$\text{Mode 2:}\quad V_2 = -V_s - V_d \tag{4}$$

$$\text{Mode 3:}\quad V_3 = -V_{dc} + V_s + V_s = -V_{dc} + 2V_s \tag{5}$$

$$\text{Mode 4 :}\quad V_4 = V_s + V_d \tag{6}$$

The sign of each above voltage varies depending on the direction of the current. As can be seen from the above four formulas, the on-voltage generated in the semiconductor component clearly affects the load voltage as compared to the ideal state. Figure 4 panel (a) shows the voltage waveforms when the pulse-width modulation (PWM) control technique (fundamental frequency: 50 Hz, carrier frequency: 1 kHz, and modulation index: 0.5) is used and the on-voltage is zero.

2.3 Output Voltage Waveform of Inverter and Minor Loop Shape

The behavior of the voltage waveform in the OFF mode depicted in Fig. 4 panel (b) is highly related to the formation of minor loops during magnetic hysteresis. In consideration of the on-voltage of the semiconductor device, the voltage does not become zero in the OFF mode. For example, in the enlarged view of the region A as described in the right part of Fig. 4, a negative voltage is applied to the load in the

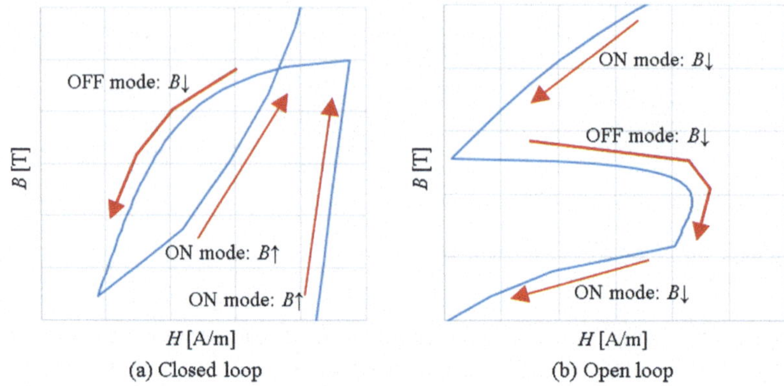

Fig. 5 Two types of minor loop

OFF mode, and a positive voltage is applied in the ON mode immediately before and after this OFF mode. When the voltage changes as this (positive → negative → positive), a phenomenon in the magnetic flux density waveform generated in the load (rising → descending → rising) occurs. This can be explained from the relationship that the magnetic flux density B is proportional to the time integral of the voltage V according to Faraday's law for electromagnetic induction, as expressed in (7).

$$B = \frac{1}{NS}\int_{0}^{T} V\,\mathrm{d}t \tag{7}$$

N: Number of turns of coil wrapped around the magnetic material, S: cross sectional area of the load (magnetic material), T: One cycle.

In this way, once the magnetic flux density waveform rises → descends → rises (or descends → rises → descends), a closed minor loop is formed as depicted in Fig. 5 panel (a). This closed minor loop is called "closed loop" hereafter. On the other hand, in the enlarged view of the region B as shown in the right part of Fig. 4, a negative voltage is applied in the OFF mode, and a negative voltage is also applied in the ON mode immediately before and after; as a result, the magnetic flux density waveform generated in the load takes a form of descending → descending → descending. In such a case, an open minor loop is formed as shown in Fig. 5 panel (b); it is called "open loop" hereafter.

Although it has been stated that the behavior of the voltage in the OFF mode significantly affects the formation of the minor loop, the shape of the minor loop possibly changes due to the magnitude (kind of difference) of the on-voltage of semiconductor component. This is because the voltage applied to the load in the OFF mode is the sum of the on-voltages generated in the switching transistor and the freewheeling diode; the increase and decrease in the magnetic flux density change according to this voltage value as expressed in (7). From this, it can be seen that the magnetic characteristics of the excited magnetic material change depending on the characteristics of the semiconductor components in the inverter.

3 Calculation Method with Mutual Consideration on Semiconductor Characteristics and Magnetic Properties

Up to this point, it can be stated that the on-voltage characteristics of the semiconductor component affect the magnetic hysteresis characteristics of the magnetic material. In this section, we present a numerical calculation method that can take into account the interaction between these characteristics. In this method, the circuit analysis with consideration on semiconductor characteristics and the magnetic analysis with consideration on magnetic hysteresis characteristics are coupled to reflect the characteristics mutually. Although this method has a constraint that it can be applied only to the unidirectional magnetic field since it does not use the finite element method often employed for magnetic analysis, it has an advantage that its calculation result can be obtained at high speed. The calculation flowchart of this method is shown in Fig. 6, and each procedure is described in detail.

3.1 Procedure (I) Generation of Ideal PWM Voltage Waveform

In this procedure, an ideal PWM inverter voltage waveform V_{ideal} is generated. As an example, the calculation is conducted with a voltage waveform with a fundamental frequency of 50 Hz, a carrier frequency of 1 kHz, a modulation index of 0.5, and an applied DC voltage of 16 V according to the modulation method comparing to an ordinary triangular wave. At this time, the switching pattern employed for the four transistors in Fig. 2 is shown in Fig. 7, and the generated ideal voltage waveform V_{ideal} is described in Fig. 4 panel (a).

3.2 Procedure (II) Calculation of Magnetic Flux Density Waveform

From the voltage waveform generated in procedure (i) or procedure (vii), the magnetic flux density waveform B is calculated using (7). In this case, $N = 254$ turns and $S = 8.75 \times 10^{-5}$ m^2 are utilized.

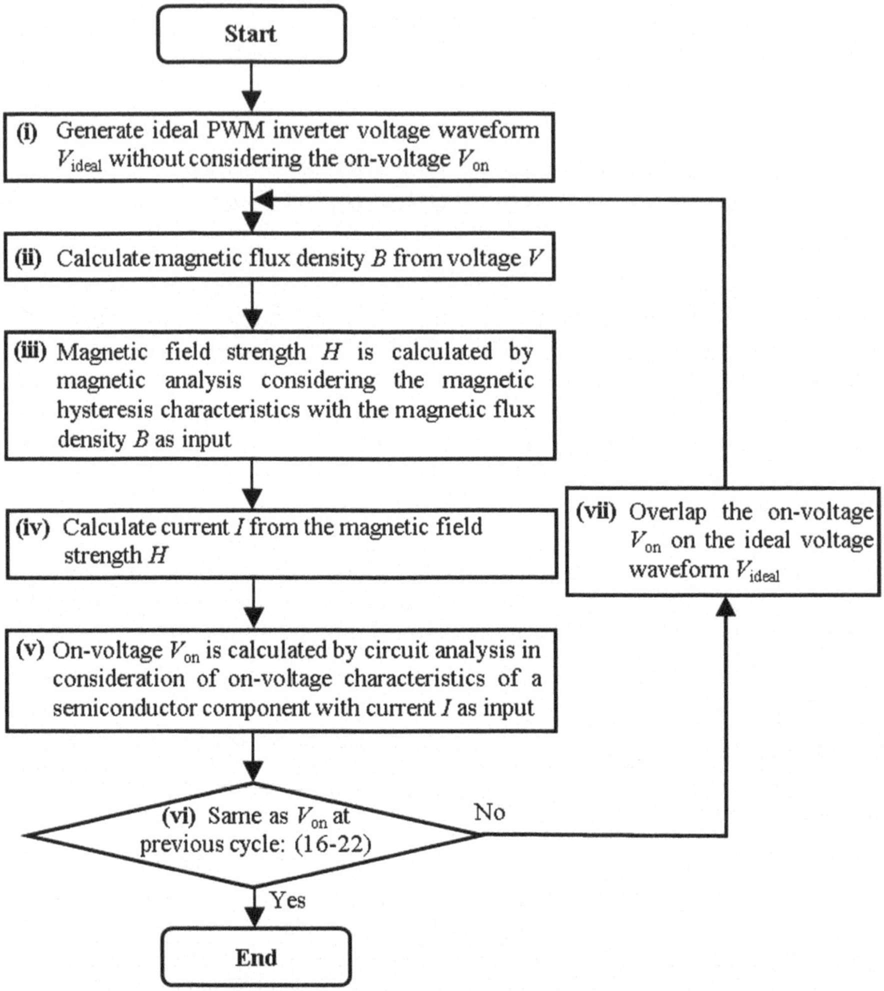

Fig. 6 Flow chart of calculation

3.3 Procedure (III) Magnetic Analysis with Consideration of Magnetic Hysteresis Characteristics

In this study, the play model [5] is adopted as a method to consider and express the magnetic hysteresis phenomenon. The play model can calculate a magnetic flux density waveform (magnetic field strength waveform) from an arbitrary magnetic field strength waveform (magnetic flux density waveform) by preparing several magnetic hysteresis loops for identification process. Since the play model is a capable method for expressing the magnetic hysteresis in the DC field, it is necessary to introduce the classical eddy current theory and the Cauer equivalent circuit theory for carefully

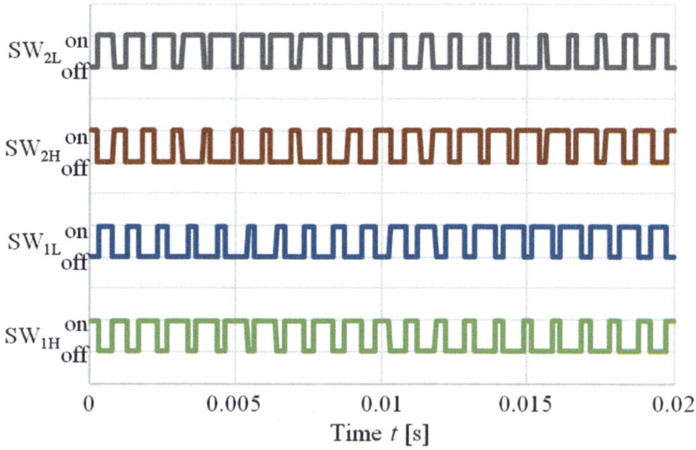

Fig. 7 Switching pattern of the four transistors used in inverter

considering the eddy current in the AC field. In the case, where the frequency is low and no reaction, magnetic field is generated (when the skin effect is negligible), it is sufficient to use the classical eddy current theory; however, in the case where the reaction magnetic field cannot be ignored (where the skin effect cannot be ignored), the magnetic hysteresis cannot be accurately drawn without using the Cauer equivalent circuit theory. In general, in the case of inverter excitation, since the carrier frequency is several ten to several hundred times of the fundamental frequency, the frequency in the region, where the reaction magnetic field cannot be ignored, is handled. In the following, a process of drawing a magnetic hysteresis loop using the play model, that inputs a magnetic flux density waveform and outputs a magnetic field strength waveform, is be described; the basic description of the play model [5] is omitted in this chapter.

(1) **With use of classical eddy current theory**
First, we describe a case for extending to the AC field using the classical eddy current theory. In general, when no reaction magnetic field is generated, the magnetic field strength $H_{AC}(B)$ of the AC field is calculated as follows.

$$H_{\mathrm{AC}}(B) = H_{\mathrm{DC}}(B) + k\frac{\sigma d^2}{12}\frac{\mathrm{d}B}{\mathrm{d}t} \tag{8}$$

$H_{DC}(B)$: magnetic field strength of DC field obtained from play model, σ: conductivity of magnetic material, d: thickness of magnetic material, and k: coefficient for considering the influence of abnormal eddy current.

In this equation, where H_{AC} represents the current I, $\mathrm{d}B/\mathrm{d}t$ represents the voltage V, $12/k\sigma d^2$ represents the resistance value R, and the nonlinear magnetic permeability μ with direct current corresponds to the inductance L, the equivalent circuit can be

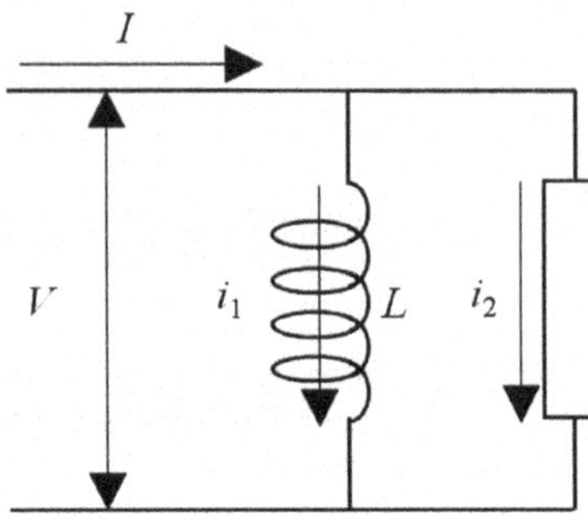

Fig. 8 Equivalent circuit of classical eddy current theory

drawn as depicted in Fig. 8. For confirmation, the circuit Eqs. (9), (10), and (11) can be obtained when substituting I for H_{AC}, V for dB/dt, R for $12/k\sigma d^2$, and L for μ in the electromagnetic Eqs. (12), (13), and (14). In addition, (14) is the same with (18) where B/μ corresponds to H_{DC}.

$$V = L\frac{\mathrm{d}i_1}{\mathrm{d}t} \tag{9}$$

$$V = Ri_2 \tag{10}$$

$$I = i_1 + i_2 \tag{11}$$

$$\frac{\mathrm{d}B}{\mathrm{d}t} = \mu\frac{\mathrm{d}i_1}{\mathrm{d}t} \tag{12}$$

$$\frac{\mathrm{d}B}{\mathrm{d}t} = Ri_2 \tag{13}$$

$$H_{AC}(B) = \frac{B}{\mu} + k\frac{\sigma d^2}{12}\frac{\mathrm{d}B}{\mathrm{d}t} \tag{14}$$

In this subsection, a concrete magnetic hysteresis loop is created. A non-oriented electrical steel sheet of 35A300 (thickness d: 0.35 mm and conductivity σ: 1.923×10^6 S/m) was used as a magnetic material of the load, with a maximum magnetic flux density $B_{max} = 1$ T under sinusoidal excitation of 50 Hz. The reproducibility of the magnetic hysteresis loop is shown in Fig. 9. The skin depth in this case is about 0.6 mm, which is larger than the thickness of the steel plate; hence, the skin effect can be ignored. Figure 9 also describes the DC magnetic hysteresis loops of 35A300 for identification process, which is used for the play model. Figure 10 shows a magnetic hysteresis loop obtained from the magnetic flux density waveform with $B_{max} = 1$ T under sinusoidal excitation and the magnetic field strength waveform calculated from (8) with input as this magnetic flux density. The magnetic hysteresis loop obtained by actual measurement is also shown in the same figure; the numerical

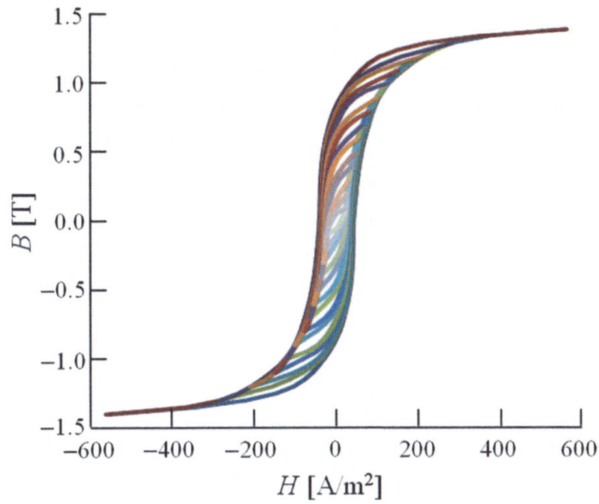

Fig. 9 DC magnetic hysteresis loops used for identification process in play model

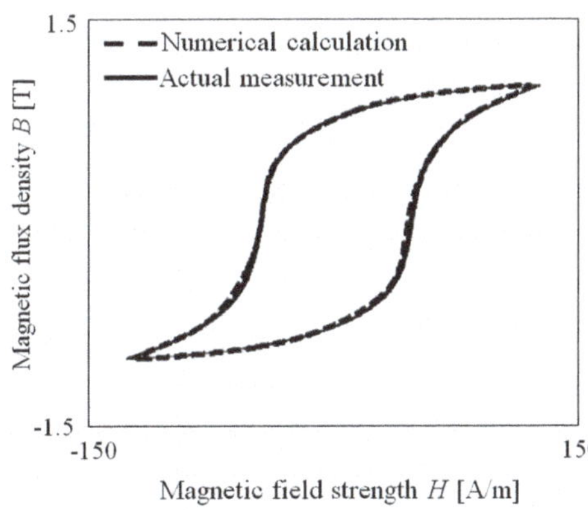

Fig. 10 Reproduction of magnetic hysteresis loop with sinusoidal excitation

calculation result demonstrates that the actual measurement result can be reproduced well.

Next, a magnetic hysteresis loop under inverter excitation is drawn; for convenience, the result of considering the on-state voltage by the circuit analysis is described as follows. The inverter excitation condition is performed so that B_{max} becomes 1 T with a fundamental frequency of 50 Hz, a carrier frequency of 1 kHz, and a modulation index of 0.5. In this case, the skin depth relative to the carrier frequency is 0.135 mm, which is smaller than the thickness of the steel plate. Similarly, Fig. 11 shows two magnetic hysteresis loops obtained from numerical calculation and actual measurement for comparison. Under sinusoidal excitation, major loops at the fundamental frequency of 50 Hz can be reproduced well, whereas minor loops are rectangular in numerical calculation results; these indicate that the minor loops are not reproduced well in this case. This is because the influence of the reaction

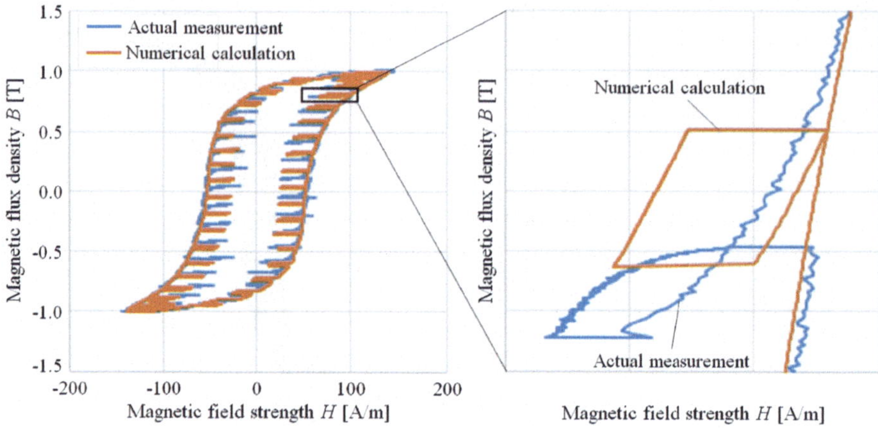

Fig. 11 Reproduction of magnetic hysteresis loop under inverter excitation (right: enlargement of minor loop)

magnetic field caused by the carrier frequency cannot be considered sufficiently, and it is difficult to draw a magnetic hysteresis loop under inverter excitation only by the classical eddy current theory, especially the minor loops.

(2) **Introduction of Cauer equivalent circuit theory**

In this subsection, we apply the Cauer equivalent circuit theory [6] in order to consider the reaction magnetic field caused by the carrier frequency. The Cauer equivalent circuit is a ladder-type circuit (ladder circuit), and the frequency can be expressed according to the rank of the ladder circuit. Here, we express the magnetic hysteresis loop under inverter excitation using an equivalent circuit called type I as shown in Fig. 12. The circuit in this figure is a second-order ladder circuit which can represent up to approximately 20 kHz; the circuit on the first floor represents for the equivalent circuit of the classical eddy current theory. The physical meaning of each component in the ladder circuit is described as follows; L represents a component for expressing a magnetic field with respect to the main magnetic flux, R represents an component for expressing an eddy current generated by the main magnetic flux, the inductor $L'/5$ which the current

Fig. 12 Cauer equivalent circuit type I (second order)

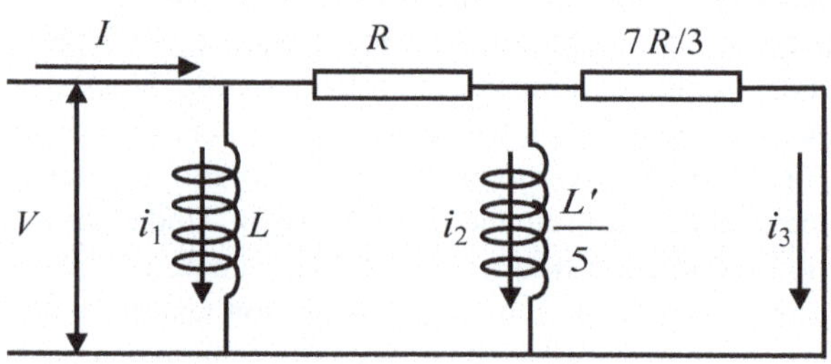

i_2 flows through is a component for expressing the reaction magnetic field generated by the eddy current, and the resistance which the current i_3 flows through is a component for expressing the other eddy current generated by this reaction magnetic field. As in the previous case of using the classical eddy current theory, when equations for representing the electric circuit in Fig. 12 are created, the electromagnetic Eqs. (15)–(18) are now obtained in which each electrical variable is replaced by a relevant magnetic variable. By rearranging these four equations, (19) is obtained. The finite difference method is used to solve these differential equations.

$$V = L\frac{\mathrm{d}i_1}{\mathrm{d}t} \tag{15}$$

$$V = R(i_2 + i_3) + \frac{L'}{5}\frac{\mathrm{d}i_2}{\mathrm{d}t} \tag{16}$$

$$\frac{L'}{5}\frac{\mathrm{d}i_2}{\mathrm{d}t} = \frac{7R}{3}i_3 \tag{17}$$

$$I = i_1 + i_2 + i_3 \tag{18}$$

$$H_{\mathrm{AC}}(B) = H_{\mathrm{DC}}(B) + \frac{7(B^k - B^{k-1}) + 2\mu' h_2^{k-1}}{7R\Delta t + 2\mu'} + \frac{3\mu'(h_2^k - h_2^{k-1})}{35R\Delta t} \tag{19}$$

In (19), Δt is the sampling interval of the waveform B, h_2 is the second term on the right side, and k is the time step. Although μ' is the permeability to the reaction magnetic field, it is reported that a modeling method matching the slope of the minor loop has been proposed and a sufficient system can be well obtained [7]. In this case, the iron loss value in numerical calculation is adjusted to be the same as the actual measurement value. When the magnetic hysteresis loop under inverter excitation is drawn using (19), it becomes as shown in Fig. 13. Minor loops caused by the carrier frequency are drawn smoothly; this clearly shows that the magnetic hysteresis under the inverter excitation can be properly expressed.

3.4 *Procedure (IV) Calculation of Magnetic Field Strength Waveform*

As expressed in (20), the current waveform I is calculated from the magnetic field strength waveform H obtained by the magnetic analysis in the procedure (iii) above.

$$I = \frac{lH}{N} \tag{20}$$

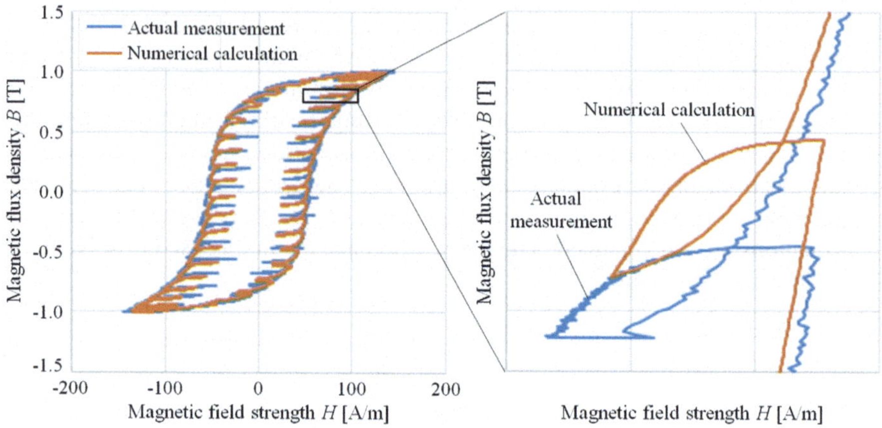

Fig. 13 Reproduction of magnetic hysteresis loop when the Cauer equivalent circuit theory is applied

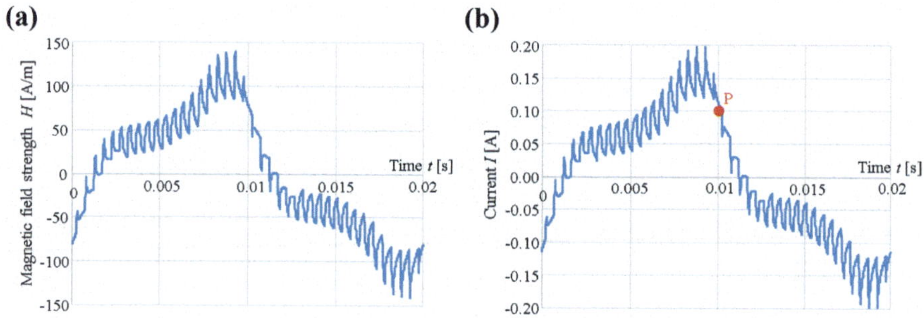

Fig. 14 Results obtained by numerical calculation; **a** Magnetic field strength waveform. **b** Current waveform

l: average magnetic path length of the magnetic material, N: number of coil turns.

In this study, $N = 254$ turns and $l = 0.36$ m are used. The magnetic field strength waveform and the current waveform obtained under the inverter excitation condition are shown in Fig. 14 panels (a) and (b), respectively.

3.5 *Procedure (V) Circuit Analysis with Consideration on On-Voltage Characteristics of Semiconductor Devices*

The circuit analysis is performed with the current waveform obtained in the procedure (iv) as the input. Figure 15 shows an example on voltage–current characteristics of the switching transistor and the freewheeling diode employed in the inverter circuit. First, the on-voltages V_s and V_d of each component corresponding to the input current

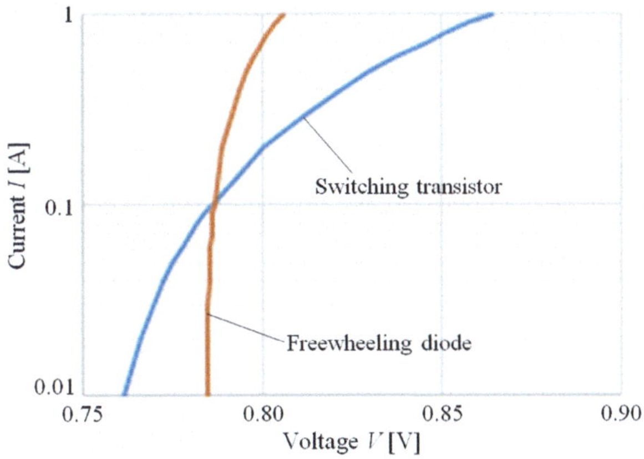

Fig. 15 Voltage–current characteristics of semiconductor devices

are calculated. For example, it can be seen that the current value at the point P (at $t = 0.01$ s) in Fig. 14 panel (b) is 0.1 A, and each on-state voltage at this time is about 0.78 V as given in Fig. 15. Next, at each time of the current waveform, it corresponds to which operation mode shown in Fig. 3, and the on-voltage value for that mode is calculated (e.g., $-2V_s$ in the case of mode 1, and $-V_s - V_d$ in the case of mode 2). The mode in use can be also determined from the switching pattern (Fig. 7) in the procedure (i). At the time $t = 0.01$ s, it is mode 2, and the on-voltage included in the output voltage is $-0.78 - 0.78 = -1.56$ V. By performing this process for one cycle, it is possible to obtain the on-voltage waveform as illustrated in Fig. 16.

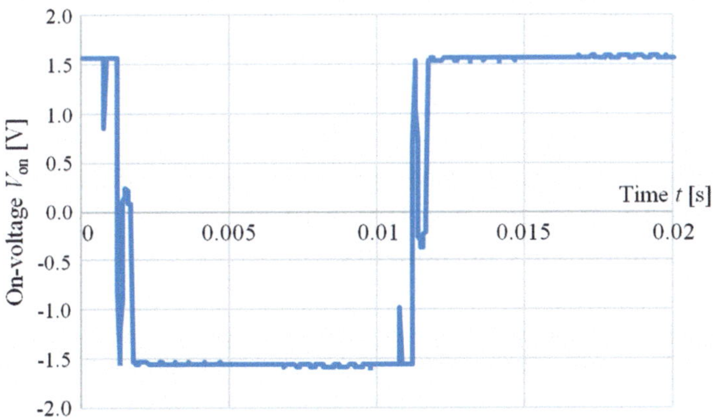

Fig. 16 On-voltage waveform

3.6 Procedure (VI) Determination of Convergence

In this method, the iterative calculation is performed; however, if the relation between the previous on-voltage (before one repetition cycle) and the present on-voltage expressed in (22) is smaller than the convergence value ε, this calculation process is terminated. Otherwise, the procedure (vii) is conducted.

$$K = \sum_{i=1}^{p} V_{\text{on},i}^{2} \tag{21}$$

$$\left| \frac{K_n - K_{n-1}}{K_{n-1}} \right| < \varepsilon \tag{22}$$

i: data point, p: number of data points per cycle, and n: number of iterations.

3.7 Procedure (VII) Overlap of on-Voltage Waveform on the Ideal Voltage Waveform

The on-voltage waveform V_{on} depicted in Fig. 16 is added to the ideal voltage waveform V_{ideal} (Fig. 4 panel (a)) generated in the procedure (i); after that, the procedure (ii) is repeated again as illustrated in Fig. 6. When this iterative calculation is completed, the final voltage waveform is presented in Fig. 4 panel (b). By mutually reflecting the magnetic hysteresis characteristics and the on-voltage characteristics of semiconductor with the above iterative calculation, it is possible to obtain the final magnetic characteristics with consideration on both the characteristics.

4 Calculation Example of Magnetic Characteristics Under Inverter Excitation Using Different Types of Semiconductor Components

By the numerical calculation method presented in the previous section, we compute the magnetic hysteresis characteristics in the case where the magnetic material is excited by the first inverter using the switching transistor and the freewheeling diode with large on-voltages, and then is excited by the second inverter using the semiconductor components with small on-voltages. Here, the first inverter with large on-voltage semiconductor components is referred to as a "large on-voltage inverter," and the other inverter with small on-voltage semiconductor components is referred to as a "small on-voltage inverter." Figure 17 shows the voltage–current characteristics of the semiconductor devices utilized in the two inverters. The final magnetic hysteresis loop obtained by numerical calculation using these characteristics of semiconductor

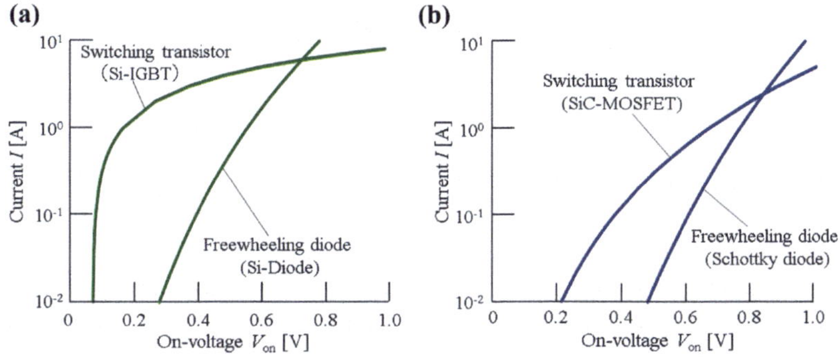

Fig. 17 Voltage–current characteristics of semiconductor devices; **a** Large on-voltage inverter. **b** Small on-voltage inverter

devices is described in Fig. 18. From this figure, it can be seen that the shapes of the minor loops (including closed loop and open loop) are dissimilar between the two inverters with different on-voltage characteristics (i.e., large on-voltage inverter: large minor loop and small on-voltage inverter: small minor loop). This tendency is similar to the actual measurement results as shown in Fig. 19; it can be said that the

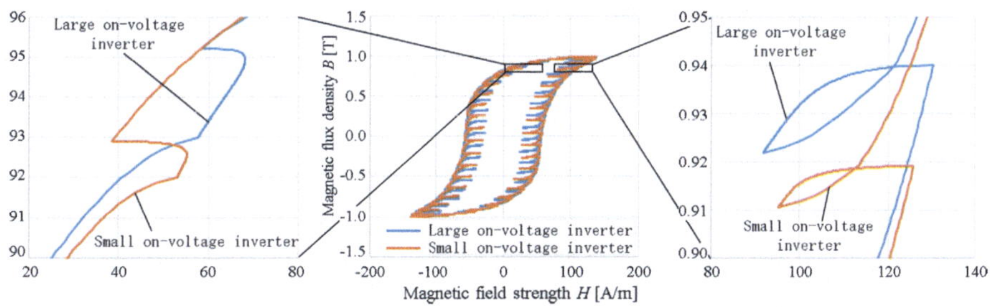

Fig. 18 Magnetic hysteresis loops obtained by numerical calculation with two types of inverter

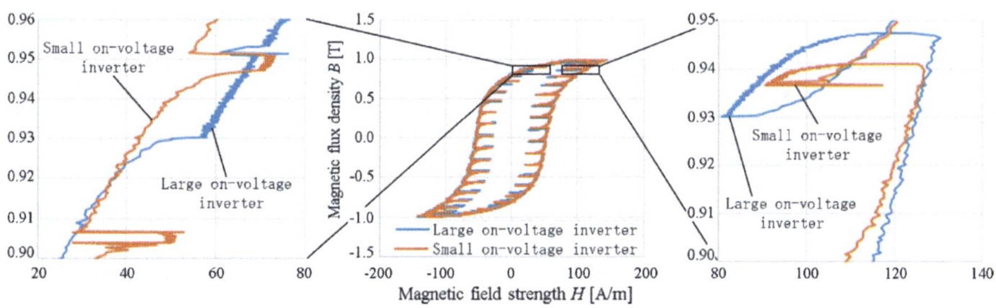

Fig. 19 Magnetic hysteresis loops obtained by experimental measurements with two types of inverter

proposed calculation method mutually considering the above-mentioned semiconductor characteristics and magnetic properties are effective for expressing an actual magnetic hysteresis loop.

References

1. Boglietti, A., et al.: Iron losses in magnetic materials with six-step and PWM inverter supply (induction motors). IEEE Trans. Magn. **27**(6), 5334–5336 (1991)
2. Kaihara, H., et al.: Effect of carrier frequency and circuit resistance on iron loss of non-oriented electrical steel sheet under single-phase full-bridge PWM inverter excitation. IEEJ Trans. IA **132**(10), 983–989 (2012)
3. Odawara, S., et al.: Iron loss characteristics of electrical steel sheet under inverter excitation by using power semiconductor with extremely low on-voltage property. IEEJ Trans. IA **134**(7), 649–655 (2014)
4. Odawara, S., et al.: Evaluation of magnetic properties considering semiconductor properties by using numerical technique coupling inverter circuit analysis to magnetic analysis. IEEJ Trans. IA **135**(12), 1191–1198 (2015)
5. Bobbio, S., et al.: Models of magnetic hysteresis based on play and stop hysterons. IEEE Trans. Magn. **33**(6), 4417–4426 (1997)
6. Shindo, Y., et al.: Cauer circuit representation of the homogenized eddy—current field based on the legendre expansion for a magnetic sheet. IEEE Trans. Magn. **52**(3), 6300504 (2016)
7. Miyazaki, T., et al.: Equivalent circuit modeling of dynamic hysteretic property of silicon steel under pulse width modulation excitation. J. Appl. Phys. **117**, 17D110 (2015)

Vector Magnetic Characteristic

Masato Enokizono

Abstract The vector magnetic characteristic was obtained by reexamining and reconstituting the conventional magnetic characteristic. The fundamental viewpoint grasps the magnetic characteristic in the vector relation. This magnetic characteristic leads to the magnetic characteristic analysis, and it is useful for the design development of electrical machinery and apparatus.

Keywords Vector magnetic characteristic · Vector locus · E & S model · Magnetic characteristic analysis

1 Basic Concept

This session provides how to decrease the core loss of the electrical machines such as motor, actuator and transformer. And it is necessary to sufficiently clarify how core loss decreases.

We have no recipe data of power magnetic material for making into low-loss design. Catalog data offered by the material company are insufficient as recipe data. Conventional numerical soft program using catalog data is not enough for magnetic characteristic analysis. Then, the vector magnetic characteristic technology has been proposed, which was reconstituted by reexamining the conventional technology.

In the past, magnetic property of non-oriented magnetic steel sheets used for motors and other parts was measured by the Epstein method, single plate magnetic test method (Single Strip Tester: SST) as shown in Fig. 1a or some other method, and the measurement data were used for design.

It is known that the data measured by these methods are generally averaged in the rolling and transverse directions, so do not precisely represent the magnetic material properties. One reason is that even "non-oriented" steel sheets have a magnetic anisotropy and their magnetic properties vary according to the direction. In addition, we have been discovering that the directions of magnetic field applied and magnetic flux generated in the magnetic material, which were believed to be the same, are

M. Enokizono (✉)
Vector Magnetic Characteristic Technical Laboratory, Oita, Japan
e-mail: enoki@oita-u.ac.jp

K. Fujisaki (ed.), *Magnetic Material for Motor Drive Systems*,
Engineering Materials, https://doi.org/10.1007/978-981-32-9906-1_17

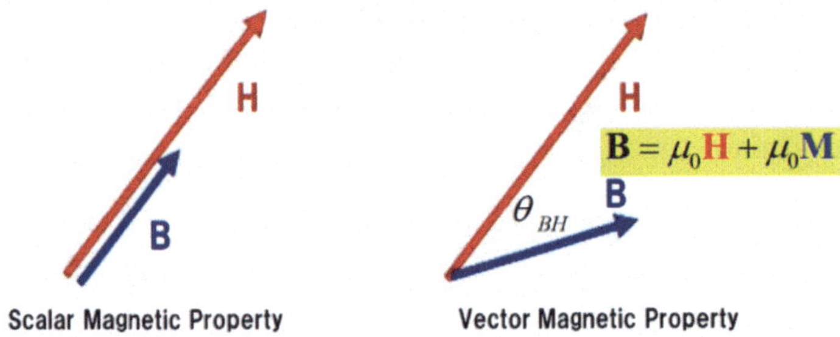

(a)Vector relation between vector **H** and vector **B**

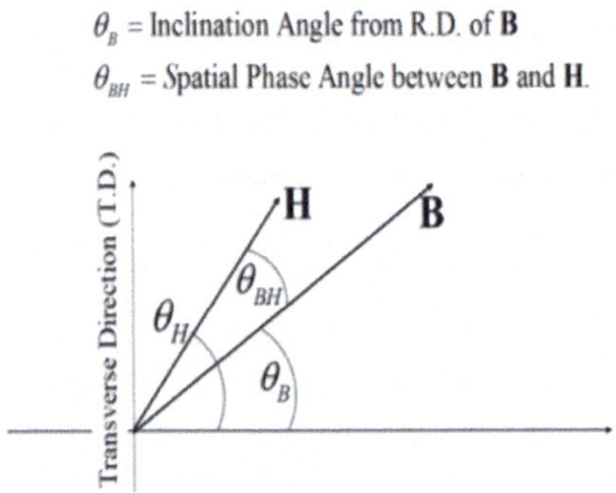

(b) Definition of vector magnetic characteristic

Fig. 1 Concept of vector magnetic characteristic

actually different. The properties that can be measured with this analyzer are called "**vector magnetic characteristic.**"

Generally, the magnetic field strength vector **H** is not always parallel to the magnetic flux density vector **B**. Therefore, the magnetic characteristics of magnetic materials have the vector relation on constitutive equation generally as shown in Fig. 1 and the following equation:

$$\mathbf{B} = \mu_0\mathbf{H} + \mu_0\mathbf{M} = \tilde{\mu}\mathbf{H} \tag{1}$$

$$\tilde{\mu} = \begin{bmatrix} \mu_{xx} & \mu_{xy} \\ \mu_{yx} & \mu_{yy} \end{bmatrix}$$

where **B** is the magnetic flux density vector, **H** is the magnetic field strength vector, **M** is the magnetization vector, and μ_0 is magnetic constant. The $\tilde{\mu}$ is magnetic permeability expressed as tensor style because it connects the relation between vector **H** and vector **B**. However, the components of this $\tilde{\mu}$ cannot be measured, which is ill-posing problems. The conventional magnetic characteristic is measured under the

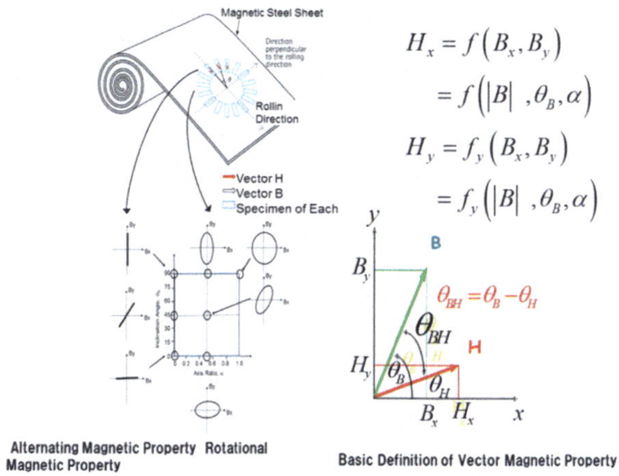

Fig. 2 Measurement specimen and measurement condition

parallel condition on both vector **H** and vector **B**. Then, it is possible to call it the scalar magnetic characteristic. The parameter shown in the figure in order to show the vector relation of the magnetic characteristic was defined as shown in the right side of the figure. The θ_B is the inclination angle of vector **B** from rolling direction, and the θ_H is the inclination angle of vector **H**. The θ_{BH} is the spatial phase difference angle between vector **B** and vector **H,** and it has the nonlinearity for magnetic flux density and angle θ_B. Conventional magnetic characteristic expresses scalar relation which handles only the size of the mapping of the measurement direction as shown in Fig. 1b.

On the other hand, the electrical steel sheet is cut down for the standard measurement as strip sheet specimen along the rolling direction only. However, this measurement results are partial and cannot grasp the characteristics of arbitrary direction as shown in Fig. 2. The electrical steel sheets are not isotropy in non-oriented electrical steel sheet. And then these have various kinds of anisotropy, which are grain orientation, crystal magnetic anisotropy, etc. Furthermore, the magnetic characteristic is measured evenly under the alternating magnetic flux condition and the rotational magnetic flux condition. On rotational magnetic characteristics, the axial ratio α is defined in the vector locus (Lissajous figure) of vector **B** in order to express magnetic characteristics under the rotational magnetic flux conditions. The $\alpha = 0$ means alternating magnetic flux, and $\alpha = 1$ means the pure circular rotational magnetic flux. The magnetic field strength becomes the function of B_x and B_y, and it can be changed in the magnitude |**B**| of vector **B**, the angle θ_B and axis ratio α of locus of vector **B**. Therefore, magnetic field strength is shown as follows in Eq. (2):

$$\begin{aligned} H_x &= f_x(B_x, B_y) = f_x(|\mathbf{B}|, \theta_B, \alpha) \\ H_y &= f_y(B_x, B_y) = f_y(|\mathbf{B}|, \theta_B, \alpha) \end{aligned} \tag{2}$$

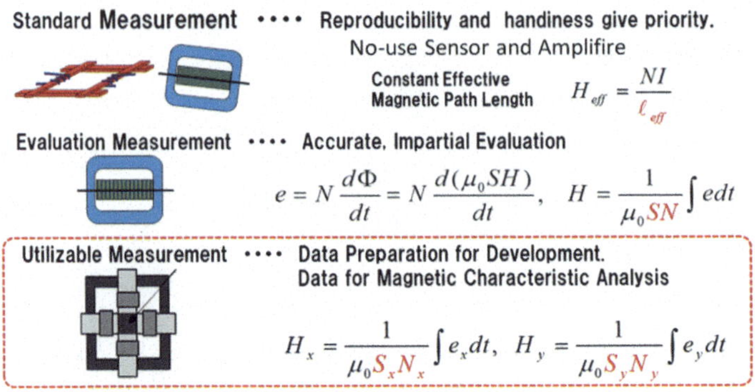

Fig. 3 Categories of magnetic measurement method

As mentioned above, the vector magnetic characteristic can propose the new characterization of arbitrary direction under alternating magnetic flux condition as follows in the expression:

$$|\mathbf{B}| - \theta_{BH} - |\mathbf{H}| \text{ characteristic: (parameter } \theta_B = 0 \sim 2\pi)$$

The measured magnetic characteristic data volume drastically increases in comparison with the conventional characteristic quantity to be measured. This vector expression can comprehensively express the whole magnetic characteristic. From the viewpoint of vector magnetic characteristic, the magnetic characteristic shows the nonlinearity in not only the magnitude but also direction. This cannot be shown by the hysteresis loop which showed the relationship only between the conventional magnitudes of magnetic property.

The magnetic characteristic measurement method can be classified into the purpose, independence at three categories about the measurement of magnetic field strength H as shown in Fig. 3.

1.1 Standard Magnetic Measurement Method

The standard magnetic measurement method takes convenience and reproducibility seriously on magnetic characteristics. This measuring method is not possible to use the sensor and the amplifier in order to avoid the calibration procedure and for keeping justice. Therefore, the effective magnetic path length is introduced in this method by Ampère's circuital law, as shown in the measurement of H in the figure. The problem of this measuring method is to keep to the constant magnetic path length by magnetic flux density. As the effective path length is a constant value, the magnetic flux density is overestimated of the B–H characteristic in the high saturation magnetization region as follows:

$$H = \frac{\mathrm{NI}}{\ell_{\mathrm{eff}}} \tag{3}$$

where ℓ_{eff} is the effective path length which is decided as IEC standard, and NI is the magnetomotive force. The measurement method is called the current method.

1.2 Evaluation Magnetic Measurement Method

The evaluate method of magnetic characteristics requires the accuracy for the research and development of magnetic materials in the laboratory. It is called *H*-coil method, which uses the *H*-coil sensor by Lenz's law and the accurate differential amplifier with high input impedance. However, the accurate calibration of NS (effective area turn) is required by using standard solenoid coil system.

$$\begin{aligned} e &= N\frac{\mathrm{d}\Phi}{\mathrm{d}t} = N\frac{\mathrm{d}(\mu_0 SH)}{\mathrm{d}t} \\ H &= \frac{1}{\mu_0(SN)}\int e\mathrm{d}t \end{aligned} \tag{4}$$

This evaluation must measure by ensuring the homogeneous region of magnetic field and magnetic flux density distribution

1.3 Utilizable Magnetic Measurement Method

The utilizable magnetic measurement method uses two kinds (x-direction and y-direction, or rolling direction and transverse direction) of *H*-coil for the measurement of vector magnetic characteristics. It is required that this two-dimensional measuring method controls the accurate angle θ_B of the vector **B** by the two-dimensional excitation by two-phase source. On the other hand, the grasp of the magnetic characteristic of the arbitrary direction of the electrical machine core material has been required for making in low-loss design of the electrical machines. This magnetic characteristic is correspondent to magnetic characteristic in the iron core of real machine such as the motor. Therefore, it is necessary to measure by two-dimensional magnetic measurement. The vector magnetic characteristics can be obtained by the following equation:

$$\begin{aligned} B_x &= \frac{1}{N_{Bx}S_{Bx}}\int e_{Bx}\mathrm{d}t \\ B_y &= \frac{1}{N_{By}S_{By}}\int e_{By}\mathrm{d}t \end{aligned} \tag{5}$$

$$H_x = \frac{1}{\mu_0 N_{Hx} S_{Hx}} \int e_{Hx} \mathrm{d}t$$
$$H_y = \frac{1}{\mu_0 N_{Hy} S_{Hy}} \int e_{Hy} \mathrm{d}t \tag{6}$$

where e_{Hx} and e_{Hy} are the output voltage of the H-coil, e_{Bx} and e_{By} are the output voltage of search coil (B-coil), and $N_{Hx}S_{Hx}$ and $N_{Hy}S_{Hy}$ are effective area turn which must be calibrated in an electromagnetic evaluation.

The core loss can solve the following equation:

$$W = \frac{1}{\rho T} \int \left(\boldsymbol{H} \cdot \frac{\partial \boldsymbol{B}}{\partial t} \right) \mathrm{d}t = \frac{1}{\rho T} \int \left(\boldsymbol{H}_x \frac{\partial \boldsymbol{B}_x}{\partial t} + \boldsymbol{H}_y \frac{\partial \boldsymbol{B}_y}{\partial t} \right) \mathrm{d}t \tag{7}$$

where ρ is the density of magnetic material. The core loss is obtained from two kinds of hysteresis characteristics, $B_x - H_x$ and $B_y - H_y$ loop characteristics.

2 Vector Magnetic Characteristics

Two-dimensional magnetic hysteresis loop of arbitrary direction obtained from vector magnetic characteristic is shown in Fig. 4 under alternating flux conditions on non-oriented electrical steel sheet and grain-oriented electrical steel sheet, respectively. Both vector **B** and vector **H** become parallel in the rolling direction and the transverse direction, and in the other directions, the vector locus produces the spatial phase difference angle between two vectors. These characteristics are composed as the hysteresis loop of two kinds of x-direction and y-direction. Though in case of the non-oriented steel sheet, the phase difference angle θ_{BH} is small, and it is very large for the grain-oriented steel sheet. In the magnetic saturation region, this angle θ_{BH} gradually decreases. The shape of the hysteresis loop of the grain-oriented steel sheet becomes abnormal for the strong anisotropy.

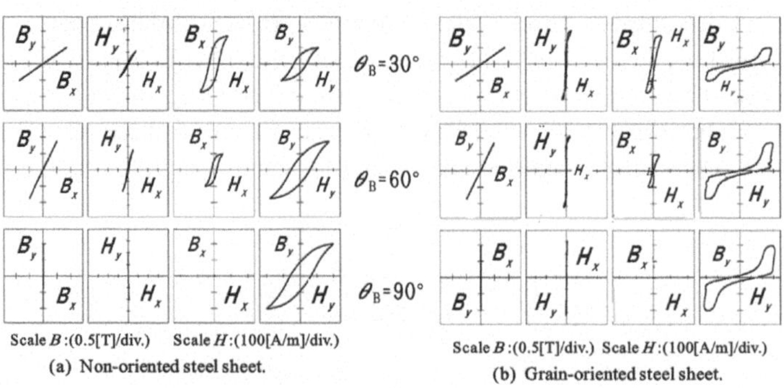

Fig. 4 Vector locus and hysteresis loop characteristics of arbitrary direction

Figure 5a, b shows the locus of vector **H** under alternating flux condition and rotational flux condition about non-oriented steel sheet and grain-oriented steel sheet, respectively. The effect of the orientation of the rolling is strong for 1.2 T for the

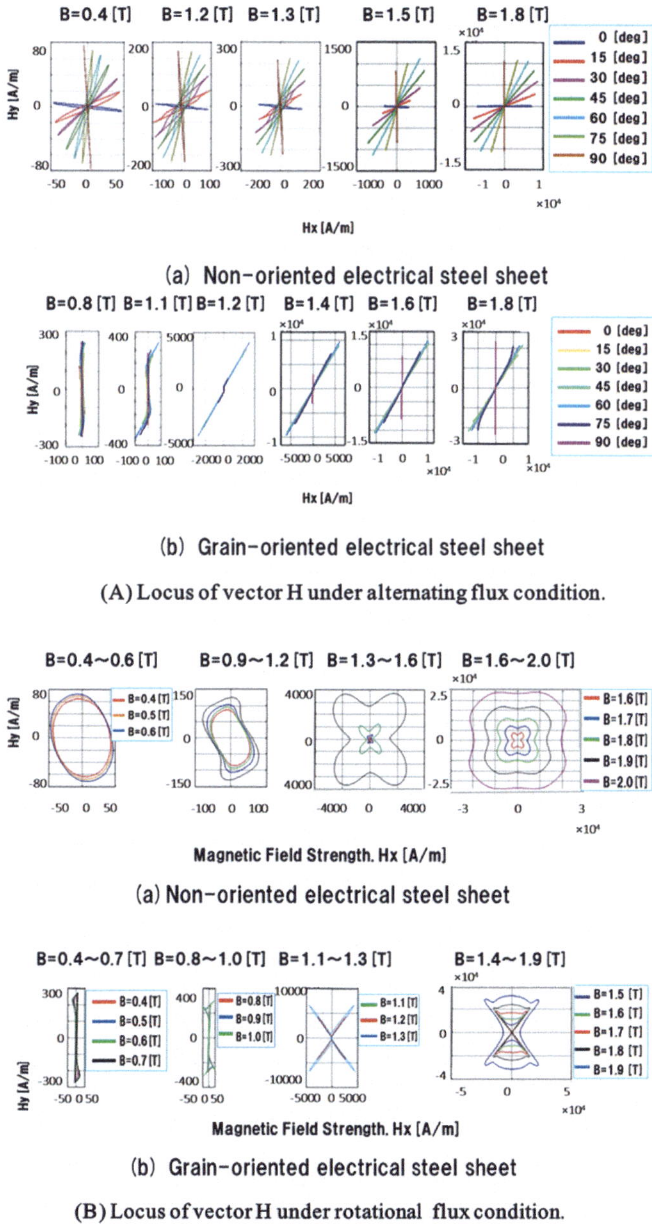

Fig. 5 Vector locus characteristics of vector **B** and vector **H**

locus characteristic, and the effect of the crystal magnetic anisotropy becomes large over 1.2 T.

3 Magnetic Characteristic Analysis

The magnetic characteristic becomes the distorted waveform in which the magnetic field strength waveform contains higher harmonic because it is measured under the magnetic flux sine waveform condition. Coefficient waveform which connects the interval of B-waveform and H-waveform is considered.

In order to introduce the above-mentioned vector magnetic characteristic into the magnetic characteristic analysis, the modeling can be proposed as follows [1–4]:

$$\begin{aligned} H_x &= \nu_{rx}(|\mathbf{B}|, \theta_B, \alpha : t)B_x + \nu_{ix}(|\mathbf{B}|, \theta_B, \alpha : t)\int B_x \mathrm{d}t \\ H_y &= \nu_{ry}(|\mathbf{B}|, \theta_B, \alpha : t)B_y + \nu_{iy}(|\mathbf{B}|, \theta_B, \alpha : t)\int B_y \mathrm{d}t \end{aligned} \tag{8}$$

where ν_{rx} and ν_{ry} are magnetic reluctivity coefficient, and ν_{ix} and ν_{iy} are magnetic hysteresis coefficient. These coefficients are obtained as a waveform.

By introducing Eq. (8) to Maxwell equation, the fundamental equation of the magnetic characteristic analysis will be shown as follows:

$$\frac{\partial}{\partial x}\left(\nu_{ry}B_y + \nu_{iy}\int B_y \mathrm{d}\tau\right) - \frac{\partial}{\partial y}\left(\nu_{rx}B_x + \nu_{ix}\int B_x \mathrm{d}\tau\right) = J_0 \tag{9}$$

Furthermore, Eq. (9) can rewrite using magnetic vector potential **A** as follow:

$$\begin{aligned} &\frac{\partial}{\partial x}\left(\nu_{ry}\frac{\partial A}{\partial x}\right) + \frac{\partial}{\partial y}\left(\nu_{rx}\frac{\partial A}{\partial y}\right) \\ &\quad + \left\{\frac{\partial}{\partial x}\left(\nu_{iy}\int \frac{\partial A}{\partial x}\mathrm{d}\tau\right) + \frac{\partial}{\partial y}\left(\nu_{ix}\int \frac{\partial A}{\partial y}\mathrm{d}\tau\right)\right\} = -J_0 \end{aligned} \tag{10}$$

The second term of Eq. (10) shows the hysteresis term.

Figure 6 shows the contrast of conventional analysis method and vector magnetic characteristic analysis method. Only the behavior of the magnetic flux density vector is solved by the conventional analytical method through vector potential; then, the conventional method cannot solve the behavior of the magnetic field strength vector. The conventional analysis method must show the loss in the function of B in order to solve only B.

By utilizing the measurement data on magnetic field strength corresponding to the different elliptical locus of magnetic flux density vector, the core loss in Eq. (7)

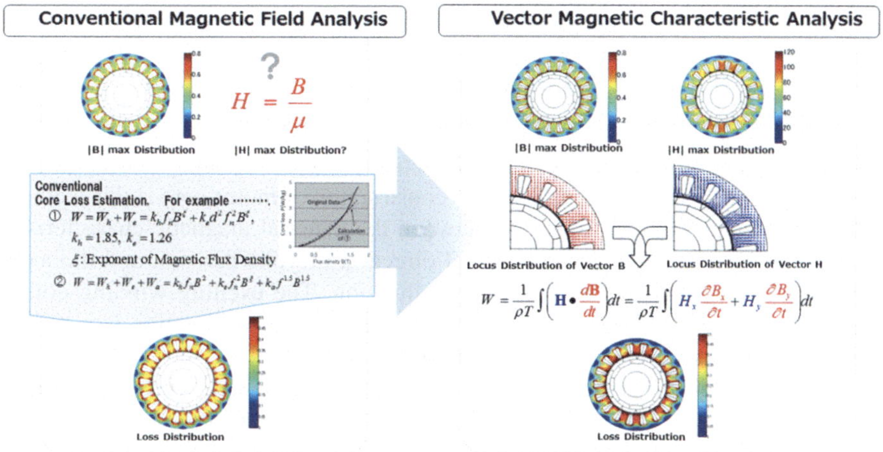

Fig. 6 Magnetic characteristic analysis

of the electric steel sheet can be analyzed same as a measurement. The analyzed core loss results include the high harmonic terms of H-waveform.

The core loss W can be calculated by the following equation [5].

$$W = \frac{1}{\rho \cdot T}\left(\int_{0}^{T} H_{kx} \cdot \mathrm{d}B_{kx} + \int_{0}^{T} H_{ky} \cdot \mathrm{d}B_{ky}\right) \tag{11}$$

The magnetic field strength H_x, H_y and magnetic flux density B_x, B_y are analyzed as well as the magnetic measurement as shown in Eq. (7), because it is solved as each waveform of H_x, H_y, B_x and B_y. It can be calculated on the sum of the area of the hysteresis loop of x-direction and y-direction.

The magnetic characteristic analysis method was applied to the analysis of the iron core which is composed of three kinds of electrical steel sheet, in order to clarify the advantage of the magnetic characteristic analysis introducing vector magnetic characteristics.

(a) Non-oriented Electrical Steel Sheets
These steel sheets are used as motor core materials and show the easy magnetized characteristic to the rolling direction as shown in Fig. 4a.
(b) Grain-oriented Electrical Steel Sheets
These steel sheets usually have (110) Goss texture. The magnetic properties are developed in the rolling direction, due to a tight control of the crystal orientation relative to the sheet as shown in Fig. 4b. The magnetic flux density increases in the rolling direction. It is used for the cores of power and distribution transformers.
(c) Double-oriented Electrical Steel

These steel sheets have secondarily recrystallized out {100} ⟨001⟩ oriented grains. These magnetic properties are developed in the rolling direction and transverse direction. The magnetization is difficult to the direction of about 50 from the rolling direction for the magnetic characteristic.

Figure 7 shows the results of the magnetic characteristic analysis introducing vector magnetic characteristics. The core construction of motor core is the segment core of understanding the difference between the material magnetic characteristic as shown in Fig. 7a. The circumferential direction of yoke division and the axial direction of teeth division are the rolling direction. The exciting winding coil is three-phase distributed winding structure. In these results, y-direction is a rolling direction of core in the evaluation region for the extraction.

The upper stage division of Fig. 7b shows the locus of vector **B** and vector **H**, and the lower division shows the $B_x - H_x$ and $B_y - H_y$ hysteresis loop. The rotating magnetic field is being generated for the three-phase excitation in the motor stator core. The magnetic characteristic in the core under a rotational magnetic flux is shown by the rotating magnetic field excitation. This vector magnetic characteristic is different from three kinds of electrical steel sheets. This vector behavior is determined in excitation direction and direction by the anisotropy of the steel sheets.

In the case of grain-oriented electrical steel core, the locus of vector **B** becomes large to the rolling direction since excitation direction of winding structure agrees with the rolling direction, and by receiving effect of the rotating magnetic field, the behavior of vector **H** to the transverse direction becomes large. The shape of hysteresis loop reflects the vector locus. It can be indicated that the spatial phase angle θ_{BH} becomes large from the vector behavior. The increase of the angle θ_{BH} is connected with the increase in vector **H**. The double-oriented steel sheet has the smaller angle θ_{BH}. In the case of double-oriented electrical steel core, the locus of vector **B** in the rolling direction becomes large, and the locus of vector **H** becomes smaller to transverse direction. Therefore, the hysteresis loop $B_x - H_x$ characteristic is very small.

From an above-mentioned knowledge, the magnetic characteristic analysis and vector magnetic characteristics offer the clue which is useful for the design of the motor iron core.

4 Summarize

Conventional magnetic characteristic measuring technique is reexamined, and vectors magnetic characteristic technology which leads to the material guideline utilization is necessary for the performance improvement of the motor.

The advantage of vector magnetic characteristic is summarized in the following.

(1) The parameter of vector magnetic characteristic is the amplitude of vector **B** and vector **H**: |**B**|.|**H**|, the inclination angle of rolling direction: θ_B, the spatial

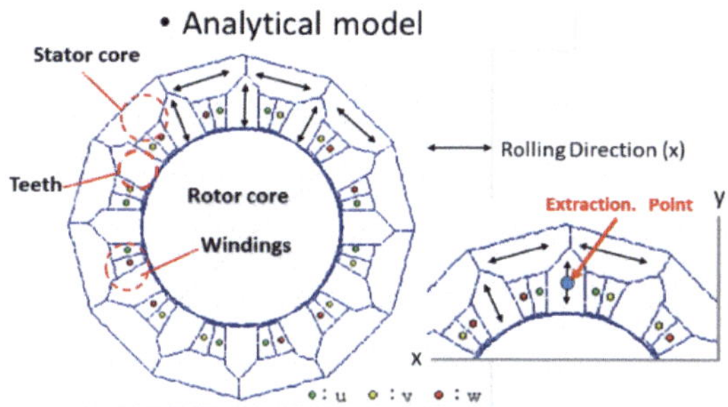

(A) Analysis model.

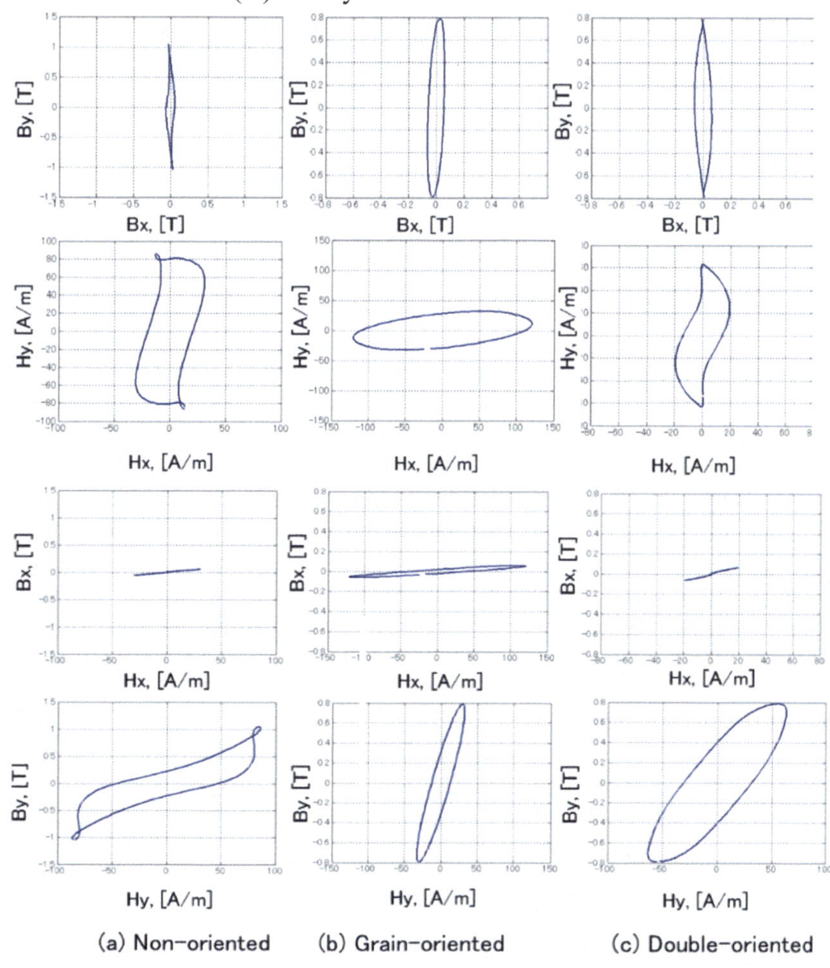

(B) Vector locus and hysteresis characteristic.

Fig. 7 Analytical results of the magnetic characteristic analysis by E and S model considering vector magnetic characteristics differential phase angle between vector **B** and vector **H**: θ_{BH}, and axis ratio of vector locus: α.

(2) In addition to the conventional hysteresis characteristic, the new characteristic can be proposed.

$|\mathbf{B}| - \theta_{BH} - |\mathbf{H}|$ characteristics
$|\mathbf{H}| - \theta_B - |\mathbf{B}|$ characteristic
$\theta_{BH} - \theta_B - |\mathbf{B}|$ characteristics.

(3) The advantage of the magnetic characteristic analysis is as follows:

(a) The distribution of both vector **B** and vector **H**, especially the behavior of vector **H** is useful because it depends on exciting current.
(b) The locus distribution of both vector **B** and vector **H**.
(c) Vector hysteresis loop ($B_x - H_x$, $B_y - H_y$) characteristics.
(d) The distribution of core loss.
(e) The parameter evaluation of vector magnetic characteristics: θ_B, θ_H, θ_{BH}, α, the loss separation between hysteresis loss and eddy current loss.

(4) In order to the reduction of the core loss of motors, the following can be examined for controlling the vector magnetic characteristics: Motor core structure and effect of the stress.

The magnetostrictive phenomenon of magnetic material controls magnetic characteristic, and it becomes a cause of noise and vibration of electrical machinery and apparatus. Reviews are also similarly necessary to the magnetostriction characteristic with the vector magnetic characteristic. The direction in which the magnetic material by a magnetic field is extended does not always agree with the direction of the magnetic flux density vector. The grasp of the magnetostriction characteristic of the arbitrary direction is required in order to catch the magnetostrictive phenomenon as a change of the shape in magnetic materials. The magnetostriction characteristic of the arbitrary direction is called the two-dimensional magnetostriction instead of the conventional magnetostriction characteristic known as a butterfly loop. Such measurement is measured under the two-dimensional excitation by the strain gauge of triaxial structure [6–8].

Furthermore, the effect of the stress on the magnetic characteristic is also important [9, 10]. The vector magnetic characteristic by being stressed from the arbitrary direction is a means for being useful for the control of the magnetic characteristic too [11, 12].

References

1. Enokizono, M., Soda, N.: Magnetic field analysis by finite element method using effective anisotropic field. IEEE Trans. Magn. VIL **31**(3), 1793–1796 (1995)

2. Enokizono, M., Soda, N.: Direct magnetic loss analysis by FEM considering vector magnetic properties. IEEE Trans. Magn. VIL **34**(5), 3008–3011 (1998)
3. Soda, N., Enokizono, M.: Hysteresis model of two-dimensional magnetic properties using an E&S model. J. Magn. Soc. Jpn. **24**(4–2), 659–662 (2000)
4. Urata, S., Shimoji, H., Todaka, T., Enokizono, M.: Measurement of two-dimensional vector magnetic properties on frequency dependence of electrical steel sheet. Int. J. Appl. Electomagn. Mech. **20**, 155–162 (2004)
5. Bozorth, R.M.: Ferromagnetism, pp. 145–149. Wiley, IEEE Press, New York (1998)
6. Yamagashira, M., Ueno, S., Wakabayashi, D., Enokizono, M.: Vector magnetic properties and two-dimensional magnetostriction of various soft magnetic materials. Int. J. Appl. Electromagnet. Mech. **44**(3, 4), 387–400 (2014)
7. Yamagashira, M., Wakabayashi, D., Enokizono, M.: Vector magnetic properties and two-dimensional magnetostriction of various electrical steel sheets under rotating flux condition. IEEE Trans. Magn. **50**(4), 4 (2014)
8. Wakabayashi, D., Enokizono, M.: Numerical representation of two-dimensional magnetostriction in electrical machines. Int. J. Appl. Electromagnet. Mech. **45**, 939–947 (2014)
9. Kai, Y., Enokizono, M.: Measurement of two-dimensional magnetostriction of a non-oriented electrical steel sheet under shear stress. Int. J. Appl. Electromagnet. Mech. **48**(2), 233–238 (2015). https://doi.org/10.3233/jae-151993. IOS Press
10. Kai, Y., Enokizono, M.: Influence of biaxial stress on vector magnetic properties and 2-D magnetostriction of a non-oriented electrical steel sheet under alternating magnetic flux conditions. IEEE Trans. Magn. **50**(4), 1–4 (2014)
11. Vector Magnetic Characteristic Technical Laboratory. http://www.vector-magtec.jp
12. Enokizono, M.: Vector magnetic characteristic technology and design method, p. 262. Kagaku-jyoho Shuppan Co., Ltd. ISBN: 978-4-904774-36-6. Apr 2015 (Japanese Book)

Future Magnetic Material

History and Future of Soft and Hard Magnetic Materials

Satoshi Sugimoto

Abstract Magnetic materials are essential in modern daily life as they are used in numerous areas, such as electronics, industrial equipment, and automobiles, as energy conversion materials that drive devices such as generators, which convert kinetic to electrical energy, and motors, which convert electrical to kinetic energy. This chapter introduces the technological history and representative examples of magnetic materials. Further details can be found in Refs. [1–12].

Keywords Magnetic anisotropy · Magnetostriction · Permeability · Maximum energy product · Coercivity

1 Soft and Hard Magnetic Materials

Soft magnetic materials generally have high magnetic permeability. For this, it is preferable for the material to have low coercivity, high magnetization, and high electrical resistance. A magnetic structure with a low magnetic anisotropy constant K and magnetostriction constant λ is required because this makes domain-wall motion and magnetization rotation easier. These properties are obtained in magnetic materials by increasing their purity and aligning their easy axes of magnetization (the direction that is most easily magnetized), and in the case of metals, this is accomplished by making particles smaller than and thin films thinner than the skin depth to suppress eddy currents.

Soft magnetic materials can be broadly categorized as metals and alloy materials, or as metal oxides, called ferrites. Examples of metals include pure Fe, Si steel (isotropic, anisotropic), permalloy, supermalloy, Sendust, permendur, amorphous metals, and nanocrystals. Examples of oxide materials include spinel ferrites (Mn–Zn, Ni–Zn), ferox planar ferrite, and garnet-type ferrite.

Hard magnetic materials are also called permanent magnets, and they have high coercivity, residual magnetization, and maximum energy product $(BH)_{max}$. Therefore, these materials should have large magnetic anisotropy and magnetostriction,

S. Sugimoto (✉)
Graduate School of Engineering, Tohoku University, Sendai, Japan
e-mail: sugimots@material.tohoku.ac.jp

K. Fujisaki (ed.), *Magnetic Material for Motor Drive Systems*,
Engineering Materials, https://doi.org/10.1007/978-981-32-9906-1_18

which requires excellent stability, including high mechanical strength, stability, and heat and corrosion resistance.

Permanent magnet materials can be broadly categorized as metal magnets or oxide magnets. Metal magnets are divided into powerful rare earth magnets, such as Nd–Fe–B magnets, Sm–Co magnets, and alloy magnets, which have good temperature properties and are exemplified by alnico magnets and Fe–Cr–Co magnets. Oxide magnets use Fe oxides and are very inexpensive, and thus ferrite magnets (Ba- and Sr-based magnetoplumbite ferrite magnets) are the magnets that people are most familiar with in daily life. Soft magnetic materials and hard magnetic materials are explained in more detail in the following sections.

2 Soft Magnetic Materials

2.1 History of the Development of Soft Magnetic Materials [5–7, 12]

The history of the development of soft magnetic materials is summarized in Table 1. Pure Fe and low-carbon mild steel were used up until around 1900. Hopkinson et al. showed that Fe loss could be reduced by the addition of Si to Fe [5, 12], and Barret et al. [13] fabricated Si steel sheet containing 1.5–2.5% Si in 1900 and established a

Table 1 History of the development of soft magnetic materials

Year	Inventor	Country	Material
1900	Barret, Brown, Hadfield	UK	Si steel plate (Fe-1.5 to 2.5%Si)
1923	Arnold, Elemen	USA	Permalloy (Fe–Ni)
1926	Honda, Kaya	Japan	Measurement and discovery of magnetocrystalline anisotropy
1929	Elemen	USA	Permendur (Fe-50%Co)
1932	Kato, Takei	Japan	Cu–Zn ferrite
1934	Goss	USA	Oriented Si steel plate
1932	Masumoto (Hakaru), Yamamoto	Japan	Sendust (Fe–Si–Al)
1947	Boothby, Bozorth	USA	Supermalloy (Fe–Ni–Mo)
1947	Snoek	Netherlands	Mn–Zn ferrite, etc.
1972	Takahashi (Migaku),Kim	Japan	$Fe_{16}N_2$
1974	Fujimori, Masumoto (Tsuyoshi), Obi, Kikuchi	Japan	Amorphous soft magnetic properties
1975	Egami, Flanders, Graham	USA	Amorphous soft magnetic properties
1988	Yoshizawa, Oguma, Yamauchi	Japan	Nanocrystalline soft magnetic materials (Fe–Si–B–Cu–Nb)

fabrication method in 1903. Permalloy, which is an Fe–Ni alloy, was developed by Arnold and Elemen [14, 15] in 1923, and permendur, which is a Fe-50%Co alloy, was developed in 1929 [16]. In 1926, Honda and Kaya [17] made the important discovery of magnetocrystalline anisotropy when they used Fe single crystals in research into magnetic materials, and this resulted in the development of oriented Si steel sheet in 1934 [18]. Furthermore, in 1932, Kato and Takei [19–21] discovered ferrites based on Fe oxides that were completely different from previous materials. This discovery led to soft magnetic materials called soft ferrites, which are generally spinel ferrites and are used, for example, as magnetic core materials, and to hard magnetic materials such as oxide powder magnets based on $CoFe_2O_4$ and magnetoplumbite-type ferrite magnets [22].

Other alloy materials were also emerging around this time. In 1932, Masumoto and Yamamoto [23] created Sendust, which is an Fe–Si–Al alloy, and in 1947, Boothby and Bozorth [24] developed supermalloy, which is an Fe-79%Ni-5%Mo alloy. At the beginning of the 1970s, amorphous alloy soft magnetic materials appeared based on different concepts from previous materials, and they consisted of non-crystalline phases formed by rapid cooling, such as liquid quenching methods [7, 25–28]. In 1988, nanocrystalline alloys [5, 29–31] were developed by precipitating nanocrystals from an amorphous phase, and these materials offered both high saturation magnetic flux density and high magnetic permeability. During the development of these amorphous alloys and nanocrystalline alloys, new fabrication techniques, such as sputtering, were used, which also led to the development of magnetic thin films [32] and granular thin-film materials [33].

However, new developments were also seen in Si steel sheet from the latter half of the 1980s, with both oriented Si steel sheet and Fe-6.5%Si alloy [34–36] becoming mass produced as a highly functional material. The following sections give an overview of representative soft magnetic materials.

2.2 *Permalloy [4, 12]*

In 1913, Elemen reported that high magnetic permeability could be obtained in Fe–Ni alloys containing 30% to 90% Ni, and this was given the name permalloy. Heat treatment methods for the Fe-80%Ni alloy were investigated in 1923, and rapid cooling from 600 °C improved the initial permeability [13]. In Fe–Ni alloys, a Ni_3Fe order phase exists at compositions of 50–85% Ni, and the order-disorder transformation occurs at 503 °C. Therefore, if the material is not quenched from a high temperature, the order phase appears which exhibits crystal magnetic anisotropy and decreases magnetic permeability. There is a strong relationship between the generation of this order phase and the permeability.

The effects of adding a third element, such as Mo or Cr, were investigated, and they were found to suppress the appearance of the regular phase and produced high permeability with good reproducibility even with slow cooling. This research led to the development of Mo permalloy and Cr permalloy. The effects of other additives, such as Cu, were also investigated, resulting in μ-metals. The Fe-72%Ni-14%Cu-3%Mo alloy published by Neumann et al. in 1934 had an initial permeability, μ_i, of 4000 and a maximum permeability, μ_{max}, of 10,000. In 1947, Boothby and Bozorth [24] developed supermalloy, which exhibits good magnetic properties of $\mu_i = 120{,}000$ and $\mu_{max} = 600{,}000$, by using high-purity materials and applying a dissolution method and heat treatment to the Fe-79%Ni-5%Mo alloy, which is a Mo permalloy.

The alloy composition of permalloy must be selected to suit its use because the magnetic properties depend on the amount of Ni. Alloys with 35–40% Ni have low permeability but high saturation magnetic flux density, B_s, making them suitable for use in alternating magnetic fields. The alloy with 36% Ni has a low thermal expansion coefficient and is called Invar. Alloys with 40–50% Ni have high B_s, electrical resistivity, ρ, and μ_{max}, which makes them suitable for use in high alternating magnetic fields, for example, in transformer magnet cores and magnetic shielding. Alloys with 70–80% Ni exhibit high permeability but have low ρ, and are used in direct current magnetic fields.

2.3 Sendust [5, 6, 12]

In 1932, Masumoto and Yamamoto [23] developed Sendust. The name Sendust was coined by combining the words "Sendai" and "dust" because the alloy readily forms a powder and was developed at Tohoku University in Sendai. Sendust is an Fe–Si–Al alloy that contains 5–11% Si and 3–8% Al and exhibits high magnetic permeability. In particular, the Fe-9.6%Si-5.6%Al alloy exhibits extremely high magnetic properties with $\mu_i = 35{,}000$ and $\mu_{max} = 116{,}000$. However, as the amounts of Si and Al are increased, intermetallic compounds Fe_3Si and Fe_3Al precipitate, and the alloy becomes hard and brittle. The alloy has a high-fixed permeability; thus, it found use as a material for various applications such as magnetic heads in the subsequent development of magnetic recording technology.

2.4 Si Steel (Magnetic Steel) [12]

Adding Si to Fe decreases the magnetocrystalline anisotropy, K, and increases the electrical conductivity, ρ. However, large amounts of Si make the steel brittle and difficult to form into thin sheets. Hadfield developed Si steel in 1900 and worked on establishing a manufacturing method. In 1934, Goss [12, 18] developed an oriented Si steel sheet (Goss texture) in which the $\langle 100 \rangle$ direction, which is the easy axis of magnetization of Fe, is aligned in the rolling direction. This is achieved by creating a secondary recrystallization texture with a combination of hot and cold working and the addition of inhibitors, such as MnS. Furthermore, Si steel in which the $\langle 100 \rangle$ direction is oriented in the two directions, parallel and perpendicular to the rolling direction, was fabricated by using AlN as an inhibitor. The Fe loss in these oriented

Si steel sheets was then reduced by methods including controlling crystal orientation, forming thin sheets, and controlling domain widths. Non-oriented Si steel sheet is typically used for applications such as rotary machines, whereas oriented Si steel sheet is used for applications such as transformers.

The addition of Si to Fe decreases the magnetocrystalline anisotropy and decreases the magnetostriction constant, λ. In particular, it was predicted that the permeability could be increased by reducing λ to near zero by adding around 6% Si. However, the addition of a large amount of Si made the steel brittle and hard. In 1988, Takada et al. [34] established a method of manufacturing Fe-6.5%Si alloy with low λ. They fabricated Fe-6.5%Si Si steel sheet by dispersing Si in Fe-3%Si Si steel sheet by chemical vapor deposition. The process technology was also developed in 1994 that made mass production possible, and the alloy is used as a low-noise magnet core material that can be used at high frequencies [35]. Recently, an even higher Si content has been achieved, and a siliconizing technique for increasing the Si concentration only on the surface of thin sheets [36] has been developed. These alloys are used in applications such as reactors as materials that offer high magnetic flux density and low Fe loss and support high-frequency ranges.

2.5 *Amorphous Alloys and Nanocrystalline Alloys [7]*

Amorphous alloys exhibit low magnetocrystalline anisotropy because they do not have a crystal phase and high permeability because they do not have pinning at domain walls due to crystal grain boundaries. They are fabricated by the liquid quenching method, which consists of ejecting a molten metal from a small nozzle and quenching it onto a rapidly rotating metal roll made of Cu or Cr. In 1974, Fujimori et al. [25] and Egami et al. [26] discovered soft magnetism in amorphous alloys, and in 1975, Kikuchi et al. [27] reported that a Co-Fe–Si–B alloy with a zero magnetostriction composition is a high permeability material. Subsequent research showed that Fe systems are suitable as high saturation magnetization amorphous alloys, and Co systems are suitable as high permeability amorphous alloys [3].

In 1975, Masumoto and Kimura [28] investigated the crystallization process by thermal annealing from the amorphous phase, and they reported that 10–20 nm nanocrystals precipitated during low-temperature annealing to form aggregates. These nanocrystals are extremely stable and do not grow at temperatures over several hundred degrees Celsius. This is because as the nanocrystals precipitate, the elements of the amorphous phase are ejected from the crystal grains and form a stable amorphous layer at the grain boundaries that suppresses the nanocrystal growth [7].

In 1988, Yoshizawa et al. [29] developed a nanocrystalline magnetic material that exhibits high saturation magnetic flux density and high permeability by adding Cu and Nb to Fe–Si–B to precipitate nanocrystals. Figure 1 [30] shows the relationship between the saturation magnetic flux density and permeability in soft magnetic materials. Higher saturation magnetic flux density B_s and higher permeability μ_r were

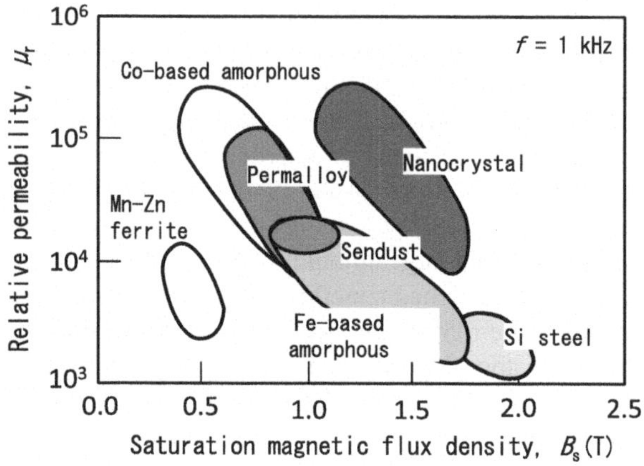

Fig. 1 Relationship between saturation magnetic flux density and permeability in soft magnetic materials (modified from Ref. [30], Copyright 2018, with permission from © Hitachi Metals, Ltd.)

obtained compared with the materials that we have discussed previously. The saturation magnetic flux density and permeability were high despite the precipitation of a crystal phase: for the following reasons: [1] the remaining amorphous phase is ferromagnetic [2], the magnetocrystalline anisotropy of the Fe is averaged out because of exchange coupling between nanocrystals [3], the magnetostriction of the nanocrystals and that of the amorphous phase cancel each other to give a low magnetostriction, and [4] the sizes of the nanocrystals are several tens of nanometers, which is smaller than the magnetic domain width, and thus does not form pinning sites for domain walls [4].

The effects of additive elements have also been investigated. Cu forms clusters at the stage before crystallization, which promotes uniform precipitation of nanocrystal grains. Furthermore, although Nb increases the concentrations of Nb and B around crystal grains during crystallization, this decreases the crystal grain size because it stabilizes the remaining amorphous phase and suppresses the crystal grain growth [5]. Si forms a solid solution inside α-Fe, and this may consequently decrease the magnetocrystalline anisotropy [4].

Makino et al. [31] reported a material with high magnetic flux density comparable to Si steel and lower Fe loss than conventional soft magnetic materials by creating a heteroamorphous structure that contains a large amount of α-Fe nanocrystal particles several nanometers in size. They achieved this by adding P and Cu to an Fe–Si–B alloy with a higher Fe content, quenching, and then forming a uniform nanocrystal texture by annealing. The texture was obtained because there are many α-Fe nanocrystal particles in the heteroamorphous structure that act as nuclei during annealing. Currently, higher magnetic flux densities and permeabilities are being obtained in this kind of nanocrystal alloy.

Methods such as sputtering are also used for fabricating amorphous and nanocrystal alloys, and since an Fe–B–N magnetic thin film was reported in 1985 [32], these materials have been developed as ribbons and as thin-film materials. The research into magnetic thin films is currently focusing on nanogranular thin films, in which ferromagnetic nanoparticles are separated by a thin amorphous insulating layer several nanometers thick of a nitride or oxide that has high electrical resistance, excellent soft magnetic properties, and exhibits tunneling effects. In particular, (Fe, Co)–Al–O thin films [33] have attracted attention for their soft magnetic properties.

2.6 *Soft Ferrite*

The soft magnetic materials that we have introduced up to this point are based on metals. However, because metals have low electrical resistance, at high frequencies, losses from eddy currents and heat generation become large. To prevent this, materials with thin bodies or small particles that are less than the skin depth have been developed. However, soft ferrites, which contains Fe oxide such as Fe_2O_3, are soft magnetic materials that do not suffer from the problem of eddy currents owing to their high electrical resistance and can be used for applications at high frequencies even though with low magnetization. The main ferrites are given by the general formula MFe_2O_4 or $MO \cdot Fe_2O_3$ (M: metal with valence 2) and are MnZn ferrite ($(MnZn)O \cdot Fe_2O_3$) and NiZn ferrite ($(NiZn)O \cdot Fe_2O_3$), which have a spinel structure.

MnZn ferrite is used at frequencies of several megahertz or less for applications such as wideband transformers and noise filters. Because MnZn ferrite has a composition such that the magnetocrystalline anisotropy and magnetostriction are zero, a high permeability is obtained with large crystal grains that give a morphology in which the domain walls move easily. Furthermore, spinel ferrites are also characterized by having the highest saturation magnetic flux density. However, because the electrical resistance is lower than other spinel ferrites, the magnetic losses are larger when the frequencies exceed several megahertz. A method for overcoming this problem is to precipitate a phase with a high electrical resistance in the grain boundaries.

In the frequency range up to several hundred megahertz, which is higher than the frequencies at which MnZn ferrite is used, NiZn ferrite, which has a high electrical resistance, is used as a magnetic core material in transformers and inductors.

3 Hard Magnetic Materials [1, 2, 8–10]

3.1 *Guidelines for Increasing Performance in Permanent Magnets [9]*

The strength of a permanent magnet is evaluated by $(BH)_{max}$, which is calculated from the second quadrant of the magnetization curve (demagnetization curve). Let us consider a permanent magnet that has an ideal demagnetization curve. In other words, if the magnetization curve is square such that the coercivity, H_{cJ}, is large ($H_{cJ} \geq J_s/2\mu_0$, where J_s is the saturation magnetic polarization, and μ_0 is the permeability of vacuum) and magnetic reversal does not occur until H_{cJ} ($J_s = J_r = B_r$, where J_r is the remanent magnetic polarization, and B_r is the remanent flux density), then the ideal value of $(BH)_{\max}$ is given by

$$(BH)_{\max} = \frac{J_s}{2} \cdot \frac{J_s}{2\mu_0} = \frac{J_s^2}{4\mu_0} \tag{1}$$

Therefore, if J_s, B_r, and H_{cJ} are large and the squareness is good, then a high $(BH)_{\max}$ is obtained, which gives a powerful permanent magnet. It is also important that the Curie temperature, T_c, is high. This is because the direction of the magnetic moment becomes random in the paramagnetic state due to the effects of thermal vibrations at or above T_c, and a high J_s cannot be obtained.

To increase B_r, it is important to use a material with a high J_s, align the easy axes of magnetization, and apply the magnetic field in the direction of the magnetization axes when the material is used. This is achieved by methods including using a magnetic field press, which presses a powder while applying a magnetic field, and thermomagnetic treatment, where a magnetic field is applied during annealing. Permanent magnets, where the easy axes of magnetization are aligned, are called anisotropic magnets, and they exhibit excellent magnetic properties in the alignment direction. In contrast, permanent magnets, where the easy axes of magnetization are randomly distributed, are called isotropic magnets, and although the magnetic properties are inferior to those of anisotropic magnets, they have the advantage of equal magnetic properties in all directions.

Squareness is evaluated based on the expansion factor of the demagnetization curve; thus, to increase this measurement, it is the same as for B_r. It is also important to control the microstructure to reduce locations in the material where the magnetic polarization is reversed at magnetic fields lower than the coercivity, H_{cJ}.

Because H_{cJ} is the strength of the magnetic field required to reverse the magnetic polarization, a high coercivity can be obtained by preventing the reversal of the magnetic polarization. Therefore, the mechanism of coercivity is affected by the rotation magnetization of single-domain particles and the magnetization of the material via domain-wall motion. For rotation magnetization, ferrite magnets, Nd–Fe–B magnets, and $SmCo_5$ [1–5] magnets use the magnetocrystalline anisotropy of compounds to produce the coercivity, whereas alnico magnets and Fe–Cr–Co magnets use magnetic shape anisotropy. When magnetocrystalline anisotropy is used, the materials have magnetic anisotropy in a single direction (uniaxial magnetic anisotropy) and high magnetocrystalline anisotropy constant K_u (or high anisotropy magnetic field H_A given by formula 2.

$$H_A = \frac{2K_u}{J_s} (= H_{cJ}) \tag{2}$$

Furthermore, to maintain a single-domain structure, the texture must be controlled to suppress the formation of reverse magnetic domain nuclei.

When magnetic shape anisotropy is used, the coercive field strength can be increased by precipitating single-domain-size particles of a material with high saturation magnetic polarization in a non-ferromagnetic phase and producing elongation in a single direction.

For domain-wall motion, even if the microstructure is not a single-domain structure and there are some reverse domains, if there are locations where the change in magnetic domain wall energy is large, the coercivity will be high. This is because the domain walls stop in these locations (domain-wall pinning) and a large magnetic field is required to overcome the pinning and move the domain walls. Magnets that have this coercive field strength mechanism include biphasic-separated Sm–Co [2–17] magnets, Pt–Co, and Pt–Fe.

To obtain high-performance permanent magnets in all cases, the magnetic materials must be a bulk material composed of single-domain particles with high saturation magnetic polarization and high magnetic anisotropy and also with the easy axis of magnetization directions aligned.

3.2 *History of the Development of Hard Magnetic Materials [8, 10, 11]*

Humans first discovered permanent magnets before the birth of Jesus Christ, and the first magnets were probably magnetite. Magnetite was produced in the Magnesia Prefecture of ancient Greece; thus, the word "magnet" may originate from the name "Magnesia." However, humans did not artificially create permanent magnetic materials until the start of the twentieth century, beginning with the development of KS steel [8, 37, 38] with a composition of Fe- [30–40] %Co- [5–9] %-(1.5–3)%Cr-(0.4–0.8)%C (wt%) by Honda and Takagi in 1917.

Table 2 shows the history of the development of hard magnetic materials. Figure 2 [11] shows that the strength of hard magnetic materials, namely, $(BH)_{max}$, which is the maximum energy products, has increased rapidly over the course of around 100 years. For example, $(BH)_{max}$ of KS steel is 7.6 kJ m^{-3}, whereas the strength of the latest Nd–Fe–B magnets is around 60 times that of KS steel.

Japanese researchers have played a prominent role in the development of hard magnetic materials. In 1931, Mishima invented MK steel [8, 37, 39], which is an Fe–Ni–Al alloy [representative composition: Fe-25%Ni-13%Al-4%Cu (wt%)] with two-phase separation morphology. It formed the basis of current alnico magnets and had $(BH)_{max}$ of 9.6 kJ m^{-3}. In 1933, Kato and Takei [21] invented oxide powder magnets, which are based on oxides, unlike conventional alloy magnets. These magnets consisted of a mixture of $CoFe_2O_4$ and Fe_3O_4 at a molar ratio of 75:25 and are the basis for modern ferrite magnets [8]. Although $(BH)_{max}$ at the start of development

Table 2 History of the development of hard magnetic materials

Year	Inventor	Country	Material
1917	Honda, Takagi	Japan	KS steel (Fe–Co–Cr–C)
1931	Mishima	Japan	MK steel (Fe–Ni–Al)
1933	Kato, Takei	Japan	Oxide powder magnet ($CoFe_2O_3$–Fe_3O_4)
1934	Honda, Masumoto, Shirakawa	Japan	NKS steel (Fe–Co–Ni–Ti)
1942	Jonas, Emden	Netherlands	Alnico magnet
1952	Went, Rathenau, Gorter, van Oosterhout	Netherlands	Ba ferrite
1960	Koch, Hokkeling, Steeg, de Vos	Netherlands	MnAl magnet
1966	Hoffer, Strnat	USA	Magnetic properties of rare earth (R)–Co compounds (YCo_5)
1968	Nesbitt, Willens, Sherwood, Buehler, Wernick	USA	SmCoCu magnet (biphasic-separated Sm_2Co_{17} magnet)
1968	Tawara, Senno	Japan	CeCoCu magnet (biphasic-separated Ce_2Co_{17} magnet)
1969	Buschow, Naastepad, Westendorp	Netherlands	$SmCo_5$ sintered magnet
1971	Kaneko, Homma, Nakamura	Japan	Fe–Cr–Co magnet
1983	Hirosawa, Fujimura, Sagawa, Yamamoto, Matsuura	Japan	Nd–Fe–B sintered magnet
1984	Croat, Herbst, Lee, Pinkerton	USA	Nd–Fe-B quenched magnet
1987	Ohashi, Yokoyama, Osugi, Tawara	Japan	R(Fe, M)$_{12}$ magnet
1990	Coey, Sun	Ireland	R–Fe–N magnet ($R_2Fe_{17}N_x$)

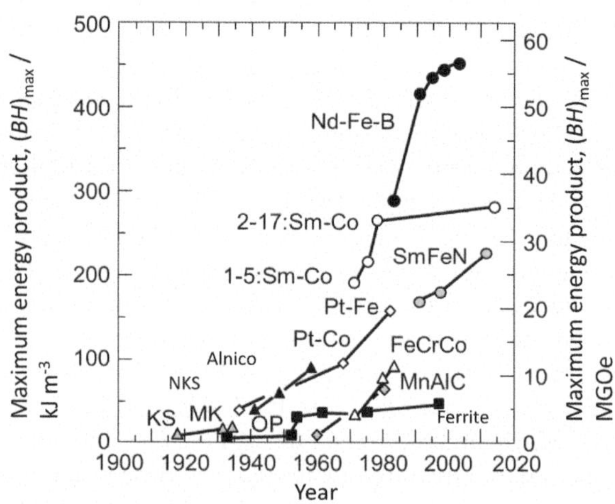

Fig. 2 Trends in $(BH)_{max}$ in hard magnetic materials (modified from Ref. [11], Copyright 2018, with permission from IOP Publishing)

was around 8 kJ m^{-3}, H_{cJ} reached around 80 kA m^{-1}, which was high compared with previous alloy magnets and was a groundbreaking discovery [40, 41].

The magnetic properties of MK steel greatly surpassed the performance of KS steel and shocked the researchers in the Institute for Materials Research at Tohoku University to which Honda belonged. In 1934, Honda, Masumoto and Shirakawa developed NKS steel, which is an Fe–Co–Ni–Ti alloy containing Co and Ti, increasing $(BH)_{max}$ to around 16 kJ m^{-3} [41, 42]. The inventions of MK steel and NKS steel led to the development of alnico magnets [43]. Furthermore, Mn-based magnets, such as MnBi, MnAl [44], and Mn–Al–C, and magnets such as Pt–Co and Pt–Fe alloys which exhibit coercive field strength were reported in 1935 [8, 41]. In 1971, Kaneko et al. [8, 45] produced an Fe–Cr–Co magnet with a lower Co content that had coercivity arising from magnetic shape anisotropy by using two-phase separation, as in alnico magnets. This is the only current magnetic material that can be cold worked.

However, Sm–Co magnets, which are superior to these permanent magnets, were first developed in 1966, and this ushered in the era of rare earth magnets. In 1966, Hoffer and Strnat [46] measured the magnetic properties of YCo_5 [1–5] and reported that it has high magnetocrystalline anisotropy. Since then, the magnetic properties of intermetallic compounds of rare earth elements and transition metal elements have been extensively investigated, and $SmCo_5$ magnets with high saturation magnetic polarization J_s and extremely high magnetocrystalline anisotropy constant K_u have been mass produced [47]. In 1968, Tawara et al. [48, 49] and Nesbitt et al. [50] discovered that a coercive force could be produced even in ingots by adding Cu to $CeCo_5$ and $SmCo_5$ alloys. This led to two-phase-separated Sm–Co magnets that have a main phase of Sm_2Co_{17} [2–17], which has excellent temperature properties and exhibits higher $(BH)_{max}$ than $SmCo_5$ magnets. However, these Sm_2Co_{17} magnets were surpassed by Nd–Fe–B magnets, which were discovered by Sagawa et al. [51] and Croat et al. [52] in 1983. During the discovery of these Nd–Fe–B magnets, the research target moved from two-element compounds, such as Sm–Co magnets, to three-element compounds, resulting in the discovery of $ThMn_{12}$ compounds. Examples include $R(Fe,M)_{12}$ (R: rare earth element, M: stabilization element such as Ti) with compositions such as $SmFe_{11}Ti$ and $NdFe_{11}TiN$ [8, 9, 48, 53], and Sm–Fe–N compounds such as $Sm_2Fe_{17}N_x$ and $SmFe_9N_x$ [8, 9, 48, 54, 55].

The number of different types of magnets that have been discovered now reaches the teens. During this time, older materials have sadly disappeared with the appearance of new higher performance materials, as with other materials. The history of permanent magnets is also the history of the discovery of new magnetic materials that produce coercive force among materials with high saturation magnetic polarization [8–10].

The current permanent magnets that have had the largest impact on the lives of people and on technology are ferrite magnets and Nd–Fe–B magnets. Ferrite magnets have a low $(BH)_{max}$, but a high performance for their price, and are currently the magnets with the highest production volume. Nd–Fe–B magnets are used in a wide variety of applications due to their strength and are still the best permanent magnets. Next, we provide an introduction to each of these materials.

3.3 *Ferrite Magnets [8, 10]*

Ferrite magnets were developed by Kato and Takei [21] from oxide powder magnets based on $CoFe_2O_4$. The composition was modified by Philips to create permanent magnets using magnetoplumbite compounds, as exemplified by $MO \cdot 6Fe_2O_3$ (M = Ba, Sr). Because of cheap raw material price and chemical stability, ferrite magnets have an excellent cost-performance ratio and are currently the largest volume mass-produced permanent magnet. The magnetism is ferrimagnetism from the super-exchange interaction via Fe^{3+} and O^{2-} in the magnetoplumbite structure, and the magnetic polarization is generated by the net total of four upward-facing magnetic moment Fe^{3+} ions in the single chemical equation.

The magnetic properties of ferrite magnets have improved. In the 1990s, improvements were made to the manufacturing and orientation of sub-micron microparticles. Around 2000, the saturation magnetic polarization, J_s, was increased by substituting Sr^{3+} with La^{4+} and substituting Fe^{3+}, which has downward magnetic moments in 4-coordinate sites ($4f_1$), with Zn^{2+} [56]. In addition, H_{cJ} was increased by replacing Zn^{2+} with Co^{2+} [57]. Furthermore, magnets have also been developed that increase B_r by substituting some La with Ca [58].

3.4 *Nd–Fe–B Magnets [8, 10]*

The Nd–Fe–B magnets developed by Sagawa et al. [51] and Croat et al. [52] in 1983 are currently the most powerful permanent magnets. These permanent magnets exhibit excellent magnetic properties because the $Nd_2Fe_{14}B$ main phase [59], which has a tetragonal crystal structure, has J_s of 1.6 T, K_u of 4.4 MJ m^{-3}, and H_A of 5.36 MA m^{-1}. Nd–Fe–B magnets can be classified as sintered magnets, hot-worked magnets, and bonded magnets.

Sintered magnets are fabricated by using a composition that is slightly Nd-rich compared with the stoichiometric composition of the $Nd_2Fe_{14}B$ main phase. Consequently, a Nd-rich phase other than the main phase appears in sintered magnets. The Nd-rich phase promotes the sintering reaction because it forms a liquid phase at temperatures of 600 °C or higher and removes reverse domain nucleation sites by removing particle surface defects. Thus, the Nd-rich phase is required to obtain the coercive force in Nd–Fe–B magnets.

The fabrication process for Nd–Fe–B sintered magnets has improved with the development of the following methods: strip casting [60]; hydrogen embrittlement, which uses hydrogen decrepitation in coarse grinding [61]; jet milling in fine grinding [48]; and various plating methods [62] to improve corrosion resistance because the magnetic material is Fe based. These improvements have resulted in $(BH)_{max}$ = 474 kJ m^{-3} [63]. Applications now include drive motors in electric vehicles and hybrid vehicles, leading to a need for heat resistance because Nd–Fe–B magnets have a low Curie temperature of 312 °C. Conventionally, the coercivity at room temperature has been increased by adding heavy rare earth elements, such as Dy or Tb, which have high coercivities in $Dy_2Fe_{14}B$ and $Tb_2Fe_{14}B$, to ensure some coercivity at high temperatures. However, it is necessary to reduce the use of heavy rare earth owing to their increasing scarcity. There are various alternative methods for increasing the coercivity. The sizes of crystal grains can be refined, for example, by He jet milling [64, 65]. Grain boundary diffusion process allows heavy rare earth elements to diffuse into the grains through the grain boundaries by coating heavy rare earth compounds on the magnet surface and annealing. It leads to increase the anisotropy field of the part near grain boundaries in the main phase resulting in high coercivity [66].

Hot-worked magnets [67] are fabricated by hot extrusion of quenched amorphous powder fabricated by liquid quenching methods. The *c*-axis, which is the easy axis of magnetization of the main phase $Nd_2Fe_{14}B$ crystal grains, becomes aligned in the rolling direction during hot working, creating an anisotropic magnet. The texture is composed of $Nd_2Fe_{14}B$ crystal grains of around 50 nm in the *c*-axis direction and 150 nm in the *c*-plane direction, and there is a Nd-rich phase at the grain boundaries. Because the grain particle size is small, the local demagnetization field is low, and these magnets have better temperature properties than sintered magnets and have been used in hybrid automobile motors as rare earth-free materials [68].

Bonded magnets [8, 48] are permanent magnets in which a powder is turned into a bulk material by mixing with resin. The powder is fabricated by liquid quenching methods [52] or hydrogenation disproportionation desorption recombination [69–71], which reduces the size of crystal grains by absorption and desorption reactions of hydrogen. This type of magnet is used in applications that require high dimensional precision, such as small motors.

4 Future of Magnetic Materials

In the future, people will increasingly use electric vehicles and robots. These products and machines have automatic control technology, such as artificial intelligence, and the role of magnetic materials in energy conversion is expected to grow. In soft magnetic materials, this requires higher efficiency and higher power, creating a demand for materials with higher saturation magnetic polarization and permeability. Furthermore, the frequencies that are used are also increasing due to the demand for high-capacity communications and high-speed switching, creating a need for materials that support high-frequency ranges.

There are no commercially available hard magnetic materials yet that are better than Nd–Fe–B magnets, and the applications and usage volume of Nd–Fe–B magnets are expected to increase. However, $ThMn_{12}$ compounds with a small amount of stabilizing element (M), such as $Sm(Fe, Co)_{12}$, $(Sm, Zr)(Fe, Co, M)_{12}$, and (Nd, Zr) $(Fe, Co, M)_{12}N_x$ compounds, in which some of the rare earth element is replaced by Zr and some of the Fe is replaced by Co [72] have been reported. These materials have better magnetic properties than $Nd_2Fe_{14}B$, and they are expected to become commercially available. If we could overcome the disadvantage that Sm–Fe–N compounds must be used as a powder for bonded magnets because they decompose into SmN and Fe at high temperature and cannot be sintered, and if technology for turning them into bulk materials advanced, then their applications would expand. Furthermore, the development of these Sm magnets, which use a rare earth element other than Nd, is strongly desirable from the perspective of efficient use of rare earth resources. The search for new magnetic materials [73] is continuing, and we can expect continued research and development of these materials and their process technologies.

References

1. Kaneko, H., Homma, M.: Materials Engineering Series 8: Magnetic Materials. The Japan Institute of Metals and Materials (1972)
2. Homma, M., Higuchi, A.: Magnetic Materials Reader. Kogyo Chosakai Publishing (1998)
3. Technical Committee on Magnetics, IEEJ: Magnetic Engineering Basics and Applications. Corona Publishing (1999)
4. Expert Committee on Functional Research into Nanoscale Magnetic Materials, IEEJ: Nanostructural Magnetic Materials: Physical Properties, Function, and Design. Kyoritsu Shuppan (2010)
5. Fukuda, M.: Well-known magnetic materials, I introduction, 1. History of magnetic materials. Spec. Steel **63**(5), 2–6 (2014)
6. Ikeuchi, S.: High permeability magnetic materials. Vac. Ind. **4**(3), 80–84 (1957)
7. Masumoto, T.: Flow in studies of amorphous metal materials. Mater. Jpn. **37**(5), 339–346 (1998)
8. Sagawa, M., Hamano, M., Hirabayashi, M. (eds.): Permanent Magnets: Materials Science and Applications. Chapter 2 (M. Hamano, Editor), Chapter 3 (S. Sugimoto and H. Kato, Joint Editors), Chapter 5 (S. Sugimoto, Editor). Agne Gijutsu Center (2007)
9. Homma, M., Sugimoto, S.: Fundamentals of permanent magnets I. J. Magn. Soc. Jpn. **25**(10), 1529–1534 (2001). Homma, M., Sugimoto, S.: Fundamentals of permanent magnets II. J. Magn. Soc. Jpn. **25**(11), 1580–1588 (2001). Homma, M., Sugimoto, S.: Fundamentals of permanent magnets III. J. Magn. Soc. Jpn. **25**(12), 1625–1640 (2001).
10. Sugimoto, S.: Recent progress in the research of permanent magnets. Mater. Jpn. **56**(3), 181–185 (2017)
11. Sugimoto, S.: Current status and recent topics of rare-earth permanent magnets. J. Phys. D Appl. Phys. **44**(6), 064001 (2011)
12. Bozorth, R.M.: Feromagnetism. Wiley-IEEE Press (1993)
13. Barret, W.F., Brown, W., Hadfield, R.A.J.: Electrical conductivity and magnetic permeability of various alloys of Fe. Sci. Trans. Roy. Dublin Soc. **7**, 67–126 (1900)
14. Arnold, H.D., Elmen, G.W.: Permalloy, an alloy of remarkable magnetic properties. J. Frank. Inst. **195**(5), 621–632 (1923)
15. Elmen, G.W.: Magnetic alloys of iron, nickel, and cobalt. Electr. Eng. **54**, 1292–1299 (1935)
16. Elmen, G.W.: Magnetic material and appliance. US patent no. 1739752 (1929)
17. Honda, K., Kaya, S.: On magnetization of single crystals of iron. Sci. Rep. Tohoku Univ. **15**, 721 (1926)
18. Goss, N.P.: Electrical sheet and method and apparatus for its manufacture and test. US patent 1965559 (1934)
19. Kato, Y., Takei, T.: Japanese patent no. 98844 (1932)
20. Kato, Y., Takei, T.: Japanese patent no. 110822 (1932)

21. Kato, Y., Takei, T.: Permanent oxide magnet and its characteristics. J. IEEJ **53**(3), 408–412 (1933)
22. Snoek, J.L.: New Development in Ferromagnetic Materials. Elsevier (1947)
23. Masumoto, H., Yamamoto, T.: On a new alloy "Sendust" and its magnetic and electric properties. J. Jpn. Inst. Met. Mater. **1**(3), 127–135 (1937)
24. Boothby, O.L., Bozorth, R.M.: A new magnetic material of high permeability. J. Appl. Phys. **18**(2), 173–176 (1947)
25. Fujimori, H., Masumoto, T., Obi, Y., Kikuchi, M.: On the magnetization process in an iron-phosphorus-carbon amorphous ferromagnet. Jpn. J. Appl. Phys. **13**(11), 1889–1890 (1974)
26. Egami, T., Flanders, P.J., Graham Jr., C.D.: Low-field magnetic properties of ferromagnetic amorphous alloys. Appl. Phys. Lett. **26**(3), 128–130 (1975)
27. Kikuchi, M., Fujimori, H., Obi, Y., Masumoto, T.: New amorphous ferromagnets with low coercive force. Jpn. J. Appl. Phys. **14**(7), 1077–1078 (1975)
28. Masumoto, T., Kimura, H.: Crystallization process of iron-based amorphous alloy (Fe–P–C) quenched from liquid. J. Jpn. Inst. Met. Mater. **39**(3), 273–280 (1975)
29. Yoshizawa, Y., Oguma, S., Yamauchi, K.: New Fe-based soft magnetic alloys composed of ultrafine grain structure. J. Appl. Phys. **64**(15), 6044–6046 (1988)
30. Products Catalog, Hitachi Metals, Ltd.: "FINEMET" Nanocrystalline Soft Magnetic Material and Website: http://www.hitachi-metals.co.jp/products/elec/tel/p02_21.html
31. Makino, A., Men, H., Kubota, T., Yubuta, K., Inoue, A.: FeSiBPCu nanocrystalline soft magnetic alloys with high B_s of 1.9 Tesla produced by crystallizing hetero-amorphous phase. Mater. Trans. **50**(1), 204–209 (2009)
32. Karamon, H., Masumoto, T., Makino, Y.: Magnetic and electrical properties of Fe-B-N amorphous films. J. Appl. Phys. **57**(1), 3527–3532 (1985)
33. Onuma, S., Mitani, S., Fujimori, H., Masumoto, T.: Soft magnetic properties at high frequencies of Co–Al–O films with granular structure. J. Magn. Soc. Jpn. **20**(2), 489–492 (1996)
34. Takada, Y., Abe, M., Masuda, S., Inagaki, J.: Commercial scale production of Fe-6.5 wt. % Si sheet and its magnetic properties. J. Appl. Phys. **64**(10), 5367–5369 (1988)
35. Okami, Y., Abe, M., Yamaji, T., Takada, Y., Ninomiya, H.: Continuous siliconizing technology of 6.5% silicon steel sheet. J. ISIJ **80**(10), 777–782 (1994)
36. Hiratani, T., Oda, Y., Namikawa, M., Kasai, S., Ninomiya, H.: Development of Si gradient steel sheet with high saturation magnetic flux density and low iron loss at high frequency. Mater. Jpn. **53**(3), 110–112 (2014)
37. Koiwa, M.: Metallurgy promenade: pursuing serendipity. Agne Gijutsu Center (2004)
38. Honda, K., Saito, S.: On K. S. magnet steel. Phys. Rev. **16**(6), 495–501 (1920)
39. Mishima, T., Makino, N.: Studies on anisotropic ML permanent magnet (I) (on the chemical composition and the additional elements). J. ISIJ **42**(11), 1063–1066 (1956)
40. Society of Non-Traditional Technology (ed.): Magnetic materials for a new age. Kogyo Chosakai Publishing (1981)
41. Ida, S., Sakurai, Y., Iwasaki, S., Nagashima, T., Iwama, Y., Watanabe, S., Kobayashi, H. (ed.): Course in magnetic engineering: hard magnetic materials. Maruzen (1976)
42. Honda, K., Masumoto, H., Shirakawa, Y.: On new K.S. permanent magnet. Sci. Rep. Tohoku Univ. **23**, 365 (1934)
43. Jonas, B., Emden, H.J.M.V.: New kinds of steel of high magnetic power. Philips Tech. Rev. **6**, 8–11 (1941)
44. Koch, A.J.J., Hokkeling, P., Steeg, M.G.V.D., de Vos, K.J.: New material for permanent magnets on a base of Mn and Al. J. Appl. Phys. **31**(5), S75–S77 (1960)
45. Kaneko, H., Homma, M., Nakmamura, K.: New ductile permanent magnet of Fe–Cr–Co system. AIP Conf. Proc. Magn. Magn. Mater. **5**, 1088–1092 (1971)
46. Hoffer, G., Strnat, K.J.: Magnetocrystalline anisotropy of YCo5, and Y_2Co_{17}. IEEE Trans. Magn. **MAG-2**(3), 487–489 (1966)
47. Buschow, K.H.J., Naastepad, P.A., Westendorp, F.F.: Preparation of SmCo5 permanent magnets. J. Appl. Phys. **40**(10), 4029–4032 (1969)
48. Tawara, Y., Ohashi, T.: Rare Earth Permanent Magnets. Morikita Publishing (1999)

49. Tawara, Y., Senno, H.: Cerium, cobalt and copper alloy as a permanent magnet material. Jpn. J. Appl. Phys. **7**, 966–967 (1968)
50. Nesbitt, E.A., Willens, R.H., Sherwood, R.C., Buehler, E., Wernick, J.H.: New permanent magnet materials. Appl. Phys. Lett. **12**(11), 361–362 (1968)
51. Sagawa, M., Fujimura, S., Togawa, N., Yamamoto, H., Matsuura, Y.: New material for permanent magnets on a base of Nd and Fe (invited). J. Appl. Phys. **55**(6), 2083–2087 (1984)
52. Croat, J.J., Herbst, J.F., Lee, R.W., Pinkerton, F.E.: Pr–Fe and Nd–Fe-based materials: a new class of high-performance permanent magnets (invited). J. Appl. Phys. **55**(6), 2078–2082 (1984)
53. Ohashi, K., Yokoyama, T., Osugi, R., Tawara, Y.: The magnetic and structural properties of R-Ti–Fe ternary compounds. IEEE Trans. Magn. **MAG-23**(5), 3101–3103 (1987)
54. Coey, J.M.D., Sun, H.: Improved magnetic properties by treatment of iron-based rare earth intermetallic compounds in ammonia. J. Magn. Magn. Mater. **87**(3), L251–L254 (1990)
55. Iriyama, T., Kobayashi, K., Imaoka, N., Fukuda, T., Kato, H., Nakagawa, Y.: Effect of nitrogen content on magnetic properties of $Sm2Fe17N_x$ ($O < x < 6$). IEEE Trans. Magn. **28**(5), 2326–2331 (1992)
56. Taguchi, H.: New magnetic materials: development of higher magnetism and new applications: trend towards higher performance of ferrite magnets. Eng. Mater. **46**(7), 53–57 (1998)
57. Ogata, Y., Kubota, Y., Takami, T., Tokunaga, M.: Improvements of magnetic properties of Sr ferrite magnets by substitutions of La and Co. IEEE Trans. Magn. **35**(5), 3334–3336 (1999)
58. Kobayashi, Y., Hosokawa, S., Oda, E., Toyota, S.: Magnetic properties and composition of Ca–La–Co M-type ferrites. J. Jpn. Soc. Powder Powder Metall. **55**(7), 541–546 (2008)
59. Herbst, J.F., Croat, J.J., Pinkerton, F.E., Yelon, W.B.: Relationships between crystal structure and magnetic properties in $Nd_2Fe_{14}B$. Phys. Rev. B **29**(7), 4176–4178 (1984)
60. Hirose, Y., Hasegawa, H., Sasaki, S.: Microstructure of strip cast alloys for high performance NdFeB magnets. In: Proceedings of 15th Workshop on Rare-Earth Magnets & Their Applications, pp. 77–86. Dresden (1998)
61. Harris, I.R., McGuiness, P.J.: Hydrogen: its use in the processing of NdFeB-type magnets and in the characterization of NdFeB-type alloys and magnets. In: Proceedings of 11th International Workshop on Rare Earth Magnets and Their Applications, pp. 29–48. Pittsuburgh, USA (1990)
62. Minowa, T., Yoshikawa, M., Honshima, M.: Improvement of the corrosion resistance on Nd–Fe–B magnet with nickel plating. IEEE Trans. Magn. **25**(5), 3776–3778 (1989)
63. Harimoto, D., Matsuura, Y.: Development of high performance Nd–Fe–B sintered magnets. Hitachi Met. Tech. Rev. **23**(3), 69–72 (2007)
64. Une, Y., Sagawa, M.: Enhancement of coercivity of Nd–Fe–B sintered magnets by grain size reduction. J. Jpn. Inst. Met. Mater. **76**(1), 12–16 (2012)
65. Nakamura, M., Matsuura, M., Tezuka, N., Sugimoto, S., Une, Y., Kubo, H., Sagawa, M.: Preparation of ultrafine jet-milled powders for Nd–Fe–B sintered magnets using hydrogenation–disproportionation–desorption–recombination and hydrogen decrepitation processes. Appl. Phys. Lett. **103**(2), 022404-1–022404-4 (2013)
66. Nakamura, H., Hirota, K., Shimao, M., Minowa, T., Honshima, M.: Magnetic properties of extremely small Nd–Fe–B sintered magnets. IEEE Trans. Magn. **41**(10), 3844–3846 (2005)
67. Lee, R.W.: Hot-pressed neodymium-iron-boron magnets. Appl. Phys. Lett. **46**(8), 790–791 (1985)
68. Daido Steel Co. Ltd. Website: http://www.daido.co.jp/about/release/2016/0712_freemag_hevmotor.html
69. Takeshita, T., Nakayama, R.: Magnetic properties and microstructures of the Nd–Fe–B magnet powders produced by the hydrogen treatment-(III). In: Proceedings of the 11th International Workshop on RE Magnet and Their Applications, pp. 49–71. Pittsburgh (1990)
70. Sugimoto, S., Nakamura, H., Kato, K., Book, D., Kagotani, T., Okada, M., Homma, M.: Effect of the disproportionation and recombination stages of the HDDR process on the inducement of anisotropy in Nd–Fe–B magnets. J. Alloys Compd. **293–295**, 862–867 (1999)
71. Mishima, C., Hamada, N., Mitaraia, H., Honkura, Y.: Dependence of the hydrogen pressure on the magnetic properties of NdFeB anisotropic magnet powders produced by the HDDR method. J. Magn. Soc. Jpn. **24**(4–2), 407–410 (2000)

72. Hirayama, Y., Takahashi, Y.K., Hirosawa, S., Hono, K.: Intrinsic hard magnetic properties of $Sm(Fe_{1-x}Co_x)_{12}$ compound with the $ThMn_{12}$ structure. Scr. Mater. **138**, 62–65 (2017)
73. Shinaji, Keita, Mase, Tsuyoshi, Isogai, Keita, Matsuura, Masashi, Tezuka, Nobuki, Sugimoto, Satoshi: Influence of heat treatment on the microstructure and magnetic properties of Mn–Sn–Co–N alloys. Mater. Trans. **54**(10), 2007–2010 (2013)

Low-Loss Soft Magnetic Materials

Shin Nakajima

Abstract Features of amorphous and nanocrystalline alloys which recently draw attention as low-loss soft magnetic materials for motor cores are presented. As an example, experimental results of an energy-efficient 11-kW axial-gap motor using amorphous alloy stator cores are shown.

Keywords Amorphous alloy · Nanocrystalline alloy · Axial-gap motor · Core loss

1 Typical Soft Magnetic Materials and Positioning

Figure 1 shows typical practical soft magnetic materials. Magnetic materials are classified into two categories: soft and hard magnetic materials. Soft magnetic materials are used for cores of motors, transformers, and reactors, etc. On the other hand, hard magnetic materials are used as permanent magnets. In these two categories of magnetic materials, there are metallic materials and ferrite materials. Soft metallic magnetic materials are classified into crystalline metallic materials and amorphous alloys. In amorphous alloys used for soft magnetic materials, there are two classes, one of which is Fe (iron)-based and the other is Co (cobalt)-based amorphous alloys. Though Ni (nickel)-based alloys are magnetic in crystalline states, they are non-magnetic in amorphous states. They are used as brazing filler alloys to join stainless steels and the like. Co-based amorphous alloys are expensive, because they contain approximately 80 wt% of Co. Therefore, they are used in niche applications such as sensors and the amount of production is small compared to that of Fe-based amorphous alloys. Generally speaking, therefore, an amorphous alloy means a Fe-based amorphous alloy. On the other hand, nanocrystalline soft magnetic alloys are the newest crystalline soft magnetic alloys which are obtained by annealing amorphous alloys at temperatures above alloys' crystallization temperatures as discussed below.

Figure 2 shows the relationship between B_S (saturation induction) and core loss. In this figure, core losses $W_{1/1k}$, at 1 kHz of operating frequency and 0.1 T of B_m

S. Nakajima (✉)
Hitachi Metals, Ltd., Tokyo, Japan
e-mail: shin.nakajima.oj@hitachi-metals.com

K. Fujisaki (ed.), *Magnetic Material for Motor Drive Systems*,
Engineering Materials, https://doi.org/10.1007/978-981-32-9906-1_19

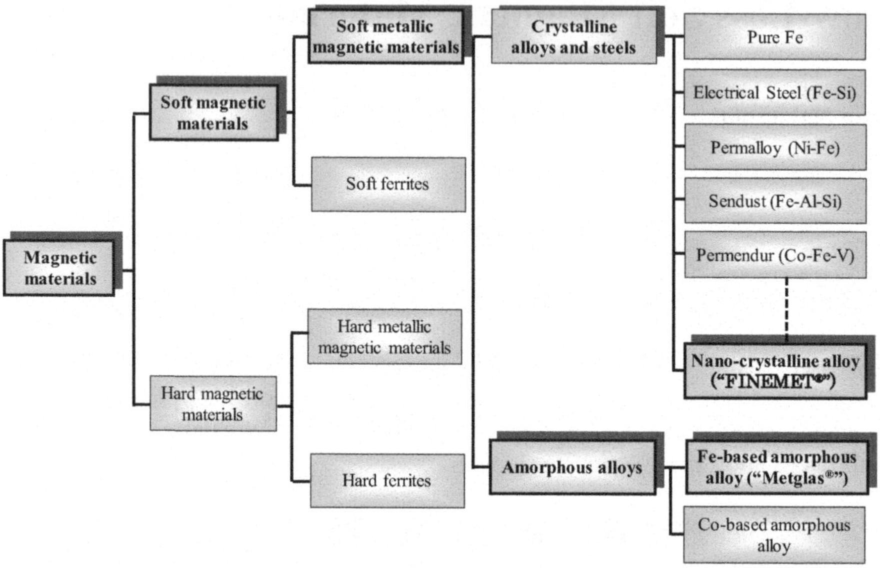

Fig. 1 Typical soft magnetic materials

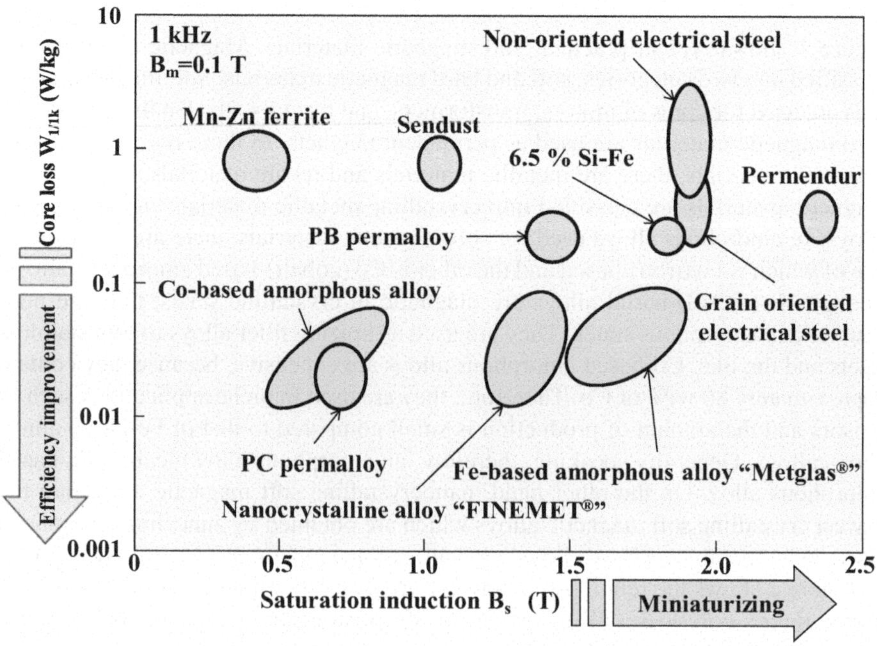

Fig. 2 Relationship between saturation induction B_S and core loss $W_{1/1k}$ of typical soft magnetic materials

(maximum operating induction) under sinusoidal flux excitation, are shown, for typical soft magnetic materials. If there is no issue of the temperature rise caused by core loss, it is desirable to select a higher B_S soft magnetic material in order to miniaturize electrical equipment. On the other hand, it is necessary to select a lower core loss soft magnetic material in order to improve the energy efficiency of electrical equipment. Therefore, it is necessary to select a soft magnetic material at lower right section of Fig. 2 in order to achieve both miniaturization and high energy efficiency. However, there are no soft magnetic materials with both high B_S and low core loss as shown in Fig. 2. Fe-based amorphous and nanocrystalline alloys are promising soft magnetic materials with relatively high B_S and low core loss.

2 Amorphous Soft Magnetic Alloys

2.1 History of the Development of Amorphous Soft Magnetic Alloys

Figure 3 shows a brief history of the development of amorphous soft magnetic alloys. In 1960, the first amorphous alloy was discovered by Prof. Duwez and others at the California Institute of Technology by rapid quenching Au (gold)–Si (silicon) alloy from the liquid state [1]. In 1969, Fe–P (phosphorus)–C (carbon) amorphous alloy,

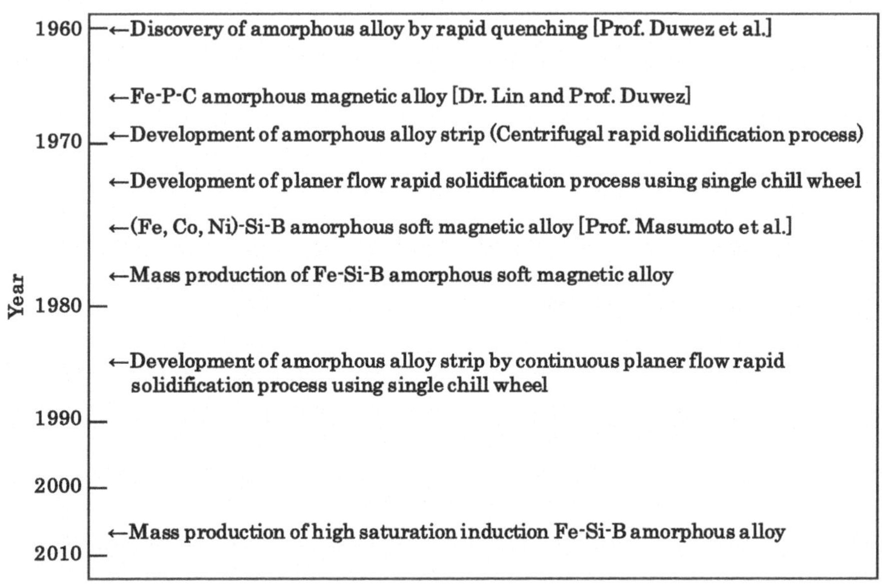

Fig. 3 History of the development of amorphous soft magnetic alloys

the first-ever soft magnetic amorphous alloy, was invented by Lin and Duwez at the same institute [2]. In the 1970s, the following two fundamental technologies for amorphous soft magnetic alloys were developed. Fe–Si–B (boron) amorphous alloy, currently the mainstream soft magnetic amorphous alloy, was invented by Prof. Masumoto and others at the Tohoku University [3]. Rapid solidification process such as planar flow casting process using a single chill wheel has been developed [4, 5]. In the 1980s and later, amorphous soft magnetic alloy was generally used for distribution transformers, medium- and high-frequency transformers and reactors, etc. In the middle of 1980s, its price per unit weight was almost the same of that of high permeability grain-oriented electrical steel [6] by developing a continuous rapid solidification process using a single chill wheel and putting it to practical use at Allied Corporation in USA. In the middle of 2000s and later, production quantity of amorphous distribution transformers drastically increased in developing countries such as China and India, because amorphous distribution transformers became one of the effective equipment in order to reduce CO_2 emission for the prevention of global warming [7]. In 2014, the annual total sales amount of amorphous alloy was almost 150,000 metric ton which was the largest amount behind electrical steel among soft metallic magnetic materials except pure Fe. Now, amorphous alloy is no longer a special soft magnetic material. Though more than 90% of the sales amount of amorphous alloy is used in distribution transformers, it is also used for medium- and high-frequency transformers and reactors. Recently, its application in motors of practical use has begun [8].

2.2 *Production Method for Amorphous Soft Magnetic Alloy [4, 5]*

Amorphous alloy is obtained by rapid solidification of molten alloy, of which cooling rate is more than 10^6 K/s. There are not many amorphous alloys in nature, because such high cooling rate is difficult to achieve in nature. The resultant alloy forms obtained by such high cooling rates are limited to thin strips, fine powders, fine wires, and thin films. Among these alloy forms, thin strips are most widely used because of their good productivity and good economic performance. Figure 4 shows a schematic diagram of typical production equipment for amorphous alloy strips with a continuous planar flow rapid solidification method using a single chill wheel. In case of typical amorphous soft magnetic Metglas® 2605SA1 [9] alloy, more than 10^6 K/s of cooling rate, inhibiting creation of crystal nucleus, is obtained by flowing liquid Fe–Si–B alloy on a high-speed rotating water-cooled wheel of which outer peripheral speed is approximately 100 km/h. Thus, amorphous alloy strips (nominal thickness: 0.025 mm, nominal width: 142.2, 170.2, and 213.4 mm) shown in Fig. 5a are obtained. Continuously cast amorphous alloy strip is wound around a paper core (inner diameter: 406 mm) mounted in a winder. When the outer diameter of amorphous alloy strip wound coil reaches a predetermined number (maximum:

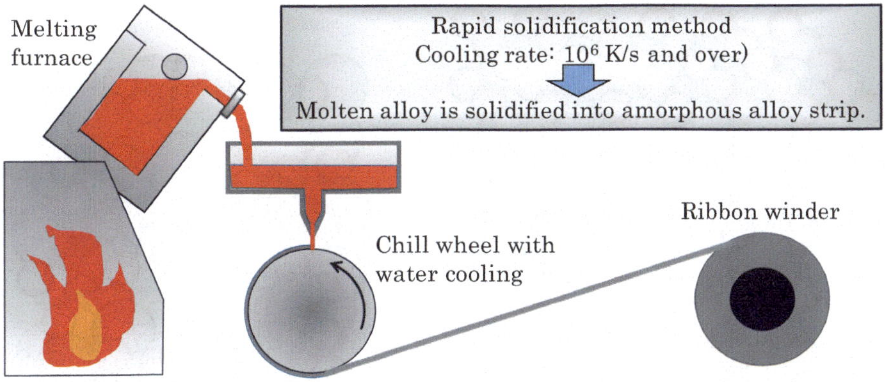

Fig. 4 Continuous planar flow rapid solidification equipment using a single chill wheel

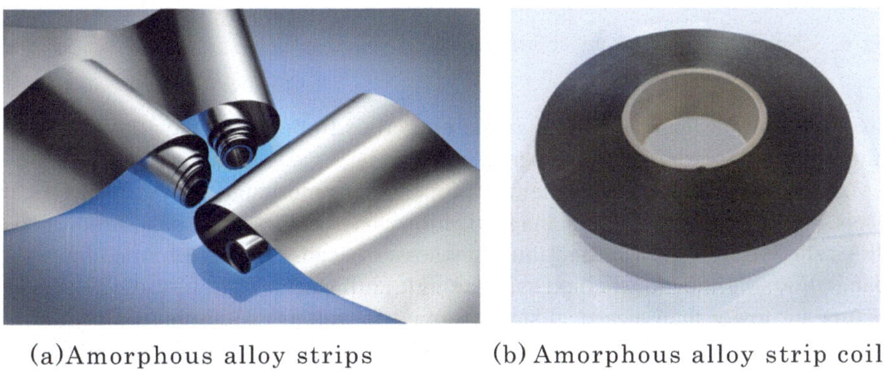

Fig. 5 Amorphous alloy strip

1,120 mm), amorphous alloy strip is cut. An amorphous alloy strip coil shown in Fig. 5b is thus produced. After an amorphous alloy strip coil is completed, continuously and speedily producing amorphous alloy strip is wound on another paper core to complete another amorphous alloy strip coil. These operations are continuous. When molten alloy is solidified into amorphous alloy strip, approximately 10 nm of surface oxidation layer is formed. Surface insulation coating is, therefore, not necessary for amorphous alloy strip to be used in motors and transformers.

2.3 *Features of Amorphous Soft Magnetic Alloy*

Figure 6 shows schematic diagrams of atomic arrangements of crystalline and amorphous alloys. Crystalline alloys have regular atomic arrangements, whereas amorphous alloys are characterized by irregular atomic arrangements. Amorphous alloys

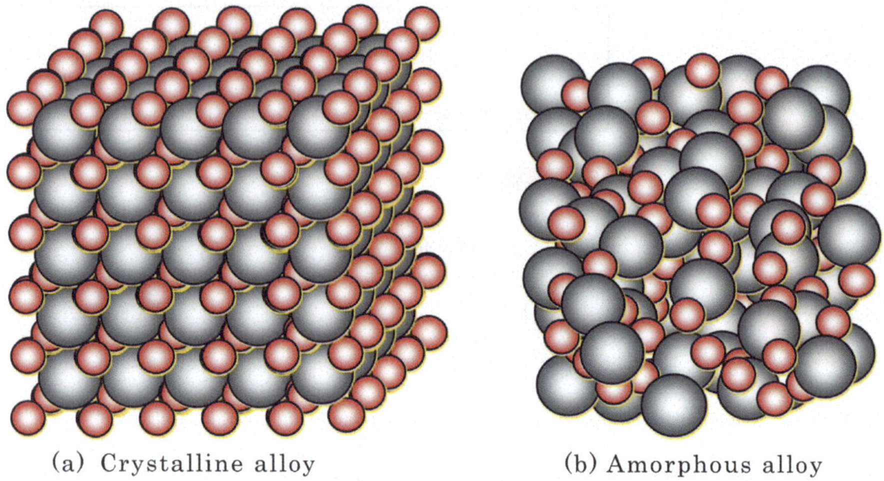

Fig. 6 Schematic diagrams of atomic arrangements

have different mechanical and magnetic properties from crystalline alloys because of their irregular atomic arrangements.

Table 1 shows physical and magnetic properties of amorphous alloys and electrical steels. Figure 7 shows DC initial magnetization curves of these materials. Amorphous alloys show rough surfaces and thin thicknesses because they are formed by rapid planar flow solidification method using a single chill wheel. Therefore, their stacking factors are lower than electrical steels formed by rolling. Amorphous alloys have about four times the tensile strength and about five times the Vickers hardness of those of conventional non-oriented electrical steel 35A300.

Amorphous alloys have large internal stresses because they are formed by rapid planar flow solidification. Therefore, annealing for reducing their internal stresses is necessary in order to achieve excellent magnetic properties. For example, in the case of 2605SA1 alloy, structural relaxation [15] is obtained by non-field anneal at 360 °C, which is lower than its crystallization temperature of 510 °C, for 60 min soak in nitrogen atmosphere. The isotropic magnetic properties shown for the non-field anneal case of Table 1 and DC initial magnetizing curve shown in Fig. 7 are thus obtained. And the magnetic properties for the longitudinal field anneal case in Table 1 and DC initial magnetizing curve shown in Fig. 7 can be obtained by applying more than 800 A/m of DC magnetizing field in the direction of magnetic path length during the anneal. Amorphous alloys show lower B_s because of their lower Fe content compared to electrical steels. However, amorphous alloys, with irregular atomic arrangements, have no magnetocrystalline anisotropies which crystalline alloys with regular atomic arrangements have. Therefore, amorphous alloys exhibit lower coercive forces and lower hysteresis losses compared with crystalline alloys. And amorphous alloys show about two times higher electric resistivity because of their irregular atomic arrangements. Moreover, thicknesses of amorphous alloys are

Table 1 Physical and magnetic properties of amorphous alloys and electrical steels [9–14]

Material		Amorphous alloy Metglas® 2605SA1		Grain-oriented electrical steel	Non-oriented electrical steel	
		Longitudinal field anneal	Non-field anneal	23P090	35A300	6.5% Si–Fe 10JNEX900
Nominal thickness (mm)		0.025		0.23	0.35	0.10
Stacking factor LF (%)		86 min.		94.5 min.	95 min.	90 min.
Density ($\times 10^3$ kg/m^3)		7.18		7.65	7.65	7.49
Electrical resistivity ($\mu\Omega$-m)		1.30		0.50	0.52	0.82
Tensile strength (N/mm^2)		2000		388	530	650
Vickers hardness		900		188	187	395
Core loss (W/kg)	$W_{10/50}$	0.051 typ.	0.056 typ.	0.28 typ.	1.11 typ.	0.5 typ.
	$W_{10/400}$	1.2 typ.	0.81 typ.	8.0 typ.	18.2 typ.	5.7 typ.
	$W_{10/1k}$	4.3 typ.	3.0 typ.	28 typ.	78 typ.	18.7 typ.
Induction (T)	B_{50}	1.56 typ.		1.88 min.	1.64 typ.	1.52 typ.
	B_S	1.56		2.03	2.03	1.80
DC relative permeability at 1.0 T		200,000 typ.	31,000 typ.	76,000 typ.	7,200 typ.	19,000 typ.
Magnetostriction (ppm)		27[a]		8[a]	8[a]	0.1[b]
Curie temperature (°C)		399		750	750	700
Crystallization temperature (°C)		510		–	–	–

[a]Saturation magnetostriction
[b]Magnetostriction at 1.0 T, 400 Hz

about 1/10 compared with conventional crystalline alloys. The eddy current losses of amorphous alloys are, therefore, considerably lower than those of crystalline alloys. Thus, amorphous alloys with lower hysteresis losses and lower eddy current losses show extremely lower core losses and higher permeabilities compared with crystalline alloys such as non-oriented electrical steels.

Magnetic properties of amorphous alloys are easily affected under mechanical stress because of their large magnetostriction. Annealed amorphous alloys tend to become brittle because of the clusters formed by annealing to reduce internal stress of non-annealed amorphous alloys [16].

Reliability test of amorphous alloys as practical materials is needed because amorphous alloys are metastable. General method for reliability test of amorphous alloys

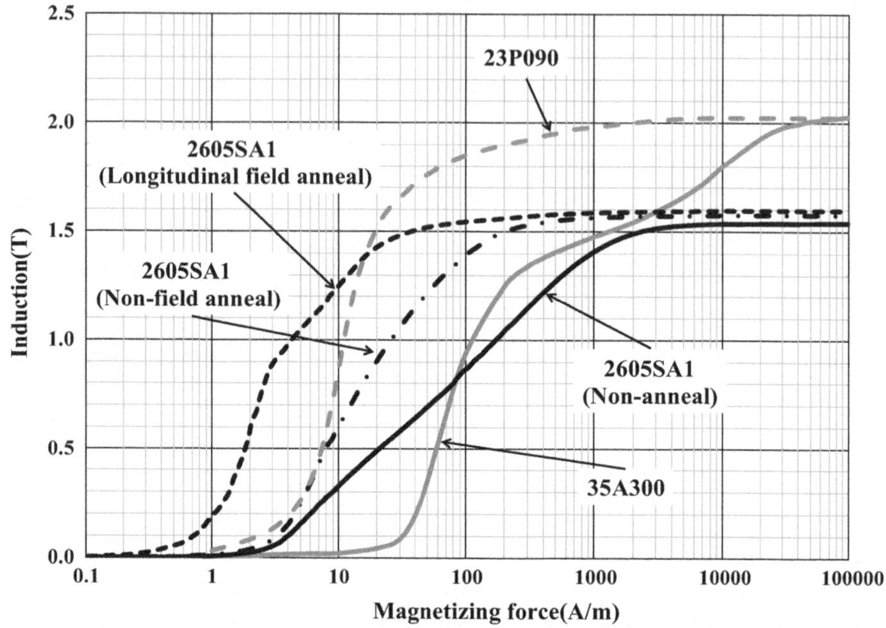

Fig. 7 DC initial magnetizing curves of amorphous alloys and electrical steels [10, 11]

is an accelerated test of magnetic properties under high-temperature atmosphere. Regarding 2605SA1 alloy, it is confirmed that its estimated lifetime at 150 °C is more than 1200 years [17]. Therefore, there is no practical issue when applying them to transformers, reactors and motors, etc.

2.4 *Application of Amorphous Soft Magnetic Alloy to Motors*

(1) Core Loss Reduction Techniques for Motor Core

On loss of a motor, there are four types of losses as follows. First one is core loss in motor cores caused by change of induction in them. Second one is conductor loss in coil caused by carrying current in it. Third one is mechanical loss at bearings and fans, etc. And forth one is stray load loss at a metallic chassis caused by leakage flux flowing in it.

Non-oriented electrical steels are generally utilized in conventional motors, and there are three types of techniques in order to reduce the core loss.

(1) Eddy current loss reduction by applying thin non-oriented electrical steels.
(2) Building factor (BF) reduction by reducing internal stresses of motor cores caused by manufacturing motor cores from soft magnetic materials.

(3) Core loss reduction by adopting lower core loss soft magnetic materials over non-oriented electrical steels.

Regarding above (1), 0.15-, 0.20-, or 0.25-mm-thick non-oriented electrical steels have become commonplace for high rotation speed motors used in electric vehicle (EVs) or Hybrid Electric Vehicle (HEVs) recently instead of conventional 0.35-mm-thick or 0.50-mm-thick non-oriented electrical steels [11]. Though a small amount of 0.10-mm-thick non-oriented electrical steels are available, their prices are high because of large milling cost involved. 6.5% Si–Fe shows a lower core loss compared with conventional non-oriented electrical steels because it has a thinner 0.1 mm thickness and near-zero magnetostriction. However, it is only used in specialized motors because its price is high, caused by a chemical vapor deposition (CVD) process needed for its production [12].

Regarding above (2), for example, adhesive lamination process becomes commonplace for motor core production instead of swaging lamination process for conventional motor core production [18].

Regarding above (3), physical and magnetic properties of practical soft magnetic materials for motor cores are shown in Table 2. In this table, the lowest core loss material is non-field annealed amorphous alloy. And core loss of other materials increases in the order of the following: non-annealed amorphous alloy, PB Permalloy, 6.5% Si–Fe, Permendur, and non-oriented electrical steel 35A300. PB Permalloy is an alloy of 50 wt% Ni and Fe. Permendur is an alloy of 48 wt% Co, 2 wt% V (vanadium), and 50 wt% Fe. These two kinds of alloys are high priced because they contain a large content of expensive metals such as Ni and Co. They are, therefore, utilized in specialized motors. The latter shows excellent performance in light of reducing size of motors, because it has the highest B_s among practical soft magnetic materials. On the other hand, amorphous alloys draw attention as low core-loss soft magnetic materials for energy-efficient motors, because their prices are not much high compared to conventional non-oriented electrical steel 35A300.

(2) Features and Issues of Amorphous Motor Core

Figure 8 shows the DC initial magnetization curves of amorphous 2605SA1 alloy and non-oriented electrical steel 35A300. By non-field annealing amorphous alloy at 360 °C for 60 min soak in nitrogen atmosphere, its internal stress is reduced and the gradient (permeability) of its DC initial magnetization curve is increased as shown by the dotted line in Fig. 8. However, the amorphous alloy becomes brittle and cannot keep its mechanical strength enough to be used as a motor core. Therefore, vacuum-pressure resin impregnation treatment using epoxy or acrylic resin is applied to an annealed amorphous alloy core in order to increase its mechanical strength. However, the gradient (permeability) of its DC initial magnetization curve is decreased, as shown by the dot-and-dash line in Fig. 8, because of internal stress caused by the resin impregnation. On the other hand, a non-annealed amorphous alloy shows several times higher core loss compared to an annealed amorphous alloy, but it shows much lower core loss compared to the conventional non-oriented electrical steel 35A300

Table 2 Physical and magnetic properties of typical soft magnetic materials for motors [9, 11–14, 19, 20]

Material		Amorphous alloy Metglas® 2605SA1		PB Permalloy	Permendur	Non-oriented electrical steel	
		Non-field anneal	Non-anneal	PB-1	YEP-2 V	35A300	6.5% Si–Fe 10JNEX900
Nominal thickness (mm)		0.025		0.10	0.2	0.35	0.10
Stacking factor LF (%)		84 min.		90 min.	93 min.	95 min.	90 min.
Density ($\times 10^3$ kg/m^3)		7.18		8.25	8.28	7.65	7.49
Electrical resistivity (μΩ-m)		1.3		0.52	0.47	0.52	0.82
Tensile strength (N/mm^2)		2000		640	728	530	90
Vickers hardness		900		130	200	187	350
Core loss (W/kg)	$W_{10/50}$	0.056 typ.	0.17 typ.	0.18 typ.	0.6 typ.	1.11 typ.	0.5 typ.
	$W_{10/400}$	0.81 typ.	2.2 typ.	2.7 typ.	8.6 typ.	18.2 typ.	5.7 typ.
	$W_{10/1k}$	3.0 typ.	7.4 typ.	10 typ.	35 typ.	78 typ.	18.7 typ.
Induction (T)	B_{50}	1.56 typ.	1.52 typ.	1.16 typ.	2.25 typ.	1.64 typ.	1.52 typ.
	B_s	1.56		1.51	2.37	2.03	1.80
DC relative permeability at 1.0 T		31,000 typ.	4700 typ.	57,800 typ.	25,000 typ.	7,200 typ.	19,000 typ.
Magnetostriction (ppm)		27[a]		25[a]	70[a]	8[a]	0.1[b]
Crystalline temperature (°C)		399		400	960	750	700

[a]Saturation magnetostriction
[b]Magnetostriction at 1.0 T, 400 Hz

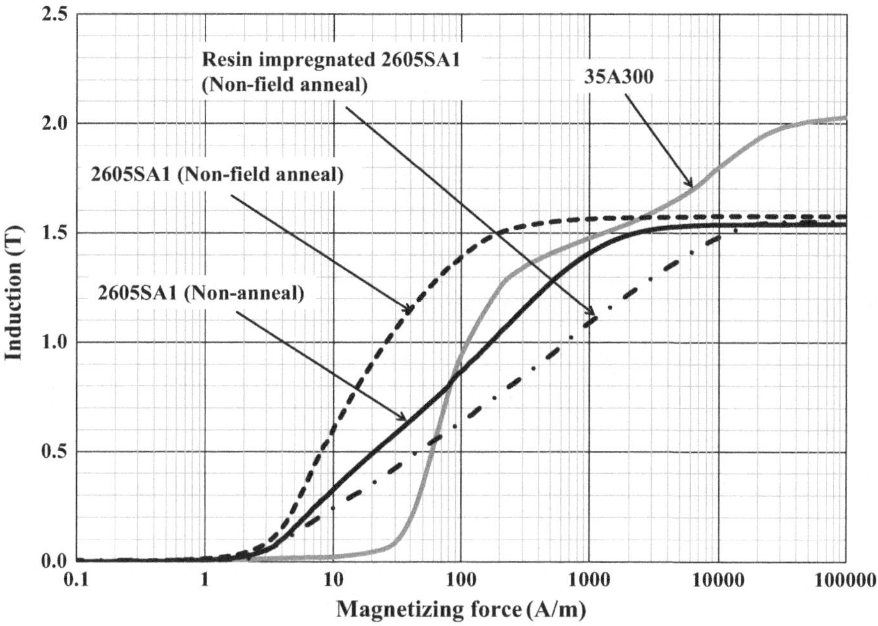

Fig. 8 DC initial magnetizing curves of amorphous alloys and non-oriented electrical steel 35A300 [11]

as shown in Fig. 9. It is thus considered to apply non-annealed amorphous alloy to motors at various laboratories and manufacturers [8, 21–23].

Nominal thickness of amorphous alloy is 0.025 mm, and its Vickers hardness is about 900. The former is less than 1/10, and the latter is about five times of those of a conventional non-oriented electrical steel 35A300. And an amorphous alloy has a feature that it is very difficult to deform plastically. Therefore in the case of making cores from amorphous alloys, there are issues as shown in Table 3.

Most of the conventional motor cores are stamp-laminated cores. It was reported that stamping of an amorphous alloy is possible by using near-zero clearance carbide tools, but the lifetime of carbide tools is very short [24]. And it was reported that stamping a heated amorphous alloy (about 200 °C) by using Polycrystalline Diamond (PCD) tools is effective in order to solve the issue regarding the lifetime of tools. However, it has not been reported how much increase the actual lifetime of tools is [25]. In the case of applying stamping process to laminated motor cores, of which the number of laminated layers is several thousands, it is necessary to drastically increase the lifetime of tools.

Etching of amorphous alloy is relatively easy. BF of etched laminated cores is low because internal stress by etching is very low. Therefore, etched laminated cores are practical and used for small-sized electronic components. However, in the case of applying etching process to laminated motor cores of which the number of laminated layers is very large, there are issues associated with the complicated and long process.

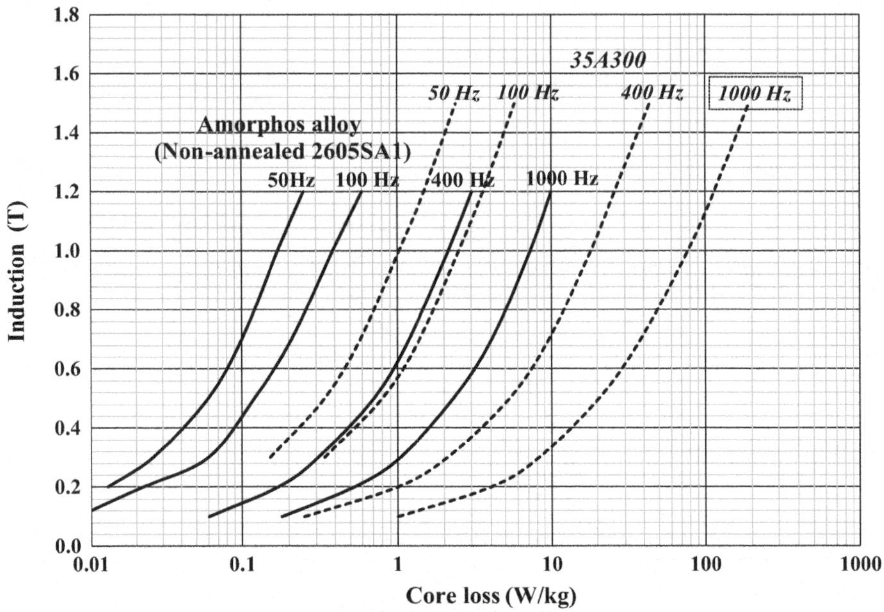

Fig. 9 Core losses of amorphous alloy and non-oriented electrical steel 35A300

Wire electrical discharge machining laminated cores are made by wire electrical discharge machining and resin impregnated amorphous laminated block cores [26]. Its process has issues of long machining time and short circuit of worked surface. In order to solve the latter issue, the short-circuited surface is removed by etching treatment. Though this process has an issue of complicated processing, it is applied to trial manufacturing or small lot production for motors, because it does not require exclusive tools such as punches and dice.

Water-jet machining wound cores are made by the following process [27]. A slit amorphous alloy strip is wound into cores. And the wound cores are impregnated with epoxy or acrylic acid resin. Then, the resin impregnated wound cores are formed into prescribed shape motor cores by water-jet machining. In these cores, there are issues of the complicated and long process. And they are limited to applicable core shapes.

Slitting and cutting laminated cores are easily manufactured because slitting and cutting processes are already put to practical use in electric components or distribution transformers. However, they are limited to applicable core shapes [8].

(3) Axial-Gap Motor using Amorphous Alloy Cores [22]

As already mentioned, the most important subject on putting motors using amorphous alloy cores to practical use is development of cost-effective and efficient manufacturing process for amorphous alloy motor cores. As an example of the solution, a

Table 3 Manufacturing methods for motor cores and their issues

Core type	Outline of manufacturing process	Issues
Stamping laminated cores	Stamping ⇒ Laminating with swaging	Lifetime of tools
Etching laminated cores	Cleaning ⇒ Drying ⇒ Photoresist application ⇒ Exposure ⇒ Unwanted photoresist stripping ⇒ Etching ⇒ Photoresist stripping ⇒ Cleaning ⇒ Drying ⇒ Laminating ⇒ Resin impregnating	Complicated processing
With wire electrical discharge machining laminated cores	Laminating ⇒ Resin impregnating ⇒ Wire electrical discharging ⇒ Cleaning ⇒ Drying ⇒ (Surface finishing)	Complicated processing, short circuit of worked surface
Water-jet machining wound cores	Slitting ⇒ Winding ⇒ Resin impregnating ⇒ Water-jet machining ⇒ Cleaning ⇒ Drying	Complicated processing, difficult to apply complicated shape core
Slitting and cutting laminated cores	Slitting ⇒ Cutting ⇒ Laminating ⇒ Resin impregnating or Inserting to bobbin	Difficult to apply complicated shape core

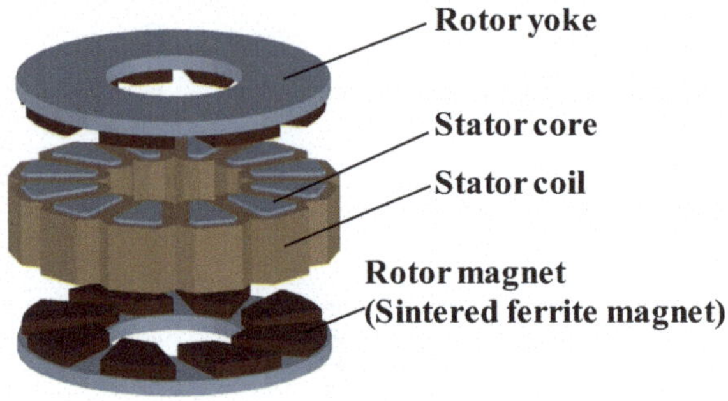

Fig. 10 Structure of axial-gap motor. Reprinted from [22], Copyright 2014, with permission from IEEJ

prototype axial-gap motor using the above-mentioned slitting and cutting laminated amorphous alloy cores is described below.

Figure 10 shows the structure of the prototype axial-gap motor. It has a stator with coils and cores in the center portion. And it has two rotors with ferrite magnets and a rotor yoke in up- and downside of the stator. By arranging the magnets facing up and down as shown, the flux variation caused by the field variation accompanied by the rotation of magnets is mainly generated in the cores of the stator. Core losses of other parts such as that of rotor yokes except that of stator cores are decreased. Table 4 shows the specifications of the prototype axial-gap motor. Its rated rotation speed is 3600 rpm.

Two types of stator cores of which appearance shown in Fig. 11 have been made with conventional non-oriented electrical steel 35A300 and non-annealed amorphous 2605SA1 alloy. Height of the stator core corresponds to the width of the slit strip. Resin impregnated lamination core shown in Fig. 11 is made by the following processes: cutting the slit strips by changing its cutting distance; laminating these cut strips in the direction of the arrow in Fig. 11; impregnating the laminated strips with resin.

Figures 12 and 13 show the core losses at frequency 400 Hz. At frequency 400 Hz and operating induction 1.0 T, BF of non-annealed amorphous alloy 2605SA1 core

Table 4 Specification for an axial-gap motor

Phase number	3
Number of stator coils	12
Pole number of rotor	8
Outer diameter of rotor (mm)	200
Rated rotation speed (rpm)	3600
Rated torque (N/m)	30
Magnet for rotor	11 mm depth sintered ferrite magnet NMF-12F ($B_r = 0.45$ T) made by Hitachi Metals, Ltd.
Stator core	Height: 50 mm, Area of base: 1020 mm^2 (trapezoid shape)

Reprinted from [22], Copyright 2014, with permission from IEEJ

Fig. 11 Stator core for 11-kW axial-gap motor. Reprinted from [22], Copyright 2014, with permission from IEEJ

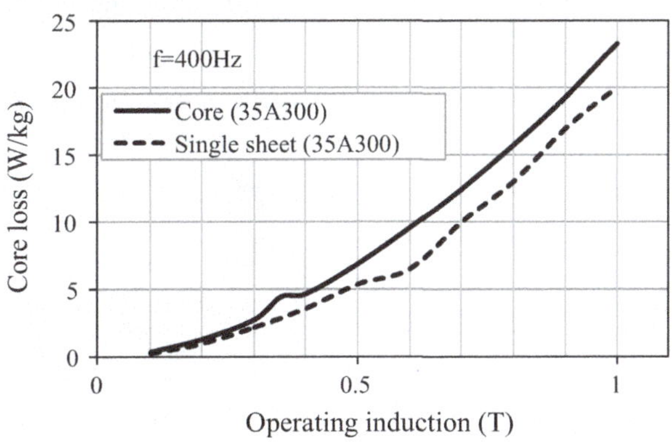

Fig. 12 Core loss of non-oriented electrical steel 35A300. Reprinted from [22], Copyright 2014, with permission from IEEJ

is 1.6. It is higher than that of a non-oriented electrical steel 35A300 core, of which BF is 1.15.

However, core loss at frequency 400 Hz and operating induction 1.0 T of a non-annealed amorphous alloy 2605SA1 core is less than 1/5 of that of a non-oriented electrical steel 35A300 core.

Figure 14 shows the appearance of parts of the prototype 11-kW axial-gap motor. Figure 14a shows the rotor with eight sintered ferrite magnets, a yoke, and a structure holding these magnets and the yoke. Figure 14b shows a stator coil with 155 turns of 1.4 mm × 1.4 mm rectangular enameled copper magnet wire. Figure 14c shows a stator made by the following process. Inserting the stator cores is shown in Fig. 11 in

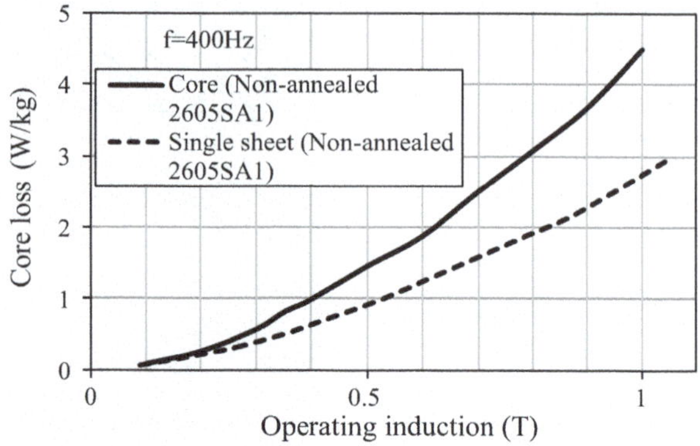

Fig. 13 Core loss of amorphous alloy non-annealed 2605SA. Reprinted from [22], Copyright 2014, with permission from IEEJ

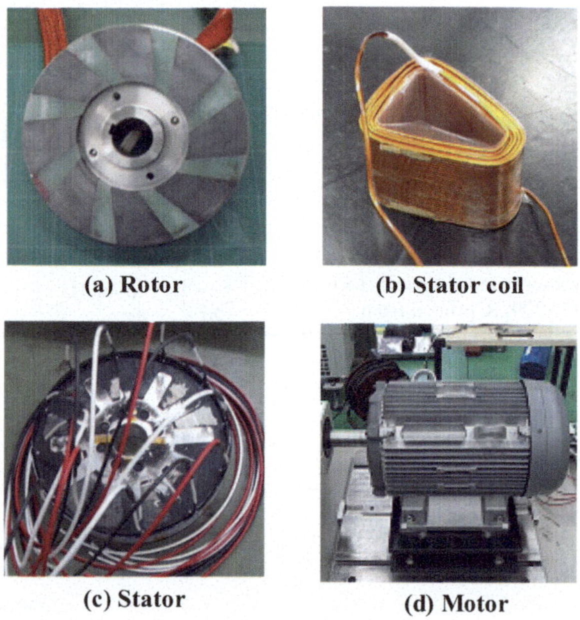

Fig. 14 Appearances of parts of prototype 11-kW axial-gap motor. Reprinted from [22], Copyright 2014, with permission from IEEJ

the center hole of the stator coils. Molding stator core coil (12 altogether) assembly with resin is shown in Fig. 10. Figure 14d shows the motor with these parts inserted in the chassis. There is useless large space in the direction of rotating axis because a commercially available chassis for an induction motor is used.

Figure 15 shows the result of no-load loss of the prototype 11-kW axial-gap motor measured by driving by an external motor. The measurement results include windage loss and mechanical loss caused by bearings, etc., in addition to core loss. However, the difference of the no-load losses is equal to that of the core losses, because windage loss and mechanical loss do not depend on core loss. Thus, the difference of the core losses at rotation speed 3,600 rpm is 65 W.

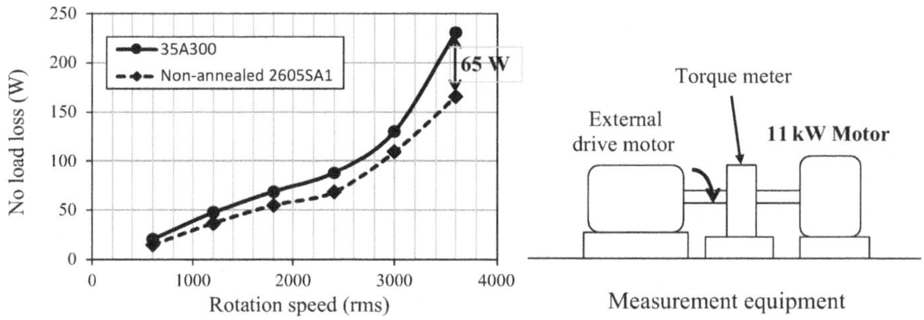

Fig. 15 No-load loss comparison of 11-kW axial-gap motor. Reprinted from [22], Copyright 2014, with permission from IEEJ

Fig. 16 Experimental result of torque characteristic of 11-kW axial-gap motor Reprinted from [22], Copyright 2014, with permission from IEEJ.

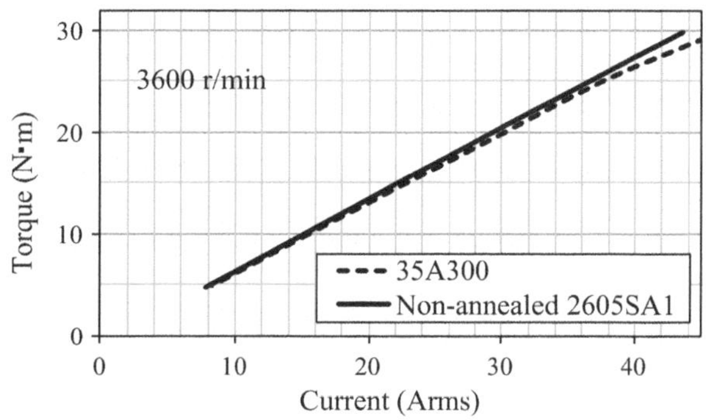

Figure 16 shows the experimental result of the torque characteristics at rotation speed of 3,600 rpm. In the high torque area, current of the motor with non-oriented electrical steel 35A300 stator cores is a little higher than that of the motor with non-annealed amorphous alloy 2605SA1 stator cores. At 30 N m of rated torque, current of the motor with non-annealed amorphous alloy 2605SA1 stator cores is 43.5 Arms and that of the motor with non-oriented electrical steel 35A300 stator cores is 46.6 Arms. The difference of the currents is caused by the higher excitation power of non-oriented electrical steel 35A300 stator cores compared to that of non-annealed amorphous alloy 2605SA1 stator cores.

Figure 17 shows the experimental results of energy efficiency of the 11-kW axial-gap motor at rotation speed of 3600 rpm. At 30 N m of rated torque, the energy efficiency of the motor with non-oriented electrical steel 35A300 stator cores is 91.7%, whereas that of the motor with non-annealed amorphous alloy 2605SA1 stator cores is 93.2%. Thus by using non-annealed amorphous alloy 2605SA1 stator cores, 1.5% higher energy efficiency is achieved. At 5.0 N m of low torque, the ratio of the core loss within the total loss increases because the currents are decreasing. Therefore in this case, the difference of energy efficiency increases to 2.3%.

Figure 18 shows the result of the loss attribution analysis of the 11-kW axial-gap

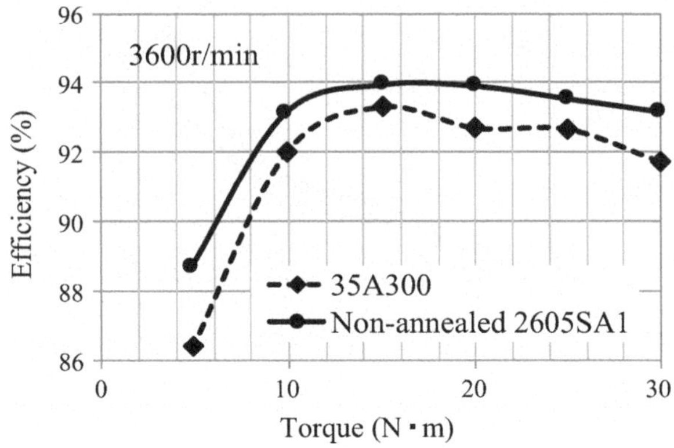

Fig. 17 Energy efficiency comparison of 11-kW axial-gap motor Reprinted from [22], Copyright 2014, with permission from IEEJ.

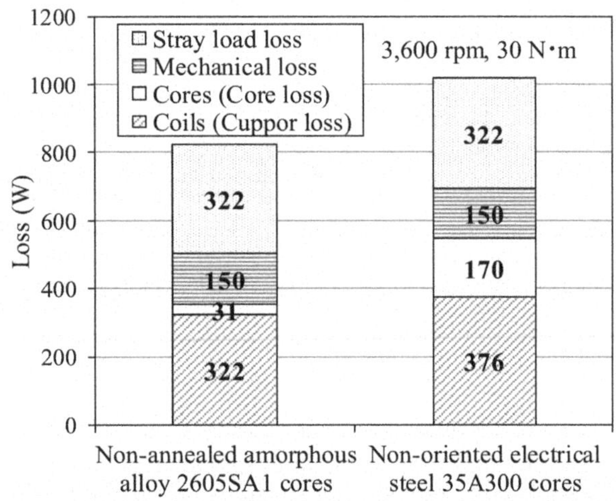

Fig. 18 Attribution analysis of loss of 11-kW axial-gap motor [22]

motor at rated power. In the case of using non-annealed amorphous alloy 2605SA1 stator cores, core loss decreases to less than 1/5 compared to using non-oriented electrical steel 35A300 stator cores. And in the case of using non-annealed amorphous alloy 2605SA1 stator cores, copper loss decreases because excitation current is lower, caused by lower excitation power compared to the case of using non-oriented electrical steel 35A300 stator cores.

In the case of an axial-gap motor, copper loss is in inverse proportion to the height of a stator core because cross section of coil can be increased in proportion to the height of a stator core. On the other hand, core loss increases in proportion to the height of a stator core. Therefore, energy efficiency can be increased by optimizing stator core height. By using this simple model, in the case of using non-annealed amorphous alloy 2605SA1 stator cores, the energy efficiency of the 11-kW axial-gap motor can be increased by approximately 1.2% by increasing the stator core height to 3.2 times. On the other hand, in the case of using the non-oriented electrical steel

35A300 stator cores, the energy efficiency of the 11-kW axial-gap motor can be only increased by about 0.3% by optimizing stator core height.

As mentioned above, it is clear that the axial-gap motor with amorphous alloy laminated stator cores made by slitting and cutting process can achieve considerably higher energy efficiency compared to that with conventional non-oriented electrical steel stator cores. This has led to energy-efficient small-size compressors with the same type motors having been commercialized [28].

3 Nanocrystalline Soft Magnetic Alloys

3.1 History of Development of Nanocrystalline Soft Magnetic Alloys

Figure 19 shows a brief history of development of nanocrystalline soft magnetic alloys. Fe–Si–B–Nb (niobium)–Cu series nanocrystalline "FINEMET®" alloy was reported by Yoshizawa and others in 1988, providing the first-ever nanocrystalline soft magnetic alloy against common knowledge that soft magnetic properties of amorphous alloys were completely damaged by crystallization [29]. By developing this alloy, it was clear that excellent soft magnetic properties were achieved by forming nanocrystalline structures which were uniform in fine bcc Fe–Si phases as shown in Fig. 20. Then, many kinds of nanocrystalline alloys such as Fe–Zr (zirconium)–B series [30], Fe–Si–B–Cu series [31, 32], and Fe–Si–B–P–Cu series [33, 34] have been developed. In the case of nanocrystalline soft magnetic alloys such as Fe–Si–B–Nb–Cu series and Fe–Zr–B series as represented by FINEMET® alloy, the content of Nb, Zr, etc., must be increased to several at% in order to form nanocrystalline structures which are uniform and in fine bcc Fe–Si phases. Their

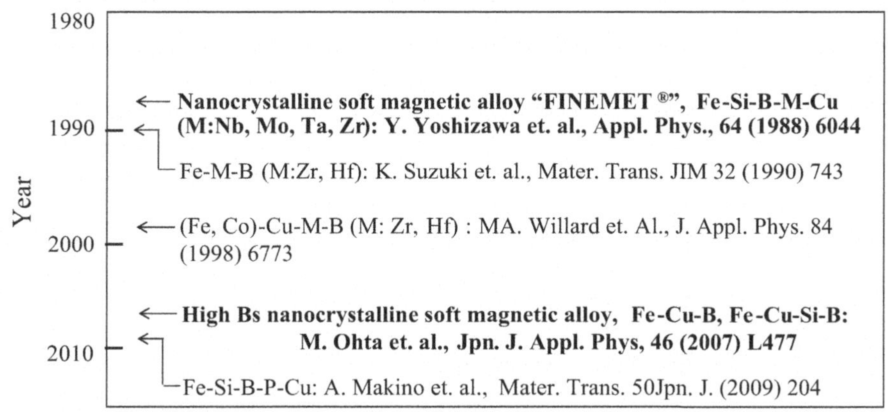

Fig. 19 History of the development of nanocrystalline soft magnetic alloy

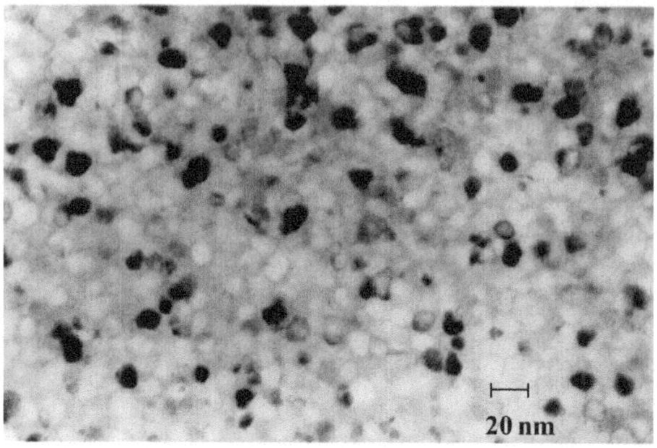

Fig. 20 TEM micrograph of nanocrystalline soft magnetic alloy

saturation inductions, B_S, are less than 1.7 T, caused by the decrease in the Fe content. Fe–Si–B–Nb–Cu series nanocrystalline alloys as represented by FINEMET® alloy have excellent magnetic properties and good productivity, becoming mainstream.

On the other hand, Fe–Si–B–Cu series nanocrystalline alloys having high B_s values with increase in the Fe content were reported by Ohta and Yoshizawa in 2007 [31]. Then, other high B_s nanocrystalline alloys such as Fe–Si–B–P–Cu series [33, 34] have been developed. However, these high B_s nanocrystalline alloys have not been put to practical use, because they have issues such as castability or annealing process as discussed in the following.

3.2 Production Method for Nanocrystalline Soft Magnetic Alloys

(1) Production Method for Fe–Si–B–Nb–Cu series Nanocrystalline Alloys [29, 35]

Fe–Si–B–Nb–Cu series nanocrystalline alloys are obtained by annealing Fe–Si–B–Nb–Cu amorphous alloys at temperatures above alloys' crystallization temperatures. Figure 21 shows schematic diagrams of crystallization mechanisms by annealing Fe–Si–B–Nb–Cu series nanocrystalline alloys and conventional amorphous alloys. As shown in Fig. 21a, in the case of a nanocrystalline alloy, Cu-rich clusters are formed in the amorphous phase in the annealing process. Then, initial fine crystals of bcc Fe–Si are formed by them as nucleation sites. The content of Fe decreases and that of Nb increases in the remaining amorphous phase as crystallization progresses. Therefore, the remaining amorphous phase is stabilized and growth of crystal grains is controlled. Thus, a nanocrystalline alloy with uniform grain sizes is obtained. As shown in Fig. 21b, in the case of annealing conventional amorphous alloys with high Fe contents and without Nb and Cu, crystal grains are coarsened because of lack of crystal nucleuses and excessive growth of crystal grain sizes.

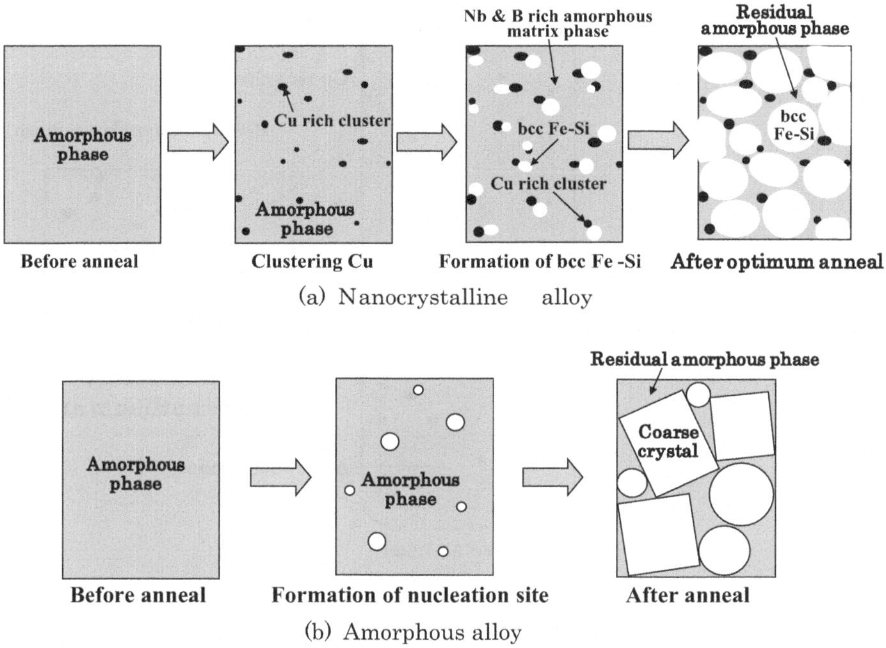

Fig. 21 Schematic diagram of crystallization mechanism by anneal [35]

In short, both forming crystal nucleuses densely and uniformly and controlling growth of crystal grain sizes are needed in order to achieve high-performance nanocrystalline alloys. Therefore, in the case of Fe–Si–B–Nb–Cu series and Fe–Zr–B series nanocrystalline alloys, slightly higher contents of Nb or Zr must be present in order to obtain nanocrystalline alloys with uniform fine bcc Fe phases by controlling growth of crystal grain sizes. However, their saturation inductions are less than 1.7 T because their content of Fe is decreased, influenced by Nb or Zr. And in the case of use in cores, mechanical strength must be provided by being inserted in plastic cases or being impregnated with epoxy resin, because the alloys becomes brittle caused by nanocrystallization.

In the case of FINEMET® FT-3 alloy which represents Fe–Si–B–Nb–Cu series nanocrystalline soft magnetic alloys, cores are produced by annealing, at 550 °C for 60 min soak in a nitrogen atmosphere, the wound cores using the amorphous alloy strip cast by the continuous planar flow rapid solidification mentioned in Sect. 2.2. The nominal thicknesses of the amorphous FT-3 alloy are 0.014 and 0.018 mm. Their standard widths are 53, 63, and 142 mm. The strip is supplied in the form of a coil which is wound around a paper core (inner diameter: 457 mm), of which maximum outer diameter is 850 mm.

(2) Production Method for Primary Crystal-Type High B_s Nanocrystalline Soft Magnetic Alloys [31, 32, 34, 35]

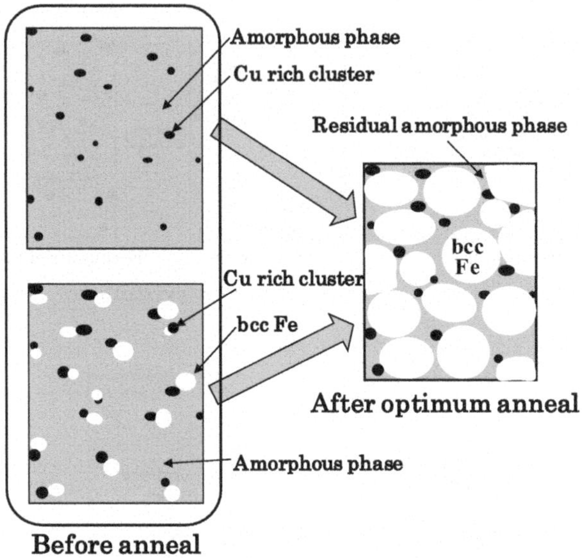

Fig. 22 Schematic diagram of crystallization mechanism by annealing for primary crystal type [35]

High B_s nanocrystalline soft magnetic alloys with high Fe content are classified into two categories, of which production methods are different. In this section, a production method for the primary crystal-type high B_s nanocrystalline alloys is described.

Figure 22 shows a schematic diagram for the crystallization mechanism by annealing a primary crystal-type high B_s nanocrystalline soft magnetic alloy. Considering that Cu does not form a solid solution with Fe, an amorphous alloy is prepared by rapid solidification of molten Fe–B or Fe–B–Si alloy with higher contents of Cu and Fe. This results in an amorphous alloy having Cu-rich clusters or nucleation sites for primary fine crystals of bcc Fe–Si. By annealing this alloy at temperatures above alloy's crystallization temperature, primary fine crystals of bcc Fe are formed. The content of B in the remaining amorphous phase is increased by the increase of the fine bcc Fe crystal phase. The remaining amorphous phase, therefore, becomes stabilized, controlling the growth of the crystal grains. Thus, a nanocrystalline alloy having high B_s, low $H_{c,}$ and low core loss is obtained. In short, the remaining amorphous phase with a high content of B is provided to stabilize the amorphous phase without Nb or Zr found in conventional nanocrystalline alloys.

As an example of primary crystal-type high B_s nanocrystalline alloy, 15-mm-wide and 0.022-mm-thick $Fe_{85}B_{15}$ and $Fe_{83.7}Cu_{1.5}B_{14.8}$ alloy strips cast by planar flow rapid solidification are annealed at 390 °C for 60 min soak in a nitrogen atmosphere. Figure 23 shows TEM micrographs of $Fe_{85}B_{15}$ and $Fe_{83.7}Cu_{1.5}B_{14.8}$ alloy before and after annealing. As shown in the figure, $Fe_{85}B_{15}$ alloy before anneal is completely amorphous, whereas $Fe_{83.7}Cu_{1.5}B_{14.8}$ alloy before anneal is amorphous with primary crystals. Uniform fine nanocrystalline structures are formed at the nucleation sites (primary crystals) in the alloys with Cu additive and a high Fe content.

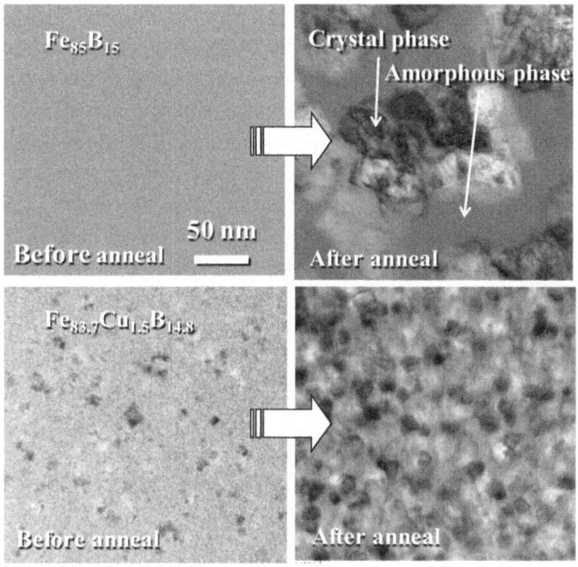

Fig. 23 TEM micrographs of $Fe_{85}B_{15}$ and $Fe_{83.7}Cu_{1.5}B_{14.8}$ alloy before and after anneal

As shown in Fig. 24, surface areas of an amorphous alloy strip with primary crystals made in contact with chill wheel during planar flow rapid solidification become amorphous without primary crystals because of its higher quench rate. Upper right portion of Fig. 24 shows a TEM micrograph of the chill wheel contacted surface area of the alloy strip annealed at a heating rate of 0.3 K/s at 400 °C for 60 min soak. The surface area has high ratio of amorphous phase, and its inner side area has coarse

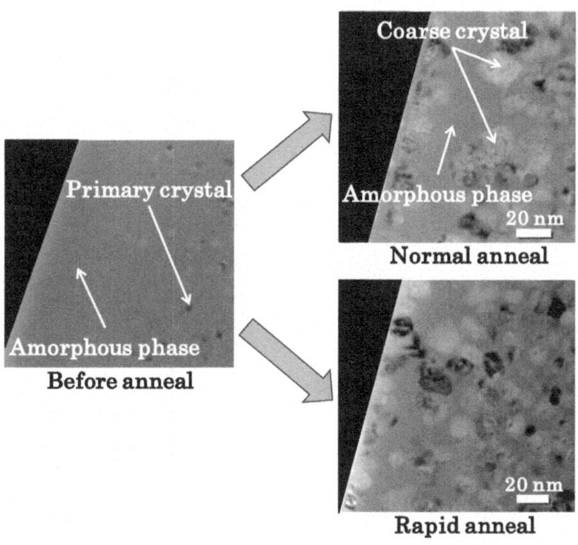

Fig. 24 TEM micrographs of $Fe_{80.5}Cu_{1.5}Si_4B_{14}$ alloy by normal anneal and rapid anneal

crystals. Lower right portion of Fig. 24 shows a TEM micrograph of the chill wheel contacted surface area of the alloy strip rapidly annealed at heating rate of 3 K/s at 400 °C for 5 min soak. A uniform fine nanocrystalline structure is obtained by the rapid anneal. Figure 25 shows the core losses of high B_s nanocrystalline alloys obtained by these two kinds of anneal. Core loss at high operating induction area is reduced by rapid annealing.

As already mentioned, it is necessary to form Cu-rich clusters or nucleation sites for primary bcc Fe–Si crystals in order to obtain primary crystal-type nanocrystalline alloys with uniform fine crystalline structures. It is, however, difficult to cast it stably for a long period of time because of its poor castability. It is a serious issue in order to put it to practical use. Also, the material is brittle because it has Cu-rich clusters or nucleation sites of primary bcc Fe–Si crystals. Its mechanical processing such as slitting or stamping is, therefore, difficult. As such, primary crystal-type nanocrystalline alloys have not been used currently as practical materials.

(3) Production Method for Rapid Anneal-Type High B_s Nanocrystalline Alloys [34, 36]

As shown in Fig. 24, it is clear that a uniform nanocrystalline structure can be formed from an amorphous $Fe_{80.5}Cu_{1.5}Si_4B_{14}$ alloy with primary crystals by rapid annealing (heating rate 3 K/s at 400 °C for 5 min soak). Similarly, rapid anneal-type nanocrystalline alloys are the alloys with uniform nanocrystalline structures formed from amorphous alloys with high Fe contents by rapid annealing.

Figure 26 shows a schematic diagram of crystallization mechanism by annealing for a rapid anneal-type high B_s nanocrystalline alloy. An amorphous Fe–B or Fe–Si–B alloy with a certain content of Cu and high Fe content is produced by planar flow rapid solidification by utilizing the low solid solubility limit of Cu in

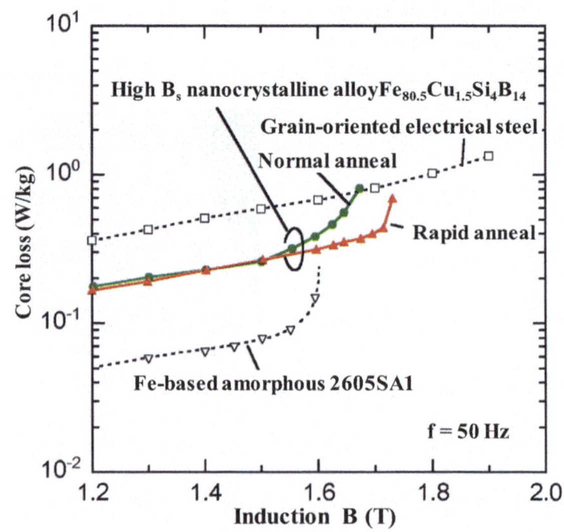

Fig. 25 Core loss comparison of normal and rapid annealed amorphous $Fe_{80.5}Cu_{1.5}Si_4B_{14}$ alloy with primary crystals [34]

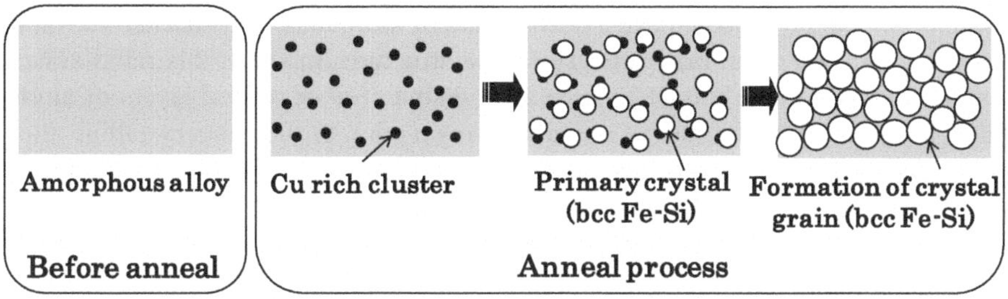

Fig. 26 Schematic diagram of crystallization mechanism of rapid anneal type

Fe. Primary fine crystals of bcc Fe–Si are formed uniformly and densely from Cu-rich clusters caused by rapid annealing. At the same time, a uniform nanocrystalline structure is formed because B-rich remaining amorphous phase controls coarsening of nanocrystal grains.

As an example of rapid anneal-type nanocrystalline alloy, Fig. 27 shows TEM micrographs of non-annealed and rapid annealed (heating rate 3 K/s at 450 °C for 5 min soak) amorphous $Fe_{82}Cu_1Nb_1Si_4B_{12}$ alloy (5 mm wide and 0.021 mm thick) cast by planar flow rapid solidification. More than 3 at. % of Nb is needed in order to form a uniform nanocrystalline structure from an amorphous Fe–Cu–Nb–Si–B alloy by conventional annealing. On the other hand, an amorphous alloy with high Fe content, in which Nb content is reduced to 1 at. %, can be formed into a uniform nanocrystalline structure by rapid annealing.

An amorphous alloy for rapid anneal-type nanocrystalline alloy can be cast stably for a long period of time. This amorphous alloy strip, however, must be annealed in

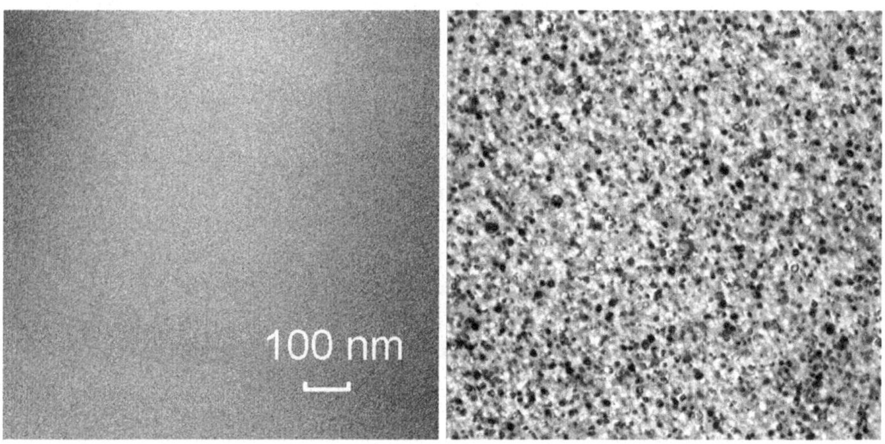

(a) Before anneal (b) After rapid anneal

Fig. 27 TEM micrographs of amorphous $Fe_{82}Cu_1Nb_1Si_4B_{12}$ alloy before and after rapid anneal (3 K/s at 450 °C for 5 min soak)[36] a single sheet in order to achieve a heating rate of more than 3 K/s needed for rapid annealing to form a uniform nanocrystalline structure. In short, this rapid anneal cannot be applied to a bulk state such as a wound core or several layers of amorphous alloy strips. In the case of manufacturing a core, brittle nanocrystalline alloy strips formed from an amorphous alloy by rapid annealing must be used for core manufacturing. This is a serious issue in practical use.

In order to solve this issue, the following process [37] has been tried. Not too brittle nanocrystalline alloy strip is formed from a single sheet of amorphous alloy strip by rapid annealing. This somewhat brittle alloy strip is wound to a core. Then, the core is normally annealed. This process is not generally used for practical products at present, as cores cannot be industrially manufactured stably by the above process.

3.3 Magnetic Properties and Motor Application of Nanocrystalline Soft Magnetic Alloys

Table 5 shows the magnetic properties of typical nanocrystalline alloys compared with an amorphous alloy and non-oriented electrical steels. Though FINEMET® FT-3M alloy which is a typical Fe–Si–B–Nb–Cu series nanocrystalline alloy has relatively low 1.23 T of B_s, its core loss is considerably low compared with other soft magnetic materials. Experimental results of low-loss prototype interior permanent-magnet synchronous motor (IPMSM) using a FINEMET® FT-3M stator core focused on its low core loss have been reported [38]. Though other five kinds of high B_s nanocrystalline alloys have more than 1.75 T of B_s, which is more than 85% of B_s of a non-oriented electrical steel, they have approximately 1/10 of core loss compared to non-oriented electrical steels. Therefore, they are exceptionally high-performance soft magnetic materials used for motor applications. It is reported that a small-size motor using Fe–Si–B–P–Cu series nanocrystalline alloy NANOMET® ($B_s = 1.84$ T, $W_{17/50} = 0.7$ W/kg) stator core, of which diameter and height are about 70 mm and about 50 mm, respectively, has been made experimentally. This report has shown 70% of core loss reduction and 6% of energy efficiency improvement compared to the same size motor using a non-oriented electrical steel stator core [39]. However, it takes time to achieve practical use of high B_s nanocrystalline alloy stator cores because there are some important issues to be solved as mentioned above. Having the properties of higher B_s and the same level of core loss compared to amorphous alloys, the high-performance soft magnetic materials discussed above are attractive as next-generation soft magnetic materials in energy-efficient motors.

Table 5 Magnetic properties of high B_s nanocrystalline soft magnetic alloys [9, 11–14]

Materials		B_S (T)	$P_{15/50}$ (W/kg)	$P_{10/400}$ (W/kg)	$P_{10/1k}$ (W/kg)
Nanocrystalline alloy	FINEMET® FT-3M	1.23	–	0.12 typ.	0.57 typ.
	$Fe_{82}Cu_1Nb_1Si_4B^a_{12}$ [34]	1.78	0.20	1.3	4.4
	$Fe_{82}Cu_1Nb_1Si_2B_{12}P^a_2$ [34	1.76	0.20	1.5	4.2
	$Fe_{80.8}Cu_{1.2}Si_5B_{11}P^a_2$ [34]	1.79	0.18	1.8	6.8
	$Fe_{80.5}Cu_{1.5}Si_4B^b_{14}$ [34]	1.80	0.27	1.6	5.8
	$Fe_{81.8}Cu_1Mo_{0.2}Si_4B^a_{14}$ [37]	1.75	0.28	1.5	5.0
Amorphous alloy Metglas® 2605SA1		1.56	0.13 typ.	1.2 typ.	4.3 typ.
Non-oriented electrical steel	35A300	2.03	2.4 typ.	18 typ.	78 typ.
	6.5% Si–Fe 10JNEX900	1.8	–	5.7 typ.	18.7 typ.

[a]Rapid anneal type
[b]Primary crystal type

References

1. Jun, W.K., Willens, R.H., Duwez, P.: Nature **187**, 869 (1960)
2. Lin, S.H.C., Duwez, P.: Phys. Status Solidi **34**, 469 (1969)
3. Mitera, M., Ohnuma, S., Watanabe, K., Masumoto, T.: Jpn. Inst. Metals Mater. **77**, 341 (1975)
4. Narasimhan, M.C.: U.S. patent no. 4142571 (1979)
5. Masumoto, T., Suzuki, K., Fujimori, H., Hashimoto, K.: Materials science of amorphous metals.Ohmusha, 7 (1982)
6. Japan Industrial Standard: Cold-rolled grain-oriented electrical steel strip and sheet delivered in the fully-processed state. JIS C 2553:2012, pp. 4 (2012)
7. Electric Power Research Institute: An EPRI white paper "amorphous metal transformer: next steps" (2009)
8. Enomoto, Y., Tokoi, H., Imagawa, T., Suzuki, T., Obata, T., Souma, K.: Hitachi Rev. **64**(8), 60–66 (2015)
9. Hitachi Metals Catalogue: Amorphous alloy ribbon Metglas®. No. HJ-B10-C (2015)
10. Nippon Steel and Sumitomo Metals Catalogue: Grain oriented electrical steel sheets. D004je_04_201509f (2015)
11. Nippon Steel and Sumitomo Metals Catalogue: Non-oriented electrical steel sheets. D005je_03_201308f (2013)
12. Kasai, S., Namikawa, M., Hiratani, T.: JFE technical report no. 36, p. 12 (2015)
13. JFE Steel Catalogue: Super Core™. Cat.No.F1J-002-05 (2014)
14. NKK Catalogue: Magnetic properties of NK super E core. Cat. No. 133-087-01 (1999)
15. Masumoto, T., Suzuki, K., Fujimori, H., Hashimoto, K.: Materials science of amorphous metals. Ohmusha, 69 (1982)
16. Masumoto, T., Suzuki, K., Fujimori, H., Hashimoto, K.: Materials science of amorphous metals. Ohmusha, 89 (1982)
17. Sato, A., Terada, H., Nagata, T., Kurita, S., Matsuda, Y., Fukui, K., Azuma, D., Hasegawa, R.: CIRED2009 session 1, Paper no. 0474 (2009)
18. Kuroda Precision Industries Catalogue: KURODA FASTEC® system. CAT.KD101-15.01 (2015)
19. Hitachi Metals Neomaterial Catalogue: NEOMAX permalloy (PB-1, 2). CAT. No. E1102 (2005)
20. Hitachi Metals Technical Report: YSS permendur YEP—2V. Technical report no. 311 (1985)
21. Fukao, T., Matsui, M., Chiba, A.: IEEJ Trans. Ind. Appl. **108**(4), 403 (1988)
22. Sugiyama, Y., Enomoto, Y., Imagawa, T., Itabashi, H., Tokoi, H.: IEEJ Trans. Ind. Appl. **134**(8), 760 (2014)
23. Okamoto, S., Denis, N., Fujisaki, K.: IEEE Trans. Ind. Appl. **52**(3), 2261 (2016)
24. Okada, S., Yamaguchi, T., Koga, N.: The proceedings of 67th Japanese Joint Conference for the Technology of Plasticity, p. 37 (2016)
25. Koga, N., Okada, S., Yamaguchi, T.: The Proceeding of the Spring Conference of 2016 Annual for The Japan Society for Technology of Plasticity, p. 257 (2016)
26. Inoue, M., Odawara, S., Fujisaki, K.: The Papers of Technical Meeting on Linear Drives, IEE Japan, LD-16-106 (2016)
27. Ertugrul, N., Hasegawa, R., Soong, W.L., Gayler, J., Kloeden, S., Kahourzade, S.: IEEE Trans. Magn. **51**(7), 8106006 (2015)
28. Kanemoto, Y., Ren, Z., Hase, M., Tanaka, H., Iyozumi, A., Tsuchiya, T., Chiba, K., Tashiro, K.: Hitachi Rev. **66**(5), 102–107 (2017)
29. Yoshizawa, Y., Oguma, S., Yamauchi, K.: J. Appl. Phys. **64**(10), 6044 (1988)
30. Suzuki, K., Kataoka, N., Inoue, A., Matsumoto, T.: Mater. Trans. JIM **31**(8), 743 (1990)
31. Ohta, M., Yoshizawa, Y.: Jpn. J. Appl. Phys. **46**(20), L447 (2007)
32. Ohta, M., Yoshizawa, Y.: J. Phys. D: Appl. Phys. **44**(6), 064004 (2011)
33. Makino, A., Men, H., Kubota, T., Inoue, A.: Mater. Trans. JIM **50**(1), 204 (2009)
34. Ohta, M., Yoshizawa, Y.: Appl. Phys. Exp. **2**, 023005 (2009)

35. Ohta, M., Yoshizawa, Y.: Mater. Jpn. **48**(3), 126–128 (2009)
36. Ohta, M., Yoshizawa, Y.: J. Magn. Magn. Mater. **321**, 2220 (2009)
37. Ohta, M., Hasegawa, R.: IEEE Trans. Magn. **53**(2), 2000205 (2017)
38. Denis, N., Inoue, M., Fujisaki, K., Itabashi, H., Yano, T.: IEEE Trans. Magn. 2700471 (2017)
39. Press Release by Tohoku University and Panasonic Corporation: Success in demonstrating energy efficient motor at the highest global standards (2014)

Nd–Fe–B-Based Sintered Magnet

Takeshi Nishiuchi

Abstract Sintered magnets based on neodymium (Nd), iron (Fe) and boron (B) were invented in 1982. They exhibit the highest performance among the industrially manufactured magnets. Nd–Fe–B magnets not only show excellent magnetic properties but are suitable for mass production as well. Therefore, they are used for many applications, especially for high-performance motors. This chapter offers basic knowledge for understanding Nd–Fe–B-based sintered magnet, including general features, an $Nd_2Fe_{14}B$ compound, phase diagram of Nd–Fe–B magnet, manufacturing process and efforts for realising high performance.

Keywords Nd–Fe–B · Sintered magnets · Manufacturing process · Magnetic properties

1 Introduction

The magnet based on neodymium (Nd), iron (Fe) and boron (B) [1, 2] was invented in 1982. Such a magnet is an epoch-making material because its magnetic properties are even beyond the strongest magnets until then, such as the ones based on samarium (Sm) and cobalt (Co). This magnet was mainly composed of Nd and Fe that have less resourceful restrictions than Sm and Co as well.

Nd–Fe–B magnets are currently manufactured industrially by several methods, as shown in Table 1. Among them, sintered magnets exhibit the highest performance and are suitable for mass production. Sintered magnets produced by powder metallurgy method have the largest production volume and have been used for many applications, such as in high-performance motors.

This chapter offers basic knowledge for understanding Nd–Fe–B-based sintered magnets, such as general features, an $Nd_2Fe_{14}B$ compound as an example, phase diagram of Nd–Fe–B magnet, manufacturing process and efforts for realising high performance.

T. Nishiuchi (✉)
Magnetic Materials Research Department, Advanced Components & Materials Research Laboratory, Hitachi Metals, Ltd., Kumagaya, Saitama, Japan
e-mail: takeshi.nishiuchi.rh@hitachi-metals.com

K. Fujisaki (ed.), *Magnetic Material for Motor Drive Systems*,
Engineering Materials, https://doi.org/10.1007/978-981-32-9906-1_20

Table 1 Industrially manufacturing methods of Nd–Fe–B-based magnets

Manufacturing method	Grain size	Anisotropy	Notes	Ref.
Sintering	$\geqq 1\ \mu m$	Good	–	[1]
Rapid quenching	10 nm–100 nm	No	Mainly used for bonded magnet powder	[2]
Hot deformation	100 nm–1 μm[*]	Good	Rapid quenching alloy is used as starting materials	[3]
Hydrogenation–disproportionation–desorption–recombination (HDDR)	100 nm–1 μm	Yes	Mainly used for bonded magnet powder	[4]

Note Dimensions of plate-like grains in the long axis direction

2 Basic Knowledge to Understand Nd–Fe–B-Based Sintered Magnets

2.1 General Indicators on the Characteristics of Permanent Magnets from a Practical Viewpoint

The characteristics of a permanent magnet are described at each point on the magnetic hysteresis curve as shown in Fig. 1. As a method to express the magnetic hysteresis curve, a *J-H* curve and a *B-H* curve are generally used. While a *J-H* curve shows the relationship between magnetic polarisation J and external magnetic field H, a *B-H* curve shows the relation between magnetic flux density B and external magnetic field H. The equation of $B = J + \mu_0 H$ (μ_0 is permeability of vacuum) is established between J and B. Using the relationship of $J = \mu_0 M$ (M is magnetisation), the equation of $B = \mu_0(M + H)$ can be achieved. Permanent magnets are used in the second quadrant of the magnetic hysteresis curve in almost all magnetic circuits.

The magnetic flux that can be extracted to the outside of the magnet can be easily obtained from the *B-H* curve. A high-performance magnet can have a high magnetic flux density B at a working point (effective magnetic field) in a magnetic circuit. Besides, the hysteresis curve varies with temperature. It is also important to consider

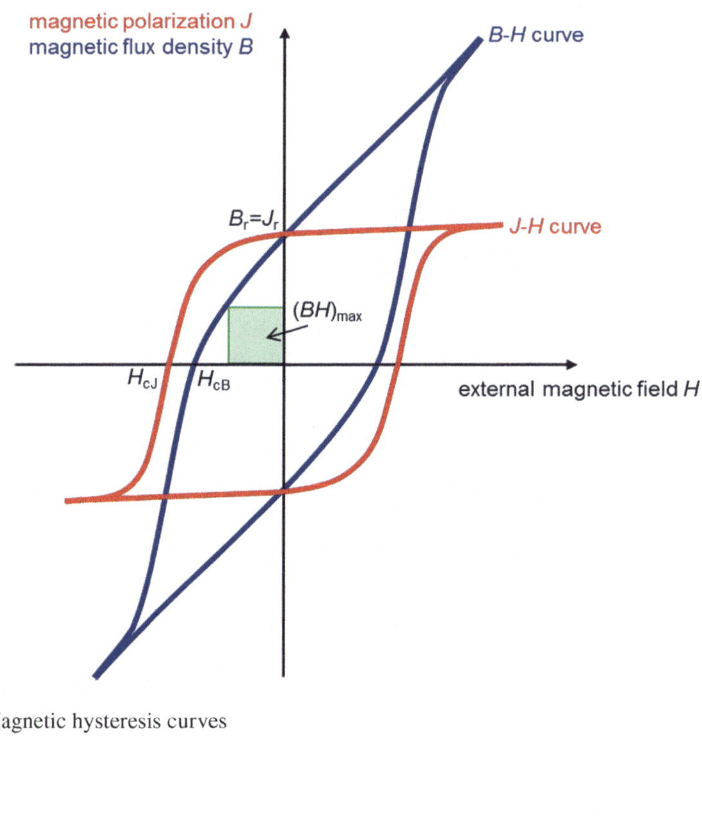

Fig. 1 Magnetic hysteresis curves

that irreversible demagnetisation does not occur in the temperature range where the magnets are exposed. Ideally, the *B-H* curve maintains linearity over the entire temperature range and the used working point (effective magnetic field).

Among the indices showing the performance of the permanent magnet, the most frequently used ones are residual magnetic flux density B_r (which is also called "remanence") and coercivity H_{cJ} (which is called "intrinsic coercivity" to be distinguished from H_{cB} on the *B-H* curve). The former one indicates the maximum value of the amount of magnetic flux to enable permanent magnet to supply the outside. The latter one shows the magnitude of the external magnetic field necessary for reversing the direction of magnetisation, i.e. the magnetic stability of the magnet. From the relational expression of J and B, the value of B_r is equal to the residual magnetic polarisation J_r. When the value of H_{cJ} is small, irreversible demagnetisation occurs. The *B-H* curve bends in the second quadrant of the magnetic hysteresis curve and becomes an effective magnetic field higher than the bending point. Furthermore, the squareness of the magnetic hysteresis curve is also important from the viewpoint of the linearity of the *B-H* curve. Several indicators can be used to show squareness. Among them, H_k is convenient to be expressed as the value of H where the value of J is 90% (or 95%) of B_r (=J_r) in the second quadrant of the hysteresis curve. H_k/H_{cJ} as squareness ratio is often used as well. In the case of two magnets having the same B_r and H_{cJ}, H_k/H_{cJ} close to one can indicate a magnet having good squareness, which is a high-performance one.

In addition to these, the maximum magnetic energy product $(BH)_{max}$ is used as an index to show the potential of the magnet. It is necessary to have sufficient H_{cJ} and squareness to obtain a high $(BH)_{max}$, in addition to high B_r. The maximum value thereof is $B_r^2/4\mu_0$.

2.2 *Factors that Determine the Characteristics of a Permanent Magnet*

As shown in Sect. 2.1, remanence B_r and coercivity H_{cJ} can act as representative indicators to show the characteristics of permanent magnets in general. When only one kind of main phase compound is responsible for the magnetic properties like the Nd–Fe–B-based sintered magnet, remanence B_r can be described by Eq. (1). J_s is the saturation magnetic polarisation of the compound. f is volume fraction of main phase to all phases consisted of the material (volume fraction of $Nd_2Fe_{14}B$ phase in the case of the Nd–Fe–B-based magnet). Besides, v is the volume ratio of voids (pores) in the material. Average value of the direction cosine component of θ in the entire material is $\langle\cos\theta\rangle$. Here, θ is an angle between the easy magnetisation direction of individual main phase crystal grains and magnetisation direction.

$$B_r = J_s \times f \times (1 - v) \times \langle\cos\theta\rangle \quad (1)$$

As can be easily understood from Eq. (1), for the purpose of achieving high B_r, it is necessary to select a compound having a high J_s as the main phase to increase the volume ratio and the orientation degree of this phase. On the other hand, with respect to the coercivity H_{cJ}, Eq. (2) is often used, which is obtained by combining anisotropic magnetic field H_a of the main phase compound and saturation magnetic polarisation J_s with coefficients α and N_{eff}.

$$H_{cJ} = \alpha H_a - N_{eff} J_s / \mu_0 \tag{2}$$

Because α in Eq. (2) usually takes a value between zero and one, the anisotropic magnetic field H_a of the main phase locally decreases and this region is pointed out as surface of grains [5]. Also, N_{eff} is called effective demagnetising field coefficient. The physical origin to determine the values of α and N_{eff} is not clear enough. Therefore, material design guidelines to improve H_{cJ} are not as clearly demonstrated as that of B_r. However, it is empirically considered that it is important to reduce the crystal grain size of the main phase [6–9] and to decouple magnetic interaction among the main phase crystal grains [6]. On the other hand, as shown in Eq. (2), when α and N_{eff} are constant, H_{cJ} is improved by increasing H_a of the main phase and decreasing J_s. The use of Dy and Tb in the Nd–Fe–B-based sintered magnet is interpreted to have an improved H_{cJ} due to such effects, which will be described later.

2.3 General Feature of Nd–Fe–B-Based Sintered Magnets

Nd–Fe–B-based sintered magnets are made by powder metallurgy method using Nd and Fe that are relatively rich in resources. Besides, such magnets are easily produced in mass. From the viewpoint of performance, Nd–Fe–B-based sintered magnets have many advantages. It is easy to obtain higher B_r than Nd–Fe–B-based magnets fabricated by other manufacturing methods. The squareness of the hysteresis curve is excellent, and magnetisation from thermal demagnetisation state is easy.

Based on these features, Nd–Fe–B-based sintered magnet is widely used in applications requiring miniaturisation as well as high power and/or high efficiency, such as voice coil motors (VCMs) of hard disc drives (HDDs), sound devices, compressor motors of air conditioners and motors for driving electric vehicles (xEV).

On the other hand, as shown in Fig. 2, the temperature change of H_{cJ} in the Nd–Fe–B-based sintered magnet is larger than that of B_r. Irreversible demagnetisation, thus, occurs on the high-temperature side depending on the working point. When the applications of this magnet spread from information devices such as HDDs to household electric appliances, air conditioners and automobiles, the temperature at which the magnet is used spreads to the high-temperature side. To prevent irreversible demagnetisation at a high temperature, several methods are generally used to ensure a sufficiently large H_{cJ} at room temperature (a concrete method will be described later).

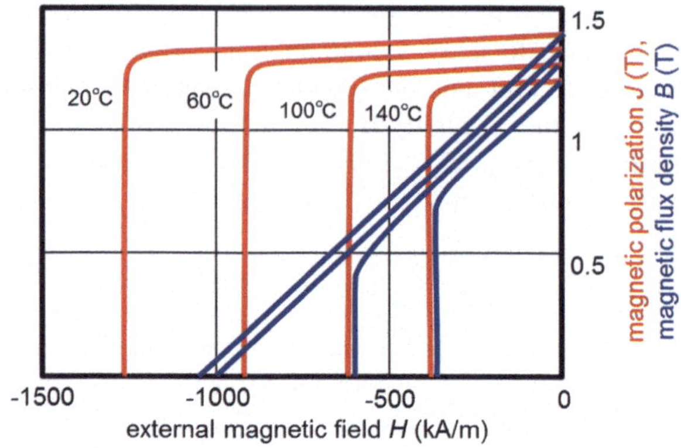

Fig. 2 Demagnetisation curves of Nd–Fe–B-based sintered magnets at various temperatures (example)

Further, the Nd–Fe–B-based sintered magnet has insufficient corrosion resistance. The corrosion progresses especially when the magnet is exposed to a harsh environment such as high temperature and high humidity for a long time. Therefore, surface treatments are performed for many cases. Depending on the use environment, plating, vapour deposition and resin coating are generally used. Simple rust prevention treatment can be applied at a small corrosion risk after installation of magnets into the product.

2.4 $Nd_2Fe_{14}B$ Compound

In many kinds of permanent magnets, excluding some magnets such as alnico magnets, compounds having high saturation magnetic polarisation J_s and anisotropic magnetic field H_a are used as the main phase. From the simultaneous rotation model of single-domain particles by Stoner and Wohlfarth [10] as well as Eqs. (1) and (2), J_s and H_a are considered as physical properties that indicate the potentials of B_r and H_{cJ}, respectively.

The main phase of Nd–Fe–B magnet is an $Nd_2Fe_{14}B$ ternary compound showing a saturation magnetic polarisation $J_s = 1.6$ T at room temperature and an anisotropic magnetic field $H_a = 5.3$ MA/m ($\mu_0 H_a = 7.0$ T, where μ_0 is vacuum permeability) [11]. The crystal structure of the $Nd_2Fe_{14}B$ compound [12] is shown in Fig. 3. This compound has a complicated crystal structure. Yttrium (Y) and lanthanides (collectively referred to as "rare earths" hereinafter) can occupy the sites of Nd in this compound. Transition metals such as cobalt (Co) and aluminium (Al) can occupy the sites of Fe, while the site of B can be substituted by carbon (C). The physical properties such as Js, H_a and T_c (Curie point) of the compound can be altered by substituting such elements.

Physical properties of the $R_2Fe_{14}B$ compound in which the Nd site is replaced with another rare earth element (R) are listed in Table 2 [11]. The values of J_s greatly change using light rare earth element (Ce to Sm) and heavy rare earth element (Gd to

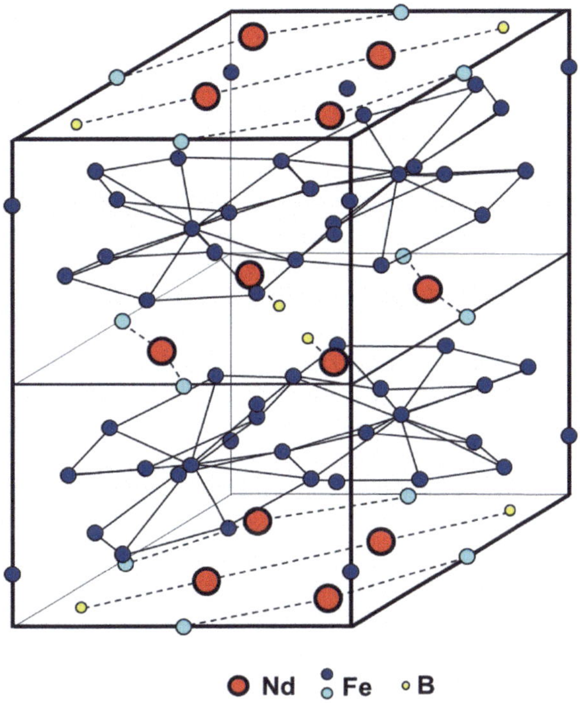

Fig. 3 Crystal structure of $Nd_2Fe_{14}B$ compound. To facilitate the recognition of the position, the colour of Fe on the same crystal plane as Nd is changed

Table 2 Lattice constant and magnetic properties of $R_2Fe_{14}B$ compounds (R:rare earth elements) (reconstruct from Ref. [11])

Element	Lattice constant		Density	Saturation magnetic polarisation (300 K)	Curie point	Anisotropic field (300 K)
R	a (nm)	c (nm)	D_x (kg/m^3)	J_s (T)	T_c (K)	H_a (MA/m)
Y	0.876	1.200	7.00	1.41	571	1.6
Ce	0.875	1.210	7.69	1.17	422	2.4
Pr	0.881	1.227	7.49	1.56	569	6.9
Nd	0.881	1.221	7.58	1.60	586	5.3
Sm	0.882	1.194	7.82	1.52	620	–
Gd	0.874	1.194	8.06	0.893	659	2.0
Tb	0.877	1.205	7.96	0.703	620	17.5
Dy	0.876	1.199	8.07	0.712	598	11.9
Ho	0.875	1.199	8.12	0.807	573	6.0
Er	0.875	1.199	8.16	0.899	551	–
Tm	0.874	1.194	8.23	1.15	549	–

Tm). The magnetic moment of the light rare earth element ferromagnetically couples with the magnetic moment of Fe in the same direction. The magnetic moment of the heavy rare earth element antiferromagnetically couples with the magnetic moment of Fe, and they counter against each other. Because yttrium (Y) has no magnetic moment, the contribution of J_s and H_a of $Y_2Fe_{14}B$ compound is interpreted by Fe sublattice. Besides, $Sm_2Fe_{14}B$ becomes in-plane magnetic anisotropy due to the distribution of the 4f electron cloud in rare earth atoms.

In the Nd–Fe–B-based magnet, a method to replace the Nd site partially with praseodymium (Pr), dysprosium (Dy) and terbium (Tb) is generally adopted. The use of Pr in Nd–Fe–B-based sintered magnets is often achieved by the mixed rare earth of Nd and Pr (called didymium) generated in the process of separation and refining of rare earths. J_s and H_a of $Pr_2Fe_{14}B$ are almost equal to that of $Nd_2Fe_{14}B$ at room temperature. However, because the decrease of H_a is large as temperature rises [11], care should be taken when Pr is added for a magnet used at high temperature. Because H_a is very high in $Dy_2Fe_{14}B$ and $Tb_2Fe_{14}B$, the value of H_{cJ} can be increased by replacing these elements with Nd in $Nd_2Fe_{14}B$ [13]. However, as mentioned above, because the magnetic moments of Dy and Tb counteract against each other with the magnetic moment of Fe, the value of J_s becomes small. As a result, it is a problem that B_r is reduced. In addition, it is worth noting that Tb at the same substitution amount has a higher H_{cJ} than Dy. Resource reserves of Tb, however, are even smaller than Dy.

2.5 *Nd–Fe–B Ternary Phase Diagram*

Nd–Fe–B-based sintered magnets can have high magnetic properties relatively easily possible due to the feature of its phase diagram. Figure 4 shows an Nd–Fe–B ternary phase diagram taken along a section where the molar ratio of Nd and B is 2:1 (stoichiometric ratio) [14]. $Nd_2Fe_{14}B$ phase as an equilibrium phase can stably exist in a wide temperature range.

In the Nd–Fe–B-based sintered magnet, a composition region having a slightly richer Nd side than the stoichiometric composition of $Nd_2Fe_{14}B$ is generally used. The primary crystal having such composition is γ-Fe phase, and the main phase is formed by a peritectic reaction of a liquid phase (L) and a γ-Fe phase at 1428 K (1158 °C), corresponding to "p_5" as shown in Fig. 4. Also, a liquid phase is formed at about 938 K (665 °C, corresponding to "E_2" as shown in Fig. 4) at which a ternary eutectic reaction described by $L \Longleftrightarrow Nd + Nd_2Fe_{14}B + Nd_{1+\varepsilon}Fe_4B_4$ takes place. In the vicinity of 1323 K (1050 °C) that is a typical sintering temperature, the $Nd_2Fe_{14}B$ phase and the liquid phase are in the equilibrium.

It is necessary to set the composition of an actual magnet so that a liquid phase having a necessary amount for sintering to proceed is obtained. On the other hand, it seems that rare earth elements such as Nd are easily oxidised. The oxidised Nd does not participate in liquid phase formation. Therefore, the amount of rare earth in the composition of the magnet is set in consideration of oxidation. When cooled

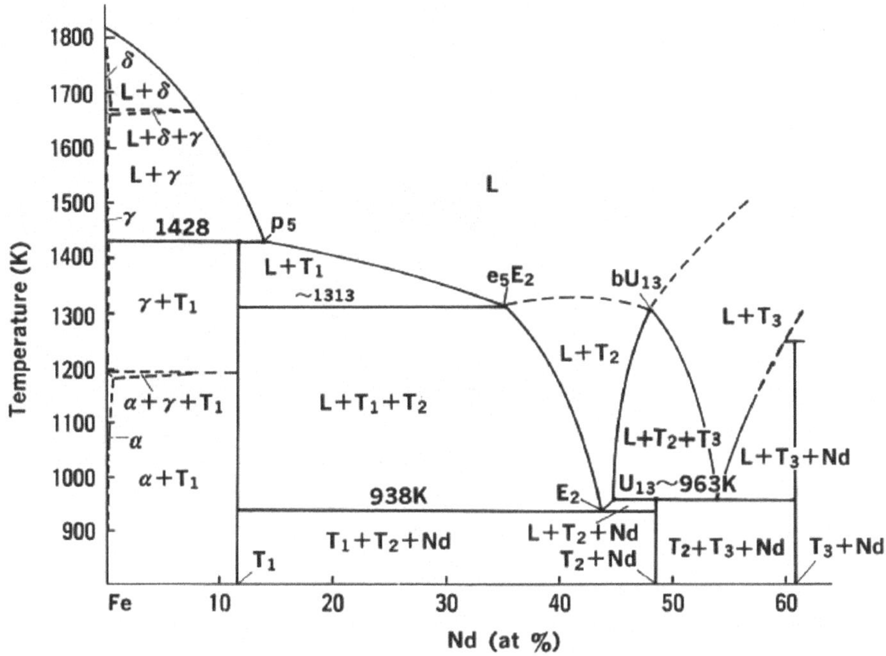

Fig. 4 Cross section of Nd–Fe–B phase diagram through the region of Nd:B = 2:1 [14]. L: Liquid, T_1: $Nd_2Fe_{14}B$, T_2: $Nd_{1+\varepsilon}Fe_4B_4$, p_5: peritectic reaction (γ-Fe + L $\Longleftrightarrow$ $Nd_2Fe_{14}B$), E_2: eutectic reaction (L $\Longleftrightarrow$ Nd + $Nd_2Fe_{14}B$ + $Nd_{1+\varepsilon}Fe_4B_4$), e_5: eutectic reaction (L $\Longleftrightarrow$ $Nd_2Fe_{14}B$ + $Nd_{1+\varepsilon}Fe_4B_4$), U_{13}: transition reaction (L + Nd_2FeB_3 $\Longleftrightarrow$ Nd + $Nd_{1+\varepsilon}Fe_4B_4$). Reprinted from [14], Copyright 1987 The Japan Society of Applied Physics

after sintering, the liquidus line moves towards the above-mentioned ternary eutectic point E_2. The ternary eutectic composition is $Nd_{67}Fe_{26}B_7$ (mol%), which is Nd-rich side [14]. Due to these features, despite the production of a liquid phase in an amount sufficient for the densification to proceed at the sintering temperature, the volume ratio of main phase becomes very high after cooling. It shows that the main phase ratio f is improved and the porosity v is reduced, which are important for obtaining high B_r as shown in Eq. (1). Besides, it is reported that the ratio of main phase at the laboratory level achieved 98% [15].

A typical microstructure of an Nd–Fe–B-based sintered magnet is schematically shown in Fig. 5. Because it is also expected from the ternary eutectic reaction formula as described above, a relatively Nd-rich phase is formed. This is a region between two main phase crystal grains, which enables to obtain a practical coercivity. In addition to the main phase and the Nd-rich phase, Nd oxide phases and B-rich phases are found in the typically Nd–Fe–B-based sintered magnets.

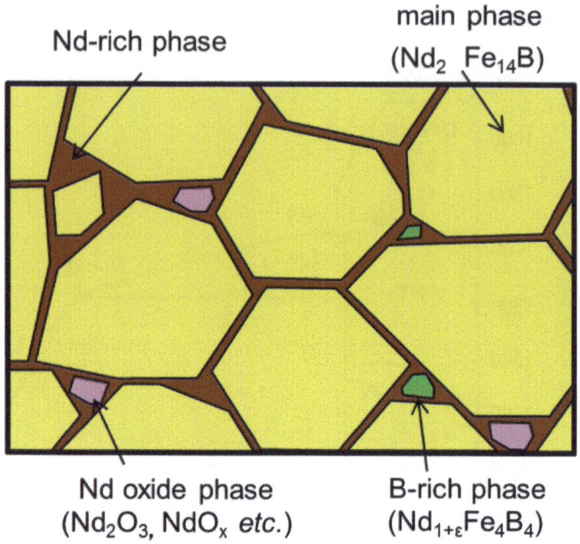

Fig. 5 Schematic illustration of microstructure in typical Nd–Fe–B-based sintered magnets. Actually, the crystal grain size of the main phase is in the order of a few micrometres, and the intergranular phase is very thin in the order of several nanometres

2.6 Manufacturing Process of Nd–Fe–B-Based Sintered Magnet

A general manufacturing process of the Nd–Fe–B-based sintered magnet is shown in Fig. 6. Each step will be described in detail below.

Raw materials

For a practical Nd–Fe–B-based sintered magnet, Nd, Fe and B are used as main components. Cobalt (Co), copper (Cu) or the like is generally added as an additive element. Among them, rare earth elements such as Nd are generally obtained by electrolysing oxides or fluorides that are separated and purified from mines by molten salt electrolysis. In addition, recently, rare earth elements are recovered from processing waste collected from the process and magnets in waste products.

Alloy Casting

A starting alloy is prepared by melting the raw material and casting. These processes are carried out in vacuum or inert atmosphere because rare earth elements such as Nd are susceptible to oxidation. As described in Sect. 2.5, the primary phase from the liquid phase is γ-Fe phase in a typical composition of an Nd–Fe–B-based sintered magnet. The alloy is obtained by pouring it into a mould, before cooling and solidifying. The α-Fe phase often exists in the alloy, which is formed by the phase transformation of the remaining γ-Fe phase with insufficient progress of peritectic reaction. The α-Fe phase present in the alloy is difficult to completely disappear even by heat treatment immediately below the peritectic temperature, which adversely affects subsequent processes such as pulverisation. To avoid such a problem, as schematically shown in Fig. 7, a strip cast (SC) method is developed and practically

Fig. 6 Typical manufacturing process of Nd–Fe–B-based sintered magnets

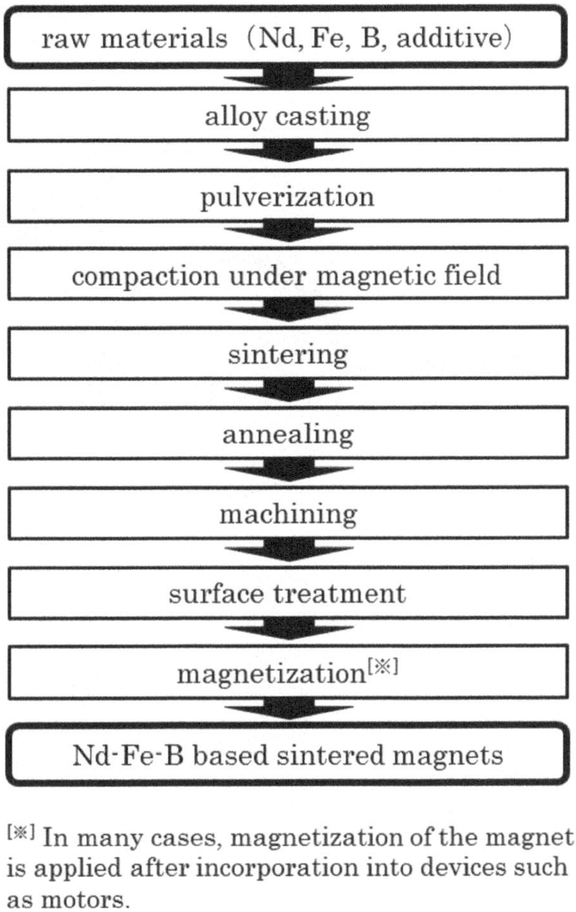

used. Molten metal is supplied onto a rotating cooling roll and solidified at a relatively high cooling rate [16, 17]. The alloy obtained by SC method has a thickness of several hundreds of micrometres. The α-Fe phase is greatly reduced compared to the casting method to previous moulds, and the Nd-rich phase appropriately dispersed. With these features, the SC alloy can easily produce single crystal-like powder or fine powder in the subsequent pulverisation step. As a result, excellent magnetic properties can be easily obtained.

Pulverisation

The Nd–Fe–B alloy is generally difficult to pulverise. However, the use of hydrogen decrepitation (HD) method enables the easy embrittlement of the alloy and the improvement in pulverisability [18, 19]. In HD method, the raw material alloy is held in a hydrogen atmosphere and hydrogen is occluded in the main phase and the Nd-rich phase. As a result, a brittle phase is generated by hydrogenation. Cracks that

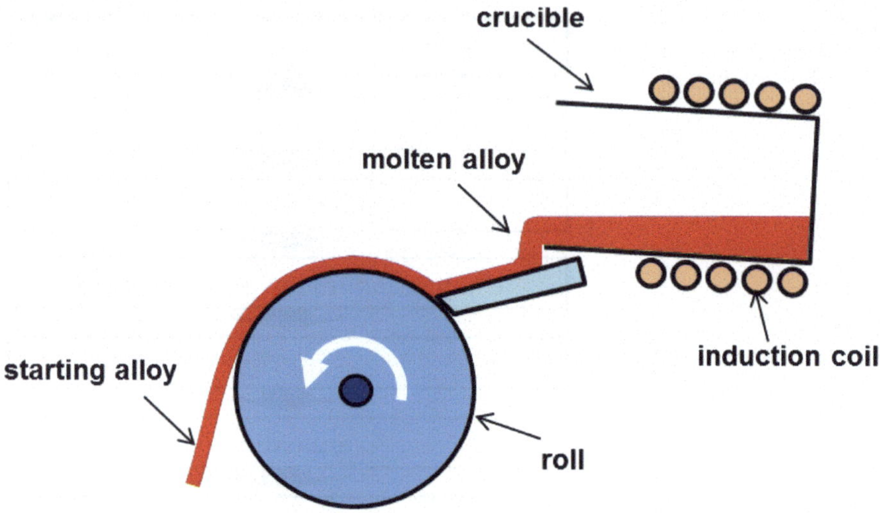

Fig. 7 Schematic illustration of strip casting (SC)

serve as starting points of fracture at the time of pulverisation are introduced into the alloy via the difference in volume expansion during hydrogen occlusion.

The alloy after HD treatment becomes fine powder through a subsequent pulverisation process. The particle size of a typical industrially adopted powder is 3–10 μm. A jet mill is often used as a pulverisation method. In this process, the alloy to be treated is charged into the high-speed airflow, before it is crushed using the collision energies of the alloy or the parts constituting the pulveriser and the alloy. A classifier may be incorporated in a pulverising apparatus to control the particle size distribution of the obtained powder.

Compaction in Magnetic Field

The obtained powder is compression-moulded while being oriented in a magnetic field to prepare a green compact. As a method to manufacture a green compact having magnetically oriented in one direction, transverse magnetic field moulding (Fig. 8a) or longitudinal magnetic field moulding (Fig. 8b) is generally used. Generally, high B_r can be obtained by transverse magnetic field moulding where turbulence of powder orientation accompanying compression is small.

Furthermore, a magnet having a characteristic orientation distribution can be manufactured by appropriately setting the orientation magnetic field distribution to be imparted during moulding. For example, radially and polar oriented anisotropic magnets having orientation distributions as shown in Fig. 9a, b are manufactured in cylindrical green compacts. To manufacture magnets having these characteristic orientation distributions, it is necessary to set moulding conditions in consideration of shrinking behaviours of the green compact in the sintering step.

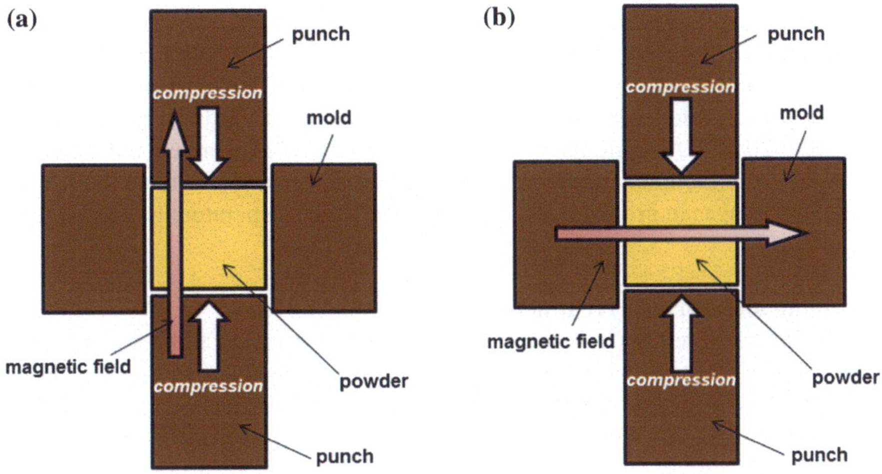

Fig. 8 Schematic illustrations of compaction under magnetic field. **a** Transverse magnetic field moulding and **b** longitudinal magnetic field moulding

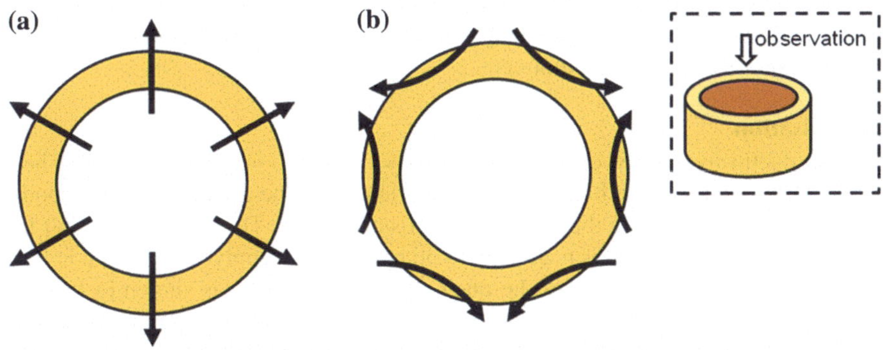

Fig. 9 Schematic illustration of distribution of orientation magnetic field for cylindrical magnets. **a** Radially oriented anisotropic magnet and **b** polar oriented anisotropic magnets

Sintering and Annealing

The obtained green compact is sintered to prepare a sintered body. As shown in Sect. 2.5, in the basic composition of the Nd–Fe–B-based sintered magnet, a liquid phase is formed at least at 938 K (665 °C). The amount of the liquid phase rapidly increases by increasing the temperature. Typical liquid phase sintering proceeds at a relatively low temperature of 1373 K (1100 °C) or less, and the densification is completed. Vacuum or inert gas atmosphere is applied during sintering to prevent oxidation of alloy.

Thereafter, heat treatment is performed at a temperature lower than the sintering temperature to optimise the magnetic properties, particularly the coercive force of the obtained sintered body. The variation of the optimised conditions for heat treatment depends on the composition. However, in the final case, heat treatment is often carried out at a relatively low temperature of around 773 K (500 °C) in many cases. Generally, Cu is added as an additive element to an industrially produced sintered magnet. The eutectic temperature on the Nd-rich side in the Nd–Cu binary system is 793 K (520 °C). A liquid phase containing Nd or Cu is generated and improves the coercivity. Recent advancement of the electron microscope technology enables to observe changes in grain boundary phase due to low-temperature heat treatments [20].

Machining and Surface Treatment

The obtained sintered body is subjected to cutting or grinding to have a predetermined product shape and size. Furthermore, surface treatment is performed to ensure corrosion resistance and/or other purposes. Surface treatment is selected according to the environment in which magnets are used. Techniques such as plating of nickel or other metals/alloys, vapour deposition such as aluminium, resin coating and other methods have been developed as well. The exposure to corrosive environment may not occur due to incorporation into parts in the case of magnets for internal permanent magnet (IPM) motors, which have become mainstream in recent years. In such case, a simple and low-cost surface treatment is applied for the purpose of rust prevention only in a relatively gentle environment from the moment the product is shipped until being incorporated into the component.

Magnetisation

The manufacturing process of the Nd–Fe–B-based sintered magnet includes the heat treatment at a temperature higher than the Curie point. The obtained sintered body, thus, is fundamentally in a thermal demagnetisation state. Therefore, to exhibit the performance as a magnet, it is necessary to magnetise it by applying a magnetic field from the outside. An example of the magnetisation properties is shown in Fig. 10. If the strength of the magnetising magnetic field is insufficient, the magnetic flux that should be generated by the magnet cannot be obtained. In addition, irreversible thermal demagnetisation may become apparent in the incompletely magnetised state.

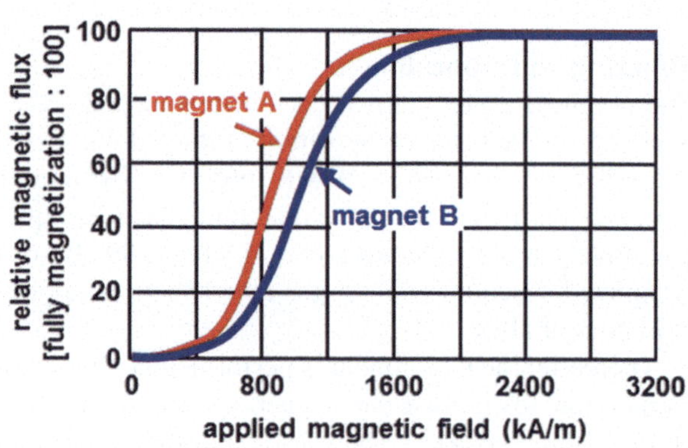

Fig. 10 Example of magnetisation properties. In this figure, the magnetisation property of magnet A is better than that of magnet B

In many cases, magnetisation of a non-magnetised state magnet is incorporated in a magnetic circuit such as a motor or like to facilitate the handling of the manufacturing process of the device. In some cases, because it is not possible to apply a sufficiently high external magnetic field depending on the magnetic circuit, easy magnetisation having a low magnetic field (good magnetisation) is a great merit in practical use. From this viewpoint, it can be said that the Nd–Fe–B-based sintered magnet is excellent in magnetisation.

3 Technology for High-Performance Nd–Fe–B-Based Sintered Magnet

3.1 Technology for Obtaining High $\mathbf{B_r}$

Generally, the factor for determining B_r of a permanent magnet is described by the aforementioned Eq. (1). However, there are no additional elements that can drastically increase the saturation magnetisation J_s of the main phase. Therefore, to obtain a high B_r, it is necessary to reduce the porosity v as well as to increase the volume fraction of the main phase f and the degree of orientation of the main phase $\langle \cos \theta \rangle$, as suggested by Eq. (1). It is necessary to bring the composition close to the stoichiometric composition of $Nd_2Fe_{14}B$ to increase the volume of the main phase f. On the other hand, an excess of Nd is required over the stoichiometric composition to reduce the porosity v by advancing the densification by sintering and to secure the contribution of Nd-rich phase to the coercivity. Because rare earth elements such as Nd are easily oxidised, it is important to suppress the formation of Nd oxide which cannot contribute to liquid phase sintering and coercive force expression [14]. A process is often adopted to suppress oxidation, where each step is blocked from oxygen (often referred to as "low oxygen process"). Besides, as a distinctive low oxygen process, pulverised powder is recovered in oil and then wet-moulded (HILOP^{TM}: Hitachi Low Oxygen Process) [21]. Furthermore, a process of filling a container with finely pulverised powder is reported, orienting in a magnetic field and sintering as it is in the same container in an oxygen-blocking atmosphere (PLP: Plessless Process) [22].

On the other hand, to improve the main phase orientation degree $\langle \cos \theta \rangle$, it is necessary to align the easy magnetisation directions of individual finely pulverised powders in the same direction as much as possible during moulding in a magnetic field. In the compaction process, as shown in Sect. 2.6, higher orientation degree is achieved by applying the transverse magnetic field moulding than by applying the longitudinal magnetic field moulding. To improve the degree of orientation, a method of isotopically compressing powder after orientation by isostatic pressing [23, 24] is used. Another method of promoting rotation and flow of individual particles is also reported by utilising an alternating pulse magnetic field and gradient magnetic field [25] to reduce the friction between the powders in the magnetic field [25, 26].

Similar methods have been reported as well. It is necessary to consider productivity for a method in the industry. It is also reported that the coercivity H_{cJ} decreases as the orientation degree increases [27]. This point also needs attention.

3.2 *Technology for Obtaining High* H_{cJ}

As mentioned in Sect. 2.3, the most effective method to increase H_{cJ} is to replace Nd partially with Dy or Tb to increase the anisotropic magnetic field H_a of the main phase compound. This method has been generally adopted. However, by increasing the substitution amount of Dy and Tb, the saturation magnetic polarisation J_s of the main phase decreases. As a result, B_r of the obtained magnet also decreases. In addition, Dy and Tb are high in resource risks due to reserves and the uneven distribution of regions to be produced.

A method to effectively utilise Dy and Tb is unevenly distributing these elements in an effective area for H_{cJ} improvement. Because the local anisotropy reduction of the main phase indicated by α in Eq. (2) is interpreted as the outermost part of the main phase crystal grains, high H_{cJ} is obtained when Dy and Tb can be concentrated in this region. Grain boundary diffusion (GBD) method has been developed as an effective way to realise the concept. A schematic diagram of the GBD process is shown in Fig. 11. After preparing the sintered body, Dy and Tb are supplied to the surface of the sintered body, which are heat treated at a temperature lower than the sintering temperature to introduce them from the surface of the sintered body through the grain boundary to form the main phase and concentrate only to the outermost part [28]. A method capable of mass production was established in the 2000s [29–31]. Although GBD has restrictions on the applicable size of the magnets, it is adopted mainly for applications of automotive and compressor for air conditioner because it can realise both high B_r and high H_{cJ} at the same time and greatly reduces heavy

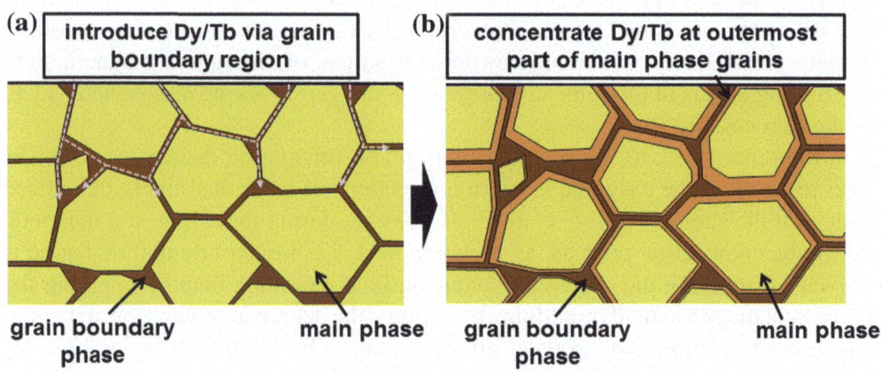

Fig. 11 Schematic illustration of the grain boundary diffusion (GBD) method. **a** During GBD and **b** after GBD

rare earth element usage. In addition, a distribution of H_{cJ} is formed in the magnet obtained by GBD. Dy (or Tb) is introduced into a specific region using GBD and increases the coercivity. The demagnetising field is high, and magnetisation reversal tends to occur (generally, it becomes a specific part in the vicinity of the magnet surface) in the magnetic circuit design. To ensure heat resistance, a small amount of Dy (or Tb) has also been proposed [32, 33].

On the other hand, as an H_{cJ} improving method that does not depend on heavy rare earth element, two approaches are developed, such as refining crystal grain size and optimising the structure near grain boundary. Grain size refinement can be realised by reducing the particle size of pulverised powder before sintering. In recent years, to impart high pulverisation energy, finely pulverised powder of about 1 μm could be prepared using helium as the crushing gas of the jet mill. The sintered magnet using this powder shows high H_{cJ} [34]. The reduction in the particle size of the finely pulverised powder is an effective way to increase the coercivity as described above. However, there are problems, including that the powder is active and difficult to handle as well as productivity is deteriorated.

On the other hand, the breakthrough progress of the tissue analysis technology since the 2000s [20, 35–39] and new findings have been obtained for the optimisation of the grain boundary structure near the grain boundary. For example, several analyses revealed that the intergranular grain boundary phase shows ferromagnetic, although it has been thought to be non-magnetic for a long time [36, 40–42]. R&D focused on the control of grain boundary region is in progress as well. For example, in actual material development, it is reported that when a specific composition including gallium (Ga) is applied, an intergranular grain boundary phase having a thickness exceeding 10 nm can be formed. The magnet having such microstructure shows high H_{cJ} [43].

4 Summary

As shown in this chapter, Nd–Fe–B-based sintered magnets not only present high performance but are suitable for mass production, a high degree of freedom to shape and orientation, excellent magnetisation properties as well, which give great benefits to apply for users. Even now, more than 30 years have passed since the announcement of this magnet, and material development is continuously in progress in terms of both scarcity resource reduction and high performance. This magnet is expected to contribute continuously to further miniaturisation, high efficiency and high power of motors.

References

1. Sagawa, M., Fujimura, S., Togawa, N., Yamamoto, H., Matsuura, Y.: New material for permanent magnets on a base of Nd and Fe (invited). J. Appl. Phys. **55**, 2083–2087 (1984)

2. Croat, J.J., Herbst, J.F., Lee, R.W., Pinkerton, F.E.: Pr-Fe and Nd–Fe-based materials: a new class of high-performance permanent magnets (invited). J. Appl. Phys. **55**, 2078–2082 (1984)
3. Lee, R.W.: Hot-pressed neodymium-iron-boron magnets. Appl. Phys. Lett. **46**, 790–791 (1985)
4. Nakayama, R., Takeshita, T., Itakura, M., Kuwano, N., Oki, K.: Magnetic properties and microstructures of the Nd–Fe–B magnet powder produced by hydrogen treatment. J. Appl. Phys. **70**, 3770–3774 (1991)
5. Kronmüller, H., Durst, K.-D., Sagawa, M.: Analysis of the magnetic hardening mechanism in RE-FeB permanent magnets. J. Magn. Magn. Mater. **74**, 291–302 (1988)
6. Ramesh, R., Thomas G., Ma, B.M.: Magnetization reversal in nucleation controlled magnets. II. Effect of grain size and size distribution on intrinsic coercivity of Fe–Nd–B magnets. J. Appl. Phys. **64**, 6416–6423 (1988)
7. Nothnagel, P., Müller, K.-H., Eckert, D., Handstein, A.: The influence of particle size on the coercivity of sintered NdFeB magnets. J. Magn. Magn. Mater. **101**, 379–381 (1991)
8. Scott, D.W., Ma, B.M., Liang, Y.L., Bounds, C.O.: The effects of average grain size on the magnetic properties and corrosion resistance of NdFeB sintered magnets. J. Appl. Phys. **79**, 5501–5503 (1996)
9. Uestuener, K., Katter, M., Rodewald, W.: Dependence of the mean grain size and coercivity of sintered Nd–Fe–B magnets on the initial powder particle size. IEEE Trans. Magn. **42**, 2897–2899 (2006)
10. Stoner, E.C., Wohlfarth, E.P.: A mechanism of magnetic hysteresis in heterogenious alloys. Phil. Trans. Roy. Soc. **240**, 599–642 (1948)
11. Hirosawa, S., Matsuura, Y., Yamamoto, H., Fujimura, S., Sagawa, M., Yamauchi, H.: Magnetization and magnetic anisotropy of $R_2Fe_{14}B$ measured on single crystals. J. Appl. Phys. **59**, 873–879 (1986)
12. Harbst, J.F., Croat, J.J., Pinkjerton, F.E., Yelon, W.B.: Relationships between crystal structure and magnetic properties in Nd2Fe14B. Phys. Rev. B **29**, 4176–4178 (1984)
13. Sagawa, M., Fujimura, S., Yamamoto, H., Matsuura, Y., Hiraga, K.: Permanent magnet materials based on the rare earth-iron-boron tetragonal compounds. IEEE Trans. Magn. **20**, 1584–1589 (1984)
14. Sagawa, M., Hirosawa, S., Yamamoto, H., Fujimura, S., Matsuura, Y.: Nd–Fe–B permanent magnetic materials. Jpn. J. Appl. Phys. **26**, 785–800 (1987)
15. Harimoto, D., Matsuura, Y.: Development of high performance Nd–Fe–B sintered magnets. Hitachi Metals Tech. Rev. **23**, 69–72 (2007). (in Japanese)
16. Hirose, Y., Hasegawa, H., Sasaki, S., Sagawa, M.: Microstructure of strip cast alloys for high performance NdFeB magnets. In: Proceeding on 15th Rare-Earth Magnets and their Applications, Dresden, pp. 77–86 (1998)
17. Okada, C., Miyake, Y., Yamamoto, K., Shibamoto, T.: Development of new production method (strip-casting method) for rare-earth magnetic alloys. J. Jpn. Soc. Powder Metall. **55**, 517–521 (2008). (in Japanese)
18. Harris, I.R., Noble, C., Bailey, T.: The hydrogen decrepitation of an Nd15Fe77B8 magnetic alloy. J. Less-Common Met. **106**, L1–L4 (1985)
19. McGuiness, P.J., Harris, I.R., Rozendaal, E., Ormerod, J., Ward, M.: The production of a Nd–Fe–B permanent magnet by a hydrogen decrepitation/attritor milling route. J. Mater. Sci. **21**, 4107–4110 (1986)
20. Vial, F., Joly, F., Nevalainen, E., Sagawa, M., Hiraga, K., Park, K.T.: Improvement of coercivity of sintered NdFeB permanent magnets by heat treatment. J. Magn. Magn. Mater. **242–245**, 1329–1334 (2002)
21. Uchida, K., Takahashi, M., Tanighchi, F., Mikamoto, T., Sasaki, K.: High-performance Nd–Fe–B permanent magnets by wet process. HILOP. Hitachi Metals Tech. Rev. **13**, 59–64 (1997). (in Japanese)
22. Sagawa, M., Une, Y.: A new process for producing Nd–Fe–B sintered magnets with small grain size. In: Proceedings on the 20th International Workshop on Rare-Earth Permanent Magnets and their Applications, Crete, pp. 103–105 (2008)

23. Sagawa, M., Nagata, H.: Novel processing technology for permanent magnets. IEEE Trans. Magn. **29**, 2747–2751 (1993)
24. Kaneko, Y., Ishigaki, N.: Recent developments of high-performance NEOMAX magnets. J. Mater. Eng. Performances **3**, 228–233 (1994)
25. Kaneko, Y., Tokuhara, K., Sasakawa, Y.: Developing and mass-producing of super high performance 400 kJ/m^3 Nd–Fe–B Magnets. Jpn. Soc. Powder Metall. **47**, 139–145 (2000) (in Japanese)
26. Kuniyoshi, F., Nakahara, K., Kaneko, Y.: Developing of 460 kJ/m^3 Nd–Fe–B magnets. Jpn. Soc. Powder Powder Metallurgy **51**, 698–702 (2003). (in Japanese)
27. Matsuura, Y., Hoshijima, J., Ishii, R.: Relation between $Nd_2Fe_{14}B$ grain alignment and coercive force decrease ratio in NdFeB sintered magnets. J. Magn. Magn. Mater. **336**, 88–92 (2013)
28. Park, K.T., Hiraga, K., Sagawa, M.: Effect of metal-coating and consecutive heat treatment on coercivity of thin Nd–Fe–B sintered magnets. In: Proceedings on the 16th International Workshop on Rare-Earth Magnets and Their Applications, Sendai, pp. 257–264 (2000)
29. Hirota, K., Nakamura, H., Minowa, T., Honshima, M.: Coercivity enhancement by the grain boundary diffusion process to Nd–Fe–B sintered magnets. IEEE Trans. Magn. **42**, 2909–2911 (2006)
30. Yoshimura, K., Morimoto, H. Odaka, T.: R-Fe-B rare earth sintered magnet and method for producing same. WO2007/102391 A1 (2007)
31. Kunieda, R.: Saving methods of heavy rare-earth elements for Nd–Fe–B sintered magnets. Mater. Integr. **24**, 270–274 (2011). (in Japanese)
32. Natsumeda, M.: Motor design method using dy diffused magnets and effect of its application. Hitachi Metals Tech. Rev. **28**, 8–13 (2012). (in Japanese)
33. Watanabe, N., Ito, M., Doi, Y., Hirota, K., Nakamura, H., Minowa, T.: Investigation of demagnetization process of the coercivity distributed Nd–Fe–B sintered magnets. In: Proceedings on the 22nd International Workshop on Rare-Earth Permanent Magnets and Their Applications, Nagawaki, pp. 359–362 (2012)
34. Une, Y., Sagawa, M.: Enhancement of coercivity of Nd–Fe–B sintered magnets by grain size reduction. J. Jpn. Inst. Metals. **76**, 12–16 (2012). (in Japanese)
35. Li, W.F., Ohkubo, T., Hono, K., Sagawa, M.: The origin of coercivity decrease in fine grained Nd–Fe–B sintered magnets. J. Magn. Magn. Mater. **321**, 1100–1105 (2009)
36. Sepehri-Amin, H., Ohkubo, T., Shima, T., Hono, K.: Grain boundary and interface chemistry of an Nd–Fe–B-based sintered magnet. Acta Mater. **60**, 819–830 (2012)
37. Itakura, M., Watanabe, N., Nishida, M., Daio, T., Matsumura, S.: Atomic-resolution X-ray energy-dispersive spectroscopy chemical mapping of substitutional Dy atoms in a high-coercivity neodymium magnet. Jpn. J. Appl. Phys. **52**, 050201 (2013)
38. Sasaki, T.T., Ohkubo, T., Hono, K., Une, Y., Sagawa, M.: Correlative multi-scale characterization of a fine grained Nd–Fe–B sintered magnet. Ultramicroscopy **132**, 222–226 (2013)
39. Sasaki, T.T., Ohkubo, T., Hono, K.: Structure and chemical compositions of the grain boundary phase in Nd–Fe–B sintered magnets. Acta Mater. **115**, 269–277 (2016)
40. Murakami, Y., Tanigaki, T., Sasaki, T.T., Takeno, Y., Park, H.S., Matsuda, T., Ohkubo, T., Hono, K., Shindo, D.: Magnetism of ultrathin intergranular boundary regions in Nd–Fe–B permanent magnets. Acta Mater. **71**, 370–379 (2014)
41. Kohashi, T., Motai, K., Nishiuchi, T., Hirosawa, S.: The magnetism at the grain boundaries of NdFeB sintered magnet studied by spin-polarized scanning electron microscopy (spin SEM). Appl. Phys. Lett. **104**, 232408 (2014)
42. Nakamura, T., Yasui, A., Kotani, Y., Fukagawa, T., Nishiuchi, T., Iwai, H., Akiya, T., Ohkubo, T., Gohda, Y., Hono, K., Hirosawa, S.: Direct observation of ferromagnetism in grain boundary phase of Nd–Fe–B sintered magnet using soft X-ray magnetic circular dichroism. Appl. Phys. Lett. **105**, 202404 (2014)
43. Sasaki, T.T., Ohkubo, T., Takada, Y., Sato, T., Kato, A., Kaneko, Y., Hono, K.: Formation of non-ferromagnetic grain boundary phase in a Ga-doped Nd-rich Nd–Fe–B sintered magnet. Scripta Mater. **113**, 218–221 (2016)

Bonded Rare Earth Permanent Magnets

Kenji Ohmori

Abstract A bonded magnet is a generic term for composite permanent magnets obtained by solidifying and shaping various magnetic powders with a binder. Plastic magnets or "Pla-Mag" are synonyms. Magnetic properties themselves are inferior to sintered magnets because they contain a binder in addition to the magnet powder and fine pores remain after forming. However, because of its high degree of freedom of shape, it is possible to manufacture various near-net shape magnets with high accuracy such as thin-walled ring. It has also the characteristics that the toughness of the molded body is high, cracking is less likely to occur, and integral molding with a shaft or the like is possible. It is superior to sintered magnets in terms of being able to reduce core wobbling, etc., and many are used for small motors for electronic devices and the like.

Keywords Bonded magnet · Rare earth · Thermosetting resin · Thermoplastic resin · Compression molding · Injection molding

1 Basic Characteristics of Magnets

Table 1 shows the basic characteristics of typical rare earth alloys for magnets. The requirement for becoming a high-performance magnet material is that the saturation magnetization J_s, the anisotropic magnetic field H_A, and the Curie temperature T_c are high respectively. Because the maximum energy product that can be possessed when magnetized is proportional to the square of the saturation magnetization when the coercive force satisfies the necessary condition, it is desired the saturation magnetization is as large as possible.

As shown in this table, the rare earth magnet powder currently used for the bonded magnet is mainly Nd type and Sm type.

K. Ohmori (✉)
The Japan Association of Bonded Magnetic Materials, Tokyo, Japan
e-mail: JDY04537@nifty.com

K. Fujisaki (ed.), *Magnetic Material for Motor Drive Systems*,
Engineering Materials, https://doi.org/10.1007/978-981-32-9906-1_21

Table 1 Basic characteristics of typical rare earth compounds for magnets

	Saturation magnetization J_s (T)	Anisotropy field H_a (MA/m)	Curie temp. T_c (K)	Crystalline anisotropy K_1 (MJ/m^3)	Density d (Mg/m^3)	Crystal structure	References
$SmCo_5$	1.14	22.3	1000	17.2	8.4	$CaCu_5$	[1]
Sm_2Co_{17}	1.25	4.8	1193	3.3	8.4	Th_2Zn_{17}	[1]
$Sm_2Fe_{17}N_3$	1.54	20.7	749	8.6	7.67	Th_2Zn_{17}	[1]
$NdFe_{11}TiN$	1.45	9.5	729	–		$ThMn_{12}$	[1]
$Nd_2Fe_{14}B$	1.6	5.3	585	4.9	7.56	$Nd_2Fe_{14}B$	[2]
$Sm_{10.6}Fe_{89.4}N_x$	1.4	6.8	743	0.2		$TbCu_7$	[3]
$(Sm_{0.75}Zr_{0.25})(Fe_{0.7}Co_{0.3})_{10}N_{1.5}$	1.7	6.2	>600	–	7.7	$TbCu_7$	[4]

Although the coercive force generating mechanism of the magnet is closely related to the magnetic anisotropy energy, there are a reverse domain nucleus type and a magnetic domain wall pinning type and the like. Particularly in the case of the nucleation type, it is believed that a linear relationship is empirically established between the anisotropic magnetic field obtained by dividing the magnetic anisotropy energy by the saturation magnetization and the coercive force. For $SmCo_5$ and $Sm_2Fe_{17}N_3$ having a large anisotropic magnetic field, coercive force of the magnitude required for the bonded magnet can be obtained by pulverizing the alloy ingot until the crystal grain has a size of several μm.

On the other hand, the anisotropic magnetic field of $Nd_2Fe_{14}B$ is considerably smaller than $SmCo_5$ and $Sm_2Fe_{17}N_3$, so it cannot be used only by pulverization.

In the case of the Nd sintered magnet since the crystal grain boundary is surrounded by the nonmagnetic phase, the necessary coercive force can be obtained even when the crystal grains are several μm. But in the case of the bonded magnet, it is necessary to pulverize the crystal grain into sub μm in order to get the necessary coercive force. As a result, even if a large coercive force can be obtained, since the rare earth alloy is an active compound, there is a high risk that the rare earth alloy will ignite as soon as it touches the air. For this reason, it is necessary to form a powder having a size that is difficult to ignite by forming a microcrystal aggregate composed of crystal grains of sub μm. As a method for manufacturing such a magnet powder, a rapid quenching method [5] and an HDDR (hydrogenation, decomposition, desorption, recombination) method using hydrogen [6] were developed.

On the other hand, as a pinning type, Sm_2Co_{17} magnet powder is representative. The ingot of the alloy is heated and controlled so as to form a two-phase structure surrounding the microcrystalline phase with another microcrystalline phase. Due to the difference in the magnetic anisotropy energy between the surrounding phases, the domain wall is pinned and the required coercive force is obtained.

2 Types of Magnet Powder

2.1 Isotropic Magnet Powder

The world's most used rare earth magnet powder is Nd type manufactured by rapid quenching method. The rapid quenching method is a method of melting a rare earth alloy and ejecting the molten metal from a nozzle onto a metal wheel that rotates at high speed to bring it into contact quenching [5]. A thin piece having a thickness of about 35 μm and a width of 1–3 mm containing microcrystal of several tens of nanometers can be obtained. The flakes prepared by this method are called isotropic magnet powders because the direction of the crystal axes of the contained fine crystals is disordered in principle. Since it has a structure in which fine crystals of submicron order are gathered densely, the effect of magnetically strengthening between crystal grains works (remanence-enhancement), and a large remanence can be obtained compared to the case where single crystal grains are simply not oriented. As a result, even if it is isotropic, a higher energy product than expected can be obtained.

In order to realize higher energy product in isotropic, it is necessary to increase the remanence. Several kinds of nanocomposite magnets have been developed which is composed of a fine soft magnetic phase having large saturation magnetization such as α-Fe and Fe_3B and a hard magnetic phase having large anisotropic magnetic field such as $Nd_2Fe_{14}B$ and $Sm_2Fe_{17}N_3$ [7, 8]. A strong exchange interaction acts between the fine crystal grains of the soft magnetic phase and the hard magnetic phase to obtain hard magnetic properties and possibility to obtain a large energy product. However, the remanence and the coercive force have not yet been improved at the same time. Representative of the magnet powder manufactured and sold in the above way is NEQ Magnequench's MQP, and Fig. 1 is its characteristic diagram [9]. The horizontal axis represents the coercive force and the vertical axis represents the residual magnetic flux density, and the size of each square indicates the characteristic width of each magnet powder.

Although not shown in this figure, a simple Nd–Fe–B-based magnet powder called MQP-A was first developed. The coercive force was as large as 1030–1350 kA/m, and it was expected to be used for a generator or a motor for automobiles, but it did not reach practical use. Subsequently, as a material capable of multipolar magnetization with compact magnets for PC spindle-motor-related applications, there was a demand for materials with reduced coercive force, and targeting high Curie temperature and high residual magnetic flux density magnet powder, Nd–Fe–Co–B (MQP-B+, etc)

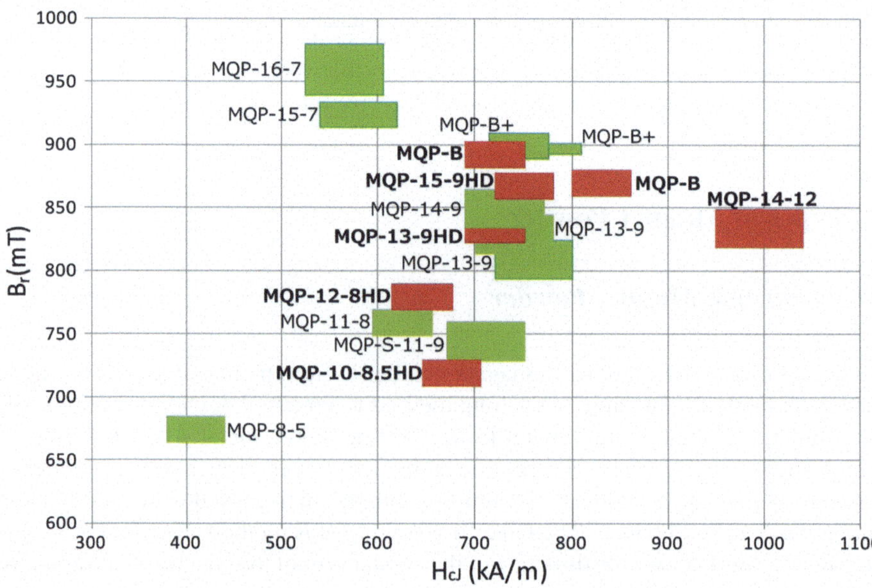

Fig. 1 Characteristic diagram of representative isotropic magnet powder MQP (MQP is a registered trademark of Neo Magnequench)

was developed. A ring magnet manufactured by a compression molding method using an epoxy resin as a binder formed a large market as an industry standard for spindle motors such as FDD, HDD, and CD-ROM. MQP-15-7 and MQP-16-7 have been developed by replacing a part of Nd with Pr in order to increase the residual magnetic flux density and squareness, and to make Fe-rich composition, Pr–Fe–Co–Nb–B-type powder. It is a nanocomposite magnet powder with low coercive force but the highest residual magnetic flux density in MQP system. This is suitable as ferrite substitution for optical disk drives and stepping motors used in office automation products.

MQP-14-12 is an Nd–Fe–Nb–B-based powder aiming for high heat resistance. The addition of Nb improves the demagnetization rate when kept at a temperature of about 150 °C, and it is a magnet that can be used at 180 °C by selecting a resin. It is specially designed for high temperature applications such as motors and sensors in automobile hood.

MQP-13-9 is a magnet powder developed by prioritizing price even if the residual magnetic flux density is lowered slightly, and it is based on the alloy of Nd–Pr–Ce–Fe–B composition. In addition to spindle motors, as price-sensitive ferrite replacement, it is used for CPU cooling fan motor, box fan, and so on.

In addition, MQP-8-5 in which Nd is substituted by Ce at about 80% has been developed. Also, MQP-S-11-9 of Nd–Pr–Fe–Co–Ti–Zr–B type for the purpose of improving fluidity in injection molding has been developed by the atomization method.

The material in bold in this figure is currently recommended by NEO Magnequench. Even though the amount of magnet powder was about 77.5 vol.% in conventional compression molding, the spring back of compression-molded products after compression molding was small even if the amount of magnet powder was increased to 80 vol.%, devised so that the withdrawing pressure could be lowered.

Table 2 shows the magnetic properties of typical isotropic magnet powder [9].

On the other hand, Daido Electronics manufactures and sells $TbCu_7$-type SmFeN-based magnetic powder. Since the anisotropic magnetic field of this alloy is about the same as that of the NdFeB alloy, the necessary coercive force cannot be obtained unless the crystal grain is made submicron or less. For this reason, the development was progressed by rapid quenching method. At the beginning of the development, it was necessary to raise the linear speed of the wheel to about 70 m/s, and the thickness of the formed thin piece was 10 μm or less, which had problems in terms of productivity, but development such as adding Zr and Co has advanced and high coercive force became possible even at low speed of around 40 m/s [4, 10]. Although it is an isotropic magnet, it is characterized by a higher energy product than the conventional MQP, excellent in corrosion resistance and less decrease in magnetic force at high temperature.

Table 2 Magnetic properties of representative isotropic magnet powder (MQP is a registered trademark of Neo Magnequench)

	MQP-10-8.5HD	MQP- 12-8HD	MQP-13-9HD	MQP-14-12	MQP-15-9HD	MQP-B	MQP-B
B_r (mT)	710–730	770–790	823–833	820–850	858–878	860–880	883–903
(BH)max (kJ/m^3)	76–86	90–100	104–112	107–120	115–123	116–124	116–124
H_{cJ} (kA/m)	645–705	615–675	690–750	940–1050	720–780	800–860	690–750
Magnetizing field to ≧95% Saturation Hs (kA/m)	≧1275	≧1350	≧1600	≧1600	≧1600	≧1600	≧1600
α (B_r) (%/°C)	−0.2	−0.17	−0.15	−0.13	−0.14	−0.14	−0.14
α (H_{cJ}) (%/°C)	−0.36	−0.39	−0.40	−0.40	−0.42	−0.44	−0.44
T_c (°C)	219	246	273	305	289	298	298
Density (Mg/m^3)	7.6	7.6	7.6	7.62	7.6	7.59	7.59

2.2 Anisotropic Magnet Powder

In order to realize a high energy product with bonded magnets, it is better to use anisotropic magnet powder. A single particle is one in which it is a single crystal or even if it is a polycrystal, the easy axis of the fine crystal is aligned in one direction. Those having a long history as an anisotropic powder are Sm–Co-based magnet powder. Although it is still used, there is a problem that it is expensive and price fluctuation is large compared with Fe-based magnets. Currently, it is used for limited applications where high heat resistance and magnetic flux stability are required.

As a method for producing Nd-type anisotropic magnet powder, the following two methods are currently devised.

The first method is to increase the density to a state where there is no vacancy while heating the powder composed of the microcrystal produced by the above-mentioned rapid quenching method and to compress the lumps in one direction to orient the crystal axis. It is called die up-set method [11]. Development of a method for producing an anisotropic powder using such thermoplastic processing has been attempted for many years. Currently, it is sold as MQA from NEO Magnequench [12].

The second method is generally called HDDR method. By subjecting the alloy ingot to a high-temperature treatment in a hydrogen atmosphere, the alloy ingot is pulverized, and the compound constituting the crystal is once decomposed into rare earth hydride and α-Fe. Thereafter, in the process of removing hydrogen, it is a technique of once again producing a powder-containing crystal of a fine compound. Since it contains fine crystals, it is possible to increase the coercive force. However, without special control, it tends to be isotropic and its properties are inferior to isotropic ones produced by rapid quenching.

Even if anisotropic particle is obtained in small-scale production, it is still difficult to control by mass production scale. Aichi Steel has realized the d-HDDR process to control the process of microcrystal formation on mass production scale. The principle for aligning the *c*-axis of the crystal has been discussed variously so far, please refer to the other report [13]. Microcrystalline orientation became possible, but since there is a distribution of about $\pm$ 20 degrees, research to increase orientation degree continues.

In producing the Sm-type anisotropic magnet powder, a method of preparing a mother alloy powder of Sm_2Fe_{17}, reacting it with nitrogen to obtain $Sm_2Fe_{17}N_3$ alloy powder, and thereafter pulverizing it to a size of several μm is used.

As an example of a method for producing Sm_2Fe_{17} mother alloy powder, there is a reduction diffusion method in which samarium oxide powder and iron powder are mixed with particulate metallic calcium and heated and reacted in an inert gas atmosphere [14]. Samarium oxide is inexpensive as a raw material, as compared with metal samarium separated and purified by distillation. Also, by choosing the reaction temperature lower than the peritectic temperature and choosing the Fe particle size in consideration of the diffusion length of Sm to Fe, a single phase of Sm_2Fe_{17} with little residual α-Fe phase can be obtained. Further, heat treatment indispensable for homogenization in the melting and casting method is unnecessary. Furthermore, since the Sm_2Fe_{17} alloy is directly obtained as a powder, the coarse pulverization step before nitriding is unnecessary. Therefore, this method is advantageous in terms of cost as compared with the melting and casting method.

In the nitriding step in which nitrogen is introduced into the Sm_2Fe_{17} phase, it is important to obtain a uniform $Sm_2Fe_{17}N_3$ phase. Therefore, as a nitriding atmosphere, a mixed gas of NH_3 and H_2 is used, and after nitriding, heat treatment is performed in Ar gas for the purpose of removing H_2 in the powder and homogenizing the nitrogen concentration distribution.

In order to obtain a high coercive force, the coarse powder obtained in the nitriding step is finely pulverized to an average particle size of 2–3 μm with a jet mill, a ball mill or the like. In addition, since the particle size after fine pulverization is very small, a treatment for inactivating the particle surface is required by various methods.

Currently, review of the manufacturing process has progressed, and magnet powder with $B_r = 1.44$ T and $H_{cJ} = 915$ kA/m has been reported [15].

As another example, there is a way to obtain a small particle size upon compound formation. A hydroxide was precipitated from an aqueous solution of sulfuric acid of Sm and Fe and a microcrystalline oxide calcined in the air was used to obtain a $Sm_2Fe_{17}N_3$ alloy powder having a particle size of about 3 μm which is almost unnecessary for fine pulverization.

The magnetic properties of the powder thus obtained were $B_r = 1.08$ T, $H_{cJ} = 1138$ kA/m. And, $B_r = 1.24$ T and $H_{cJ} = 1074$ kA/m were also obtained by mechanical disintegration [16]. In addition, $B_r = 1.21$ T and $H_{cJ} = 1530$ kA/m have been reported by adding TiO_2 ultrafine powder at the time of production to suppress grain growth of magnetic powder [17].

In addition, when a part of Fe in $Sm_2Fe_{17}N_3$ alloy is substituted with Mn and subjected to excessive nitrogen treatment, an amorphous phase containing Mn and a large amount of nitrogen precipitates inside the single crystal phase. As a result, high coercive force can be obtained due to domain wall pinning [18]. There is a possibility that the temperature change of the coercive force is small and it can withstand use at a relatively high temperature.

Table 3 shows the magnetic properties of representative anisotropic magnet powder [9, 19, 20].

2.3 Recent Development Trends

Development of magnet powder for bonded magnets with $ThMn_{12}$-type alloys is underway. Up to now, in order to stabilize the structure, it was necessary to replace a part of Fe with Ti or V. It was difficult to increase the magnetization as expected, but substituting a part of Sm or Nd with Zr, it has succeeded in alloying with less reduction of Fe and researches on increasing coercive force have been advanced [21].

In addition, since SmFeN-based magnet powder contains nitrogen, there is a problem of decomposition at a temperature at which it can be sintered, but researches

Table 3 Magnetic properties of typical anisotropic magnet powder (Wellmax is a registered trademark of Sumitomo Metal Mining Co., Ltd.) (MAGFINE is a registered trademark of Aichi Steel Corporation) (MQP is a registered trademark of Neo Magnequench)

	WELLMAX SFN	MAGFINE MF15P MF18P		37-11	MQA 38-14 37-16		36-18
B_r (mT)	1300–1440	1270–1390	1200–1300	1300	1310	1290	1240
(BH)max (kJ/m^3)	268–347	295–350	239–302	290	300	295	285
H_{cJ} (kA/m)	796–916	1035–1274	1193–1432	840	1120	1270	1430
Magnetizing field to ≧95% Saturation H_s(kA/m)	>1350	>2000*	>2000*	>1600	>1600	>2000	>2500
α (B_r) (%/°C)	−0.07	−0.11	−0.14	−0.27	−0.09	−0.13	−0.14
α (H_{cJ}) (%/°C)	−0.52	−0.56	−0.49	−0.75	−0.66	−0.65	−0.64
T_c (°C)	478	310	310	339	369	347	343
Recoil permeability	1.1	1.1–1.2	1.1–1.2	1.1	1.14	1.11	1.05
Density (Mg/m^3)	7.67	7.6*	7.6*	7.47	7.51	7.51	7.54

*Estimated value

are underway to increase the density by using spark plasma sintering (SPS) or the like [22, 23].

3 Binder

The main binders for solidifying and shaping the magnet powder are thermosetting and thermoplastic resins. As the thermosetting resin, EP (epoxy) resin is generally used. Examples of the thermoplastic resin are PA (polyamide, nylon) resin, PPS (polyphenylene sulfide) resin and elastomers such as NBR (acrylonitrile butadiene rubber), CPE (chlorinated polyethylene), and EVA (ethylene vinyl acetate).

4 Molding Method

When producing bonded magnets using isotropic magnet powder, it is important how to increase the density of the compact with good productivity. Depending on the magnetization characteristics of the magnet powder, arbitrary magnetization is theoretically possible as long as the molded body can be formed, and a required magnetization pattern can be realized.

4.1 Compression Molding Method

Compression molding is a simple shape, but it can increase the density among bonded magnets, so a high-energy product can be realized. Mix the raw material magnet powder and EP resin together with additives to prepare a compound, which is put into a mold and compression molded. Thereafter, it is cured by heating and then excess powder is washed and surface coated. Important elements at the time of compound manufacture are selection of magnet powder, surface treatment of powder, selection of EP resin, selection of kneading conditions, etc. [24].

As is well known, it is possible to increase the density by optimizing the particle size distribution. In order to raise the density, it is effective to increase the slipperiness between the magnet powders by using a liquid resin. On the other hand, there is a dilemma that in order to put a stable amount in the mold within a short time, the surface of the compound should be dried.

Although the content of the magnet powder is about 98% in terms of weight, pores and the like remain so that it is only about 80% in volume conversion. It is an important technique to uniformly distribute about 2% by weight of EP resin on the surface of the magnet powder beforehand, and it is necessary to select EP resin in accordance with powder feedability of compound and densification of molded body.

In the case of forming an isotropic magnet powder, it is not necessary to apply a magnetic field, it is sufficient that the density change and the weight change of the molded body are small and the powder supply property to the mold is good, but in the case of forming an anisotropic magnet powder, the degree of orientation of the magnet powder by application of a magnetic field is added as an important condition. So what kind of resin is used is an important know-how. In other words, considering the orientation of the magnetic powder, it is important to create a state in which the individual magnet powder can overcome the viscosity of the binder and move freely when a magnetic field is applied, and a liquid resin is advantageous, while on the other hand, considering the supply of the compound to the mold, the surface of the compound needs to be dried, and a solid resin is advantageous. In order to overcome these conflicting conditions, a method has been devised to make the solid type EP resin melted by heating the mold when magnetic field is applied [25, 26]. Currently, practical application using d-HDDR powder is being promoted to electric tools and others. A major goal is for automotive use, and a move toward practical use as a seat motor is introduced.

Even fine powders such as SmFeN can realize a high-energy product by increasing the molding pressure, but from the viewpoint of practical use, it is preferable to use Nd-based relatively large particles to increase the density at low pressure.

Since it is necessary to add a curing agent to the compound, curing may proceed depending on storage conditions. Therefore, in general, the compound manufactured by each company on its own is used.

Table 4 shows magnetic characteristics examples of compression-compacted bonded magnets [20, 27].

4.2 *Injection Molding Method*

Injection molding is characterized by being able to mold complex shapes without post-processing. PA resin or PPS resin is used as the binder. Those using PPS resin are superior in heat resistance and chemical resistance, so they are widely used recently for heat-resistant and oil-resistant applications such as automobiles.

The raw material magnet powder is mixed with PA resin or PPS resin and additives and kneaded in a kneader to form pellets. The compound is charged into an injection molding machine, heated and melted in a cylinder, and injected into the mold. If the viscosity of the resin is too low, separation from the magnet powder occurs and it becomes impossible to inject, and if the viscosity of the resin is too high, the injection itself cannot be performed. Therefore, selection of resins and additives is important know-how of each company as well as kneading conditions. As the content of the magnet powder is increased, the magnetic properties can be enhanced. However, due to the above restriction, the amount of the magnet powder is about 93% in terms of weight and about 60% in terms of volume.

In the case of injection molding using anisotropic magnet powder, it is necessary to orient the magnet powder by configuring the necessary magnetic circuit for the mold. In order to produce a high-performance anisotropic injection-molded magnet, it is necessary to sufficiently orient the crystal axis of the magnet powder contained in the molten compound before solidifying. For this purpose, it is important that the fluidity of the compound is high and the orientation magnetic field is large. In particular, since the coercive force of the rare earth magnet powder is larger than that of the ferrite powder, a large magnetic field is necessary for the orientation, and magnetic field analysis of the mold becomes a very important factor. In addition, in the case of anisotropic injection molding, since the molten compound is rapidly cooled in the mold, the fluidity tends to decrease and it becomes difficult to orient the compound. This becomes more prominent as the magnet becomes smaller and thinner. When sufficient orientation is achieved, the higher the content of the magnet powder is, the better the magnetic properties are. However, in practice, when the content is large, the fluidity decreases and the degree of orientation decrease. As a result, although the magnetic properties of the powder itself are high, on the contrary, the magnet characteristics may deteriorate, so care is required.

Table 5 shows examples of magnetic properties of injection-molded magnets [19, 20].

Table 4 Example of magnetic properties of compression-molded bonded magnets (NEOQUENCH, NITROQUENCH are registered trademarks of Daido Denshi Co., Ltd.) (MAGFINE is a registered trademark of Aichi Steel Corporation)

Magnet	Isotropic						Anisotropic		
	NEOQUENCH-P				NITROQUENCH-P		MAGFINE		
	NP-8L	NP-8R	NP-8SR	NP-12L	SP-14	SP-14L	MF14C	MF16C	MF18C
B_r (mT)	640–710	580–650	600–680	720–770	750–820	750–830	900–070	850–1050	850–1050
$(BH)_{max}$ (kJ/m^3)	68–80	60–72	65–77	88–99	98–112	98–112	151–183	143–175	139–171
H_{cJ} (kA/m)	637–796	1035–1353	835–1075	716–836	670–800	550–670	1035–1154	1194–1354	1314–1433
α (B_r) (%/°C)	−0.1	−0.15	−0.13	−0.1	−0.05 to −0.07	−0.05 to −0.07	−0.11	−0.11	−0.11
α (H_d) (%/°C)	−0.4	−0.4	−0.4	−0.4	−0.4	−0.4	−0.56	−0.47	−0.46
Recoil permeability	1.2	1.13	1.13	1.2	1.2	1.2	1.10–1.20	1.10–1.20	1.10–1.20
Density (Mg/m^3)	5.6–6.1	5.6–6.1	5.8–6.1	6.1–6.4	5.8–6.4	5.8–6.4	6.1–6.3	6.1–6.3	6.1–6.3
Binder (resin)	Epoxy								

Table 5 Example of magnetic properties of injection-molded magnets (Wellmax is a registered trademark of Sumitomo Metal Mining Co., Ltd.) (MAGFINE is a registered trademark of Aichi Steel Corporation)

Magnet powder	Isotropic		Anisotropic			
	Wellmax		Wellmax			
	NdFeB		SmFeN	SmFeN + Ferrite	SmFeN + NdFeB	
Binder (resin)	PA12	PPS	PA12	PA12	PA12	PPS
B_r (mT)	410–730	390–600	630–810	360–570	860–940	710–780
$(BH)_{max}$ (kJ/m^3)	28–76	29–57	76–115	19–60	131–147	91–107
H_{cJ} (kA/m)	637–875	676–836	660–910	430–770	955–1155	995–1195
α (B_r) (%/°C)	−0.11	−0.11	−0.07	−0.14 to −0.04	−0.09	−0.10
α (H_{cJ}) (%/°C)	−0.40	−0.40	−0.50	−0.44 to −0.30	−0.65 to −0.56	−0.61 to −0.55
Recoil permeability	1.0–1.2	1.1	1.1		1.1	1.1
Density (Mg/m^3)	4.1–5.9	4.3–5.4	4.1–4.9	3.6–4.3	5.4–5.6	5.0–5.2
Magnet powder	Anisotropic					
	Wellmax		MAGFINE			
	SmCo (2-17)		MF15P		MF18P	
Binder (resin)	PA12	PPS	PA12	PPS	PA12	PPS
B_r (mT)	630–700	550–650	790–840	650–700	760–810	640–690
$(BH)_{max}$ (kJ/m^3)	75–88	59–72	115–123	80–88	111–119	76–84
H_{cJ} (kA/m)	763–995	710–980	1035–1115	1035–1115	1274–1354	1274–1354
α (B_r) (%/°C)	−0.03	−0.03	−0.11	−0.11	−0.14	−0.14
α (H_{cJ}) (%/°C)	−0.30	−0.30	−0.56	−0.56	−0.49	−0.49
Recoil permeability	1.1	1.1	1.10–1.20	1.10–1.20	1.10–1.20	1.10–1.20
Density (Mg/m^3)	5.5–5.7	5.3–5.7	5.0–5.2	4.6–4.7	5.0–5.2	4.6–4.7

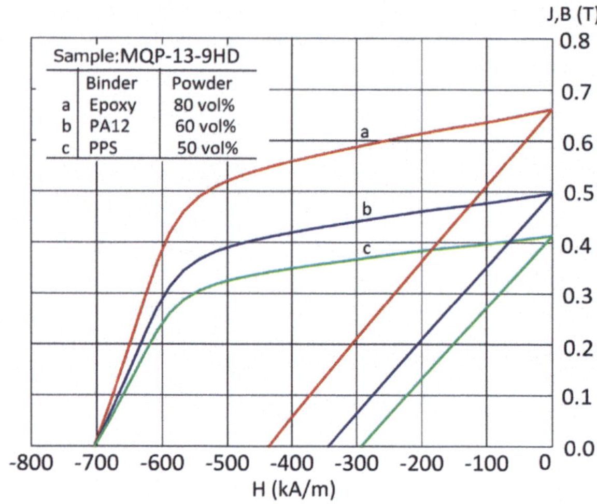

Fig. 2 Magnetization curve of MQP-13-9 HD magnet manufactured by various molding methods

4.3 *Other Molding Methods*

Another method is calendaring or extrusion. Binder is elastomers. A calendar-processed magnet is made of a roll, and the extruded magnet is made from a mouthpiece. The mixing ratio is often in the vicinity of 9:1 by weight ratio of magnet powder: elastomers. It has low magnetic properties, but it is characterized by its high flexibility.

Extruded magnets using elastomers have uniform magnetic properties in the extrusion direction and are suitable for manufacturing long and sheet-shaped bonded magnets, and since molds can be made shorter and inexpensive compared to injection molding. Cost performance is high. The magnet powder in the composition is extruded with its crystal axes aligned by the magnetic field applied to the mouthpiece portion at the tip of the molding machine. Depending on the thickness of the magnet and production conditions, maximum energy product of up to 81 kJ/m^3 can be obtained using anisotropic SmFeN fine powder [28]. Although it is possible to mold even with use of MQP which is flat and large in particle size, it is easy to peel off from the surface of the molded product, so care is required.

Figure 2 shows the magnetization curve of the MQP-13-9 HD magnet prepared by various molding methods [9].

5 Orientation Technology

When manufacturing a bonded magnet using an anisotropic magnet powder, the degree of orientation of the magnet powder is important. At the time of molding, it is necessary to orient the crystal axes of the particles in a desired direction and to increase the density like an isotropic magnet. In the case of a ring magnet, the orientation direction includes an axial direction, a radial direction, a multipole direction, and the like. Particularly, when oriented in the radial direction, the shape becomes a problem. When forming the ring, a magnetic field is applied in the repulsion direction from both ends of the shaft to orient the magnet powder in the radial direction. It is said that it is necessary to consider $f_R = A/2S$ between the area S of the inner circle and the outer surface area A here. This is known as the radial factor proposed by Hamano. For example, when orienting Sm_2TM_{17} with a coercive force on the order of 10 kOe, sufficient orientation is obtained when f_R is smaller than 1, but it is said that orientation becomes difficult when f_R is 2 or more [29].

In the case of multipole orientation, in general, sintered magnets are placed in a mold to orient crystal axis. In the case of a ferrite magnet, the coercive force is small, it is possible to orient with a relatively small magnetic field, but in the case of rare earth magnet powder, a relatively large magnetic field is required for orientation. Therefore, it is necessary to make an optimum state by performing magnetic field analysis and the like.

Since the direction of magnetization of the magnet formed by applying the magnetic field is fixed, the direction of magnetization is limited and cannot be magnetized except for the direction determined after molding. However, since the direction of the axis of easy magnetization is fixed, there is a feature that magnetization can be made relatively stable if a firm orientation is made during molding.

In the case of molding a bonded magnet using anisotropic magnet powder, it is necessary to align the crystal axes, so it is important to apply magnetic field application technique and special die technology. When taking out a molded product, it is general to demagnetize by applying a reversal magnetic field or the like. In contrast, when a bonded magnet is formed using isotropic magnet powder, application of a magnetic field is unnecessary, and the mold and the like become simple.

Consultation is often asked whether an anisotropic magnet powder can be used as an isotropic magnet by molding without applying a magnetic field.

If the powder itself is isotropic such as a rapid quenched powder, because there is a strong interaction among the crystal grains so that the residual magnetization can hold 70–80% of the saturation magnetization, the maximum energy product is obtained about one-half as large as an anisotropic magnet. In the case of the question, the residual magnetization becomes about one-half as large as the saturation magnetization. It is known that the maximum energy product is proportional to the square of the residual magnetization, so it should be noted that it is extremely small as 1/4 that of an anisotropic magnet.

Figure 3 shows the magnetization curve of the Wellmax-S3A12MA magnet fabricated by the injection molding method under the condition of the application of magnetic field [30]. The residual magnetization is 0.72 T in the case of anisotropic and 0.39 T in the case of isotropic. Although it is unknown whether the anisotropy is not perfect or whether the residual magnetization exists in the mold and affects the orientation, the residual magnetization of the isotropic is 54% of the anisotropic. In addition, the maximum energy product is 94 kJ/m^3 in the case of anisotropic, whereas it is 26 kJ/m^3 in the case of isotropic, which is 28% of the anisotropic.

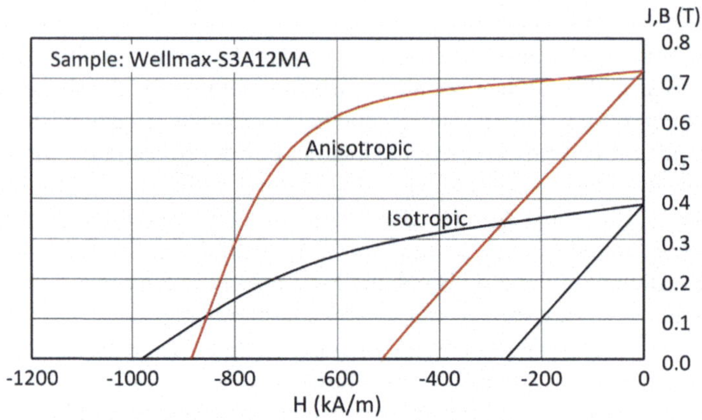

Fig. 3 Magnetization curve of Wellmax-S3A12MA magnet manufactured by injection molding under the condition of the presence or absence of a magnetic field (Wellmax is a registered trademark of Sumitomo Metal Mining Co., Ltd.)

6 Magnetizing Technique

Regardless of isotropic or anisotropic magnet powder, when used as a magnet, it is necessary to magnetize the compact. Particularly in small motor applications, it is not uncommon to use magnets with *N* poles and *S* poles multipolarly magnetized at 2–3 mm pitch. In such applications, a sufficient magnetic field cannot always be applied due to the restriction of the magnetizing yoke. Therefore, care should be taken in the difference between the magnetic characteristics of the test piece magnetized up to the saturation magnetization and that of the actual product due to the flux obtained by the magnetization characteristics. It is necessary to investigate magnetization characteristics with reference to catalogs, etc. [20, 31–35].

7 Temporal Change of the Magnetic Flux

Parts are assembled using the surface magnetic flux of the magnets, but since the surface magnetization acts as a demagnetizing field for the magnets themselves, magnetic flux changes with time depending on the holding temperature. This is called change over time. It is important to investigate the temporal change of the magnetic flux at the holding temperature of magnets with different demagnetizing factor by referring to the catalog of magnets [20, 36–39].

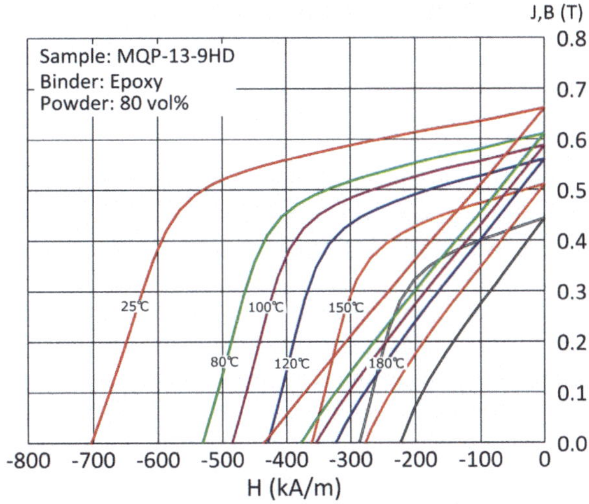

Fig. 4 Temperature change of magnetization curve of MQP-13-9 HD magnet manufactured by compression molding method

8 Temperature Characteristics

The residual magnetic flux density, the coercive force, and the maximum energy product change at the holding temperature. In the catalog of the magnet, the temperature coefficient α (B_r) of the residual magnetic flux density and the temperature coefficient α (H_{cJ}) of the coercive force are shown as the rate of change in an arbitrary temperature range. However, unlike sintered magnets, bonded magnet generally has poor squareness, so it is difficult to estimate the temperature range that can be used only with this data. These data are used as guidelines, and when actually used, it is necessary to measure the magnetization curve at the necessary temperature.

Figure 4 shows the temperature change of the magnetization curve of MQP-13-9 HD prepared by compression molding method [9].

9 Surface Coating

In the case of NdFeB-based compression-molded magnets, surface coating of the molded body is indispensable to prevent rust. In general, the coating methods using the epoxy resin are electrodeposition coating (20–35 μm), spray coating (15–30 μm), dip coating (1–5 μm), spray coating with barrel (3–18 μm). When metallic coating is required, Ni plating (10–30 μm) is applied.

On the other hand, in the case of injection-molded magnets, the blend ratio of the resin is high, so it is often used without a surface coating. The SmFeN system magnet is more resistant to rusting than the NdFeB system. When it is used in a bad environment, an epoxy-type resin coating is applied, but it is sometimes necessary for discrimination or powdering prevention degree.

10 Weather Resistance

Evaluation criteria for weather resistance vary greatly depending on the application. Corrosion resistance test, chemical resistance test, and heat resistance test, etc., are general tests.

As an environment for the corrosion resistance test, changes in appearance (in particular rust and oxide generation), changes in mass, magnetic properties and the like are evaluated by dipping for a certain period of time in tap water, pure water, salt water (5% NaCl aqueous solution) or the like. Evaluate as compared with those exposed to room temperature atmosphere as a control experiment.

As a solution for the chemical resistance test, it is preferable to use a solution of pure water (ion exchanged water), tap water, salt water, neutral detergent, hydrochloric acid, nitric acid, sulfuric acid, acetic acid, citric acid, sodium hydroxide, sodium carbonate, machine oil No. 1 (kinematic viscosity 18.71–21.05 mm^2/s), machine oil No. 2 (kinematic viscosity 19.19–21.52 mm^2/s), machine oil No. 3 (kinematic viscosity 31.96–34.18 mm^2/s), gasoline, kerosene, and others.

In the appearance test, the surface of the bonded magnet is evaluated in order to observe the occurrence of discoloration or rust, presence or absence of swelling, loss of magnet powder, presence or absence of cracks, dissolution or decomposition, etc.

The heat resistance test is carried out by exposing the test piece to a test environment (room temperature, 60, 80, 100, 125, 150, 180, 200 °C) for a certain period of time (class A: 100 h class B: 1000 h) and evaluates changes in appearance (especially rust, oxide generation), mass, etc., before and after the change, and judges whether the change is within the specified range.

The heat resistance for magnetic properties is calculated by the magnetic temperature coefficient. When it measures the changes in mechanical properties or other necessary characteristics with respect to temperature, refer to JIS K 7212 and JIS K 7226. However, the test should be carried out with respect to characteristics considered to be important for practical use, and a test using a bonded magnet product itself as a test piece is recommended. Numerical values obtained by alternative tests with common physical test pieces are effective for the evaluation and quality control of various bonded magnets, but in the case of bonded magnets, they are complex magnetic properties in the presence of magnetic field and structural stress, its value may deviate from general alternative test. Therefore, attention must be paid to the test method according to the purpose of the heat resistance test, and the test using the bonded magnet product itself as a test piece is appropriate [40].

References

1. Fujii, H., Sun, H.: Handbook of Magnetic Materials, vol. 9, p. 395 (1995) (Elsevier)
2. Buschow, K.H.J.: Handbook of Magnetic Materials, vol. 10, p. 506 (1997) (Elsevier)
3. Katter, M., et al.: J. Appl. Phys. **70**, 3188 (1991)
4. Sakurada, S., et al.: J. Appl. Phys. **79**, 4611 (1996)

5. Neo Magnequench HP: Neo powder history. http://www.mqitechnology.com/bonded-neo-powder-history.jsp. Accessed 21 Aug 2017
6. Nakayama, R., Takeshita, T.: J. Alloys Compd. **193**, 259 (1993)
7. Coehoorn, R., et al.: J. Mag. Mag. Mater. **80**, 101 (1989)
8. Kneller, E.F., Hawig, R.: IEEE Trans. Magn. **27**, 3588 (1991)
9. Neo Magnequench HP: MQP product name. http://www.mqitechnology.com/product-name-mqp.jsp. Accessed 21 Aug 2017
10. Omatsuzawa, R., et al.: Trans. Magn. Soc. Jpn. **4**, 113 (2004)
11. Kawashita, Y., et al.: Trans. Magn. Soc. Jpn. **4**, 46 (2004)
12. Neo Magnequench HP: MQA overview brochure. http://www.mqitechnology.com/downloads/brochures_PDF/MQA-Overview.pdf. Accessed 21 Aug 2017
13. Yamazaki, M., et al.: AIP Adv. **7**, 056220 (2017)
14. Kawamoto, A., et al.: IEEE Trans. Magn. **35**, 3322 (1999)
15. Ishikawa, T.: BM News **51**, 35 (2014)
16. Kume, M.: BM News **23**, 34 (2001)
17. Tada, S.: BM News **46**, 16 (2011)
18. Ishikawa, T., et al.: Ninth International Conference on Ferrites (ICF-9), p. 543 (2004)
19. Sumitomo Metal Mining Co., Ltd. HP: Properties of Wellmax. http://www.smm.co.jp/products/material/magnet/property.html. Accessed 21 Aug 2017
20. Aichi Steel Corporation HP: Properties of MAGFINE. https://www.magfine.com/downloads/ Accessed 21 Aug 2017
21. Kuno, T., et al.: AIP Adv. **6**, 025221 (2016)
22. Takagi, K., et al.: J. Magn. Magn. Mater. **324**, 2336 (2012)
23. Okada, S., et al.: AIP Adv. **7**, 056219 (2017)
24. Ohmori, K., Yoshizawa, S.: BM News **11**, 26 (1994)
25. Honkura, Y., Mitarai, H.: BM News **15**, 49 (1996)
26. Mitarai, H., et al.: Proceedings of the 16th International Workshop on Rare-Earth Magnets and Their Applications, p. 787 (2000)
27. Daido Electronics Co., Ltd. HP: Properties of neoquench & nitroquench. http://www.daido-electronics.co.jp/product/index.html. Accessed 21 Aug 2017
28. Ida, T.: BM News **26**, 36 (2001)
29. Hamano, M.: Bulletin of the 58th Topical Symposium of the Magnetic Society of Japan, 58-10, p. 67 (1989)
30. Courtesy of T. Ishikawa
31. Neo Magnequench HP: Magnetizing characteristics of MQP. http://www.mqitechnology.com/jp/downloads/research-notes/bonded-neo-magnet-saturation.pdf. Accessed 21 Aug 2017
32. Neo Magnequench HP: Magnetizing characteristics of MQP. http://www.mqitechnology.com/jp/downloads/research-notes/unsaturated-bonded-neo-magnet.pdf. Accessed 21 Aug 2017
33. MS-Schramberg GmbH & Co. KG HP: Magnetizing characteristics of various kinds of magnet. http://www.magnete.de/uploads/media/aufmagnetisierungsfeld_en.pdf. Accessed 21 Aug 2017
34. Daido Electronics Co., Ltd. HP: Magnetizing characteristics of Neoquench. http://www.daido-electronics.co.jp/product/neoquench_p/magnetizability/index.html. Accessed 21 Aug 2017
35. Daido Electronics Co., Ltd. HP: Magnetizing characteristics of Nitroquench. http://www.daido-electronics.co.jp/product/nitroquench_p/magnetizability/index.html. Accessed 21 Aug 2017
36. Neo Magnequnch HP: Aging properties of MQP. http://www.mqitechnology.com/jp/downloads/research-notes/factors-affecting-magnet-aging.pdf. Accessed 21 Aug 2017
37. Daido Electronics Co., Ltd. HP: Aging properties of Neoquench. http://www.daido-electronics.co.jp/product/neoquench_p/temperature/index.html. Accessed 21 Aug 2017

38. Nichia Corporation HP: Aging properties of bonded SmFeN magnets. http://www.nichia.co.jp/jp/product/magnet_compound_b.html. Accessed 21 Aug 2017
39. Nichia Corporation HP: Aging properties of bonded hybrid magnets. http://www.nichia.co.jp/jp/product/magnet_compound_j.html. Accessed 21 Aug 2017
40. Japan Bonded Magnet Association (ed.): Guidebook of test method for bonded magnet. Refer to BMG-4001, BMG-4002 and BMG-4003 (2000)

The Rare Earths Problem for Permanent Magnets

Satoshi Hirosawa

Abstract The rare earths criticality issue in the permanent magnets implies in a narrower sense the sudden inflation of prices of rare earth oxides and alloys that stared in 2011 and ceased in 2012 as a result of mishandling of the market in China, but more generally, it includes the issues concerning the rare earth resources and/or supply risks caused by rapid increase of the market demands coupled with increasing environmental regulations on mining or smelting operations in developing countries. Thus, there has existed R&D efforts for the development of permanent magnets that do not use or significantly reduce the usage of the rare earth metals for more than a decade, and their importance is ever increasing. In this chapter, this issue is briefly overviewed.

Keywords Elements criticality · Rare earth magnets · Dy-free Nd–Fe–B · Non-rare earth permanent magnets · Atomistic theory of coercivity · Academia-industry cooperation

1 Introduction

Abundance in the upper continental crust of important magnetic elements, namely the rare earth elements starting from praseodymium (Pr) of atomic number 59 to ytterbium (Yb) of atomic number of 70 and major transition metals of the iron family is shown in Table 1 together with the leading country of production and risk scores provided by British Geological Survey in 2015 [1].

There was a surge of prices of rare earth elements in 2011, which was caused by a sudden enforcement of regulations on exports of rare earth oxides and alloys by the Chinese government. Although the high prices kindled investments on reopening rare earth mines outside China, those movements have faded away after relaxation of the regulations on export and following downfall of rare earth prices. The rare earth prices settled at the levels more than double of previous prices and have been more

S. Hirosawa (✉)
Elements Strategy Initiative Center for Magnetic Materials, National Institute for Materials Science, 1-2-1 Sengen, Ibaraki 305-0047, Japan
e-mail: HIROSAWA.Satoshi@nims.go.jp

K. Fujisaki (ed.), *Magnetic Material for Motor Drive Systems*,
Engineering Materials, https://doi.org/10.1007/978-981-32-9906-1_22

Table 1 Abundance of typical magnetic elements in the upper continental

Element	Abundance in the upper continental crust[a]	Leading country[b]	Risk score[b]	Permanent magnets using the element
Cr	9.2×10^{-5}	South Africa	6.2	Fe–Cr–Co
Mn	7.75×10^{-4}	China	5.7	MnAlC
Fe	1.83×10^{-1}	China	5.2	Nd–Fe–B
Co	1.73×10^{-5}	Congo	8.1	Sm–Co
				Fe–Cr–Co
Ni	4.7×10^{-5}	Indonesia	5.7	Alnico
Pr	7.1×10^{-6}			Nd–Fe–B
Nd	2.7×10^{-5}			Nd–Fe–B
Sm	4.7×10^{-6}			Sm–Co
Eu	4×10^{-6}			–
Gd	4×10^{-6}			–
Tb	7×10^{-7}	China	9.5	Nd–Fe–B
Dy	3.9×10^{-6}			Nd–Fe–B
Ho	8.3×10^{-7}			–
Er	2.3×10^{-6}			–
Tm	3×10^{-7}			–
Yb	2×10^{-6}			–

Crust, the leading production countries, risk score, and permanent magnets using these elements
[a]Chronological Science Tables 2013, Ed. National Astronomical Observatory of Japan, Maruzen Publishing Co. Ltd. 2012
[b]British Geological Survey, Risk List 2015, http://www.bgs.ac.uk/mineralsuk/statistics/risklist.html/. As of 28 Aug 2017

or less stable from 2013 to 2016, but showing a slight increase toward the end of 2017 [2]. The structure of the supply chains that yielded the 2011 crisis, i.e., China being the practically monopolizing supplier, has not been changed. Therefore, there still exist risks in supply of rare earth elements at present.

The rare earth elements that are used as the indispensable ingredients in permanent magnets are, in the order of atomic numbers, praseodymium (Pr), neodymium (Nd), which are for neodymium–iron–boron (Nd–Fe–B) permanent magnets, and samarium for samarium–cobalt (Sm–Co) and samarium–iron–nitrogen (Sm–Fe–N) permanent magnets. Terbium (Tb) and dysprosium (Dy), which are used to avoid irreversible thermal demagnetization of the Nd–Fe–B permanent magnets at elevated operation temperatures by replacing a part of Nd, are regarded as important ingredients since the early stage of development of the Nd–Fe–B, especially the sintered-type materials.

There exist abundant minable resources worldwide for the elements from Pr to Sm, which are called light rare earth elements [3]. On the other hand, for Tb and Dy, which belong to the heavy rare earths, there are few ores that can be mined commercially except for those in the south China classified into "ion-adsorption clays" or "laterite ores," which have been developed because of relatively high concentrations of these heavy rare earth elements and relatively low concentrations of radioactive elements such as thorium.

An important aspect of the rare earth issue is the fact that, because of their chemical similarity, all rare earth elements are mined simultaneously. The molar fraction of each element in ores is significantly unbalanced compared to the demands in the market. The elements such as cerium (Ce) which does not have enough markets are put aside in stockyard, and their production costs are added on top of highly demanded elements such as Pr and Nd. In order to secure stability in the market prices of those highly demanded rare earth elements for production of permanent magnets, developments of applications of the unused rare earth elements such as Ce are desirable.

To reduce the problematic aspects of using rare earth elements, either one of the following technological items should be effective:

(1) Permanent magnets containing no rare earths,
(2) High performance iron-based permanent magnets that contain smaller amounts of light rare earths compared to that in Nd–Fe–B-type magnets,
(3) Nd–Fe–B-type high-coercivity permanent magnets that contain no heavy rare earths.

The only magnetic element that can be regarded as "ubiquitous" is iron (Fe). Manganese (Mn) is relatively abundant, but yields only small magnetic polarizations in magnetic alloys as a result of antiparallel magnetic coupling of magnetic moments of Mn atoms with those of ferromagnetic transition metals such as Fe or even among Mn atoms themselves when the Mn atoms have large atomic magnetic moments [4]. Abundance of other magnetic elements is by many orders of magnitudes smaller than that of Fe. For instance, the crustal abundance of Co is not different from that of Nd. Therefore, the targets of investigations aiming at new materials listed above must be Fe-based, containing no significant amount of critical elements such as Co and rare earths.

The main application of the high-performance permanent magnets is electromagnetic actuators, typically motors, and generators, which may be classified as electric energy–mechanical energy converting devices. Magnetic properties required for permanent magnets are strongly dependent on the design of those devices, opening a possibility that the rare earth issue may be considerably reduced by redesigning the devices.

2 Typical Permanent Magnets and Rare Earth Elements Used Therein

Many variations of permanent magnet materials are used in a variety of applications to which characteristic features of each material are customized. The rare earth elements are used not only in rare earth magnets such as Sm–Co, Nd–Fe–B, and Sm–Fe–N-type magnets but also in some classes of hard ferrite magnets such as lanthanum–cobalt (La–Co) and calcium–cobalt (Ca–Co) types. The permanent magnets that do not contain rare earth elements are alnico, iron–chromium–cobalt (Fe–Cr–Co), and manganese–aluminum–carbon (Mn–Al–C) magnets. Characteristics of these materials are discussed in other chapter in this book.

The typical room-temperature values of remanence (B_r) and intrinsic coercivity (H_{cJ}) of permanent magnets materials mentioned in the previous paragraph are shown in Fig. 1, data in which have been compiled from data sheets publicly provided by magnet producers except for those for MnAlC taken from [5, 6]. The room-temperature values of saturation magnetization (J_S) of the main hard magnetic compound, which is the theoretical upper bound of B_r of each type of magnets, are shown as a function of concentration of rare earth elements in the compounds in Fig. 2 which were compiled using various data sources [7–13]. It should be noted that H_{cJ} needs to exceed one half of B_r in order to obtain a large energy density ($(BH)_{max}$) close to the theoretical limit, $(1/4)\mu_0^{-1}J_S^2$, where μ_0 is the permeability of vacuum. The permanent magnet materials that satisfy this condition are limited to hard ferrites and the rare earth magnets. The rare earth magnets are selected for devices such as traction motors used in electric vehicles that require a high power density for the sake of high overall efficiency as a system.

2.1 Perspectives of the High Performance Iron-Based Permanent Magnets Containing Light Rare Earths

Hard magnetic Fe-based rare earth compounds that have less concentrations of rare earth elements are R_2Fe_{17} and RFe_{12} types of compounds, where R denotes rare earth elements. The crystal structures of these compounds are derivatives of that of $SmCo_5$ (with the $CaCu_5$-type crystal structure). R_2Fe_{17} is isomorphs of Sm_2Co_{17}, and the molar fraction of R in these compounds is about 11% smaller than that in the $Nd_2Fe_{14}B$-type compounds.

The molar fraction of rare earths in the RFe_{12} compounds is about 27% smaller than that in $Nd_2Fe_{14}B$ although RFe_{12} is unstable and needs to be stabilized by partially substituting Ti for Fe and/or Zr for R. When 25% of R is further replaced with Zr, the molar fraction of rare earth elements in the compound becomes about one half of that in $Nd_2Fe_{14}B$. Moreover, the binary RFe_{12} compounds with the Fe-sublattice partially replaced with Co have been shown to have larger J_S and higher Curie temperatures compared to $Nd_2Fe_{14}B$ [7]. Thus, the RFe_{12}-type compounds

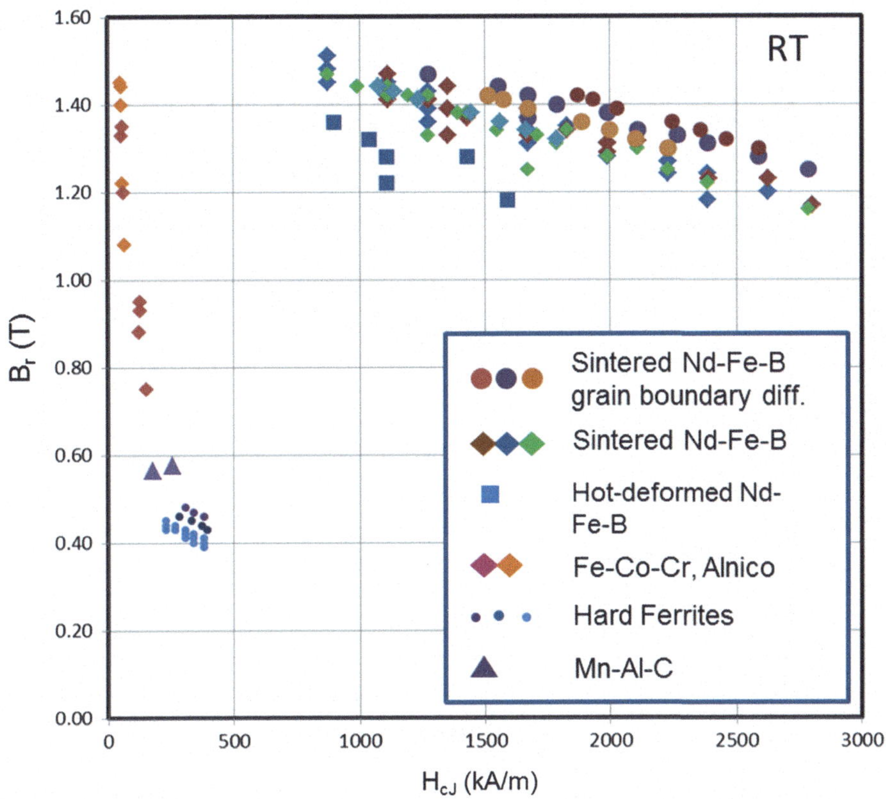

Fig. 1 Typical room-temperature values of remanence (B_r) and intrinsic coercivity (H_{cJ}) of typical permanent magnets materials compiled using various sources. *Data source* Data sheets publicly provided by magnet producers, Hitachi Metals Co. Ltd., Shin-Etsu Chemical Co. Ltd., TDK Corporation, Daido Steel Co. Ltd., except for those for MnAlC which are from Refs. [5, 6]

are strongly attractive as the base compounds for development of new Fe-based permanent magnets containing less rare earth elements than Nd–Fe–B. To generate a high enough for typical motor design coercivity with the RFe_{12}-type compounds in anisotropic magnet is a challenging technical issue, toward which intensive R&D efforts are necessary.

The 2–17 family of Sm–Co permanent magnets contains no heavy rare earth elements except for very special grades developed in order to provide nearly temperature-independent B_r characteristics for special applications [8]. Owing to their excellent stability of magnetic properties at elevated temperatures compared to that of Nd–Fe–B, sintered 2–17-type Sm–Co magnets have been recently installed in traction motors for urban tramcars [14] and also in variable flux motors used in high-end home appliance applications, by tuning the coercivity below about 500 kA/m so that the remanent flux states of the magnets can be switched by magnetic fields generated by stator windings [15]. Another 2–17-type magnet commercially

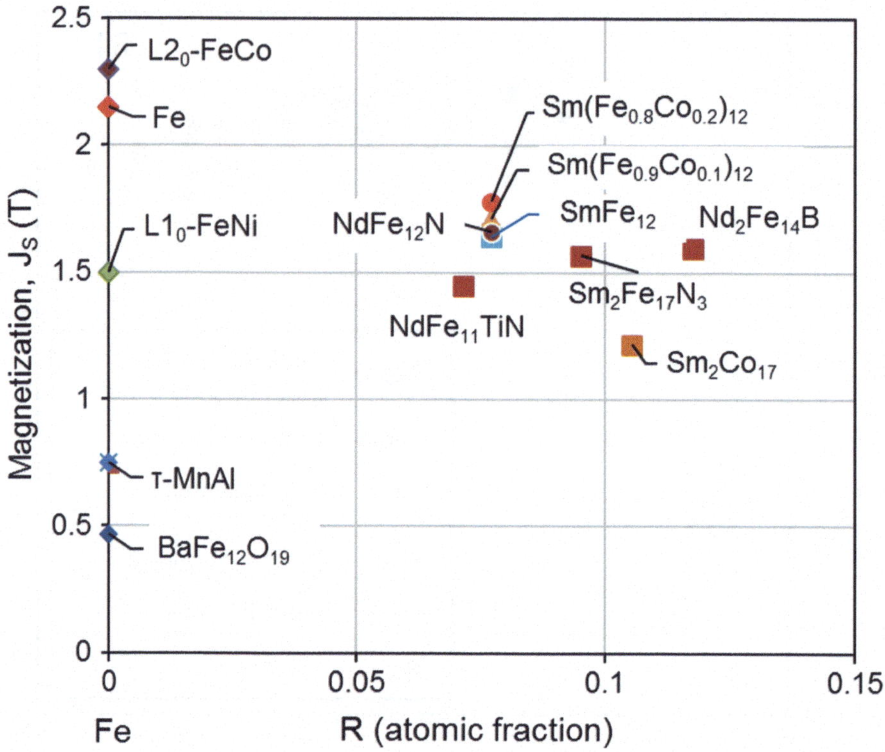

Fig. 2 Room-temperature values of saturation magnetization (J_S) of the main hard magnetic compound as a function of concentration of rare earth elements in the compound compiled using various sources [7–13]

in use is $Sm_2Fe_{17}N_3$-based bonded magnets. The $Sm_2Fe_{17}N_3$ powders are also used in some anisotropic Nd–Fe–B bonded magnets as a hard magnetic filler component to improve their magnetic performances. Those bonded magnets are used in various types of applications including motors because their high freedom in shaping can reduce the device's assembly cost, overcoming their high material costs.

2.2 *Nd–Fe–B Permanent Magnets Free from Heavy Rare Earth Elements*

Nd–Fe–B permanent magnets free from heavy rare earth elements with competitive performance against conventional sintered Nd–Fe–B containing Dy are produced commercially in both powder metallurgical (sintering) and hot-deformation processes. These Dy-free grades have the room temperature coercivity at most about 1.6 MA/m, which is significantly smaller than 2.4 MA/m obtained with Dy-containing grades conventionally used in hybrid vehicles. In order to make use of the Dy-free grades, redesigning of the motor system including not only magnetic circuits but also methods of cooling the magnets have been performed simultaneously [16]. Investments on such additional developments were justified by the high risk of using the heavy rare earth elements in such applications. However, for electric vehicles in which high power rates are constantly required, the 2.4 MA/m class grades, which are possible with using the heavy rare earth elements Tb and Dy, are still in strong demands in order to avoid the necessity of cooling the magnets. For this reason, further developments toward high-coercivity grades with smaller amount of heavy rare earths are strongly required.

The essences of technologies for reduction of the usage of Dy are grain size reduction and control of magnetic properties of the grain boundary regions in the magnet [17]. The reduction of grain sizes is realized either by decreasing the particle sizes in the powders in the case of sintered magnets or composition adjustment and hot-deformation conditions in the case of hot-deformed Nd–Fe–B magnets. The control of magnetism of the grain boundary regions is currently performed by means of grain boundary diffusion of Tb or Dy from the surface of the nearly finished pieces of permanent magnets by heat treatments [18]. Another way for improving coercivity of the Dy-free magnets has been proposed more recently, in which alloys of Nd or Pr with Al or Cu that have low melting temperatures are infiltrated into Dy-free fine-grained Nd–Fe–B magnets [19].

3 Required Properties of Permanent Magnets for Motor Applications

Required properties of permanent magnets for motor applications include not only a large remanence (B_r) but also other features such as a large enough coercivity (H_{cJ}) to withstand armature magnetic fields produced by coil windings, straight *B-H* characteristics, small temperature dependences of B_r and H_{cJ}, easiness of magnetic charging, high electric resistivity to avoid the eddy current heating, and high mechanical toughness to name a few. However, it is difficult for a material to fulfill all of these requirements. More importantly, the requirements also vary when the design concept of the motor becomes different.

One of technological proposals to solve the problem of degrading efficiency of permanent magnet type motors in high revolution ranges, in which the magnetic flux of the permanent magnets generates a large electromotive force in the stator windings, is the variable flux motors or "memory motors" [20]. While several different types of variable flux motors have been proposed, the basic idea is to reduce the amount of air gap magnetic flux generated by the permanent magnets either by changing the remanence of the magnets or by redirecting the path of the magnetic flux from the permanent magnets in the magnetic circuit [21, 22], instead of using the flux weakening operation using the stator winding current component.

4 Perspectives of Materials Research Toward Solving the Rare Earth Issue in Permanent Magnets

The central issue in materials development toward less usage of critical elements in rare earth magnets is to find a way to avoid the decrease in coercivity caused by reducing or deleting the amount of the critical elements. Since the true factor that governs the magnitude of coercivity is believed to lie in atomic-level defects in the material, it must be understood that, in the theory of coercivity that could lead to achieving high coercivity in rare earth magnets without using critical rare earths such as Dy, the conventional theory that regards the hard magnetic materials as a continuum substance is not sufficient enough. Instead, atomistic descriptions of the coercivity-structure relationships are required from the size scales of atoms that compose the material [23]. The size and time scales that the atomistic micromagnetic calculations can deal with are many orders of magnitude smaller than the microstructural features of the permanent magnets and the measurement time of their magnetic hysteresis curves, respectively. Therefore, the development of theoretical tools to bridge the atomistic micromagnetic theories based on electronic theories of magnetism and discrete spin systems to real rare earth magnet systems is important.

In order to bridge those two regimes, technological developments including, not only gap-bridging multi-scale computational methods connecting atomistic behaviors to bulk hysteresis behaviors but also atomic-resolution analyses of interfacial microstructures and visualization of magnetic interactions of magnetic domains with structures of the materials from meso- to microscopic scales are required. Furthermore, to develop materials processing technologies for realizing the microstructural engineering as inferred from such theoretical approaches, construction of both a thermodynamic data basis and a system for utilizing those data is indispensable.

The building up of such a multi-discipline academic platform required for such multi-aspects research activities ranging from physics to thermodynamics and material processing is only possible by long-term research funding schemes, which, in Japan, have been endorsed by Elements Strategy Initiative Project to form Research Centers, outsourced by Ministry of Education, Culture, Sports, Science, and Technology (MEXT) since the fiscal year of 2012 until FY 2021 [24, 25]. In such a project, edification of human resources in the industrial section, who have literacy in the outputs of fundamental research, is one of the important issues. For a long-term impact of the academic researches strengthening the competitive capability of the domestic permanent magnet industry, industrial R&D activities, formation of academia-industry cooperative schemes have a pivotal importance.

References

1. British Geological Survey: Risk List 2015: http://www.bgs.ac.uk/mineralsuk/statistics/risklist.html/. As of 28 Aug 2017
2. Price–Precious Metal–Rare Earth—Rare Earth Metals, Shanghai Metals Market. http://original.metal.com/metals/rare-earth/prices. Jishaku-Navi. As of 28 Aug 2017
3. Rare Earths—Statistics and Information, USGS: https://minerals.usgs.gov/minerals/pubs/commodity/rare_earths. As of 28 Aug 2017
4. Coey, J.M.D.: Scripta Mater. **67**, 524 (2012)
5. Tanaka, Y., Aikawa, Y.: J. Japan Soc. Powder Metall. **47**, 15 (2000)
6. Ohtani, T., Kato, N., Kojima, S., Kojima, K., Sakamoto, Y., Konno, I., Tsukahara, M., Kubo, T.: IEEE Trans. Magn. **MAG-13**, 1328 (1977)
7. Hirayama, Y., Takahashi, Y.K., Hirosawa, S., Hono, K.: Scr. Mater. **95**, 70 (2015); *ibid.* **138**, 62 (2017)
8. Strnat, K., Tauber, A.: J. Less-Common Met. **93**, 269 (1983)
9. Coey, J.M.D.: Magnetism and Magnetic Materials. Cambridge University Press, Cambridge (2010)
10. Hirosawa, S., Matsuura, Y., Yamamoto, H., Fujimura, S., Sagawa, M., Yamauchi, H.: J. Appl. Phys. **59**, 873 (1986)
11. Kojima, T., Mizuguchi, M., Koganezawa, T., Osaka, K., Kotsugi, M., Takanashi, K.: Jpn. J. Appl. Phys. **51**, 010204 (2012)
12. Lewis, L.H., Mubarok, A., Poirier, E., Bordeaux, N., Manchanda, P., Kashyap, A., Skomski, R., Goldstein, J., Pinkerton, F.E., Mishra, R.K., Kubic Jr., R.C., Barmak, K.: J. Phys.: Condens. Matter **26**, 064213 (2014)
13. Hosoda, M., Oogane, M., Kubota, M., Kubota, T., Saruyama, H., Iihama, S., Naganuma, H., Ando, Y.: J. Appl. Phys. **111**, 07A324 (2012)
14. Permanent Magnet Synchronous Motor Incorporating SmCo magnet for 305 Series EMUs of Kyushu Railway Company, Toshiba Review 2015 (2015) 32. https://www.toshiba.co.jp/tech/review/2015/high2015/high2015pdf/1504.pdf
15. Variable flux permanent magnets, Toshiba Materials Co. Ltd. https://www.toshiba-tmat.co.jp/res/theme10.htm (Japanese URL only)
16. Hioki, K., Hattori, A., Iriyama, T.: J. Magn. Soc. Jpn. **38**, 79–82 (2014)
17. Hono, K., Sepehri-Amin, H.: Scripta Mater. **67**, 530 (2012)
18. Hirota, K., Nakamura, H., Minowa, T., Honshima, M.: IEEE Trans. Magn. **42**, 2909 (2006)
19. Sepehri-Amin, H., Ohkubo, T., Hono, K.: Acta Mater. **61**, 1982 (2013)
20. Ostovic, V.: IEEE Ind. Appl. Conf. Record **4**, 277–284 (2001)
21. Kato, T., Fukushige, T., Akatsu, K., Lorenz, R.: SAE Tech. Paper **2014–01**, 2014 (1869)
22. Huang, J., Wang, X., Sun, Z.: ITEC Asia-Pacific **2014**, 1569920713 (2014)
23. Hirosawa, S.: Adv. Nat. Sci.: Nanosci. Nanotechnol. **8**, 013002 (2017)
24. MEXT, Element Strategy Initiative: To form core research centers. http://www.mext.go.jp/a_menu/kagaku/nano/genso/index.htm (in Japanese)
25. Elements Strategy Initiative Center for Magnetic Materials. http://www.nims.go.jp/eng/research/group/ESICMM/index.html

High-Frequency Magnetics

Makoto Sonehara

Abstract When the magnetic material is used as the high-frequency devices such as RF inductor, the domain walls do not move; so, a soft magnetic material with an uniaxial magnetic anisotropy is often used in the case. A magnetic field is applied in the direction of the hard axis, and magnetization rotation is used in the soft magnetic material. In the soft magnetic material, not only the higher real part and the lower imaginary part of the complex permeability but also the electrical resistivity should be high. For example, when it is used for the magnetic core of an inductor, eddy current loss increases when the electric resistivity of the material is low. In addition, we also need various resources such as reducing the eddy current loss by putting a slit in the magnetic material in order to apply it in RF devices.

Keywords RF inductor · Uniaxial magnetic anisotropy · Magnetic thin film · Power inductor · Magnetic composite material · Eddy current loss

1 Introduction

The mobile terminal has various functions which are not only called function but also GPS, wireless LAN, Bluetooth, and various radio functions [1]. Also, the fourth-generation communication standards and LTE-advanced [1], WiMAX are adopting multiple-input and multiple-output (MIMO) communication using multiple antennas, and the radio circuit section is increasing. Such an increase in the radio circuit part in the portable terminal accompanying the progress of multifunctionalization becomes a factor of increasing the volume of the portable terminal [2]. Figure 1 shows an example of complementary metal oxide semiconductor low noise amplifier (CMOS-LNA) circuit [3]. For example, according to Ref. [3], two planar spiral air-core inductors are arranged in the central part of the CMOS-LNA, and most of the CMOS-LNA circuits are occupied by inductors. Therefore, particularly in the RF circuit, it is an urgent matter to miniaturize an inductor element or integrate it into an RF-IC. If these are realized, unnecessary wiring loss as well as miniaturization

M. Sonehara (✉)
Shinshu University, Nagano, Japan
e-mail: makoto@shinshu-u.ac.jp

K. Fujisaki (ed.), *Magnetic Material for Motor Drive Systems*,
Engineering Materials, https://doi.org/10.1007/978-981-32-9906-1_23

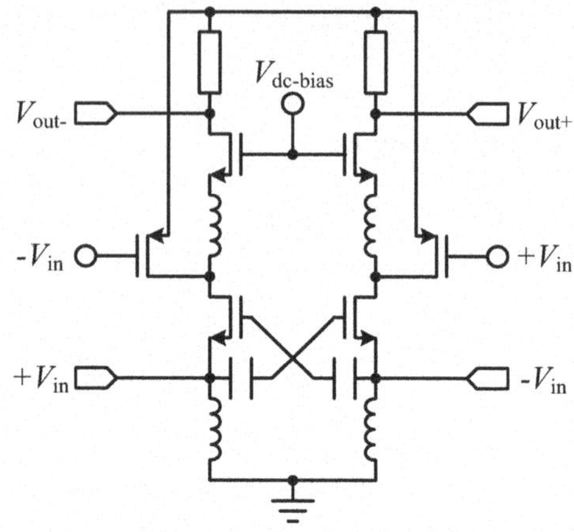

Fig. 1 Circuit diagram of a representative CMOS-LNA in the LTE-RF circuit

of the circuit can be reduced, and the LNA has lower loss. In addition, since we can amplify even weak radio waves, it is considered to contribute the expansion of the communication area.

For integration into RF-ICs, chip-size packages (CSP) and the like, as shown in Fig. 2, have been proposed. However, in the current air-core inductor, as shown in Fig. 2a, the so-called crosstalk noise in which the alternating magnetic flux leaking from the inductor induces the induction phenomenon at the neighboring element/wiring and superimposes the noise becomes a problem. As a solution to this problem, development of a magnetic thin-film inductor having a structure of sandwiching a planar spiral air-core inductor with a magnetic thin film as shown in Fig. 2b is proceeding [4].

In this section, we focus on the magnetic thin-film inductor for the UHF band (0.3–3 GHz), especially as an application to the above circuit. In addition, here "high frequency" refers to the UHF band or more. From the conclusion, as a title "high frequency magnet", (1) a real part of high complex relative permeability, (2) an imaginary part of low complex relative permeability, and (3) a soft magnetic material (thin film) having high electric resistivity are required.

2 Soft Magnetic Thin Film Having Uniaxial Magnetic Anisotropy

It is not easy to use magnetic materials in the UHF band. If it is several kHz or less, a high magnetic permeability can be obtained by applying a driving magnetic field in the direction of the easy axis of magnetization and using magnetization reversal by domain wall movement. However, as the frequency of the driving magnetic field increases, the movement of the domain wall does not follow, the magnetization reversal does not occur, and the relative magnetic permeability becomes 1 as with

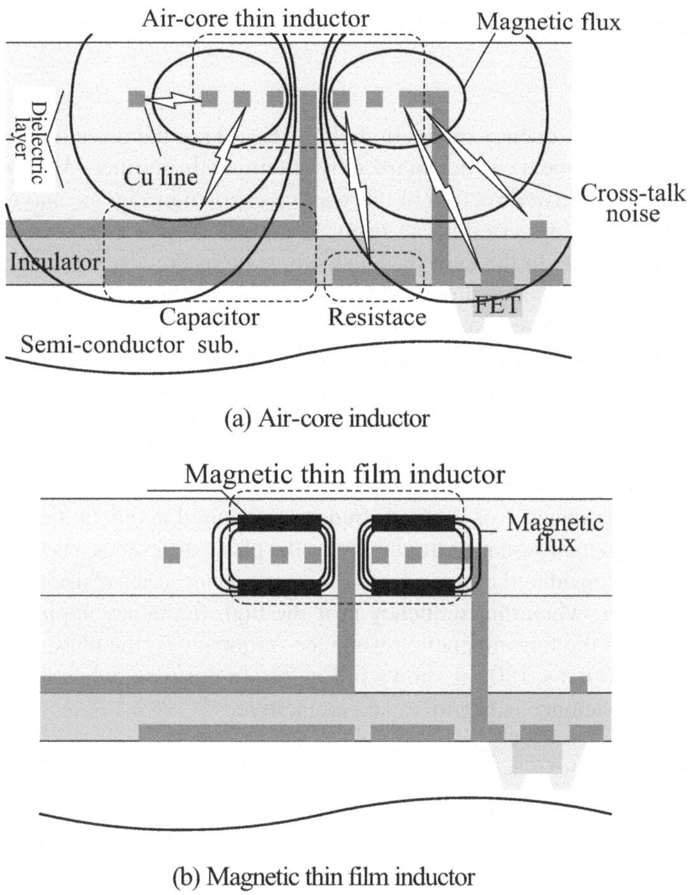

Fig. 2 Cross-sectional schematic view of chip-size package (CSP) RF-IC

air. Therefore, in the high-frequency region, a magnetic thin film having uniaxial magnetic anisotropy as shown in Fig. 3 is exclusively used, and the hard magnetic axis direction is defined as the driving magnetic field direction. In the figure, H_k represents an anisotropic magnetic field.

Fig. 3 Schematic view of soft magnetic thin film with uniaxial magnetic anisotropy

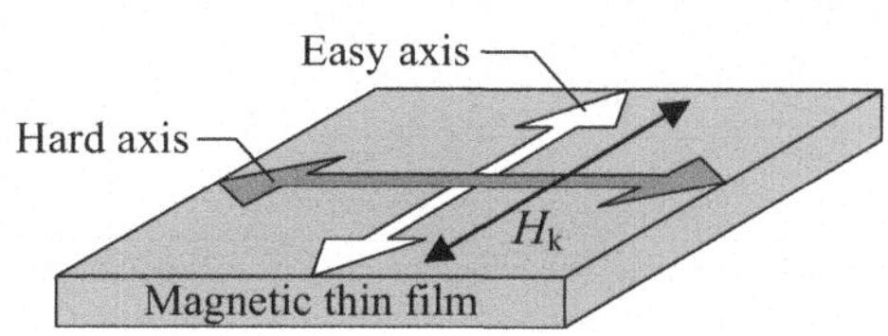

2.1 Magnetization Rotation by Magnetization Hard Axis Excitation

By applying a high-frequency magnetic field of Eq. (1) in the direction of the hard axis, the magnetic moment can be obtained by obtaining the torque—$M \times h$ from the initial state in the positive direction of the x-axis as shown in Fig. 4a, and eventually precesses while drawing an elliptical orbit around the x-axis as shown in Fig. 4b which is a steady state. In the case of the steady state in Fig. 4b, the magnetization component "m_y" in the y-axis direction has a high frequency applied magnetic field and a phase difference δ and can be expressed by Eq. (2).

When the frequency f of the high frequency applied magnetic field is lower than the ferromagnetic resonance frequency f_r, the phase difference δ and the frequency f of the applied high-frequency magnetic field are such that the phase difference δ is $0 < \delta < 90°$. At this time, when the phase difference δ is the smaller, the more reactance is positive, and the equivalent series resistance is small in an inductor. On the other hand, when the frequency f of the high frequency applied magnetic field coincides with the ferromagnetic resonance frequency f_r, the phase difference δ is 90° as shown in Fig. 5b. When considered as an inductor, the equivalent series resistance becomes very large. Further, when the frequency f of the high frequency applied magnetic field is higher than the ferromagnetic resonance frequency f_r, the phase difference δ takes a range of $90 < \delta < 180°$ as shown in Fig. 5c. In this case, when considered as an inductor, the reactance is negative, i.e., capacitive.

Fig. 4 Magnetization process when high-frequency alternating magnetic field is applied in hard axis direction

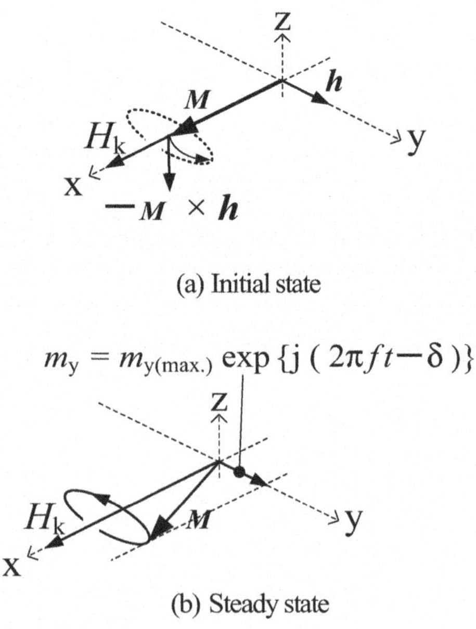

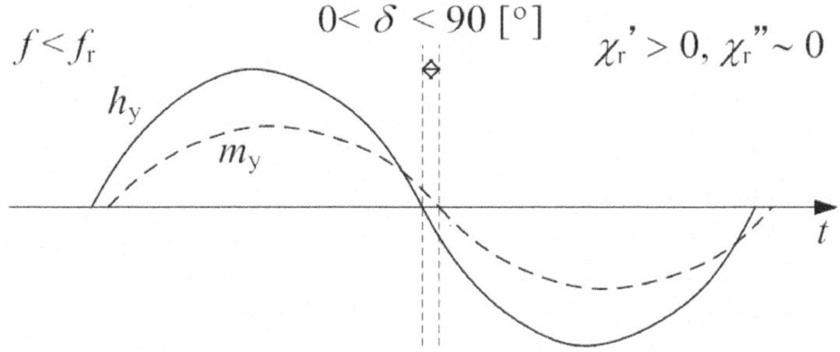

(a) $f<f_r$

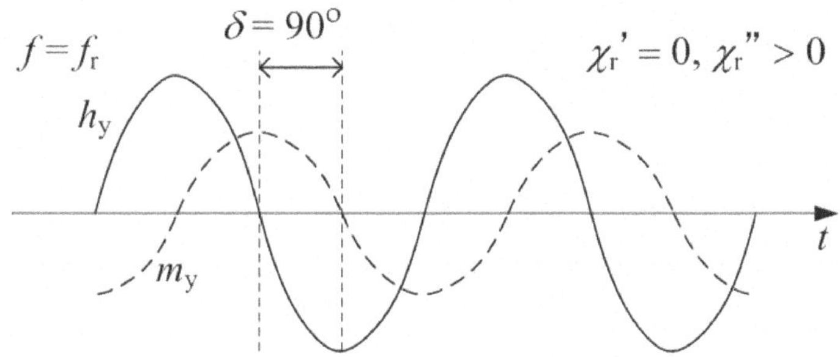

(b) $f=f_r$

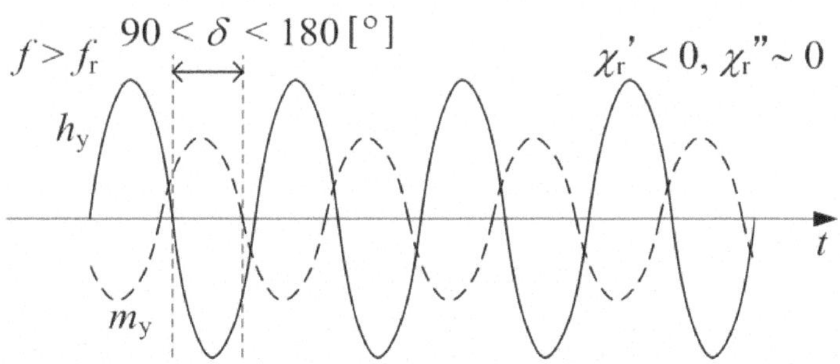

(c) $f=f_r$

Fig. 5 Relation between the frequency f of the high-frequency alternating magnetic field h_y in the magnetic thin film, the magnetization m_y and the phase difference δ

$$h_y = h_{y(\max.)} \exp(j2\pi f t) \tag{1}$$

$$m_y = m_{y(\max.)} \exp\{j(2\pi f t - \delta)\} \tag{2}$$

Here, j is an imaginary number. From Eqs. (1) and (2), the magnetic susceptibility χ_r in the direction of the hard axis can be expressed by the following equation.

$$\begin{aligned}\chi_r &= m_y/\mu_0 h_y = m_{y(\max.)} \exp\{j(2\pi f t - \delta)\}/\mu_0 h_{y(\max.)} \exp(j2\pi f t) \\ &= m_{y(\max.)} \exp(-j\delta)/\mu_0 h_{y(\max.)} = m_{y(\max.)}(\cos\delta - j\sin\delta)/\mu_0 h_{y(\max.)}\end{aligned} \tag{3}$$

Here, μ_0 is the magnetic permeability in vacuum (4×10^{-7} H/m). The real part χ_r' and the imaginary part χ_r'' of the complex relative magnetic susceptibility χ_r are expressed by the following equations.

$$\chi_r = \chi_r' - j\chi_r'' \tag{4}$$

From Eqs. (3) and (4), the real part χ_r' and the imaginary part χ_r'' of the complex relative magnetic susceptibility are expressed by the following equations.

$$\chi_r' = m_{y(\max.)} \cos\delta/\mu_0 h_{y(\max.)} \tag{5}$$

$$\chi_r'' = m_{y(\max.)} \sin\delta/\mu_0 h_{y(\max.)} \tag{6}$$

The real part μ_r' and the imaginary part μ_r'' of the complex relative magnetic permeability are expressed by the following equations.

$$\mu_r' = m_{y(\max.)} \cos\delta/\mu_0 h_{y(\max.)} + 1 \tag{7}$$

$$\mu_r'' = m_{y(\max.)} \sin\delta/\mu_0 h_{y(\max.)} \tag{8}$$

If the real part μ_r' and the imaginary part μ_r'' of the complex relative magnetic permeability are proportional to the inductance L and the equivalent series resistance R of the inductor, respectively, as in Eq. (9), the phase difference δ is small as described above desired.

$$j2\pi f \mu_r = j2\pi f(\mu_r' - j\mu_r'') = j2\pi f \mu_r' + 2\pi f \mu_r'' \propto j2\pi f L + R \tag{9}$$

Since AC loss such as eddy current loss appear as R, it is obvious that a high electric resistivity is required for the magnetic thin film.

2.2 Ferromagnetic Resonance Frequency

In the previous section, it is described that it is desired that the phase difference δ is small, but this means that the ferromagnetic resonance frequency f_r of the soft magnetic thin film must be sufficiently higher than the frequency (band) to be used.

When the ferromagnetic material is in a sufficiently large static magnetic field $\boldsymbol{H}$, the total magnetic moment begins to precess at the angular frequency ω with the static magnetic field $\boldsymbol{H}$ direction as the axis and they are directed in the direction of the static magnetic field $\boldsymbol{H}$. Considering the braking effect due to eddy current, it is expressed by Landau-Lifshitz-Gilbert (LLG) equation of Eq. (10) [5].

$$\mathrm{d}M/\mathrm{d}t = -\gamma(M \times H) + \alpha/M_s(M \times \mathrm{d}M/\mathrm{d}t) \tag{10}$$

Here, α is the damping constant ($0.01 \leqq \alpha \leqq 0.02$), γ is the gyro magnetic constant ($1.105\ g \times 10^5$ m/A s), and g is the Lande factor (=2).

When applying a high-frequency magnetic field $\boldsymbol{h}$ with an angular frequency ω_r in a direction perpendicular to the static magnetic field $\boldsymbol{H}$, the magnetic moment absorbs energy from the high-frequency magnetic field $\boldsymbol{h}$ and continues precession without attenuation. This is called ferromagnetic resonance absorption, and the frequency of the applied high-frequency magnetic field is referred to as ferromagnetic resonance frequency f_r. The equation of motion at this time can be considered as Eq. (11) which eliminates the damping term of the second term on the right side of Eq. (10). By solving this equation, the ferromagnetic resonance frequency f_r can be obtained.

$$\mathrm{d}M/\mathrm{d}t = -\gamma(M \times H) \tag{11}$$

In general, the static magnetic field $\boldsymbol{H}$ in Eq. (11) includes various elements expressed by Eq. (12).

$$H = H_{\mathrm{dc}} + H_{\mathrm{k}} + H_{\mathrm{m}} + H_{\mathrm{d}} \tag{12}$$

Here, $\boldsymbol{H}_{\mathrm{dc}}$ is a direct current external magnetic field, $\boldsymbol{H}_{\mathrm{m}}$ is a molecular magnetic field, and $\boldsymbol{H}_{\mathrm{d}}$ is a demagnetizing field.

Consider a case where the DC external magnetic field $\boldsymbol{H}_{\mathrm{dc}}$ is applied in the z-axis direction. Assume that the anisotropic magnetic field $\boldsymbol{H}_{\mathrm{k}}$ is a uniaxial anisotropic case and is oriented in the z-axis direction. Since the molecular magnetic field $\boldsymbol{H}_{\mathrm{m}}$ is parallel to the magnetization $\boldsymbol{M}$, it is 0 from the right side of Eq. (11). The demagnetizing field $\boldsymbol{H}_{\mathrm{d}}$ is due to the magnetic poles generated on the surface of the sample. For the sake of simplicity, if it is assumed to be a spheroidal sample and its principal axis is in the x- , y- , and z-directions, the demagnetizing field coefficient N is $N_{(x)}$, $N_{(y)}$, and $N_{(z)}$. The demagnetizing field $\boldsymbol{H}_{\mathrm{d}}$ can be represented by the components in the x- , y- , and z-directions as shown in Eq. (13).

$$H_{\mathrm{d}(i)} = -N_{(i)}M_{(i)}/\mu_0 \quad (i = x, y, z) \tag{13}$$

To summarize the above equations, the static magnetic field $\boldsymbol{H}$ of Eq. (13) can be expressed as the components in the x-, y- , and z-directions as shown in Eqs. (14), (15), and (16).

$$H_{(x)} = -N_{(x)}M_{(x)}/\mu_0 \tag{14}$$

$$H_{(y)} = -N_{(y)}M_{(y)}/\mu_0 \tag{15}$$

$$H_{(z)} = H_{\mathrm{dc}} + H_{\mathrm{k}} - N_{(z)}M_{(z)}/\mu_0 \tag{16}$$

Assuming that $\boldsymbol{M}(t) = \boldsymbol{M}\ \exp\,(\mathrm{j}\omega t)$, by substituting Eqs. (14)–(16) into Eq. (11) and computing each component, Eqs. (17), (18), and (19) are derived. $M_{(x)}$, $M_{(y)}$, «H_{dc}, H_{k}, $M_{(z)}$ are assumed, and the products of small values are omitted.

$$-\mathrm{d}M_{(x)}/\gamma \mathrm{d}t = -\mathrm{j}\omega M_{(x)}/\gamma = M_{(y)}\left\{H_{\mathrm{dc}} + H_{\mathrm{k}} + \left(N_{(y)} - N_{(z)}\right)M_{(z)}/\mu_0\right\} \tag{17}$$

$$-\mathrm{d}M_{(y)}/\gamma \mathrm{d}t = -\mathrm{j}\omega M_{(y)}/\gamma = M_{(x)}\left\{H_{\mathrm{dc}} + H_{\mathrm{k}} + \left(N_{(x)} - N_{(z)}\right)M_{(z)}/\mu_0\right\} \tag{18}$$

$$-\mathrm{d}M_{(z)}/\gamma \mathrm{d}t = 0 \tag{19}$$

It considers that $M_{(z)} \simeq M_{\mathrm{s}}$ because $M_{(z)}$ is invariant to time. Considering the above, Eqs. (20) and (21) are derived.

$$-\mathrm{j}\omega M_{(x)}/\gamma - M_{(y)}\left\{H_{\mathrm{dc}} + H_{\mathrm{k}} + \left(N_{(y)} - N_{(z)}\right)M_{\mathrm{s}}/\mu_0\right\} = 0 \tag{20}$$

$$M_{(x)}\left\{H_{\mathrm{dc}} + H_{\mathrm{k}} + \left(N_{(x)} - N_{(z)}\right)M_{\mathrm{s}}/\mu_0\right\} - \mathrm{j}\omega M_{(y)}/\gamma = 0 \tag{21}$$

In order for $M_{(x)}$, and $M_{(y)}$ to have a solution, when Eq. (21) is a determinant and solved, angular frequency ω is expressed by Eq. (22).

$$\omega = \gamma\left[\left\{H_{\mathrm{dc}} + H_{\mathrm{k}} + \left(N_{(y)} - N_{(z)}\right)M_{\mathrm{s}}/\mu_0\right\}\left\{H_{dc} + H_{\mathrm{k}} + \left(N_{(x)} - N_{(z)}\right)M_{\mathrm{s}}/\mu_0\right\}\right]^{0.5} \tag{22}$$

Equation (22) is the angular frequency (ferromagnetic resonance angular frequency ω_{r}) of the high-frequency magnetic field h applied in a direction perpendicular to the static magnetic field $\boldsymbol{H}$. Therefore, the ferromagnetic resonance frequency f_{r} is expressed by Eq. (23).

$$f_{\mathrm{r}} = (\gamma/2\pi)\left[\left\{H_{\mathrm{dc}} + H_{\mathrm{k}} + \left(N_{(y)} - N_{(z)}\right)M_{\mathrm{s}}/\mu_0\right\}\left\{H_{\mathrm{dc}} + H_{\mathrm{k}} + \left(N_{(x)} - N_{(z)}\right)M_{\mathrm{s}}/\mu_0\right\}\right]^{0.5} \tag{23}$$

Consider the case of a magnetic thin film used in a magnetic thin film inductor or the like. In order to consider the case where the DC external magnetic field $\boldsymbol{H}_{\mathrm{dc}}$ is not applied, $\boldsymbol{H}_{\mathrm{dc}} = 0$. In the case of a magnetic thin film, it has $N_{(x)} \approx 1$, $N_{(y)}$

$= N_{(z)} \approx 0$ since it has a demagnetizing field coefficient only in the film thickness direction. Therefore, in the case of a magnetic thin film, the ferromagnetic resonance frequency f_r is expressed by Eq. (24).

$$f_r = (\gamma/2\pi)\{H_k(H_k + M_s/\mu_0)\}^{0.5} \tag{24}$$

From Eq. (24), it is found that it is necessary to select a magnetic material having a high saturation magnetization M_s for the magnetic thin film and to impart a large magnetic anisotropy and increase the anisotropic magnetic field H_k. On the other hand, as can be seen from Eqs. (7) and (8), care should be taken because the permeability decreases when the anisotropic magnetic field H_k is too high.

3 Development Example of Magnetic Thin-film Inductor

Magnetic thin films for magnetic thin-film inductors are actively developed. For example, a method using a dynamic sputtering method such as a carousel sputtering apparatus [6], a method using a shape anisotropy by a slit pattern [7], a method of epitaxially growing Fe on a single-crystal GaAs to obtain a large crystal magnetic anisotropy method [8], a method using an exchange coupling laminated film using exchange bias magnetic field [9], a ferrite film [10–12], a nanogranular magnetic thin film [13–15], and the like. Here, we describe a development example of a magnetic thin film inductor using CoFeSiO/SiO_2 granular magnetic thin film [16]. In addition, the development examples of high Q inductor [17] using magnetic fine particle composite material will also be outlined.

3.1 CoFeSiO/SiO_2 Granular Magnetic Thin-film Inductor

For the formation of CoFe/SiO_2, an inductively coupled RF sputtering apparatus was used. It is fabricated by alternate lamination of granular layer and SiO_2 layer formed by co-sputtering of $CoFeSiO_2$ alloy and SiO_2, and it is possible to obtain extremely narrow ferromagnetic resonance half-width in laminated structure of 6 nm-CoFeSiO/1 nm-SiO_2 know. Electric resistivity is high by 1 nm-SiO_2 layer. Figure 6 shows the frequency response of the complex permeability. The resonance frequency is in the vicinity of 2.7 GHz, the imaginary part of the complex relative magnetic permeability is sufficiently small up to at least 2 GHz, and the CoFeSiO/SiO_2 granular magnetic thin film has uniaxial anisotropy.

Figure 7 shows a schematic diagram of the four-turn thin-film inductor fabricated. It is a single-sided magnetic structure composed of CoFeSiO/SiO_2 granular magnetic thin-film layer/polyimide insulating layer/Cu spiral coil conductor line layer on the glass substrate from the lower layer. Since CoFeSiO/SiO_2 has uniaxial anisotropy, the axis of easy magnetization is induced in a direction orthogonal to A-A' in Fig. 7. As

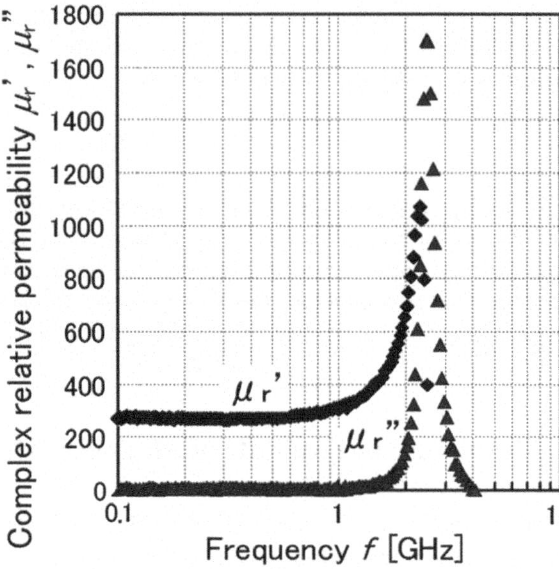

Fig. 6 Measurement results of the frequency characteristic of complex relative magnetic permeability of CoFeSiO/SiO_2 granular magnetic thin film

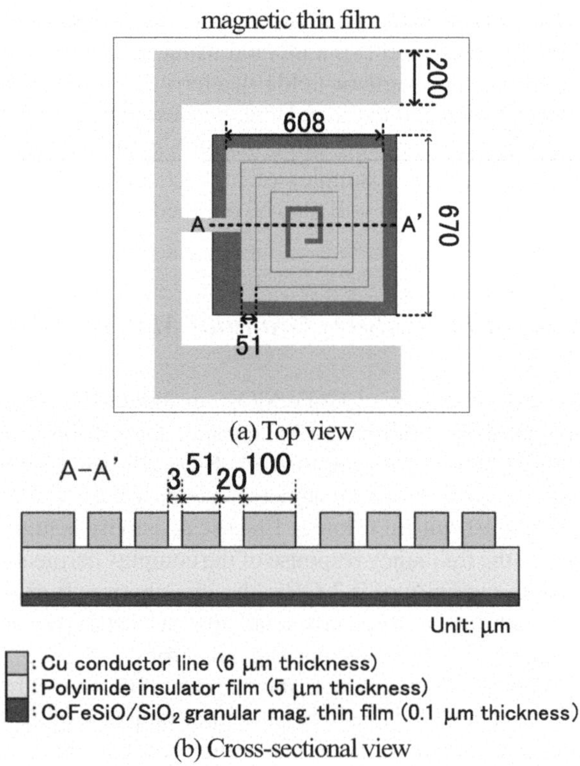

Fig. 7 Schematic view of prototype CoFeSiO/SiO_2 granular magnetic thin film inductors

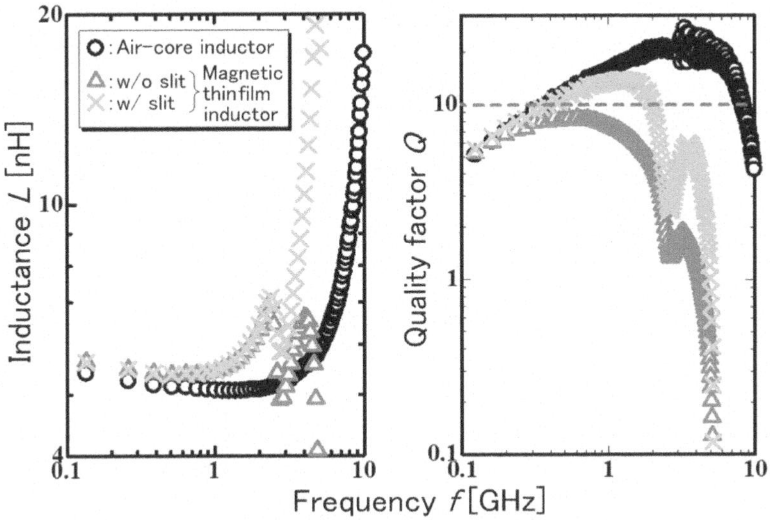

(a) *L* vs *f* (b) *Q* vs. *f*

Fig. 8 Measurement results of frequency dependence of inductance *L* and quality factor *Q* of prototype CoFeSiO/SiO_2 granular magnetic thin-film inductors

a result, the area of half of the square spiral coil becomes the magnetization rotation area. For the purpose of increasing the *Q* at high frequencies, the magnetic thin film was slitted for the in-plane eddy current suppress caused by the perpendicular magnetic flux component, and the coil conductor was slitted for the increase of the AC copper loss suppress by the coil conductor interlinkage magnetic flux [18].

Figure 8 shows the inductance *L* and the figure of merit *Q* ($=\omega L/R$) of a thin-film inductor with a one-sided magnetic structure composed of a magnetic thin film (0.1 μm thick) and three divided conductor lines (6 μm thick). In the same figure, the measurement results of thin-film inductors and air-core inductors of the same size which are not slit-processed in the magnetic thin film are also shown for comparison. In the case of adding a thin magnetic film, the inductance *L* increases from around 1 GHz and shows a characteristic of rapidly decreasing around 2 GHz. This is considered to correspond to the frequency characteristic of the real part μ_r' of the complex relative permeability of the thin magnetic film in the vicinity of the resonance frequency as shown in Fig. 6. Assuming that the *Q* value of a magnetic thin film inductor with a slit is 10 or more as the operating frequency, the operating range is from 400 MHz to 2 GHz. Also, *Q* is 7.54 for slitless magnetic thin film inductors at 1 GHz, but it increases to 13.3 for slitted magnetic thin film inductors.

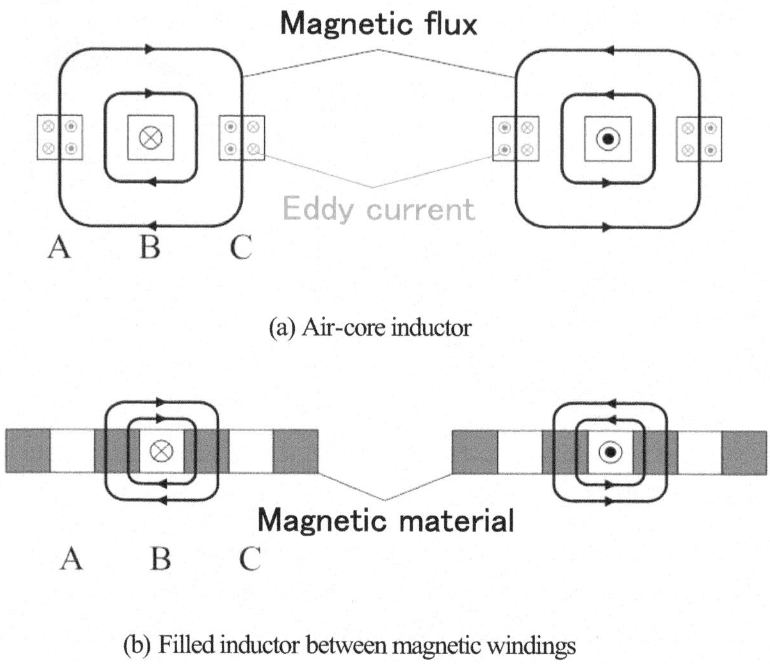

Fig. 9 Cross-sectional view of planar spiral inductor

3.2 *High Q Inductor Using Magnetic Fine Particle Composite Material*

Here, outline of the high Q inductor using the magnetic fine particle composite material is described. Figure 9 shows a cross-section of the planar spiral air-core inductor, in which the alternating magnetic flux generated by the high-frequency current flowing through the winding B is linked to the adjacent windings A and C. Then, in the windings A and C, eddy current due to the induction phenomenon is generated, i.e., loss due to the proximity effect occurs. On the other hand, in the proposed method as shown in Fig. 9b, the magnetic flux generated by the current flowing through each winding passes the magnetic material embedded between the windings with higher magnetic permeability than the adjacent windings, and it is possible to suppress the loss due to the proximity effect.

Therefore, a carbonyl iron powder (CIP)/epoxy composite material magnetic core having a structure in which magnetic fine particles are dispersed in an insulator is used as a magnetic material to be filled in a space between conductor lines. The initial state of the composite material magnetic core is in the form of a paste, and it can be filled between the lines of the inductor as shown in Fig. 10 by applying it using the metal mask and the squeegee to the upper surface of the inductor [19]. In addition,

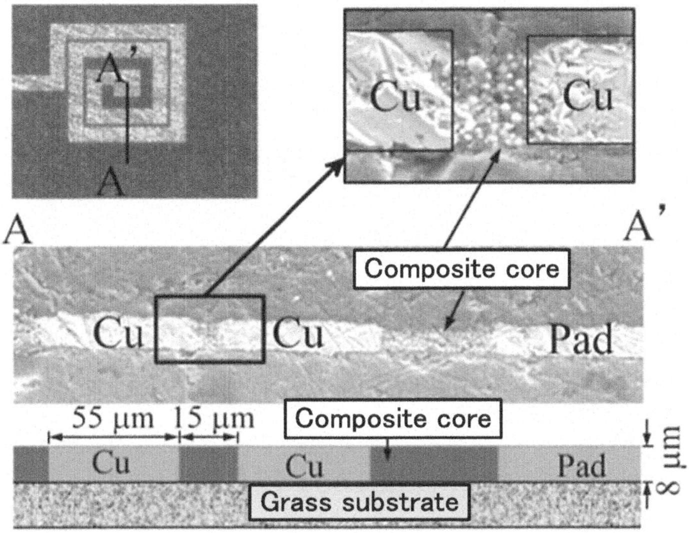

Fig. 10 Top view and cross-sectional photograph in prototype planar spiral inductor with magnetic fine particle composite material filling between windings

since it has a higher resistivity than the conventional metal magnetic thin film, the eddy current loss at high frequencies can be reduced.

The details are omitted and only the results are described. Figure 11a shows the frequency characteristics of the inductance *L* of the interwinding inductor of Fig. 10, and Fig. 11b shows the frequency characteristic of the *Q* value. Also, the air-core inductors and analysis values of the same size are shown in solid lines or broken lines in each figure. From Fig. 11a, the inductance L of the composite core loaded inductor is 2.0 nH at 1 GHz, which is 22% higher than the air-core inductor. It is thought that although the composite material is filled only between the conductor lines, the inductance *L* increases though slightly depending on the magnetic permeability of the composite material. On the other hand, from Fig. 11b, the *Q* value of the composite material loaded inductor is 29 at 1 GHz, which is 29% higher than the air-core inductor. As described above, these are considered to be due to suppression of proximity effect and reduction of AC copper loss. However, at frequencies higher than 1 GHz, the *Q* value decreases, which is considered to be the effect of high frequency loss of the composite material. Loss of material at high frequencies can be expected to be improved by using magnetic fine particles with a small particle size or coating the surfaces of the magnetic fine particles with an insulating material.

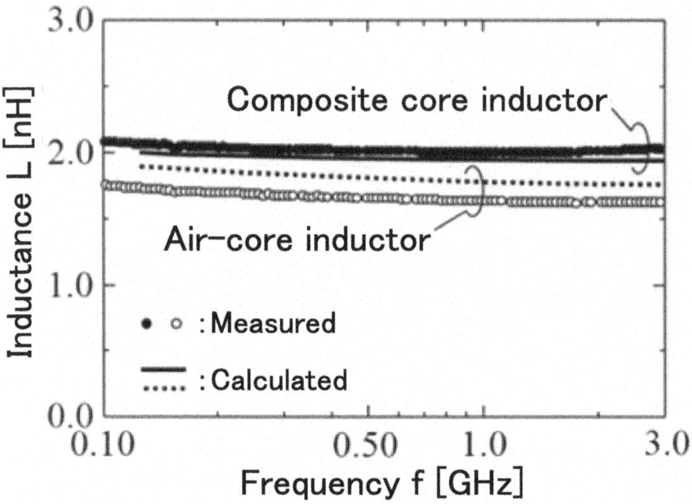

(a) *L vs. f*

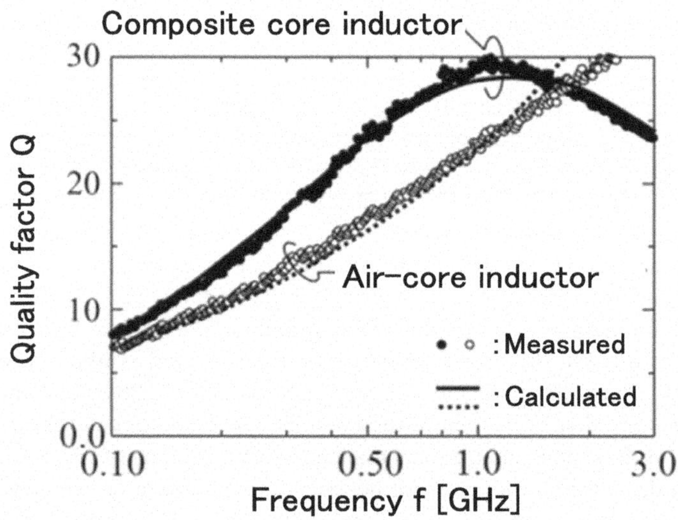

(b) *Q vs.f*

Fig. 11 Measurement results and analysis results of frequency dependence of inductance *L* and quality factor *Q* of prototype planar spiral inductor with magnetic fine particle composite material filling between windings

4 Conclusions

Lastly, the development examples of inductors using concrete magnetic thin film inductors and magnetic fine particles were introduced, but in particular, high-frequency inductors using magnetic materials used at frequencies exceeding 1 GHz have not been put to practical use. The authors believe that high-frequency inductors can not be put to practical use unless they are developed from both sides of the magnetic material and the device.

On the other hand, power supply circuits using next-generation power semiconductors SiC or GaN are being developed and marketed. Currently, there are many power supplies that use switching frequency of several MHz, but "Magnetic core material for 100 MHz switching power supply" in the JST future creation project adopted in FY2017 [20]. The switching frequency is expected to be further increased in the future. Magnetic materials for magnetic components such as inductors and transformers which are also the subject of the same task but low loss and small even at several MHz or more are bottlenecks at present. Development of magnetic materials satisfying the requirements (1) to (3) mentioned at the beginning of this chapter is an urgent task, and research and development of power supply inductors using the magnetic microparticles described in Sect. 3.2 are underway [21, 22]. Since it is expected that it will be applied to inverters and converters that make up the motor drive system, the "high-frequency magnetism" mentioned in this chapter seems to be important from now on.

References

1. 3GPP Homepage: http://www.3gpp.org/technologies/keywords-acronyms/97-lte-advanced. Accessed 21 July 2017
2. Maruyama, T., et al.: Experimental study of cross-talk suppression effect by introducing magnetic thin film to spiral inductor for CSP RF-IC. The paper of Technical Meeting on Magnetics, IEE Japan, MAG-10-201, pp. 11–16 (2010)
3. Hidayov, O.A., et al.: 0.7–2.7 GHz wideband CMOS low-noise amplifier for LTE application. Electron. Lett. **49**(23), 1433–1435 (2013)
4. Gardner, D.S., et al.: Review of on-chip inductor structures with magnetic films. IEEE Trans. Magn. **45**(10), 4760–4766 (2009)
5. Ohta, K.: Foundation of magnetic engineering II, Kyoritsu Zensho (1990)
6. Munakata, M., et al.: Magnetic properties and frequency characteristics of $(CoFeB)_x$–$(SiO_{1.9})_{1-x}$ and CoFeB films for RF application. Trans. Magn. Soc. Japan **2**(5), 388–393 (2002)
7. Ikeda, S., et al.: A design chart of effective permeability of magnetic film with parallel slit structure. Technical Report of IEE of Japan, MAG-02-111, pp. 25–29 (2002)
8. Tsai, C.S., et al.: Wideband electronically tunable microwave bandstop filters using iron film-gallium arsenide waveguide structure. IEEE Trans. Magn. **35**, 3178–3180 (1999)
9. Sonehara, M., et al.: Preparation and characterization of Mn-Ir/Fe-Si exchange-coupled film for high-frequency micromagnetic devices. J. Magn. Soc. Japan **29**(2), 132–137 (2005)
10. Kondo, K., et al.: FMR study of plated Ni-Zn-Co ferrite films. In: 28th Digest of Annual Conference on Magnetics in Japan, vol. 28, p. 479 (2004)

11. Ishida, H., et al.: Fabrication of Ni-Zn ferrite thin film using the low-target-voltage reactive ECR sputtering. method. In: 28th Digest of Annual Conference on Magnetics in Japan, vol. 28, p. 480 (2004)
12. Yanai, T., et al.: High temperature magnetic properties of Fe-based toroidal core with controlled permeability. In: 30th Digest of Annual Conference on Magnetics in Japan, vol. 30, p. 291 (2006)
13. Itoh, T., et al.: Nano-structure and magnetic properties of Ni-Fe granular thin films produced by evaporation. In: 28th Digest of Annual Conference on Magnetics in Japan, vol. 28, p. 482 (2004)
14. Ohnuma, S., et al.: Nano-granular magnetic thin films for applications as soft magnetic materials. In: Digests of the 30th Annual Conference on Magnetics, vol. 30, pp. 435–436 (2006)
15. Naoe, M., et al.: Investigation of controlling in-plane uniaxial anisotropy of CoPd-CaF_2 nanogranular films by tandem-sputtering deposition. IEEE Magn. Lett. **5**(1), 1–4 (2014)
16. Ikeda, K., et al.: IEEE INTERMAG Conference 2009, BS-07 (2009)
17. Sonehara, M., et al.: Fundamental study of high Q-factor RF spiral inductor using carbonyl-iron/epoxy composite magnetic core. IEEJ Trans. Electr. Electron. Eng. **11**(S1), S3–S8 (2016)
18. Ikeda, K., et al.: Technical meeting on magnetics. IEE Japan, MAG-09-80 (2009)
19. Sugawa, Y., et al.: Carbonyl-iron/epoxy composite magnetic core for planar power inductor used in package-level power grid. IEEE Trans. Magn. **49**(7), 4172–4175 (2013)
20. JST MIRAI Homepage: https://www.jst.go.jp/mirai/jp/project/index.html
21. Sugimura, K., et al.: Formation of high electrical-resistivity thin surface layer on carbonyl-iron powder (CIP) and thermal stability of nanocrystalline structure and vortex magnetic structure of CIP. AIP Adv. **6**(5), #055932 (2016)
22. Sato, K., et al.: Fabrication of metal composite magnetic core transformer with surface-oxidized carbonyl-iron powder and its application to the flyback-type dc-dc converter. Trans. Magn. Soc. Japan (Special Issues) **1**, 44–52 (2017)

Magnetic Application

Iron Loss Analysis of Motors

Katsumi Yamazaki

Abstract In this chapter, methods to calculate iron losses of motors by using electromagnetic field analysis are presented. First, the components of iron loss, i.e., hysteresis loss, eddy current loss, and excess loss, are explained. Next, the iron loss calculation methods that consider the effects of nonlinear characteristics of the core materials, mechanical stress, and harmonic fields are summarized. Several approximations for the practical loss estimation are introduced. Finally, an application of the iron loss calculation method to motors is indicated.

Keywords Iron loss · Motors · Mechanical stress · Harmonics · Finite element methods

1 Introduction

Owing to energy problems in recent years, the loss analysis of motors becomes more important subject because the energy consumption of motors is nearly the half of the total electricity in most of advanced countries.

The motor losses can be classified into mechanical loss, copper loss, and iron loss. Among them, the modeling of core materials and estimation of harmonic electromagnetic fields are important to calculate the iron loss accurately, as follows:

(1) Modeling of core materials

The magnetic field H and flux density B are in nonlinear relationship due to the magnetic saturation and hysteresis phenomenon. In addition, the directions of B and H are not identical to each other due to the vector magnetic properties [1]. The accuracy of the iron loss analysis is determined by the modeling of these phenomena. The modeling for the effects of mechanical stress on the magnetic properties is also required because the iron loss often increases with the stator stress caused by shrink-fitting and rotor stress caused by centrifugal force [2–8].

K. Yamazaki (✉)
Chiba Institute of Technology, Narashino, Japan
e-mail: yamazaki.katsumi@it-chiba.ac.jp

K. Fujisaki (ed.), *Magnetic Material for Motor Drive Systems*,
Engineering Materials, https://doi.org/10.1007/978-981-32-9906-1_24

(2) Harmonic electromagnetic fields in motors

The magnetic field produced by stator and rotor includes a lot of space harmonics, such as magnetomotive, phase-band, and slot harmonics [7–10]. Furthermore, time harmonics caused by inverter carrier are also generated when the motors are driven by inverters [9, 10]. The sum of the losses by these harmonics often becomes several times larger than the fundamental loss. Therefore, the consideration of these harmonic losses is indispensable.

Although many papers related to the method of iron loss calculation have been published, and there is no method that considers all the phenomena in (1) and (2) because long calculation time or vast material data are required. Several approximations should be required for practical iron loss calculation.

In this chapter, firstly, the components of iron loss are explained. Then, the calculation methods for iron loss are summarized. Finally, an application of the iron loss calculation method to motors is indicated.

2 Components of Iron Loss and Calculation Methods

The iron loss per weight w_i in electrical steel sheet used for motor cores can be regarded as the sum of eddy current loss w_e and hysteresis loss w_{h}, as follows:

$$W_{\mathrm{i}} = w_{\mathrm{h}} + w_{\mathrm{e}} \tag{1}$$

Further, w_{e} can be regarded as the sum of classical eddy current loss w_{e}' and excess loss w_{ex} that is generated by the micro movement of domain wall. In this case, w_i is expressed by the sum of three terms, as follows:

$$w_{\mathrm{i}} = w_{\mathrm{h}} + w_{\mathrm{e}}' + w_{\mathrm{ex}} \tag{2}$$

The most of iron loss calculation methods proposed in recent years are based on (1) or (2). In these methods, each term is determined by using results of the core material experiments and the electromagnetic field analysis of motors. Many iron loss calculation methods proposed in Japan are based on (1) [3–10], whereas those in western countries are typically based on (2) [11–15].

The definition of hysteresis loss in (1) and (2) are identical to each other. This loss can be directly obtained from the hysteresis loops at each point of the motor core when the hysteresis phenomenon is taken into account in the electromagnetic field analysis. On the other hand, when the hysteresis phenomenon is neglected, it can be approximately calculated from the amplitude of flux density and the corresponding hysteresis loss in the core material experiments.

The eddy current loss w_e in (1) can be directly calculated from electromagnetic field analysis that considers eddy currents. It can also be calculated by static electromagnetic field analysis from the amplitude of flux density and the corresponding eddy current loss in the core material experiments.

The classical eddy current loss w'_e in (2) is often calculated by the following expression that neglects the skin effect:

$$w'_e = \frac{\sigma_{iron}(\pi h)^2}{6D} f^2 B_m^2 \tag{3}$$

where σ_{iron}, h, D are the conductivity, thickness, density of the core material, f is the frequency, B_m is the amplitude of flux density. In this case, the excess loss w_{ex} is defined as the residual that is calculated by subtracting w'_e from the total eddy current loss w_e.

Figure 1 shows an example of components of core loss per/weight of an electrical steel sheet, which are separated due to (1)–(3) [9]. The K_e and K_h in the figure are the eddy current and hysteresis loss coefficients due to the following approximated modeling.

$$w_i = w_e + w_h = K_e f^2 B_m^2 + K_h f B_m^2 \tag{4}$$

The vertical axis of Fig. 1 corresponds to W_i/fB^2. Due to (4), this value can be expressed as $K_e f + K_h$. Therefore, the intercept of the characteristics in Fig. 1 corresponds to K_h, whereas the slope corresponds to K_e. On the other hand, the slope of the characteristics in Fig. 1 decreases with frequency f. In the loss separation due to (1), the reason for this decrease is considered to be the skin effect of the electrical steel sheet. In addition, the K_e at low frequency, where the skin effect is negligible, is larger than the slope of the classical eddy current loss. This increase is caused by the existence of the excess loss. Therefore, the modified coefficient of

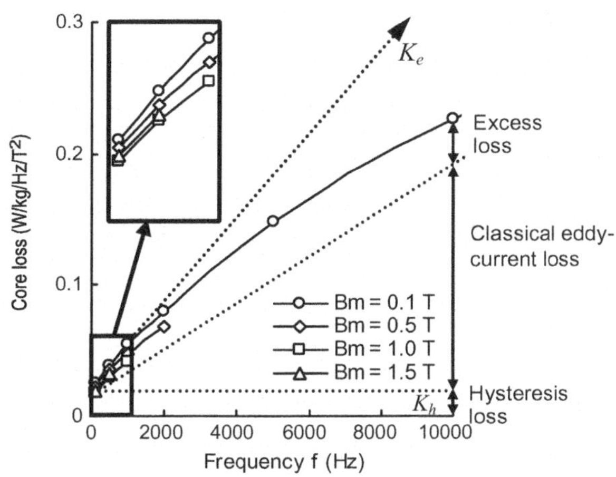

Fig. 1 Experimental iron loss characteristics of electrical steel sheet. © [2012] IEEE. Reprinted, with permission, from [9]

excess loss κ can be derived from (3) and the first term in the right side members in (4), as follows [9]:

$$\kappa = \frac{6DK_e}{\sigma_{iron}(\pi h)^2} \tag{5}$$

The total eddy current loss that consider the skin effect can be approximately obtained by multiplying the κ in (5) to the classical eddy current loss obtained by macro electromagnetic field analysis that neglects the magnetic domain.

On the other hand, in the iron loss calculation due to (2) with the assumption of the classical eddy current loss by (3), the excess loss is often approximated to be proportional to $(fB)^{1.5}$, as follows [11–15]:

$$w_{ex} = K_{excs} f^{1.5} B_m^{1.5} \tag{6}$$

where K_{excs} is the excess loss coefficient.

All the methods mentioned above are based on some approximation for the actual physical phenomena. However, there is an important difference between these methods. In the method based on (1), the coefficients K_e, K_h, and κ can be determined to be constants from the results of core material experiments, whereas K_{ex} in (2) is a complex function of frequency and flux density [9].

On the other hand, in the iron loss calculation methods that considers the vector magnetic properties, the loss separation by (1) and (2) is often not carried out. In this case, the total iron loss is directly calculated from the measured iron loss of core materials under various conditions [1]. In this case, vast measured results should be required. This is the reason why the number of application reports of the method that considers the vector magnetic property is small and limited.

3 Calculation Method of Motor Iron Loss Considering Harmonics

In this section, the calculation method of the iron loss of motors is explained by considering the harmonic electromagnetic field and mechanical stress.

The iron loss in the motors cannot be expressed by (3) and (4) with single frequency f because harmonic fluxes are included in the motor cores. Furthermore, the waveform of flux density varies at each position in the motor cores. Therefore, electromagnetic field analysis, for example, finite element method, is required to obtain the total iron loss, which is obtained by summing up the loss calculated at each element.

On the basis of (4), which neglects the skin effect, the total eddy current and hysteresis losses W_e, W_h can be calculated from the flux density waveform obtained by the electromagnetic field analysis, as follows [16]:

$$W_{\mathrm{e}} = \frac{K_{\mathrm{e}} D}{2\pi 2} \int_{\mathrm{iron}} \frac{1}{N} \sum_{k=1}^{N} \left\{ \left(\frac{B_{\mathrm{r}}^{k+1} - B_{\mathrm{r}}^{k}}{\Delta t} \right)^2 + \left(\frac{B_{\theta}^{k+1} - B_{\theta}^{k}}{\Delta t} \right)^2 \right\} \mathrm{d}v \tag{7}$$

$$W_{\mathrm{h}} = \frac{K_{\mathrm{h}} D}{T} \sum_{i=1}^{NE} \frac{\Delta V^i}{2} \left(\sum_{j=1}^{N_{\mathrm{pr}}^i} \left(B_{\mathrm{mr}}^{ij}\right)^2 + \sum_{j=1}^{N_{\mathrm{p}\theta}^i} \left(B_{\mathrm{m}\theta}^{ij}\right)^2 + \sum_{j=1}^{N_{\mathrm{pz}}^i} \left(B_{\mathrm{mz}}^{ij}\right)^2 \right) \tag{8}$$

where Δt is the time step, N is the sampling number per time period, B_{r}, $B\Delta$ are the radial and circumferential components of flux density, ΔV^i is the volume of the i-th element, N_{pr}^i, $N_{\mathrm{p}\theta}^i$, N_{pz}^i are the number of peaks of radial, circumferential, and axial components of flux density, respectively, B_{mr}^{ij}, $B_{\mathrm{m}\theta}^{ij}$, B_{mz}^{ij} are the amplitude of the major and minor hysteresis loops. Note that (7) and (8) are valid only for the laminated cores and the in-plane eddy currents caused by axial flux density are not included. In addition, the effect of the rotational magnetic field is approximately taken into account under the assumption that the total loss can be calculated by the sum of the losses which are generated by each component of flux density.

When the skin effect should be taken into account, the eddy current loss W_{e} can be calculated from the eddy current density $\boldsymbol{J}_{\mathrm{e}}$ obtained by 3D electromagnetic field analysis with κ in (5), as follows [17]:

$$W_{\mathrm{e}} = \frac{1}{T} \kappa \int_{0}^{T} \int_{\mathrm{iron}} \frac{|J_{\mathrm{e}}|^2}{\sigma_{\mathrm{iron}}} \mathrm{d}v \mathrm{d}t. \tag{9}$$

As shown in Fig. 2, W_{e} can also be calculated by using 1D analysis along the thickness of electrical steel sheet as a post process of 2D analysis that neglects the eddy currents [9].

The eddy current and hysteresis loss W_{e}, W_{h} in (7)–(9) can be approximately decomposed into harmonic components by using Fourier transformation [9, 10]. This decomposition is very useful to understand the main factors that cause iron loss of the motors. Note that, strictly speaking, the hysteresis loss cannot be decomposed into harmonic components due to its nonlinear characteristics. As a result, the sum of the decomposed harmonic hysteresis losses is often not equal to the total hysteresis loss, particularly when the magnetic saturation in the cores is considerable.

In the hysteresis loss calculation of (8), a single hysteresis loss coefficient K_h is used. In this case, the major and minor loops are assumed as similar in shape. On the other hand, more accurate methods that consider the actual shape of minor hysteresis loops by introducing hysteresis models to electromagnetic field analysis have been proposed in recent years [18].

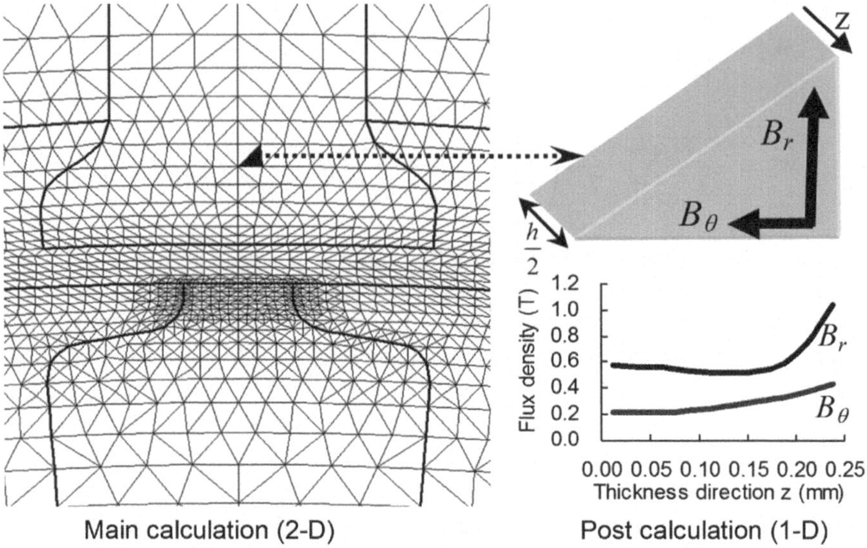

Fig. 2 Post 1D analysis for consideration of skin effect in electrical steel sheet. © [2012] IEEE. Reprinted, with permission, from [9]

4 Calculation Method of Motor Iron Loss Considering Mechanical Stress

The reluctivity and iron of the motor cores are affected by the mechanical stress caused by shrink-fitting and centrifugal force. To consider the effect of the mechanical stress on the iron loss, the combined stress-electromagnetic field analysis should be carried out with the results of core material experiments for the dependence of magnetic properties of the core on the mechanical stress.

Figure 3 shows the stress and magnetic field in rotating machines [7], where (σ_1, σ_2) are the maximum and minimum principal stresses, (B_x, B_y) are the x and y components of flux density, (h_1, h_2) and (B_1, B_2) are the principal axes components of the unit vector along the magnetic field and the flux density. Note that the directions

Fig. 3 Magnetic field and principal stress in motors. © [2014] IEEE. Reprinted, with permission, from [7]

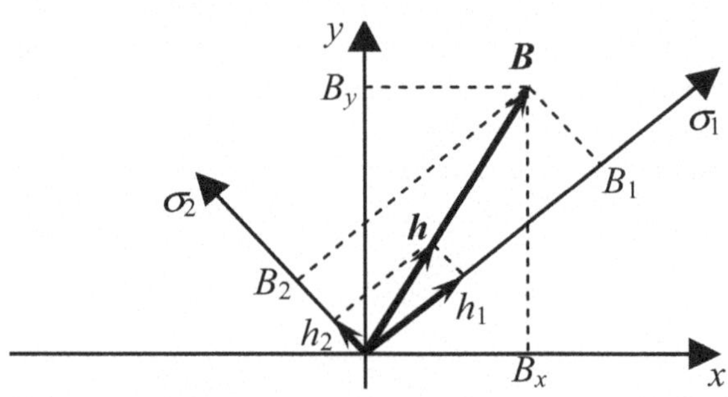

of (h_1, h_2) and (B_1, B_2) are identical to each other according to the neglect of the vector magnetic properties. In addition, 2D approximation is applied to the stress and electromagnetic field.

The reluctivity of the electrical steel sheet varies not only with the amplitude of the principal stress and flux density, but also the angle between them. Therefore, the governing equation of magnetic field in the core can be expressed, as follows:

$$\nabla \cdot \{\nu(\sigma_1, \sigma_2, B_x, B_y)\nabla A_z\} = 0 \tag{10}$$

where A_z is the axial component of magnetic vector potential. In this case, reluctivity ν is the function with 4 variables h_1, h_2, B_x, and B_y. (5 variables if the components of the stress tensor σ_x, σ_y, τ_{xy} are used instead of σ_1 and σ_2) In this case, vast measured results of core material experiments should be required to determine the function. It can be stated that the analysis based on (10) is not practical.

From this viewpoint, was often regarded as the function of von Mises stress, as follows [2, 3]:

$$\sigma_{VM} = \sqrt{\frac{1}{2}\{\sigma_1^2 + \sigma_2^2 + (\sigma_1 - \sigma_2)^2\}} \tag{11}$$

In this case, in (10) is approximated, as follows:

$$\nu(\sigma_1, \sigma_2, B_x, B_y) \cong \nu(\sigma_{VM}, |\mathbf{B}|) \tag{12}$$

σ_{VM} can be obtained by the stress analysis, whereas $|\boldsymbol{B}|$ can be obtained by the electromagnetic field analysis. Then, it is determined by the function obtained by the core material experiment. However, in this case, the effect of the angle between the principal stress and flux density vector is neglected. Furthermore, there is no evidence that the reluctivity of the electrical steel sheet is determined by σ_{VM}, which is used for estimating the material failure in the mechanical engineering.

To overcome these problems, an equivalent stress is proposed, as follows [19]:

$$\sigma_{eq} = \left(\sigma_1 - \frac{1}{2}\sigma_2\right)h_1^2 + \left(\sigma_2 - \frac{1}{2}\sigma_1\right)h_2^2 \tag{13}$$

The elastic energy when the single axial σ_{eq} is imposed along the magnetic field direction is equal to that by σ_1, σ_2, B_x, and B_y [19]. Therefore, if the magnetic properties under these conditions with the same elastic energy are assumed to be identical to each other, in (10) can be approximated, as follows:

$$\nu(\sigma_1, \sigma_2, B_x, B_y) = \nu(\sigma_{eq}, |\mathbf{B}|) \tag{14}$$

The electromagnetic field analysis that considers the angle between principal stress and the magnetic field can be carried out by using σ_{eq} and $|\boldsymbol{B}|$, which are estimated by the stress and electromagnetic field analysis [7, 8].

The iron loss can also be calculated by using σ_{eq}, as follows:

$$w_c = \sum_{k=1}^{2} \left\{ C_e\left(\sigma_{eq,k}, B_{max,k}\right) w_{e,k,\sigma 0} + C_h\left(\sigma_{eq,k}, B_{max,k}\right) w_{h,k,\sigma 0} \right\} \tag{15}$$

where w_c is the iron loss per volume, $w_{e,k,}$ σ_0 and $w_{h,k,}$ σ_0 are the eddy current and hysteresis losses when the stress is zero, respectively, C_e and C_h are the increase ratios of eddy current and hysteresis losses by the stress, respectively, $B_{max,k}$ is the amplitude of the flux density component along the principal axis B_k, and $\sigma_{eq,k}$ is the equivalent stress for B_k, as follows:

$$\sigma_{eq,1} = \left(\sigma_1 - \frac{1}{2}\sigma_2\right) \tag{16}$$

$$\sigma_{eq,2} = \left(\sigma_2 - \frac{1}{2}\sigma_1\right) \tag{17}$$

C_e and C_h in (15) can also be determined by the core material experiment, in which the single axial stress is imposed along direction of magnetic field. In addition, $w_{e,k}$, σ_0 and $w_{h,k}$, σ_0 can be obtained by the method described in the previous section.

5 Example of Iron Loss Calculation of Motors [8]

This section describes the loss calculation of a 100 kW class interior permanent magnet synchronous motor by considering the harmonics and mechanical stress [8].

Table 1 lists the specification of the motor. This motor is used for variable speed

Table 1 Motor specification © [2017] IEEE. Reprinted, with permission, from [8]

Phases and poles	3 phases, 8 poles
Maximum rotational speed	10,000 $\min^{-1}$
Maximum armature current	700 A
Continuous rated armature current	300 A
Diameters of stator/rotor	200 mm/130 mm
Core length	150 mm
Number of stator slot	48 (distributed)
Magnet type, Magnetization	Nd–Fe–B, 1.1 T
Thickness of electrical steel sheet	0.3 mm
Inverter	$f_c = 10$ kHz, $V_{dc} = 400$ V

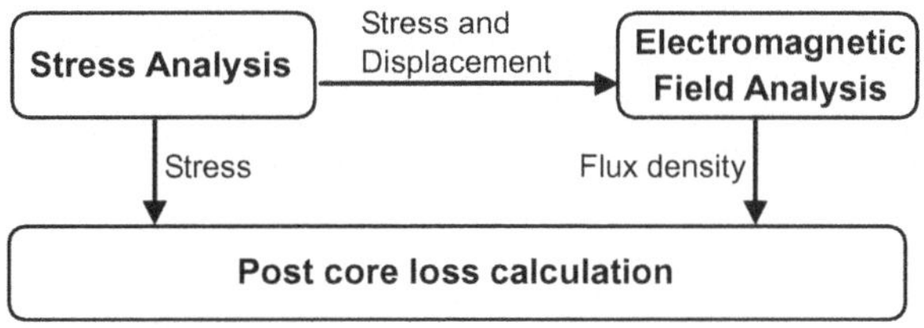

Fig. 4 Block diagram of combined stress-electromagnetic field analysis. © [2017] IEEE. Reprinted, with permission, from [8]

applications. The maximum rotational speed is 10,000 r/min. The stator is shrink-fitted by an aluminum housing.

Figure 4 shows the block diagram of the combined stress-electromagnetic field analyses. The 2D approximation is applied for both the analyses. In the stress analysis, the stator stress caused by the shrink-fitting and rotor stress caused by the centrifugal force are estimated. The reluctivity of the core used in the electromagnetic field analysis is determined by (14) due to the results of the stress analysis. In addition, the finite element mesh in the electromagnetic field analysis is deformed due to the displacement calculated by the stress analysis.

Figure 5 shows the distribution of principal stress vector. A fixed boundary condition that the stator surface is displaced to inside by 25 μm is imposed in the stress analysis according to the shrink-fitting. It is observed that the circumferential compressive stress is generated at the stator yoke by the shrink-fitting. It is also observed that large tensile stress is generated at the rotor surface and bridge between the inside magnets by the centrifugal force at the maximum speed, whereas the rotor stress becomes very small at low speeds. The figure indicates that the direction of the compressive stress at the stator yoke is nearly the same as that of the flux. In this case, the consideration of the stress effect by using von Mises stress σ_{VM} has no big problems. On the other hand, the direction of the tensile stress at the rotor surface

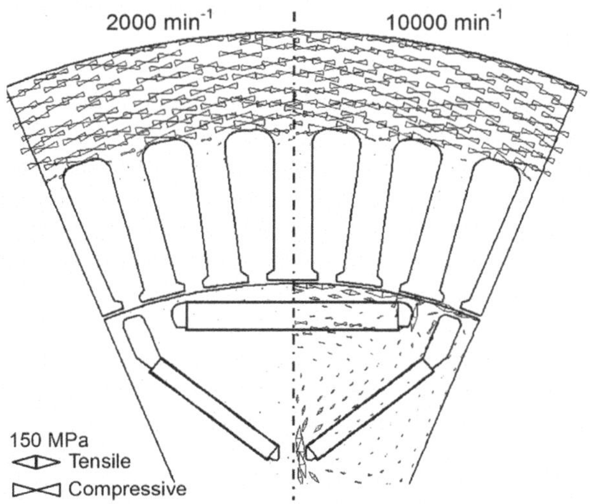

Fig. 5 Calculated principal stress vector distribution. © [2017] IEEE. Reprinted, with permission, from [8]

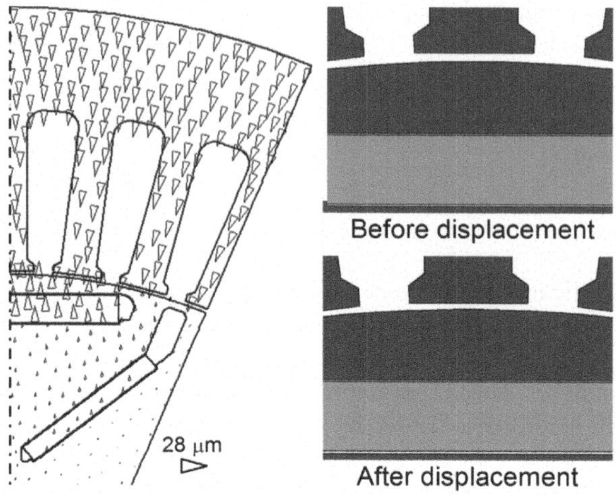

Fig. 6 Calculated displacement (10,000 r/min). © [2017] IEEE. Reprinted, with permission, from [8]

at the maximum speed is different from the direction of the flux. They are nearly in right angle. In this case, σ_{eq} in (13) implies that the effect of the tensile stress (plus sign) perpendicular to the magnetic field is equal to that of the 1/2 compressive stress (minus sign) along the magnetic field. Therefore, an increase in the harmonic iron loss is expected because the effect of the compressive stress on the magnetic properties of core material is considerably larger than that of the tensile stress.

Figure 6 shows the calculated displacement at the maximum speed. The stator is displaced to inside by the shrink-fitting, whereas the rotor is displaced to outside by the centrifugal force. As a consequence, the width of the air gap decreases. The stator displacement can be taken into account in the motor design in advance because it does not vary with the rotational speed. On the other hand, the displacement of the rotor varies with the rotational speed and it becomes nearly 5% of the air-gap width at the maximum speed. This decrease in the air gap also causes the increase in the harmonic iron losses.

Figure 7 shows the experimental and calculated iron losses of the motor. The calculated results that neglect the stress effects are also shown. It is observed that the accuracy of the iron loss calculation is improved by considering the mechanical stresses. From the comparison between the results (a) and (c), it is clarified that the rotor core loss, as well as the stator core loss increases at 10,000 r/min, whereas only the stator core loss increases at 2000 r/min. This rotor core loss increase must have been caused by the increase in the centrifugal force. The difference between the results (a) and (b) is caused by the decrease in the air-gap width. The rotor core loss increases by 6% only by this effect. On the other hand, the difference between the results (b) and (c) must have been caused by the variation in the magnetic properties by the tensile stress whose direction is different from the harmonic fluxes. The rotor core loss increases by 26% by this effect.

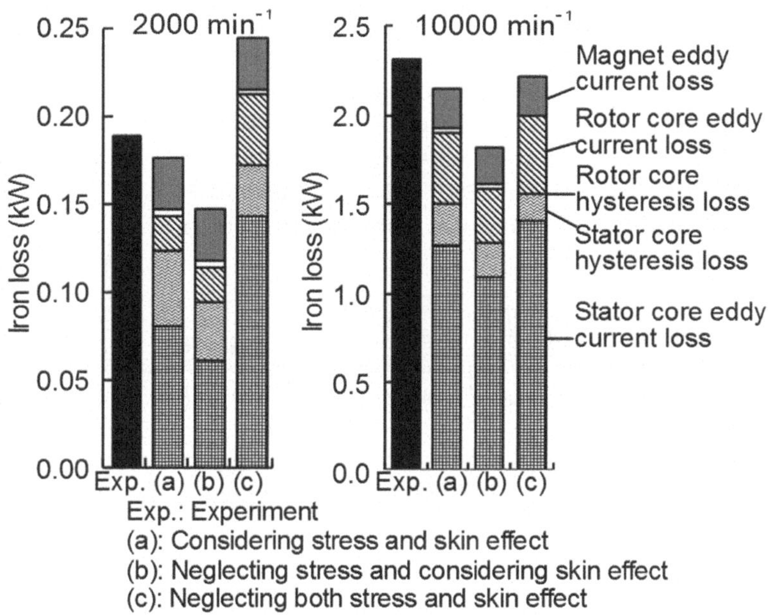

Fig. 7 Experimental and calculated iron losses. © [2017] IEEE. Reprinted, with permission, from [8]

Figure 8 shows the classified iron losses according to origins. It is clarified that the hysteresis loss mainly increases by the stress in the fundamental iron loss, whereas the eddy current loss increases by the stress in the harmonic components, in particular, the eddy current loss caused by inverter carrier.

To understand this reason, the variation in the permeability distribution and the flux density waveform at the stator yoke are investigated. Figure 9 shows the results. It is observed that the permeability of the yoke considerably decreases by the circumferential compressive stress. In this case, the skin effect in the electrical steel sheet decreases. On the other hand, when the stress is neglected, the harmonic ripples are concentrated at the surface of the electrical steel sheet due to the skin effect.

Figure 10 shows the distribution of the fundamental and carrier harmonic flux density along the thickness of the electrical steel sheet obtained by applying the Fourier transformation to the flux density waveform in Fig. 9. The figure indicates that the distribution of the fundamental flux density does not considerably vary with the stress. On the other hand, the distribution of the carrier harmonic flux density considerably varies with stress because of the decrease in the skin effect.

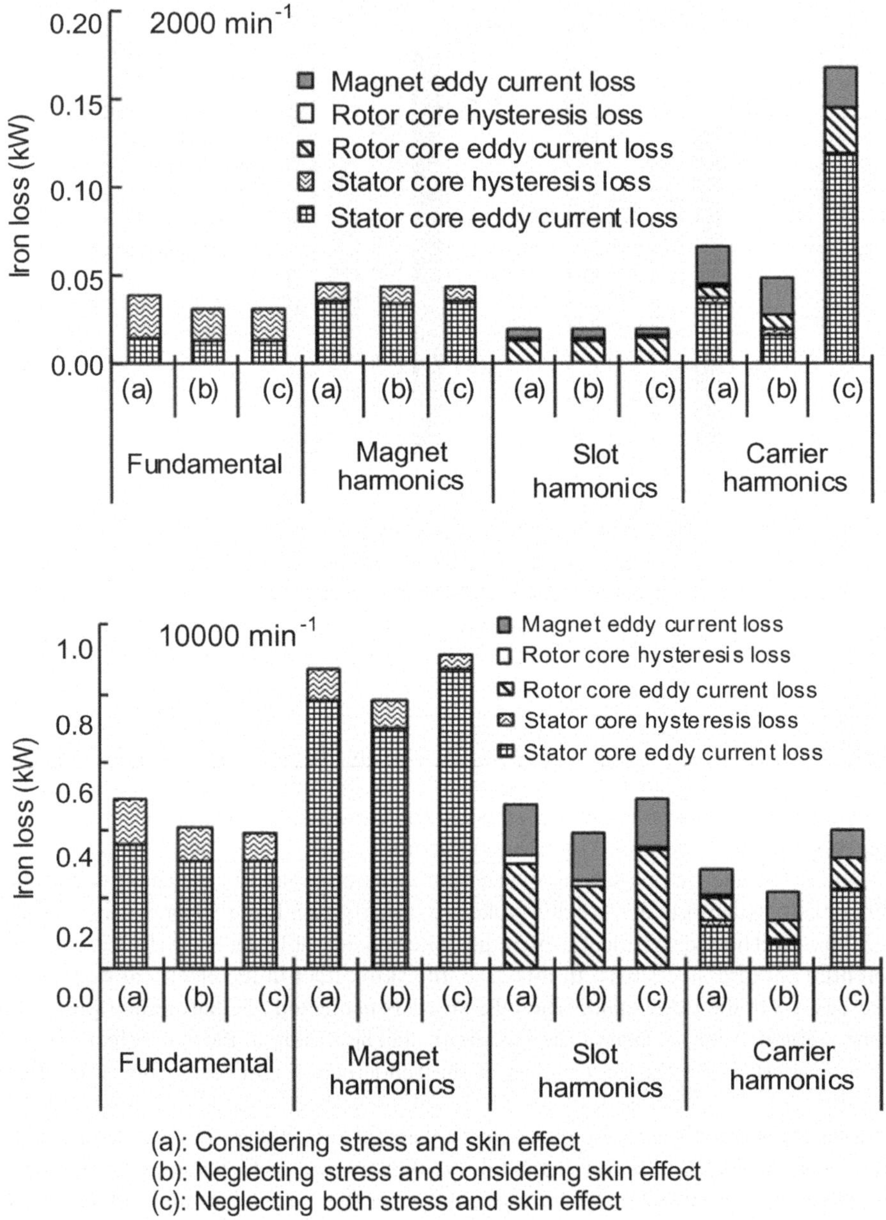

Fig. 8 Separated iron losses according to origins. © [2017] IEEE. Reprinted, with permission, from [8]

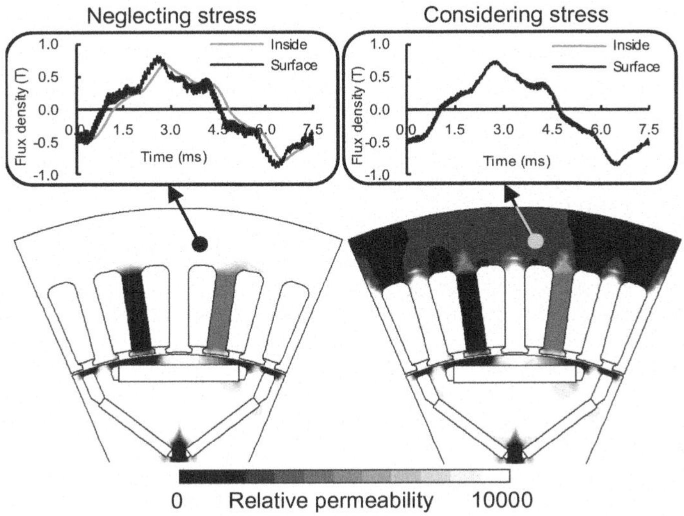

Fig. 9 Calculated permeability distribution and flux density waveform. © [2017] IEEE. Reprinted, with permission, from [8]

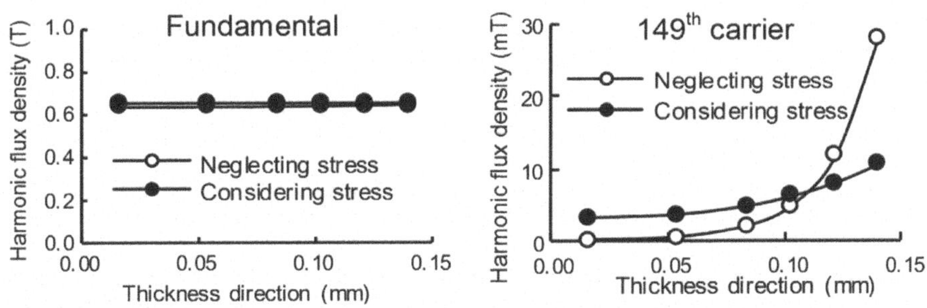

Fig. 10 Calculated harmonic flux density distributions along the thickness of electrical steel sheet. © [2017] IEEE. Reprinted, with permission, from [8]

References

1. Urata, S., Enokizono, M., Todaka, T., Shimoji, H.: "Magnetic characteristic analysis of the motor considering 2-D vector magnetic property",IEEE Trans. Magn. **42**(4), 615–618 (2006)
2. Daikoku, A., Nakano, M., Yamaguchi, S., Tani, Y., Toide, Y, Arita, H., Yoshioka, T., Fujino, C.: An accurate magnetic field analysis for estimating motor characteristics taking account of stress distribution in the magnetic core. IEEE Trans. Ind. Appl. **42**(3), 668–674 (2006)
3. Fujisaki, K., Hirayama, R., Kawachi, T., Satou, S., Kaido, C., Yabumto, M., Kubota, T.: Motor core iron loss analysis evaluating shrink fitting and stamping by finite-element method. IEEE Trans. Magn. **43**(5), 1950–2007 (2007)
4. Takahashi, N., Morimoto, H., Yunoki, Y., Miyagi, D.: Effect of shrink fitting and cutting on iron loss of permanent magnet motor. J. Magnetizm Magnetic Mater. **320**(20), e925–e928 (2008)

5. Miyagi, D., Maeda, N., Ozeki, Y., Miki, K., Takahashi, N.: Estimation of iron loss in motor core with shrink fitting using FEM analysis. IEEE Trans. Magn. **45**(3), 1704–1707 (2009)
6. Nakano, M., Fujino, C., Tani, Y., Daikoku, A., Toide, Y., Yamaguchi, S., Arita, H., Yoshioka, T.: High precision calculation of iron loss by considering stress distribution of magnetic core. IEEJ Trans. IA **11**(129), 1060–1067 (2009)
7. Yamazaki, K., Kato, Y.: Iron loss analysis of interior permanent magnet synchronous motors by considering mechanical stress and deformation of stators and rotors. IEEE Trans. Magn. **50**(2), 7022504 (2014)
8. Yamazaki, K., Takeuchi, H.: Impact of mechanical stress on characteristics of interior permanent magnet synchronous motors. IEEE Trans. Ind. Appl. **53**(2), 963–970 (2017)
9. Yamazaki, K., Suzuki, A., Ohto, M., Takakura, T.: Harmonic loss and torque analysis of high speed induction motors. IEEE Trans. Ind. Appl. **48**(3), 933–941 (2012)
10. Yamazaki, K., Seto, Y.: Iron loss analysis of interior permanent magnet synchronous motors—variation of main loss factors due to driving condition. IEEE Trans. Ind. Appl. **42**(4), 1045–1052 (2006)
11. Bertotti, G.: General properties of power losses in soft ferromagnetic materials. IEEE Trans. Magn. **24**(1), 621–630 (1988)
12. Atallah, K., Zhu, Z.Q., Howe, D.: An improved method for predicting iron losses in brushless permanent magnet motor. IEEE Trans. Magn. **28**(5), 2997–2999 (1992)
13. McClay, C.I., Williamson, S.: The variation of cage motor losses with skew. IEEE Trans. Ind. Appl. **36**(6), 1563–1570 (2000)
14. Cassat, A., Espanet, C., Wavre, N.: BLDC motor stator and rotor iron losses and thermal behavior based on lumped schemes and 3-D FEM analysis. IEEE Trans. Ind. Appl. **39**(5), 1314–1321 (2003)
15. Mthombeni, T.T., Pillay, P.: Lamination core losses in motors with nonsinusoidal excitation with particular reference to PWM and SRM excitation. IEEE Trans. Energy Convers. **20**(4), 836–843 (2005)
16. Yamazaki, K.: Torque and efficiency calculation of an interior permanent magnet motor considering harmonic iron losses of both the stator and rotor. IEEE Trans. Magn. **39**(3), 1460–1463 (2003)
17. Yamazaki, K., Fukushima, N.: Iron loss modeling for rotating machines: Comparison between Bertotti's three term expression and 3-D finite element method. IEEE Trans. Magn. **46**(8), 3121–3124 (2010)
18. Takeda, Yoshimi, Takahashi, Yasuhito, Fujiwara, Koji, Ahagon, Akira, Matsuo, Tetsuji: Iron loss estimation method for rotating machines taking account of hysteretic property. IEEE Trans. Magn. **51**(3), 7300504 (2015)
19. Daniel, L., Hubert, O.: An equivalent stress for the influence of multiaxial stress on the magnetic behavior. J. Appl. Phys. **105**, 07A313 (2009)

Iron Loss of the Inductors

Toshihisa Shimizu

Abstract Since inductors used in power electronics circuits are magnetized under completely different conditions from sinusoidal alternating voltage/current, the manner of occurrence of iron loss is often greatly different. In this chapter, we clarify the difference between the magnetizing modes of the transformer and inductor used in the power electronics circuit and introduce recent research examples on the method of measuring and calculating the iron loss of the inductor.

Keywords Iron loss · Inductor · Transformer · Pre-magnetization · Open-loop hysteresis curve

1 Magnetization of the Inductor in PWM Inverter

First, the functional differences between a transformer and an inductor in a power electronics circuit will be explained by taking DC–DC converter and PWM inverter as examples. Figures 1 and 2 are the basic circuit configurations of the DC–DC converter and the single-phase PWM inverter, respectively. In both figures, the primary winding of the transformer is connected to the AC output terminal of the inverter. In the case of a DC–DC converter, a rectifier circuit is connected to the secondary winding of the transformer, and its DC output terminal is connected to a load circuit via an inductor. On the other hand, in the case of a PWM inverter, the inductor is connected immediately after the secondary winding of the transformer, and the load circuit is connected.

Next, consider a transformer. The equivalent circuit of the transformer is described by the magnetizing inductance L_g, the leakage inductance L_{1l}, L_{2l}, the winding resistance R_w, and the ideal transformer as shown in Fig. 3. Figure 4 shows an example of the operation waveform of each part of the transformer in the DC–DC converter.

A rectangular wave inverter output voltage v_{inv} alternating positive and negative is applied to the primary winding of the transformer. The primary winding current

T. Shimizu (✉)
Tokyo Metropolitan University, Hachiōji, Japan
e-mail: shimizut@tmu.ac.jp

K. Fujisaki (ed.), *Magnetic Material for Motor Drive Systems*,
Engineering Materials, https://doi.org/10.1007/978-981-32-9906-1_25

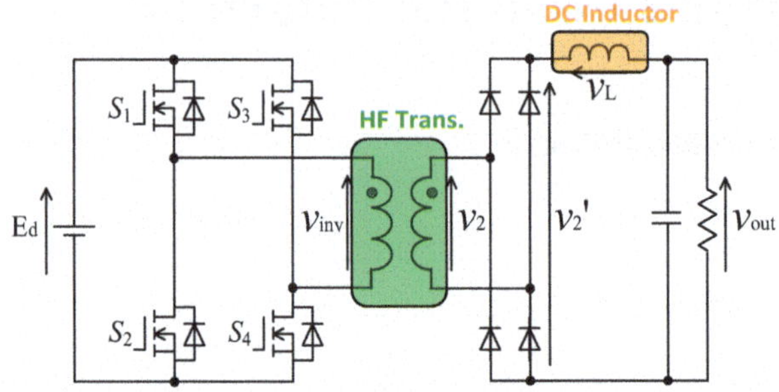

Fig. 1 Circuit configuration of the DC–DC converter

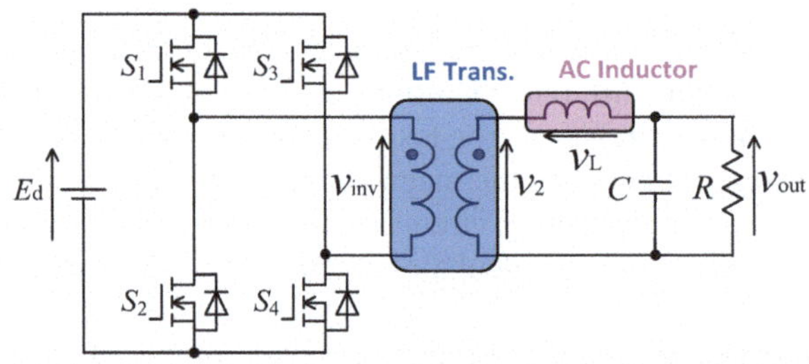

Fig. 2 Circuit configuration of the single-phase PWM inverter

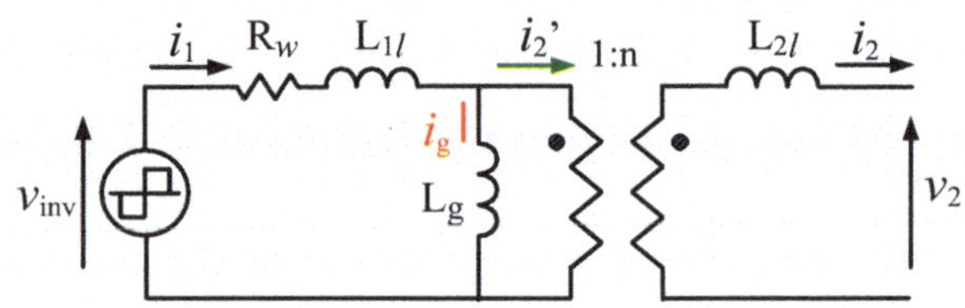

Fig. 3 Equivalent circuit of the transformer

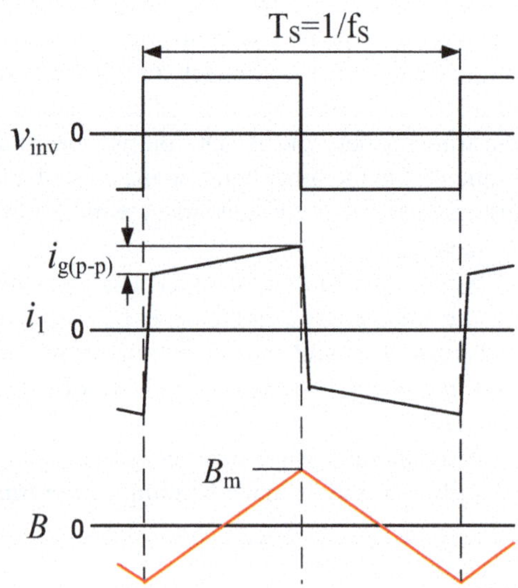

Fig. 4 Operation waveforms of the DC–DC converter

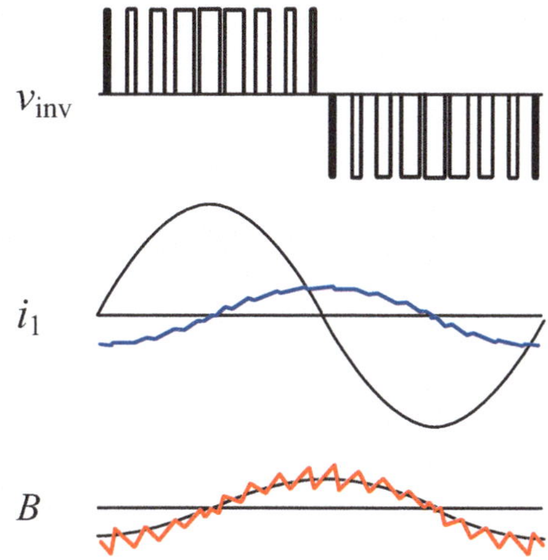

Fig. 5 Operation waveforms of the single-phase PWM inverter

i_1 of the transformer is the sum of the current i_g flowing through the magnetizing inductance L_g of the transformer and the load current i_2'. Since the magnetic material of the transformer is magnetized by the rectangular wave AC voltage, the magnetic flux density waveform has a triangular waveform. Figure 5 shows the operating waveform of each part of the transformer in the PWM inverter.

The inverter output voltage v_{inv} is applied to the primary winding of the transformer. The voltage waveform of the inverter output voltage v_{inv} follows a positive rectangular pulse voltage sequence in which the pulse width gradually changes in accordance with the AC output voltage of the inverter, it becomes a negative pulse voltage sequence. The current i_g flowing through the magnetizing inductance L_g and the load current i_2' of the inverter flow through the primary winding of the transformer.

Since the magnetizing current i_g is the integral value of the inverter output voltage, the low-frequency component gradually changing at the cycle of the load voltage v_{out} of the inverter AC output and the high-frequency current component of the approximately triangular wave shape corresponding to the switching cycle of the inverter it becomes sum. In addition, the magnetic flux density waveform also has a waveform in which a high-frequency component is superimposed on a low-frequency component. In principle, the magnetizing waveform of the transformer is determined by the inverter output voltage applied to the transformer and is not affected by the load current.

Furthermore, consider an inductor. The equivalent circuit of the transformer is described by magnetizing inductance L_g, leakage inductance L_{l1}, and winding resistance R_w as shown in Fig. 6. Figure 7 shows an example of the operating waveform of each part of the inductor in the DC–DC converter. The voltage v_L applied to the inductor is the difference voltage between the voltage v_2' obtained by rectifying the inverter output voltage v_{inv} and the load voltage v_{out}.

In addition, because the load current flows through the inductor, the inductor current i_1 is equal to the magnetizing current i_g of the inductor. Here, assuming that

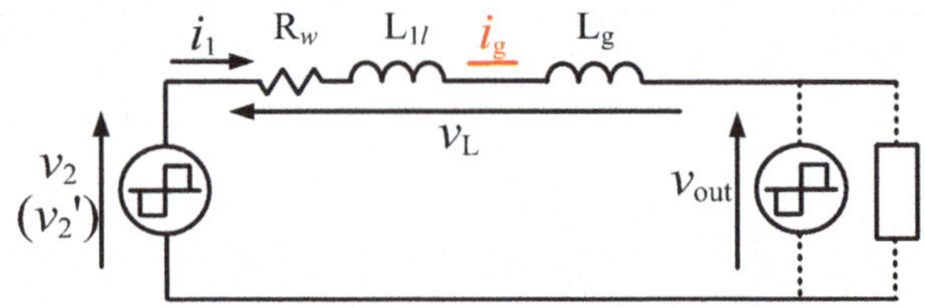

Fig. 6 Equivalent circuit of the inductor

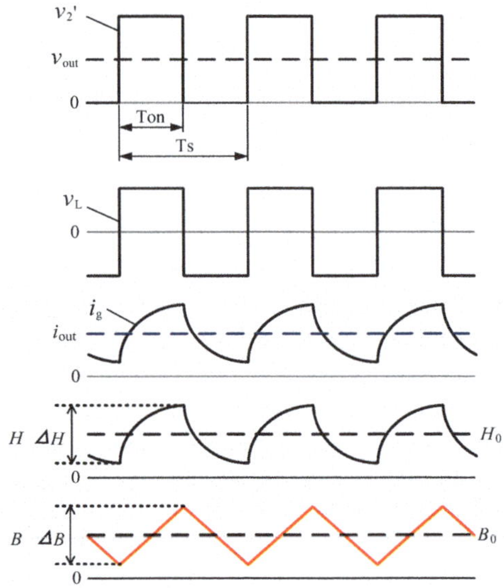

Fig. 7 Operation waveforms of the DC–DC converter

the load voltage is a smooth direct current, the magnetic flux density waveform has a triangular waveform having a frequency corresponding to the switching frequency as an integral value of the inductor voltage v_L and a bias magnetic flux density corresponding to the load current value. A waveform is added. Figure 8 shows the operation waveform of each part of the transformer in the PWM inverter. The voltage v_L applied to the inductor is the difference voltage between the inverter output voltage v_{inv} and the load voltage v_{out}. The fact that the inductor current i_l is equal to the magnetizing current i_g of the inductor is the same as in the case of the DC–DC converter, but the load current varies from moment to moment in accordance with the inverter output voltage and the like. The magnetic flux density waveform is also the sum of the triangular waveform having a frequency corresponding to the switching frequency and the AC waveform corresponding to the time change of the load voltage.

As described above, the magnetizing state of a transformer and an inductor used in an inverter, etc., are largely different. In particular, in the case of an inductor, it is necessary to pay attention that the magnetic field bias component which varies depending on the load condition has a large influence on the iron loss.

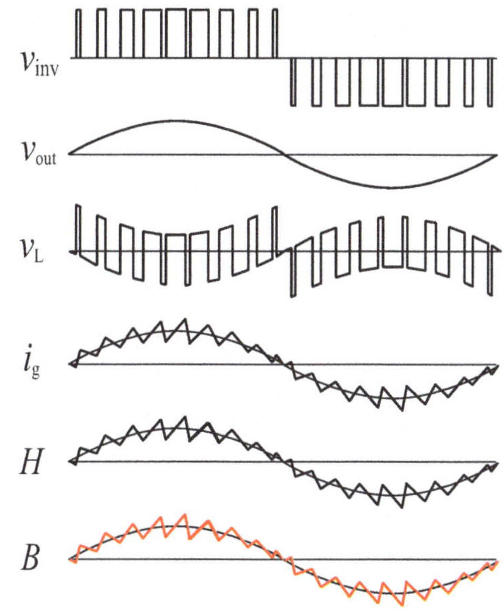

Fig. 8 Operation waveforms of the single-phase PWM inverter

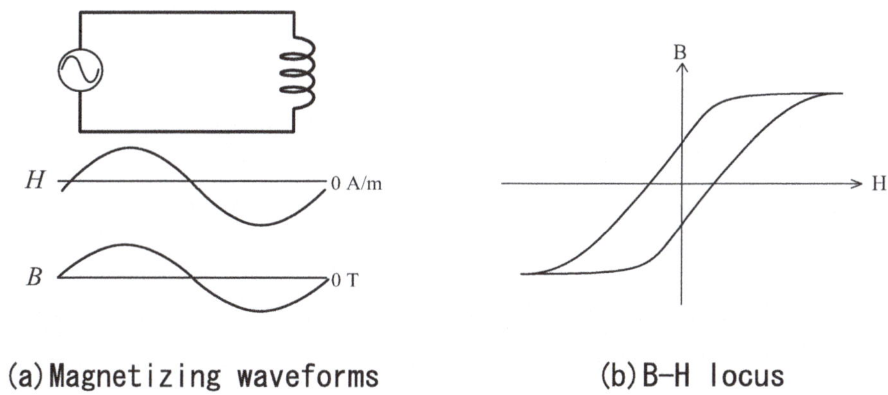

Fig. 9 Magnetizing waveforms and BH locus on a sinusoidal voltage magnetization

2 Measurement Method of Iron Loss Under Direct Magnetic Field Bias Condition

The expression of Steinmetz described by formula (1) has been used for a long time in notation of iron loss magnetized with a sinusoidal voltage as shown in Fig. 9 [1].

$$P = K_i f^{\alpha} B^{\beta} \tag{1}$$

On the other hand, in the transformer used for the inverter device, taking into consideration that it is magnetized by a rectangular wave voltage as shown in Fig. 10, improved generalized Steinmetz equation was proposed [2].

$$P = \frac{1}{T}\int_{0}^{T} K_i \left|\frac{\mathrm{d}B}{\mathrm{d}t}\right|^{\alpha} |\Delta B(t)|^{\beta-\alpha}\mathrm{d}t \tag{2}$$

Equation (2) is based on the assumption that a hysteresis curve is drawn around the origin of the BH plane, the iron loss of the transformer connected to the output of the DC–DC converter can be expressed relatively accurately. However, as in the inductor of the step-down chopper circuit shown in Fig. 11, a hysteresis curve (referred to as "Minor Hysteresis Loop") is drawn in a place other than the origin by the magnetic

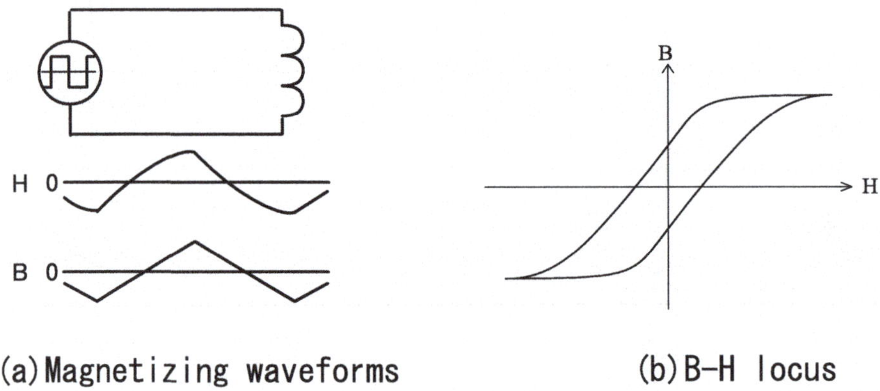

Fig. 10 Magnetizing waveforms and BH locus on a rectangular voltage magnetization

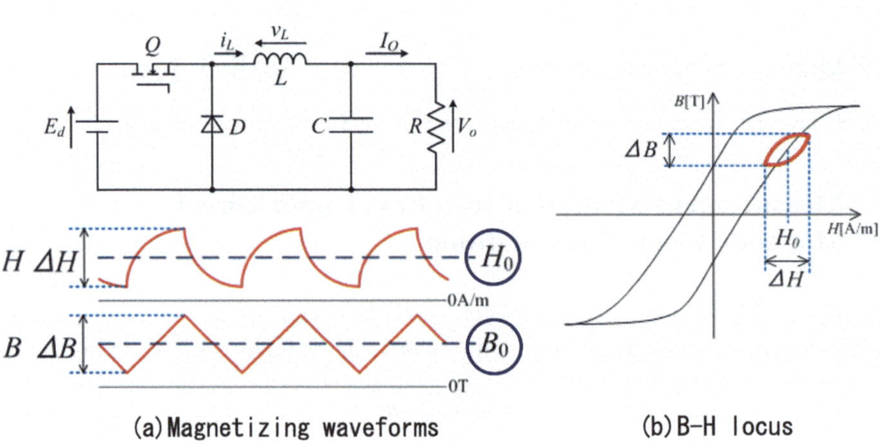

Fig. 11 Inductor magnetizing waveforms and BH locus on a buck-shopper

field bias and its position changes in the inverter operation state and the load state, it was difficult to express the iron loss in Eq. (2).

Figure 12 shows an example of a method for measuring iron loss when the magnetic field bias changes variously. In this method, the BH analyzer and the actual buck chopper circuit are used, and the buck chopper circuit is operated with the synchronization signal from the BH analyzer. The BH analyzer detects the magnetic flux density B from the integrated value of the secondary winding electromotive force of the inductor of the buck chopper and the magnetic field strength H from the primary winding current of the inductor and measures the iron loss value with the BH analyzer. The merit of this method is to be able to measure the iron loss of the inductor that operates with an actual buck chopper circuit and measures the inductor iron loss under a wide range of voltage, current, and frequency conditions by appropriately setting the specification of the buck chopper. Figure 13 shows a measurement

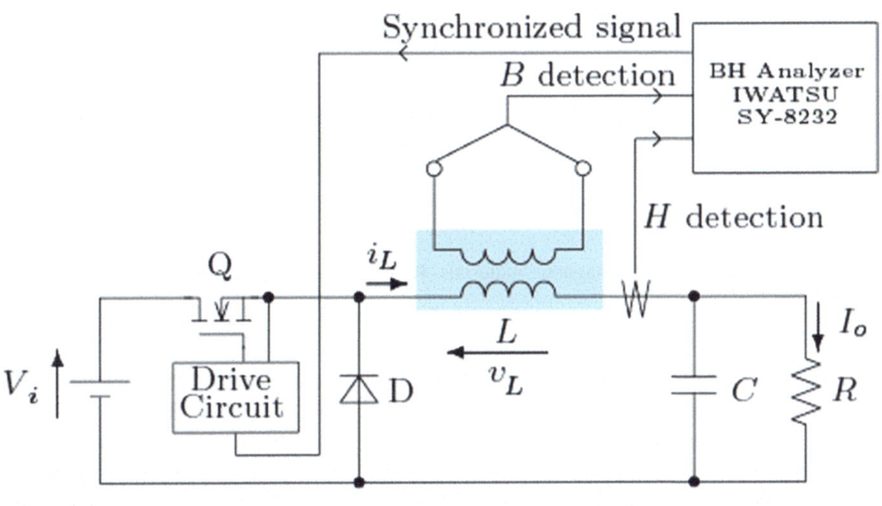

Fig. 12 Circuit configuration of the iron-loss-measuring instrument

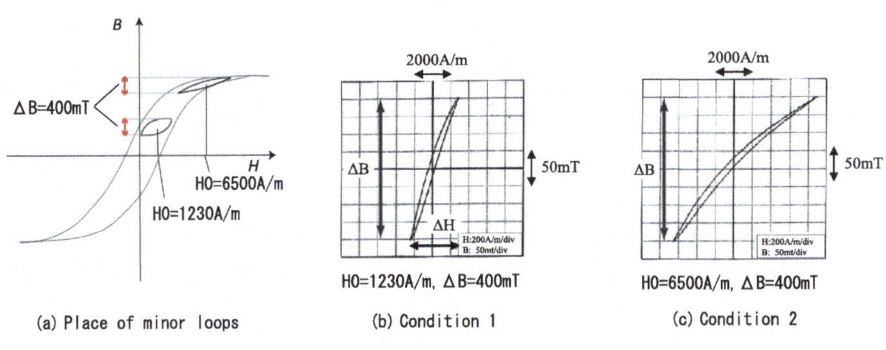

Fig. 13 Magnetic field bias and minor hysteresis loop

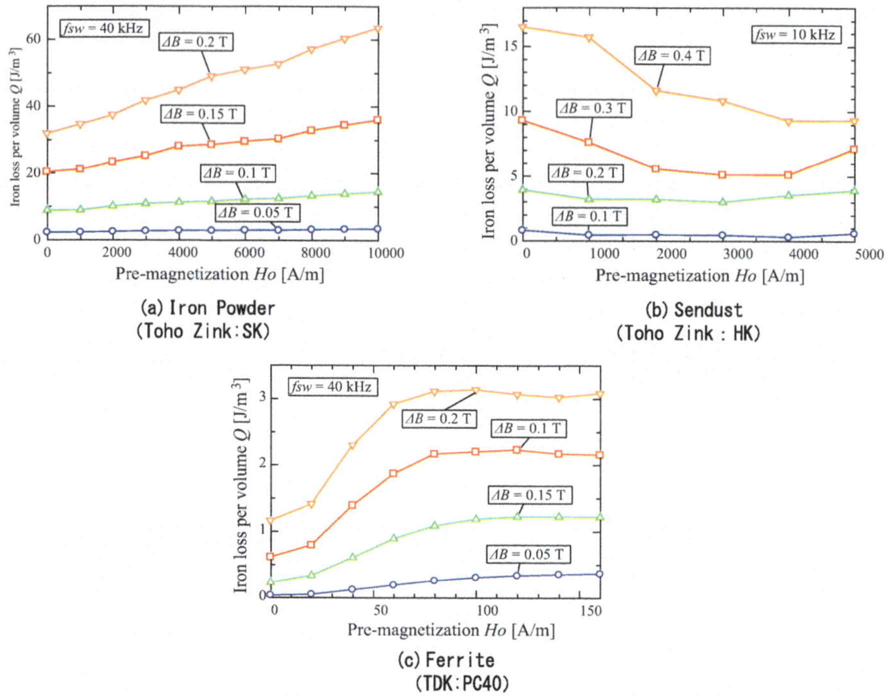

Fig. 14 Loss map on various magnetic materials. Reprinted from [6], Copyright 2017, with permission from IEEJ

example of a minor hysteresis loop of a toroidal core using iron powder material. When the magnetic field bias H_0 is changed from 1230 A/ m to 6500 A/m under the same flux density change amount ($\Delta B = 400$ mT), the shape of the hysteresis loop greatly changes and the iron loss value increases from 115 to 145 J/m^2. Since the manner of change in iron loss varies in various ways, it is difficult to write in a single mathematical expression. Figure 14 shows a graph in which the loss is described as a database (here, it is called a "loss map") [3]. As is apparent from these data, it can be seen that the iron loss value increases and decreases with increase of the magnetic field bias H_0 and that the amount of change greatly differs depending on the type of the magnetic material. It is found that designing based on these loss characteristics is important for inductors used in inverters and the like.

3 Iron Loss Calculation Method and Evaluation Method of Inductor

In the case of the inductor of the DC chopper circuit, the iron loss value of the inductor can be calculated directly from the loss map, as the minor hysteresis loop can be regarded as an almost constant closed loop shape under certain operating conditions as shown in Fig. 11.

On the other hand, in the case of the filter inductor of the PWM inverter shown in Fig. 2, its appearance varies greatly, as shown in Fig. 15. Figure 15a shows the BH curve for one period of AC output, but in addition to the large (low frequency) hysteresis curve centered on the origin of the BH plane, the hysteresis curve which changes according to the switching of the inverter is appeared. Figure 15b is an enlarged view of the hysteresis curve for three cycles of switching, but it can be seen that the hysteresis curve associated with switching does not become a closed loop.

A method for calculating iron loss for each switching cycle from such an open-loop hysteresis curve will be described with reference to Fig. 16 [3]. Figure 16 is a

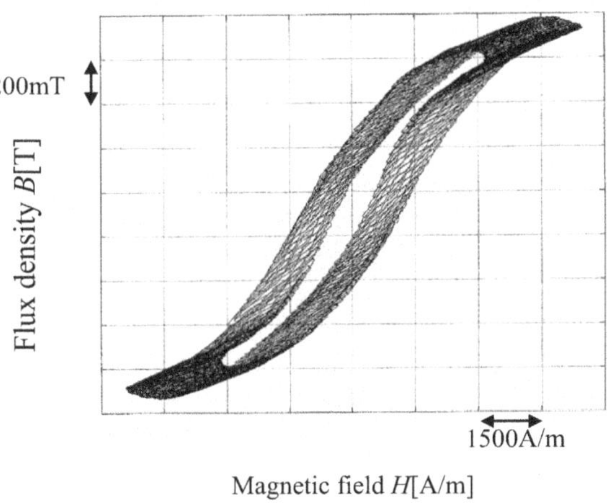

(a) B-H locus during one output cycle of inverter output voltage

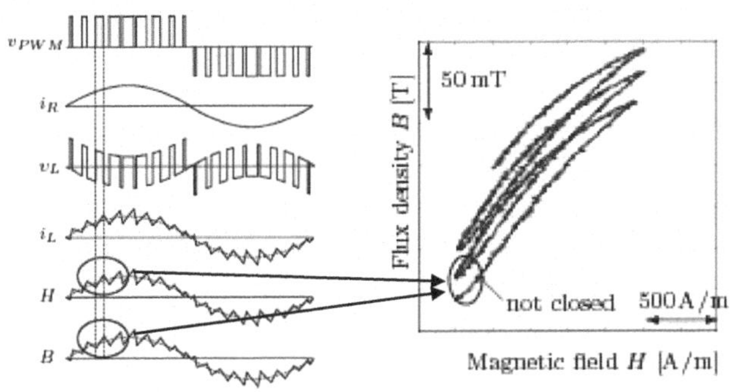

(b) B-H locus in each switching cycle

Fig. 15 BH locus of the filter inductor in a single-phase PWM inverter

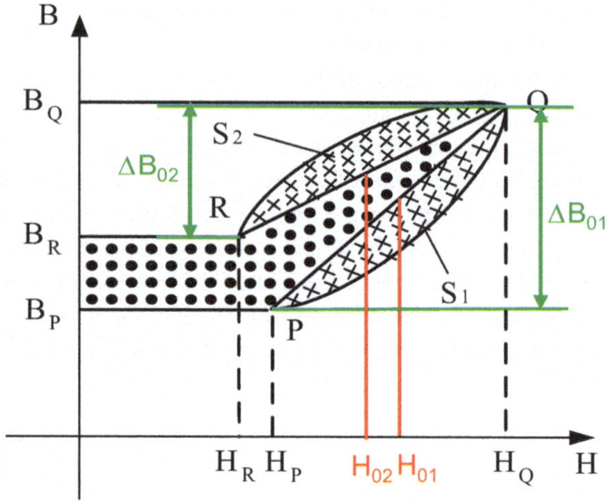

Fig. 16 Iron loss on an open-loop shape hysteresis locus. Reprinted from [7], Copyright 2013, with permission from IEEJ

hysteresis curve corresponding to one switching period of the inverter, and the magnetization state at the start of switching is assumed to be at point P. Magnetization in the positive direction is performed from this point and the magnetization state moves to point Q. When it reaches point Q, magnetization in the negative direction is performed and the magnetization state moves to R point. With the delay in magnetization corresponding to the change in magnetization between the two points, the hysteresis curve has two arc shapes. Next, consider the change in magnetic energy accumulated in the magnetic material of the inductor. By positive magnetization, magnetic energy corresponding to the area surrounded by straight $\overline{B_P P}$, arc $\widehat{PQ}$, and straight $\overline{QB_Q}$ rays is accumulated in the magnetic material. Due to the negative magnetization, magnetic energy corresponding to the area surrounded by straight $\overline{B_Q Q}$, arc $\widehat{QR}$, and straight $\overline{RB_R}$ rays is emitted from the magnetic material. Therefore, the part marked with ● and the part marked with x in the figure can be regarded as the magnetic energy remaining in this switching section. The magnetic energy of the part marked with ● corresponds to the change in the low-frequency output current of the inverter, so by taking the sum of the values of the area (positive or negative) over one period of the low-frequency output, it is thought that iron loss accompanying with low-frequency magnetization. Next, if the iron loss does not occur when the magnetization states of P to Q and Q to R change, the magnetization curve is considered to fluctuate linearly.

That is, the magnetic energy S_{pwm} of the x portion generated by drawing a circular arc can be regarded as the magnetic energy loss of the magnetic material caused by switching. This can be thought of as a graphical interpretation of the improved generalized Steinmetz equation described by Eq. (2). However, since the shape of this circular arc varies depending on the magnitude of the magnetic field bias H_0 at the center of the circular arc, Eq. (2) which does not take this into account can not be applied as it is. Now, considering that the area of the circular arc of the × portion corresponds to 1/2 of the minor loop area shown in Fig. 13, the value of 1/2 of the iron loss value at the corresponding H_0 and ΔB on the loss map data in Fig. 14, and the iron loss of the circular arc portion is required. In addition, since the magnetic field bias H_0 and the magnetic flux density ripple ΔB in each switching period can be calculated comparatively easily by using an electric circuit simulator or the like as shown in Fig. 17, if these are coupled, the inductor iron loss per switching period becomes can be calculated.

Next, consider the means to evaluate the validity of the iron loss calculation value by the loss map. Figure 18 shows the circuit configuration and appearance of the inductor loss analyzer (ILA). The inverter main circuit is a three-phase full-bridge inverter, and a filter inductor, a filter capacitor, and a load resistance are connected to the AC output. The operation signal of the inverter is supplied from the PWM signal generator based on the synchronization signal from the high-speed memory. The secondary voltage and the primary current of the inductor are detected by the

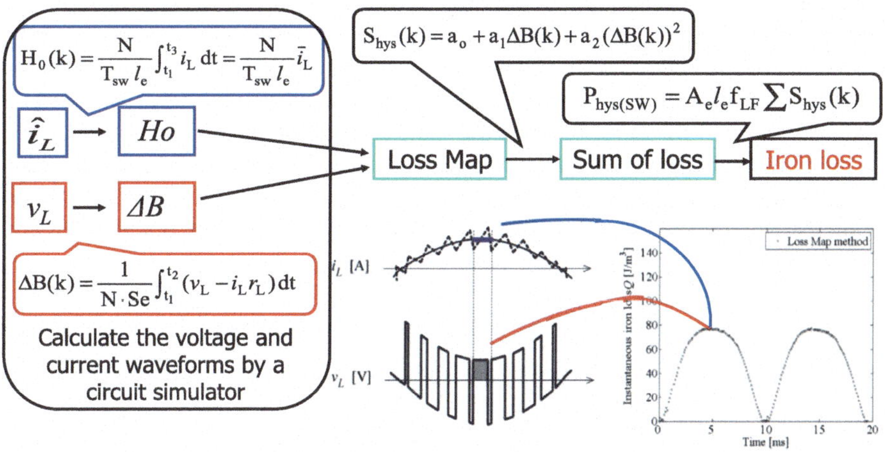

Fig. 17 Calculation procedure of the iron loss in each switching cycle

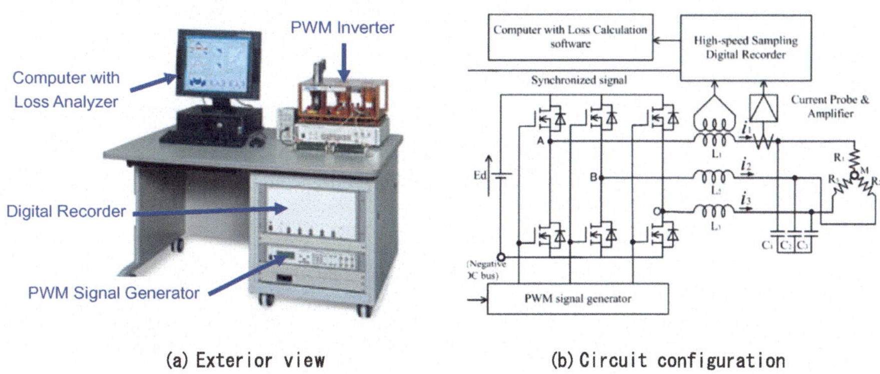

Fig. 18 Iron loss analyzer (ILA)

high-speed memory, and the iron loss value of the inductor per switching cycle of the inverter is calculated by the personal computer. Figure 19 shows the evaluation results of the iron loss of the filter inductor of the single-phase PWM inverter. Figure 19b shows the measured value of the magnetic flux density waveform (upper) and the iron loss (J/cm^3) (referred to as the instantaneous iron loss in this paper) for each switching period of the inverter and the calculated value by the loss map method). It is understood that the instantaneous iron loss varies with the time change of the magnetic flux density. Table 1 shows the average loss (W) of the inductor iron loss during one cycle of AC output of the inverter. Since the measured value and calculated value of instantaneous iron loss and average iron loss are in good agreement, the validity of the instantaneous iron loss calculation method by the loss map method is recognized. Figure 20 shows the measurement results of the iron loss of the filter inductor of the three-phase PWM inverter. As is clear from Fig. 20a, in the section a, the measured value and the calculated value are in good agreement, but in the section b, they do not match. Looking at the magnetic flux density waveform in Fig. 20b, the triangular waveform is continuous in section a, but in section b, the flux density change rate dB/dt when the magnetic flux density decreases is changing midway. In the conventional iron loss calculation method, as shown in Eq. (2), since the loss for each section of dB/dt is obtained, the same calculation method is adopted in the loss map method calculation, but considering this as the cause of calculation error. The iron loss of the magnetic material can be regarded as the difference between the increase and decrease of the magnetic flux density and the accumulation amount

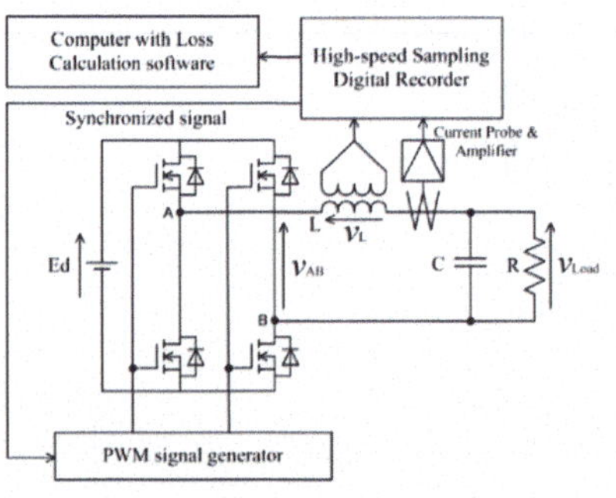

(a) Experimental circuit of single-phase PWM inverter

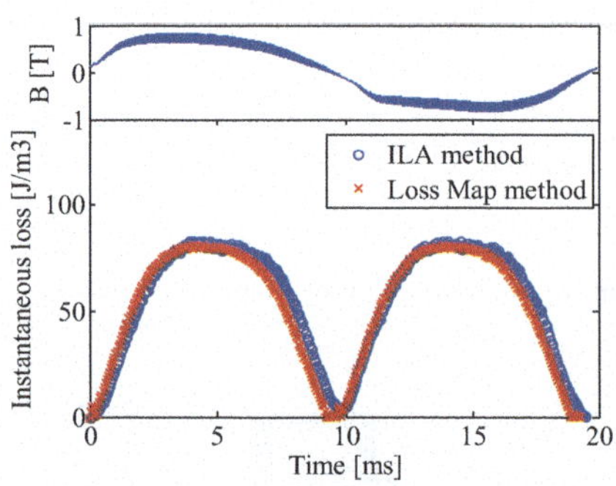

(b) Calculated and measured results of the instantaneous iron loss

Fig. 19 Evaluation of the iron loss on a single-phase PWM. Reprinted from [7], Copyright 2013, with permission from IEEJ

Table 1 Deviation of the measured and the calculated iron loss

Condition	Loss (w)	Error (%)
ILA method	2.41	–
Loss map method	2.32	3.7

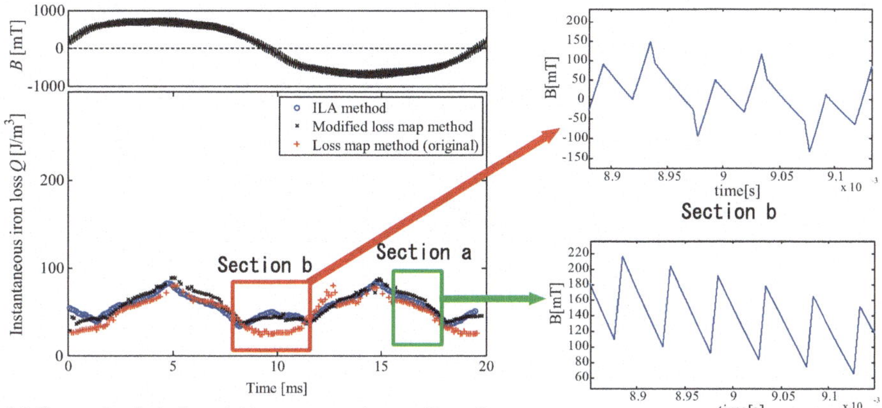

(a) The calculated and the measured results of the instantaneous iron loss

(b) Flux density waveforms

Fig. 20 Evaluated results of the inductor iron loss

and emission amount of magnetic energy at the time of descent. In Brockmeyer et al., when the rise rate (or descending rate) of the magnetic flux density changes in the middle and is represented by the magnetic flux density waveforms of n straight lines, this is collectively taken as a loss associated with a series of ascending (or descending) operations. We developed a method to calculate the iron loss using the equivalent frequency $f_{\text{eq(n-seg)}}$ given by the Eq. (3) [4].

$$f_{\text{eq(n-seg)}} = \sum_{k=2}^{K} \frac{(B_k - B_{k-1})^2}{(B\text{max} - B_{\text{min}})^2} \cdot \frac{1}{t_k - t_{k-1}} \tag{3}$$

When applying this, as shown in Fig. 21, the iron loss accompanying the series of decreases in the magnetic flux density from the point c to the point f is expressed by the magnetic flux density change amount ΔB_{cf} and the equivalent frequency $f_{\text{eq(cf)}}$ expressed by Eq. (4). Iron loss can be calculated from the loss map by using.

$$f_{\text{eq(cf)}} = \frac{B_{\text{cd}}^2 f_{\text{cd}} + B_{\text{de}}^2 f_{\text{de}} + B_{\text{ef}}^2 f_{\text{ef}}}{(B_{\text{cd}} + B_{\text{de}} + B_{\text{ef}})^2} \tag{4}$$

where, $f_{\text{cd}} = 1/t_{\text{cd}}$, $f_{\text{de}} = 1/t_{\text{de}}$, $f_{\text{ef}} = 1/t_{\text{ef}}$.

This method is called an improved loss map method [5]. The calculation result of the iron loss using the improved loss map method (Fig. 20a) is in good agreement with the measurement result over the entire range, and the validity of the calculation method can be confirmed. In addition, Fig. 22 shows a comparison between the iron-loss-measured value and the calculated value of the iron loss when the modulation factor of the PWM inverter is changed. The calculated value by the improved loss map method is in good agreement with the measured value, but in the conventional

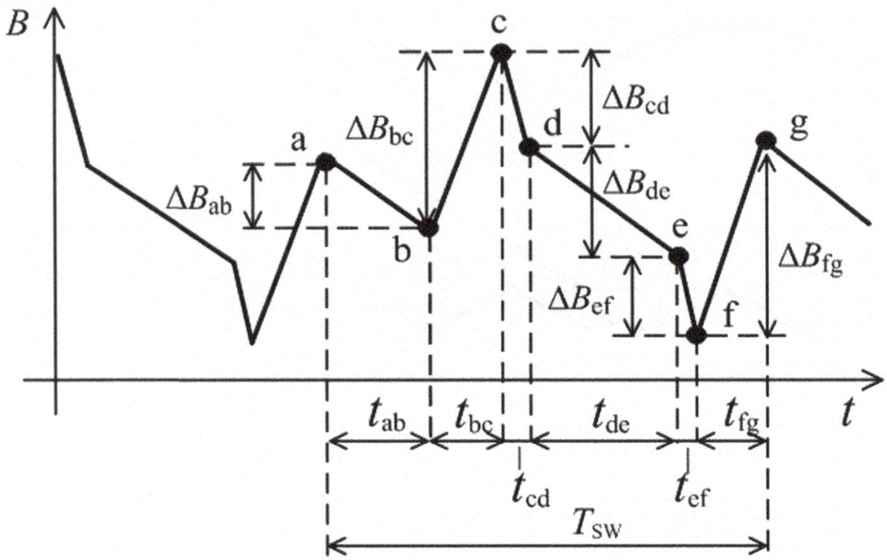

Fig. 21 Flux density waveform during section. Reprinted from [7], Copyright 2013, with permission from IEEJ

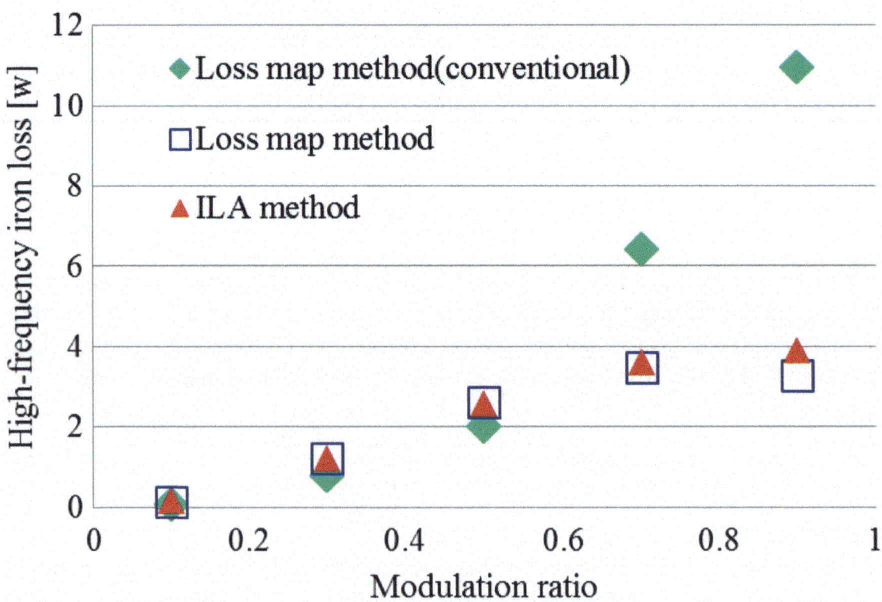

Fig. 22 Measured and calculated iron loss

iron loss calculation method, the calculation error increases as the modulation rate becomes higher.

As described above, in the filter inductor of the PWM inverter or the like, the magnetizing voltage of the PWM inverter is mainly the rectangular wave voltage, and also a magnetic field bias is induced. Hence, the shape of the magnetic flux density waveform must be considered to calculate the iron loss of the inductor. It is to be noted that the magnetic field dependence of the iron loss greatly varies depending on the magnetic material.

References

1. Steinmetz, C.P.: On the law of hysteresis. Proc. IEEE **72**(2), 197–221 (1984)
2. Li, J., Abdallah, T., Sullivan, C.R.: Improved calculation of core loss with nonsinusoidal waveforms. In: 36th IEEE IAS Annual Meeting, vol. 4, pp. 2203–2210 (2001)
3. Shimizu, T., Iyasu, S.: A practical iron loss calculation for AC filter inductors used in a PWM inverters. IEEE Trans. Ind. Electron. **56**(7), 2600–2609 (2009)
4. Albach, M., Durbaum, Th., Brockmeyer, A.: Calculating core losses in transformers for arbitrary magnetizing currents—a comparison of different approaches. In: IEEE Proceedings of 27th Applied Power Electronics Conference, vol. 2, pp. 1493–1468 (1996)
5. Matsumori, H., Shimizu, T., Takano, K., Ishii, H.: Evaluation of iron loss of ac filter inductor used in three-phase PEM inverters based on an iron loss analyzer. IEEE Trans. Power Electron. **31**, 3080–3095 (2016)
6. Miwa, Y., Shimizu, T., Takano, K., Ishii, H.: Study on low-loss design of inductor used for buck-chopper circuit based on iron loss characteristics considering pre-magnetization. IEEJ Trans. Ind. Appl. **137**(5), 385–394 (2017)
7. Matsumori, H., Shimizu, T., Takano, K., Ishii, H.: Iron loss calculation of AC filter inductor for three phase PWM Inverters. IEEJ Trans. Ind. Appl. **133**(10), 1009–1021 (2013)

Application of Magnetism to Automobiles

Tetsuya Aoki

Abstract Automobiles are designed to provide safe and comfortable movement to people all over the world. In order to accomplish this mission, car makers have focused on functional development including safety performance, environmental performance, driving performance, and provision of comfortable space. A wide variety of material technologies related to magnetic circuits is applied in the automobiles. These technologies have supported the development of the automobile industry and have become indispensable for high functionality and performance of the automobiles. This section will describe the relation between the basic functions of automobiles, namely, "run," "turn," and "stop," discuss the application of magnetism from the historical point of view, and explain the latest technologies and future prospects.

Keywords Automobile · Engine control component · Actuator and sensor products · HEV products

1 "Run"

As mentioned in the previous section, the main purpose of the automobiles is to provide safe and comfortable movements to people, with the most important function in movement being able to "run." Currently, internal combustion engines, including gasoline and diesel engines, and electric motors for electric vehicles and hybrid vehicles are used as the major sources of power to run an automobile. Improvement in efficiency is required to reduce the CO_2 emission for environmental preservation, and this, in turn, requires improvement in the accuracy and speed of engine control and motor control. As a result, there is an increased demand for the control of actuators that drive the magnetic circuits and for better magnetic properties of the materials used. Several technologies to improve the accuracy and efficiency of the internal combustion engines and the motor control for automobiles are described below from the viewpoint of magnetic applications.

T. Aoki (✉)
Denso Corporation, Kariya, Japan
e-mail: tetsuya.aoki.j5t@jp.denso.com

K. Fujisaki (ed.), *Magnetic Material for Motor Drive Systems*,
Engineering Materials, https://doi.org/10.1007/978-981-32-9906-1_26

1.1 Fuel Injection Control

The liquid fuel, including gasoline and light oil, is used for the internal combustion engines in automobiles, with the combustion style of the fuel directly affecting the fuel efficiency and the environment. Various engine control components burning the fuel employ magnetic application technologies for driving and/or for positional detection of objects and these technologies are effective in improving fuel efficiency (Fig. 1) [1].

Fuel is injected into the intake port or the cylinder of the engine by an injector. The injector is an actuator utilizing high-level magnetic circuit technologies and magnetic material application technologies (Fig. 2) [1]. As the injector is designed to operate the high-speed on-off valve at high fuel pressure, magnetic materials and circuits are used to achieve high-suction power at quick response. Regarding the specific material requirements, high-magnetic flux density is required to ensure suction power and materials with high-electric resistance suppressing the overcurrent are required to ensure quick response. Individual component manufacturers may exercise ingenuity in designing the magnetic circuits. However, it is generally assumed that the efficiency of the magnetic circuits can be improved by designing a circuit that does not cause magnetic saturation and minimizes the air gap.

Figure 3 shows an example of the magnetic circuit in the injector. When the valve opens, the movable core in the injector is magnetized and the magnetic flux density increases to generate the suction power quickly. However, the magnetic flux density

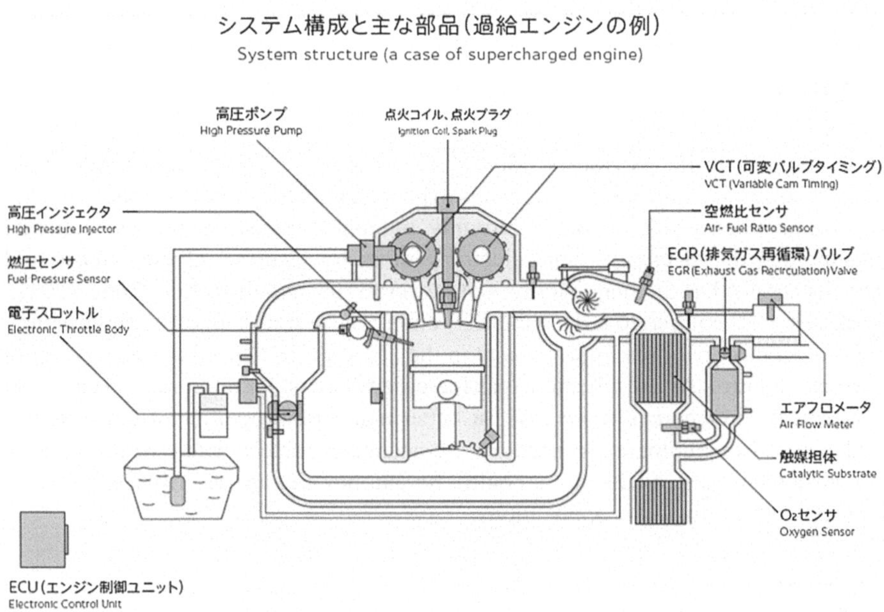

Fig. 1 Outline of the electronic fuel injection (EFI) system [1]

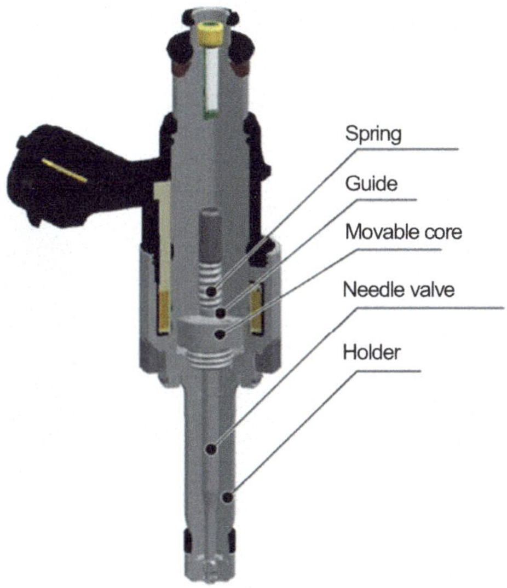

Fig. 2 Injector [1]

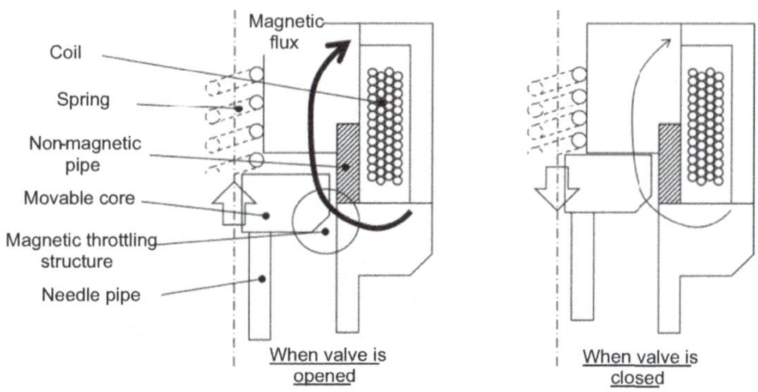

Fig. 3 Example of a magnetic circuit in an injector

needs to be reduced quickly when the valve closes. Thus, the magnetic throttling structure is employed to change the magnetic flux density according to the behaviors of the core.

The injector is structured so that the transistor built in the computer for fuel injection called the engine control unit (ECU) executes switching to control the on-off valve. However, in the earlier days (around 1995), in order to control the engine, the driver's "intention" to run applied onto an accelerator pedal was transmitted via the mechanical connection and the wires between the accelerator pedal and throttle body that controlled the air inflow volume to the engine. However, at present, the drive-by-wire has been introduced and the accelerator position information is transmitted to the

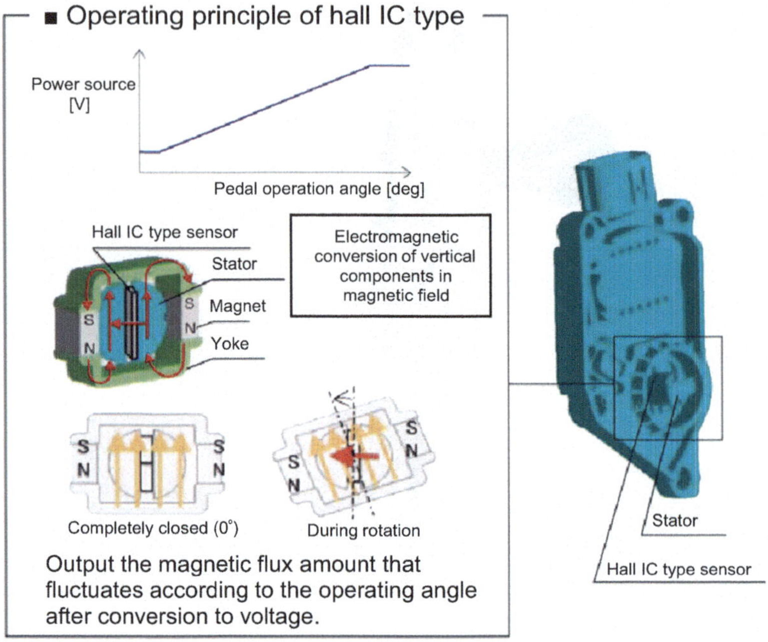

Fig. 4 Hall IC type sensor for detecting the accelerator pedal position [1]

ECU by electric signals. The magnetic application here involves the detection of the accelerator pedal position. Conventionally, the slide-type sensor equipped with the variable resistor was used. However, owing to the limited product life of the sliding section, the hall IC type sensor is used now (Fig. 4) [1]. The accelerator position is transmitted to the ECU and then to the electronic throttle (Fig. 5) [1] and that controls the air inflow volume to the engine by electric signals. The electronic throttle detects the angle of the valve that actually opens with the magnetic non-contact sensor and feeds back the rotation angle to the ECU. The rotation angle sensor is required to ensure the linearity and low magnetic hysteresis in the angle detection characteristics and satisfies these requirements by means of the well-devised magnetic circuits and/or magnetic materials selected. As discussed above, the ECU calculates, selects, and commands to supply the optimum amount of fuel and air to the engine according to the program to ensure that optimum engine control is realized. The ECU corrects the accelerator position by receiving the driving conditions on each spot and situation from various sensors, including the peripheral environment such as temperature and atmospheric pressure, driving speed, loads on the engine on an ascending slope, and temperature of the cooling water, in addition to the accelerator position information. Then, it executes optimum control to burn the fuel most efficiently.

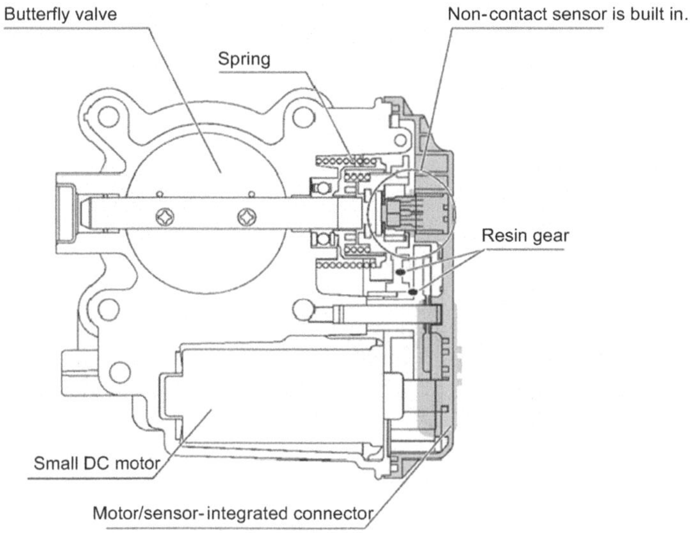

Fig. 5 Electronic throttle and non-contact sensor (magnetic type) [1]

1.2 Ignition and Combustion Control

When the air/fuel mixture is injected into the cylinder, the engine ignites it and causes combustion and explosion. In the case of a gasoline engine, compressed air/fuel mixture is ignited by the spark discharged from the ignition plug. In the case of a diesel engine, spontaneous ignition is caused by the high-pressure injection of fuel into the compressed air in the cylinder. This section describes the magnetic application technologies for ignition combustion control in the gasoline engine.

The high discharge spark energy to prevent incomplete combustion of fuel and to ensure optimum control of the ignition timing is required to operate the engine efficiently. This can be compared to a swing. The power in the swing is equivalent to the ignition energy and the timing of the swing is equivalent to the ignition timing. This means that the engine can be operated efficiently when both of them are attained appropriately, with magnetic application technologies aiding in the operation. First, the battery voltage needs to be boosted by the ignition coil to generate discharge spark that will be the ignition energy. The ignition coil is provided with a function to boost the battery voltage by electric/magnetic energy conversion. Figure 6 shows the boosting effect by the electric/magnetic energy conversion of the ignition coil. The voltage of the automobile battery is 12 V, but the output voltage of the ignition coil reaches as high as 40,000 V. The ignition coil is the inductor element itself and is structured to accumulate the magnetic energy in the iron core by flowing the electric current to the primary coil and inducing this magnetic energy to the secondary coil as electric energy when the electric current to the primary coil is shut off. A voltage is as high as 40,000 V is generated depending on the winding number ratio of the primary

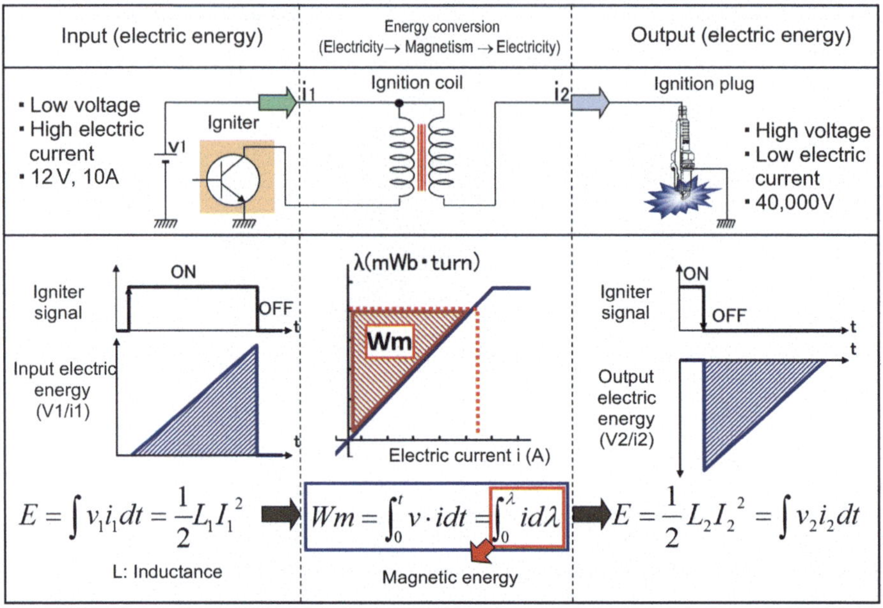

Fig. 6 Electric/magnetic energy conversion by the ignition coil

and the secondary coils. Figure 7 [1] shows an example of the ignition coil. The switching of the electric current is conducted by the insulated gate bipolar transistor (IGBT) element called the igniter, installed in the ignition coil and driven by the on-off command signal from the ignition ECU.

The outputs required for the ignition coil are high voltage and high energy, with both of them requiring the iron core material with high-magnetic flux density and low magnetic loss. Hence, electromagnetic steel sheet or powder magnetic core material

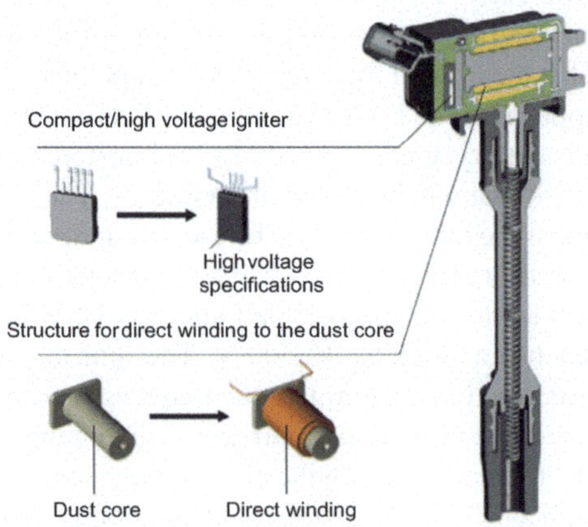

Fig. 7 Ignition coil structure [1]

that has high-magnetic flux density and low iron loss is used for the iron core. In addition, in order to relieve the magnetic saturation in the iron core and downsize the ignition coil, increase in the output is attempted by applying diamagnetism bias in the opposite direction of the excitation current magnetic field to the magnetic circuits via neodymium magnet.

The timing to ignite air/fuel mixture is considered next. This is also called the ignition timing control and is comparable to the timing of the swing in the aforementioned example. In other words, this is the pacemaker of the gasoline engine. As the emission gas needs to be always kept clean by the control that advances or retards the ignition timing according to the load condition of the engine.

The ignition ECU needs to recognize the actual position of the piston correctly. This is done by detecting the rotation angle of the crankshaft, which is the output axis of the engine with the magnetic sensor. Figures 8 and 9 [1] show the outlines of the crank angle and the cam angle sensors, respectively. The system is roughly categorized into two types. The first one is the pickup type that was mainly used around 2000. The other type is the IC magnetic sensor system, which is becoming the current mainstream. The pickup type consists of the neodymium magnet, coil, and iron core. The magnetic flux generated by the neodymium magnet is changed by the detection gear opposite to the iron core and the coil detects the voltage of such change as the rotation angles. The IC type uses the magneto-resistive element (MRE) sensor. The angle of magnetic flux generated from the ferrite bond magnet is changed by the opposing detection gear and the change of angle in the magnetic field is detected as the rotation angle. The IC type is cost effective compared to the pickup type as no rare-earth magnet is used and the winding wire of the coil is not needed. In addition, the system of this rotation angle sensor is also used for the wheel speed sensor for the brake control, as described later.

The rotation angle information detected by the magnetic signals is transmitted to the ignition ECU. Then, the ignition ECU calculates the appropriate ignition timing according to the fuel injection program and transmits the ignition command to the ignition coil.

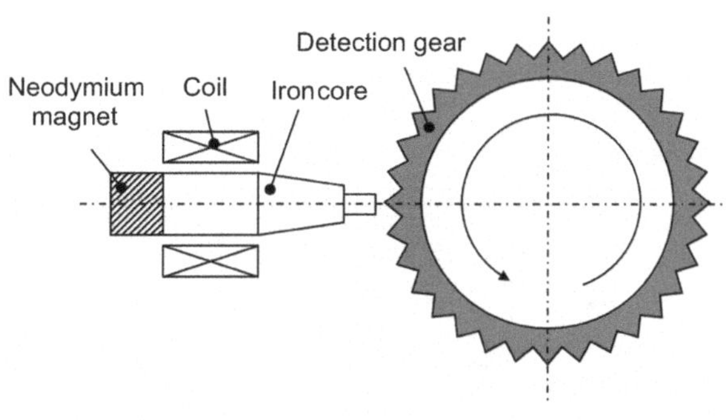

Fig. 8 Pickup type crank/cam angle sensor

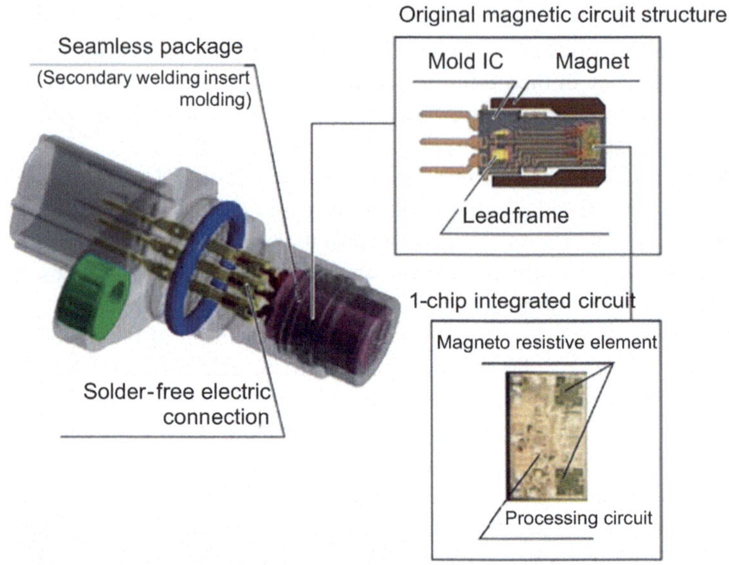

Fig. 9 IC magnetic sensor type crank/cam angle sensor [1]

1.3 Transmission Control

As the engine is an internal combustion unit and has the specific torque band, the automatic transmission (AT) is used. The AT controls the transmission gear by hydraulic pressure, while magnetic circuits are used in the solenoid valve that controls the hydraulic pressure. Figure 10 shows an example of the AT linear solenoid. The magnetic throttling structure that generates the suction power of the stator core and the magnetic shield structure made by the laser boring is provided to improve the magnetic circuits. The magnetic throttling structure ensures the linearity of suction power and the long stroke of the plunger by moving the magnetic saturation zone of

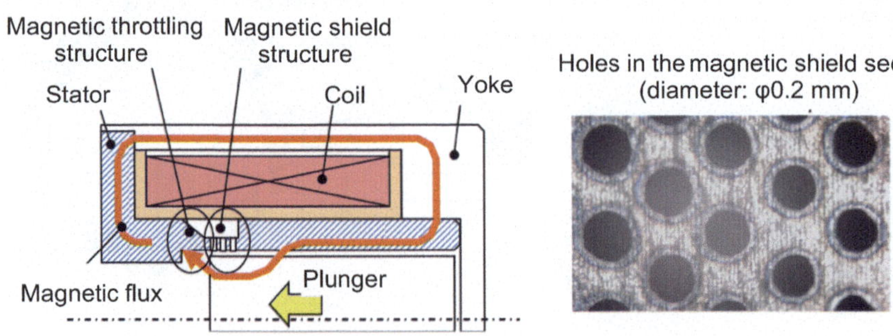

Fig. 10 Magnetic throttling/shield structures of the AT linear solenoid

the stator core, while the plunger is moved by the excitation. The magnetic shield structure is employed to reduce the cost by integrating the stator core, which may be separated into two components when an air gap is formed.

1.4 Cooling

Automobiles generate heat while running. This heat includes the fuel combustion heat generated from the engine, the heat loss from the motor, and the switching heat of an inverter in a hybrid vehicle. The temperatures of these engines, motors, and inverters are controlled by water cooling, with the cooling systems with the application of magnetism including the radiator cooling fan motors and the water pump motors. Considering the output density, cost, silence requirements, mountability, etc., the conventional DC motor or the DC brushless motor, which is common in the recent uses, is selected. In the case of the brushless motor, sensor-less control is adopted in many cases in order to reduce the control costs at the start of the rotation. The sensor-less control uses inductance component detection according to the positional relation between the rotator and the stator of the motor, this being another magnetic application technology.

1.5 Starting and Power Supply

A starter is used to start an automobile engine, and a drive motor is used for a hybrid vehicle. In addition, as the automobiles require large amount of electric power to drive the aforementioned EFI systems, electricity is also needed for vehicles other than hybrid or electric vehicles. A generator, called an alternator, is used to supply electric power in engined cars, while a drive motor equipped with the power generation function or a generator is used in the hybrid vehicles and electric vehicles. The relation between the magnetic application technology and the system to control these motors and generators is omitted here as it is already introduced in this book and this section will only describe sensing of electric current in the automobiles.

Battery charge control has been adopted in engined cars from around 2000. This is a technology that suppresses power generation from the alternator and reduces loads to the engine during acceleration or steady running to improve the fuel efficiency or output. This technology also promotes power generation by the alternator during deceleration to manage electric power balance in the entire automobile. The key aspect of this technology is the charge–discharge management of the battery, along with the adoption of the magnetic application technology to the sensor. Figure 11 [1] shows the outline of the battery current sensor. The principle used is the magnetization of the iron core component in the magnetic field generated by the Ampère's circuital law and detection of the magnetic flux density with the hall elements. Electric currents ranging from large electric current that the starter generates when starting the

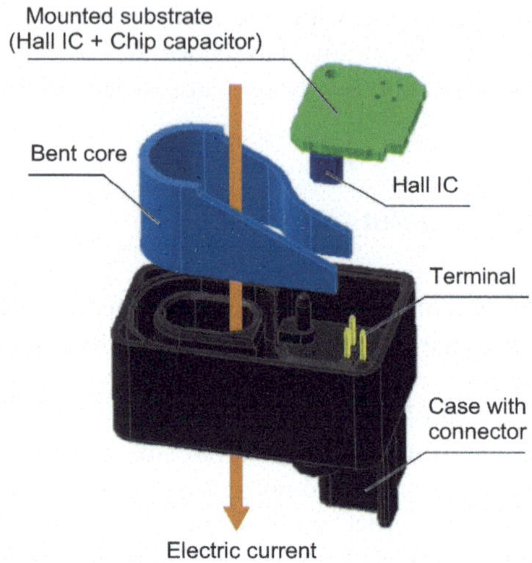

Fig. 11 Electric current sensor [1]

engine to the entire charge–discharge zone need to be detected; hence, wide detection range and high accuracy are required. Therefore, permalloy or grain-oriented electromagnetic steel sheet with low hysteresis and high-magnetic flux density is used as the iron core material for the bent core. In addition, this current detection system is also used for the vector control of motors for hybrid vehicles and electric vehicles.

1.6 Drive Motor Control

The power control unit (PCU, Fig. 12) [1], installed in hybrid vehicles and electric vehicles, is designed to generate triple-phase alternating current from the direct-

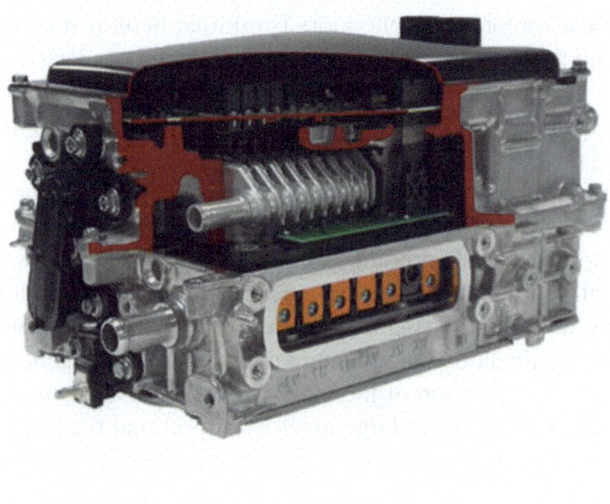

Fig. 12 Power control unit (PCU) [1]

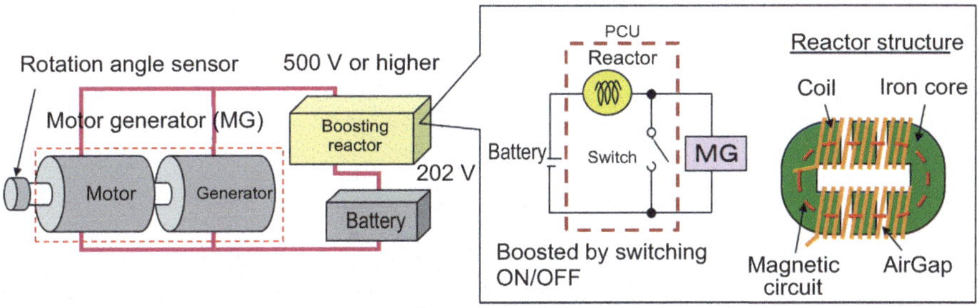

Fig. 13 Outline of the boosting reactor

current battery and supply it to the drive motor. The direct voltage of the battery is approximately 200 V in the case of a hybrid vehicle, but the voltage may be boosted in a PCU for downsizing, weight reduction, and increase in the rotation speed of the motor. Increasing the voltage is effective in downsizing and cost reduction of the motor because the wire size for the motor and wire harnesses can be reduced, thus reducing the amount of copper wires required. Moreover, as the upper limit of the voltage can be raised for motor control, the rotation speed of the motor can also be raised and the output can be improved. The magnetic application product called a reactor is used for such boosting. Figure 13 shows the outline of the boosting reactor. The reactor boosts the voltage by the conversion of electric/magnetic energy using the same principle as for the aforementioned ignition coil. The difference from the ignition coil is that the reactor has a coil only on the primary side. It can be imagined that the reactor connected in series with the battery circuit converts energy through switching by IGBT and works as a battery by itself to boost the voltage. The present reactor is driven at approximately 10 kHz and requires lower magnetic loss to ensure efficiency of the circuits and to suppress heating. In addition, higher saturation magnetic flux density is also required for downsizing. In order to satisfy these requirements, electrical steel sheets and/or powdered materials containing 6.5% silicon are used for the present reactor, which can be expected to be driven at higher frequencies in the future through further evolution, such as the application of SiC to the switching elements.

2 "Turn"

The procedure for automobiles to make a turn can be described as follows: the driver's "intention" operates the steering wheel, which then rotates the steering shaft, moves the tie rod via the pinion rack gear, and changes the traveling direction of the automobile. The hydraulic assist technology had been used in many vehicles up to around the year 2000; however, after that time, the electric power steering (EPS) system, which is an electrical assist technology, have started to be used frequently in order to improve fuel efficiency and to reduce the costs of the automobiles (Fig. 14).

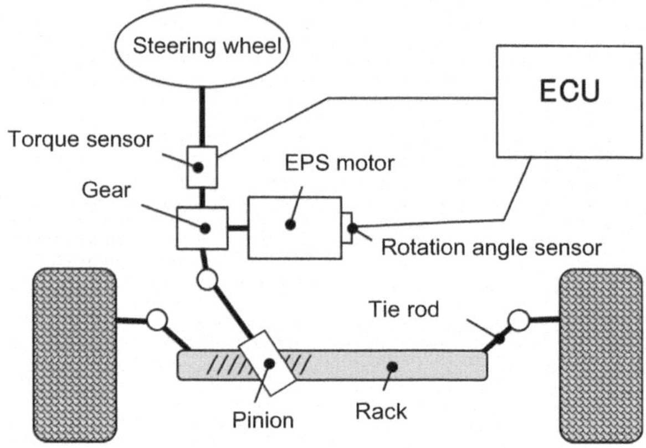

Fig. 14 Electric power steering system

This section describes the operation torque of the steering wheel and detection of the rotation angle of the EPS motor.

The torque generated to the steering shaft after the driver's "intention" to turn the steering wheel is detected by the magnetic sensor. The outline of the sensor is shown in Fig. 15. When the torque generated by the steering is applied to the yoke and magnet mounted to the shaft, the positional relation between the magnet and the yoke changes, the magnetic flux of the magnet is transmitted to the yoke equipped with the hall sensor, and the torsional change in the shaft is detected. The detected steering signal is transmitted to the ECU of the EPS, and the drive command is sent to the EPS motor.

The EPS motor generates torque to assist steering by means of the rotary drive, but the rotation angle needs to be controlled strictly in order to ensure appropriate drive control to realize the conforming of the steering angle to the command. This function is supported by a magnetic rotation angle sensor. The absolute value type rotation angle sensor called "resolver" is still used widely for controlling the rotation angle of the DC brushless motor.

The history of this sensor technology can be tracked back to the directional control of the battery during the World War II. This technology is capable of detecting

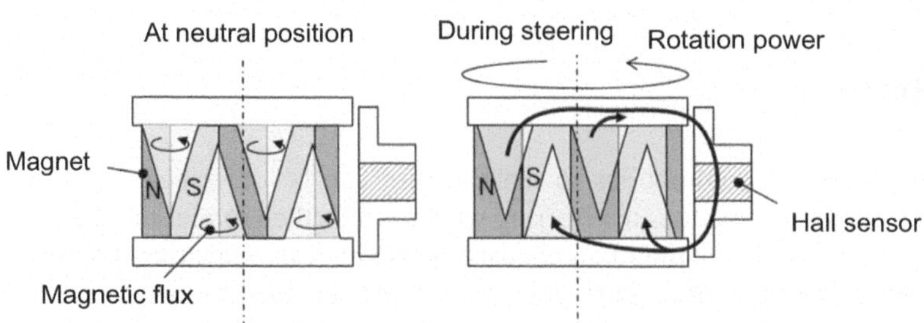

Fig. 15 Outline of the EPS torque sensor

absolute angles very precisely, but the reduction of the size and cost is difficult as this sensor consists of iron cores and coils. Thus, the IC rotation angle sensor is becoming popular nowadays. The tunnel magneto-resistance (TMR) element that ensures high sensitivity and temperature stability is used for this sensor.

3 "Stop"

The brake of the automobile works when the hydraulic cylinder receives the instruction from the driver after the brake pedal and brake pads provided for each wheel is pressed against the brake disks by hydraulic pressure. In the latter half of 1980s and later, the anti-lock brake system (ABS) was introduced for the brake system of automobiles to improve the attitude stability of the automobile upon braking by increasing and reducing the hydraulic pressure for the brake. Therefore, the ABS is equipped with the oil pump motor to increase the hydraulic pressure of the brake and the solenoid to reduce the pressure, with these units controlled by the ECU of the ABS. A general DC motor is used for the oil pump motor, while the magnetic circuit technology using composite magnetic material shown in Fig. 16 is introduced to the solenoid for cost reduction and downsizing requirements.

The properties of the composite magnetic material can be switched between the ferromagnetic and the non-magnetic by thermal treatment and processing technologies. This material was developed initially based on non-magnetic stainless material (austenitic stainless steel) and allowed the ABS solenoid at that time to be downsized by approximately 30% compared with the conventional one. However, at present, solenoids using composite magnetic material are not manufactured because of the further requirements for cost reduction.

Meanwhile, brake technologies have been evolved recently as a preventive safety technology. Technologies to detect obstacles in front of vehicles or traveling direction by using a millimeter-wave radar or image recognition camera and to automatically operate the brake through the ECU have been developed. However, even these advanced technologies are supported by high reliability and magnetic circuit technologies of the aforementioned oil pump motor and/or solenoid products.

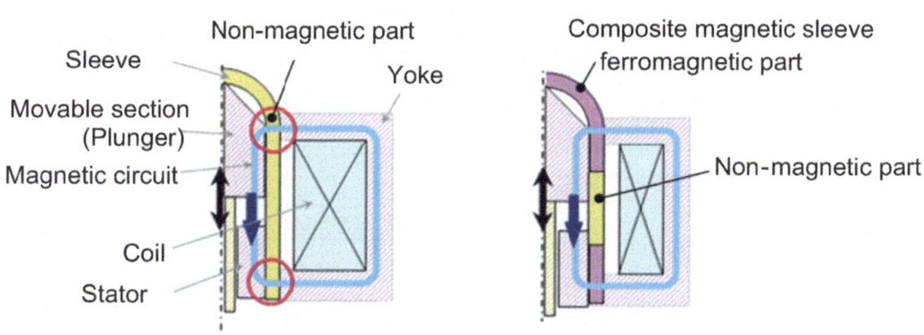

Fig. 16 Magnetic circuit using composite magnetic material (ABS solenoid)

4 Evolution of Magnetic Materials for Automobiles

The magnetic materials used for automobiles include the iron core material and the magnetic material. As described in this chapter, the iron core material is used for the motor, solenoid valve, and linear solenoid. The low-cost cold rolled steel sheet is used for the cores of auxiliary rotary products such as the conventional alternator, starter, and electric fan motor. Regarding the magnetic material, lower-cost ferrite magnet is used in many cases. On the other hand, electromagnetic stainless steel, pure iron, etc., are used for the solenoid valves in order to satisfy the requirements of corrosion resistance and magnetic hysteresis suppression. As discussed above, previously it was more likely to use the highly functional magnetic materials for engine control components that needed to satisfy high requirement values and to select lower-cost materials for auxiliary rotary products depending on the required performance. However, the neodymium magnet discussed in this chapter has allowed the development of innovative automobile technologies. Figure 17 shows the change in the performance of the neodymium magnet and the history of its application to automobile components. The neodymium magnet was mainly used for small sensors needed for the automobile components at the beginning. However, when the power electronics technology was developed at the same time and synchronous motors started to be used for the software control, its scope of application was extended instantaneously to the field of automobiles, making mass production of hybrid and electric vehicles possible. However, it is difficult to use the neodymium magnet for conventional auxiliary device products because it costs approximately ten times the ferrite magnet. The cost reduction and development of new magnet are currently promoted by the industry-academia collaboration projects or by individual magnet manufacturers.

The iron core material may be roughly classified into two classes in terms of applications. One of them is the electromagnetic steel used for the drive system.

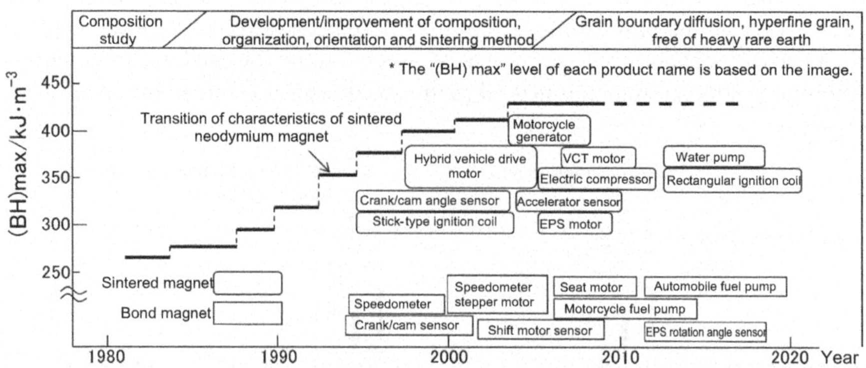

Fig. 17 Change in performance of neodymium magnet and history of its application to automobile components

However, reduction of loss will be required along with the requirement for high-speed drive to downsize the motor for hybrid and electric vehicles. The other is the application as the energy conversion material such as a reactor or non-contact power supply. For this application, it is expected that the drive frequency can be increased by approximately ten times by introducing SiC for the switching elements. Powder or nanopowder material with high silicon contents, amorphous materials, and their pressed powder materials have been studied. It is another large area to realize both costs and loss reduction requirements of these materials and the development has also been promoted by the industry-academia collaboration project or others.

The needs and seeds for development of these materials, magnetic circuits, and magnetic application technologies have started from the neodymium magnet, which is therefore called the "invention that has changed the world." We expect the development and application of new magnetic materials to take place through various activities that can demonstrate the Japanese technical strength to the maximum.

Reference

1. DENSO Corporation website: https://www.denso.com/ja/ (December 1, 2017)

Magnetic Application in Linear Motor

Hiroyuki Wakiwaka

Abstract In this chapter, small- and medium-sized linear motors used in the industrial field are considered, firstly classified into four types based on their operation principle, and their each features are described. Next, the method of characterizing the linear motor is explained, and some of the characteristic evaluation indexes are shown. Furthermore, in order to evaluate the characteristics of the linear motor, the relation between the thrust and the tangential stress is explained, and the traditional magnet array, the repulsion type array, and the Halbach magnet array are shown as the magnet arrangement in the case of permanent magnet field. In addition, recently developed several high-performance linear synchronous motors are classified into permanent magnet field type and variable reluctance type from the viewpoint of thrust source. Finally, the values of the motor constant square density of those synchronous motors are compared.

Keywords Classification of linear motors · Characteristic evaluation index · Permanent magnet field · Variable reluctance · LSM classification · Motor constant square density

1 Introduction

Linear motors [1] are electric machines which move objects linearly using electromagnetic force to directly drive movers (movable side). In order to drive objects linearly, it is not necessary to apply a rotary motor conversion mechanism driving source, and simple linear motor drive systems are possible. The movable ranges (strokes) in which linear motors are used are minute range (micrometer order) to long distance (kilometer order), and there are linear motors suitable for each stroke. Each linear motor constitutes a linear drive system. Linear drive systems are configured to control the positioning and speed of the mover. Linear drive systems [2] are

H. Wakiwaka (✉)
Shinshu University, Nagano, Japan
e-mail: wakiwak@shinshu-u.ac.jp

K. Fujisaki (ed.), *Magnetic Material for Motor Drive Systems*,
Engineering Materials, https://doi.org/10.1007/978-981-32-9906-1_27

Table 1 Classification of linear motors [3]

Linear induction motor (LIM)	• Linear induction motors in general • Magnetic induction type electromagnetic pump
Linear DC motor (LDM)	• Unipolar type (voice coil motor) • Brush commutation type
Linear synchronous motor (LSM)	• Electromagnet type • Permanent magnet type • Variable reluctance type
Linear stepping motor (LSTM)	• Variable reluctance type • Permanent magnet type

driven by linear motors with supporting mechanism, sensors, and drive circuits, with a variety of applications such as transportation and information equipment. In this chapter, small- and medium-sized linear motors mainly used in industrial application fields are described.

2 Linear Motor and Linear Drive System

2.1 Classification of Linear Motors

The construction of linear motors is equivalent to rotary electric machines which engage their rotary motor linearly, as showed by Table 1 [3], and linear motors are classified as linear induction motor (LIM), linear DC motor (LDM), linear synchronous motor (LSM), and linear stepping motor (LSTM).

Linear oscillatory actuator (LOA) and linear electromagnetic solenoid (LES) are categorized as short stroke actuators. A LOA is defined as a "linear actuator which brings linear reciprocating motion to the mover directly, without any conversion mechanism from electrical input." A LES in defined as "mechanical parts which bring linear movement to the mover iron core directly by magnetic force gained from exciting magnetizing coils by applying voltage." LOA and LES can also be included in LSM or LDM.

2.2 Linear Drive System

As showed by Fig. 1, linear drive systems use positioning control or/and constant speed to operate an object, using a combination of electronic drive power circuits (driver), position sensor, speed sensor, and control circuit (controller) with the linear motor as the driving source. Figure 1 shows a linear drive system [4] with an LSM

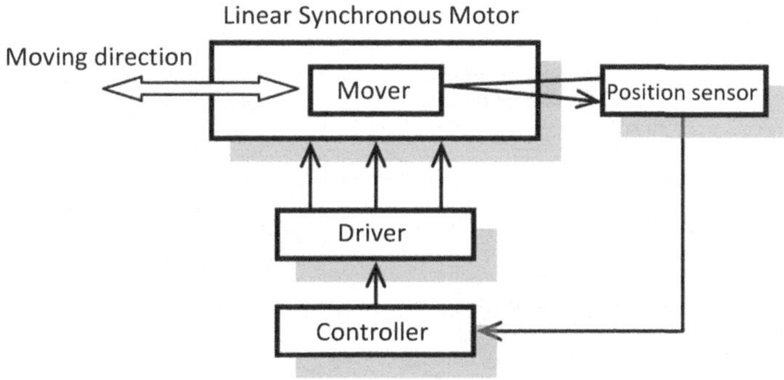

Fig. 1 Linear drive system [4]. Reprinted from [4], Copyright 2009, with permission from IEEJ

driving source. A different sensor and driver would be required if the linear drive system uses other types of linear motors as their driving source; however, the principle is basically the same.

2.3 Linear Induction Motor (LIM)

LIM is defined as "moving linear motors which cut open the stator and rotor of a rotary induction motor around a central axis and stretch linearly." The operating principle and characteristics of LIM are basically equivalent to rotary induction motors because they move asynchronously with slip speed and do not synchronize with the input frequency.

Figure 2 illustrates development of LIM structures [3]. A basic LIM consists of a primary side and a secondary side. The primary side is composed of laminated magnetic steel sheet cores and multiphase coils. The secondary side is composed of steel plates with aluminium or copper on the surface and is called a reaction plate. Both the primary side and secondary side can be either movable or fixed. When polyphase AC power is supplied to the coils on the primary side, a moving magnetic field is generated in the gap, and induced current flows in the conductor on the secondary side. Linear thrust is generated by electromagnetic force acting between the moving magnetic field and induced current.

The different types of LIM are: the single-sided LIM which positions the primary side on one side of the secondary conductor, the double-sided LIM which positions a primary side on either side of the secondary conductor, and the cylindrical LIM where the primary side has a cylindrical shape. These have a case to provide an iron plate on the back side of the secondary conductor (back iron) to concentrate magnetic flux onto the secondary conductor. Unlike an LSM with permanent magnets for field magnetism and an LSTM with grooves on the secondary conductor, a LIM can be constructed easily because the secondary side consists only of a conductor plate and an iron plate. However, in general, the power density, thrust, and efficiency of LIM

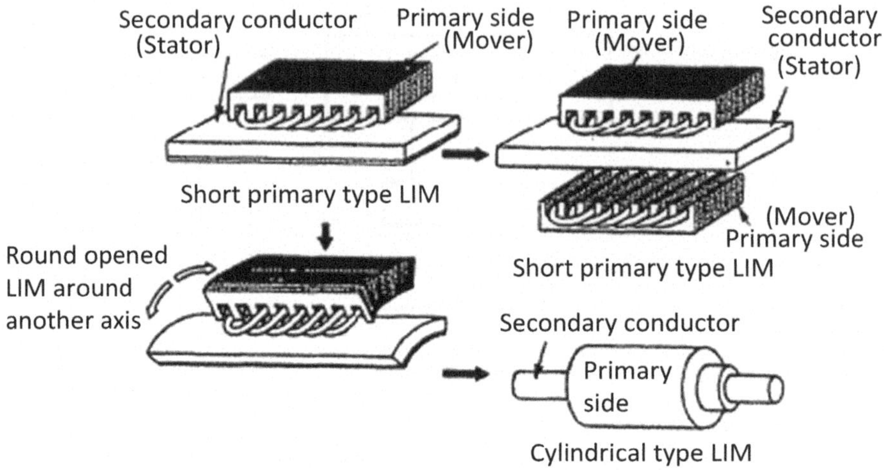

Fig. 2 Linear induction motor (LIM) [3]. Reprinted from [3], Copyright 2003, with permission from IEEJ

are lower than those of LSM. Recently, due to advancements in power electronics, LIM drives can apply vector control more efficiently.

In general, LIMs are used for transport, shipping, transfer lines in factories, and metro lines because a LIM is suitable for continuous movement over long distances (large stroke).

2.4 Linear DC Motor (LDM)

The different types of LDM are: the single unipolar type for which the magnetic field flux direction is within the range of motion (stroke) and the type which uses a multipolar brush. Multipolar LDM switching without the brush is classified as LSM. LDM is used for information equipment in many cases, because an LDM can easily be precisely positioned at high speed in combination with a sensor such as a linear encoder.

Figure 3 shows typical and basic structures of an LDM [5]. Every type generates a magnetic field in the gap between the permanent magnet and the yoke (or permanent magnet) via the permanent magnetic field, and the Lorenz force which drives current through the coil gap.

Thrust F of an LDM is calculated using the equation below.

$$F = NlBI \text{ [N]}$$

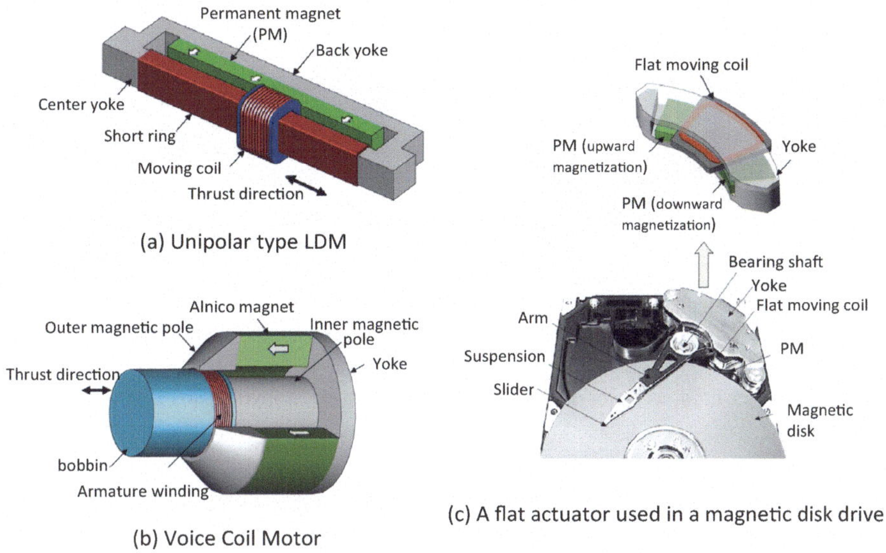

Fig. 3 Linear DC motor (LDM) [5]. Reprinted from [5], Copyright 2010, with permission from IEEJ

where N—number of coil turns, l—length of the coil where the magnetic flux intersects [m], B—magnetic flux density crossing the coil [T], and I—current in the coil [A].

Figure 3a shows a unipolar type LDM with magnetic flux and stroke working to the same direction. When external dimensions are fixed, the accompanying stroke gets larger and magnetic saturation will happen in the yoke, so magnetic flux density becomes smaller and it becomes difficult to get a large thrust in a wide stroke range [6].

Figure 3b shows a magnetic circuit of an electrodynamic speaker type, used for short stroke shaker actuators, like to focus the adjustment of a laser lens. Figure 3c shows a flat LDM embedded into a magnetic disk device where a rare-earth magnet is most frequently used.

2.5 Linear Synchronous Motor (LSM)

LSMs are originally rotary synchronous motors, "due to the interaction of the armature and the field magnetic pole, synchronous movement happens as the moving magnet field moves over the mover which has a magnetic pole." LSMs operate using a synchronous input frequency.

LSMs are classified into electromagnetic, permanent magnetic (PM), and variable reluctance (VR) types according to the configuration of their magnetic field [3].

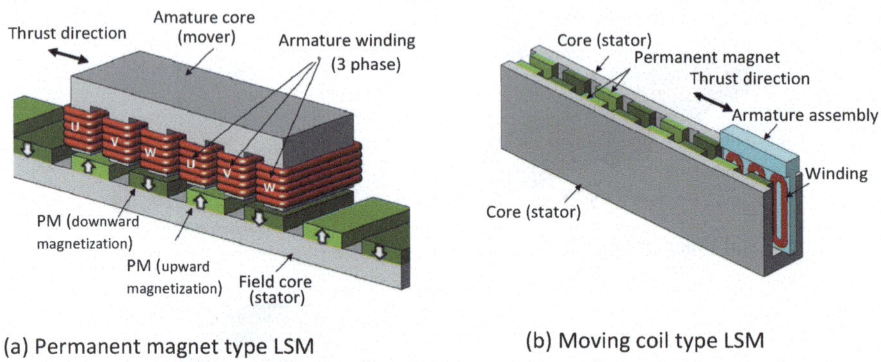

Fig. 4 Linear synchronous motor (LSM) [5]. Reprinted from [5], Copyright 2010, with permission from IEEJ

Figure 4a shows the structure of a linear synchronous motor with a permanent magnet as a field and an armature as the moving side. Figure 4b shows a linear synchronous motor in which the secondary side is double-sided, and the coil is movable.

Armature cores are made by laminating thin insulated magnetic steel sheets. Permanent magnet field LSMs do not require an exciting current for the field, so it has a simple structure, high output density, and high efficiency. Since permanent magnet LSMs cannot weaken their field on control, they are unsuitable for high-speed railway applications. However, due to the improved performance of rare-earth magnets in recent years, PM-LSM has been widely used in servos controlling speed and position in industrial semiconductor manufacturing equipment, liquid crystal manufacturing equipment, machine tools, and in general transport.

2.6 *Linear Stepping Motor (LSTM)*

LSTM is defined as "a motor which moves (steps) according to a pulse signal input," also known as a linear pulse motor (LPM). Both LSM and LSTM are generally known as linear synchronous motors. LSM and LSTM (i.e., LPM) can be regarded as linear synchronous motors in a broad definition. In particular, when performing closed-loop control for high-speed and high-precision driving of a LSTM, the LSTM cannot be distinguished from LSM in a narrow definition. LSTM is also classified into reluctance and permanent magnetic (PM) types according to the structure of their magnetic circuit.

LSTM [5] shown in Fig. 5 is excited using a permanent magnet and armature winding coil (exciting coil). The LSTM shown in Fig. 5 uses four exciting coils to pulse current in a single-phase excitation pattern, in which every 1/4 magnetic pole pitch movement correspondents to pulse current input. Permanent magnet LPM (PM-LPM)and LSTM with no magnet are called variable reluctance type LSM (VR-LSM).

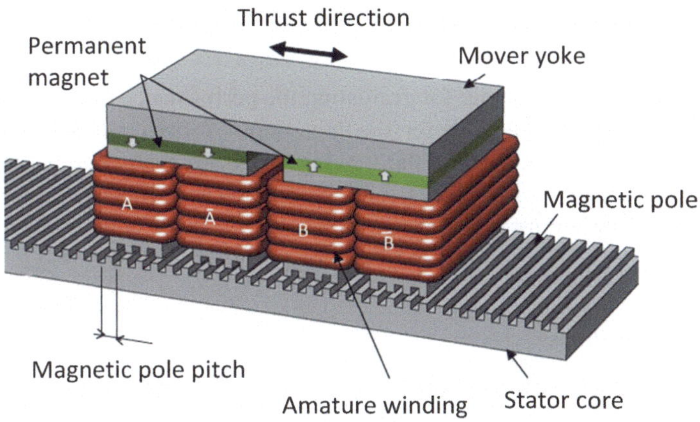

Fig. 5 Linear stepping motor (LSTM) [5]. Reprinted from [5], Copyright 2010, with permission from IEEJ

2.7 Linear Actuator

Figure 6a shows a linear oscillation actuator (LOA) [3], where a mover is fixed on the outside with a spring and often have a vibration frequency system. Figure 6b shows a linear electromagnetic solenoid (LES) commonly known as a solenoid; it cannot produce a large stroke (limited to a few mm approx.), though retains a large attraction (thrust) to its comparatively small size.

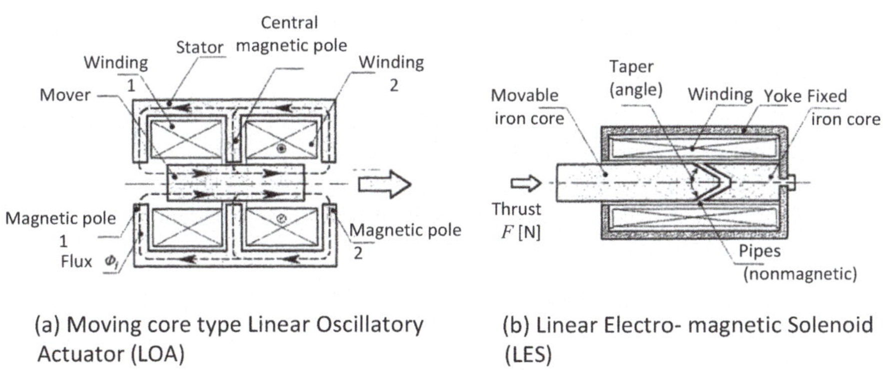

Fig. 6 Linear actuator [3]. Reprinted from [3], Copyright 2003, with permission from IEEJ

3 Characteristic Evaluation of Linear Motor

Efficiency is used as a parameter for evaluating the performance of the rotary motor. However, in the case of a linear motor, the linear motor is driven by acceleration and deceleration in a short time within the moving range, so evaluation based only on efficiency is not appropriate [7]. Therefore, the characteristic values representing the merits of a linear motor are as follows.

(1) Ratio of thrust/input.
(2) Ratio of thrust/volume.
(3) Ratio of thrust/mass.

The thrust characteristic used for these calculations refers to starting thrust, and input refers to electric power input. These characteristics are calculated based on the volume and mass in the area enclosed by the dotted line in Fig. 7.

3.1 Thrust Constant

The thrust constant of the linear motor K_f is found by measuring the static thrust of the motor, then calculating from current I [A] and the slope of the static thrust F [N].

$$K_f = \frac{F}{I} \text{ [A]}$$

3.2 Motor Constant

Motor constant evaluation is effective for high-frequency acceleration/deceleration applications. The motor constant K_m is calculated by dividing thrust by the square root of copper loss W_c [W], which is also equal to the thrust constant above divided by the square root of coil resistance R [Ω], as shown below.

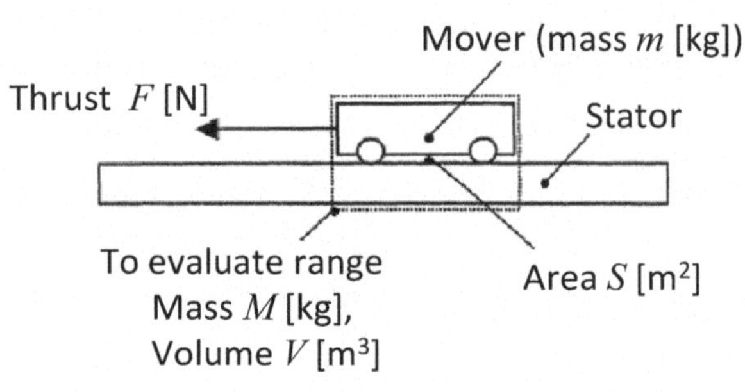

Fig. 7 Characteristic evaluation of linear motor [3]. Reprinted from [3], Copyright 2003, with permission from IEEJ

$$K_m = \frac{F}{\sqrt{W_c}} = \frac{K_f I}{\sqrt{I^2 R}} = \frac{K_f}{\sqrt{R}} \left[\mathrm{N}/\sqrt{\mathrm{W}}\right]$$

3.3 *Motor Constant Square Density*

The motor constant generally increases as volume increases. In order to compare motors with different volumes, the motor constant square density G is converted from predefined volume V [m^3] as below.

$$G = \frac{K_m^2}{V} = \frac{F}{W_c} \cdot \frac{F}{V} \left[\mathrm{N}^2/(\mathrm{Wm}^3)\right]$$

3.4 *Power Rate*

High-frequency acceleration and deceleration applications evaluate acceleration speed for objects directly loaded. Here, the rated power rate Q_r is defined as the value of a suitable servo system with predetermined load.

$$Q_r = \frac{F^2}{m} \ [\mathrm{W/s}]$$

Effective acceleration due to the driving load α is calculated as below.

$$\alpha = \frac{F^2}{m + m_L} = \frac{\sqrt{Q_r/m}}{1 + m_L/m} \ [\mathrm{m/s^2}]$$

where m is the mass of mover [kg], and m_L is the mass of load [kg].

4 Magnetic Circuit for High-Performance LSM

4.1 *Thrust and Normal Force*

The linear motor that faces the primary side and the secondary side shown in Fig. 2 has tangential force and normal force as shown in Fig. 8a. The tangential force becomes the thrust and the normal force becomes the attractive force. An effective way to increase the thrust of a linear motor is to increase the magnetic flux density in the gap. However, doing this generally increases the attractive forces on the primary

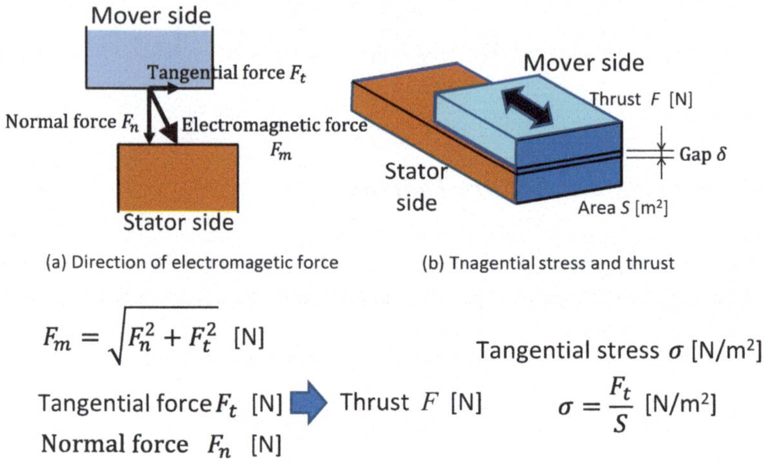

Fig. 8 Tangential force and normal force

and secondary sides. Especially for frequently used PM-LSM in industry, the normal force becomes over 10 times greater than the tangential force, such that the structural support load is too high for single-sided linear motors. These applications require the double-side type and/or the cylindrical type to offset normal forces while increasing thrust to reduce load on the support mechanism. As shown in Fig. 8b, evaluating tangential force per unit area of the magnetic force working surface shows the tangential stress σ.

$$\sigma = \frac{F}{S} \text{ [Pa]}$$

4.2 Arrangement of Permanent Magnet

Normally, the permanent magnet of a magnetic field is magnetized in the gap direction. Geometry may lend itself to magnetize in the stroke direction; however, many cases still magnetize in the gap direction regardless. In recent years, "Repulsion arrangement" and "Halbach arrangement (or Halbach array)" have been proposed as the arrangement of the permanent magnets for the field like interior permanent magnet rotary motor (IPM). Repulsion arrangement is an arrangement in which a pole tooth is sandwiched between a permanent magnet magnetized in the stroke direction and a permanent magnet magnetized in the opposite direction. Halbach arrangement is an arrangement in which a permanent magnet magnetized in the gap direction is sandwiched between permanent magnets magnetized in the stroke direction, and a yoke is basically not necessary.

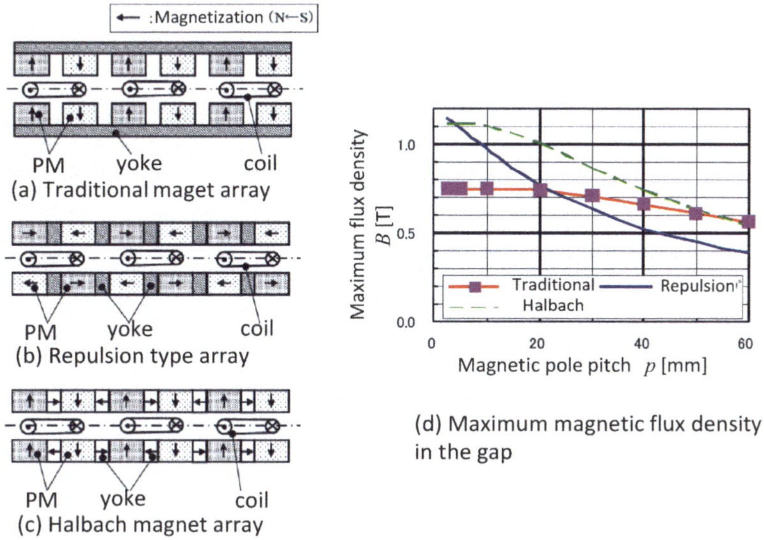

Fig. 9 Magnetic flux density by PM field [4]. Reprinted from [4], Copyright 2009, with permission from IEEJ

Figure 9a, b, and c shows the structure of each type of magnetic field, and Fig. 9d compares magnetic flux densities in the gap [4]. Magnetic field density in the gap for ordinary magnets is almost equal with magnetic field density in permanent magnets, and therefore, increasing permanent magnet thickness only increases density approx. 0.8 T. The repulsion arrangement concentrates magnetic flux into the pole teeth of the permanent magnet and which increases magnetic flux in the gap. However, as the magnetic pole pitch gets larger, the sectional area of the permanent magnet becomes smaller as compared with a normal structure, and the magnetic flux generated in the gap would be smaller. The Halbach arrangement increases magnetic flux density in the gap of the normal structure of the yoke to make magnetomotive force bigger. As the magnetic pole pitch gets bigger, the effect of the magnetized permanent magnet on stroke direction gets weaker.

4.3 Example of High-Performance LSM

(1) **Classification of magnetic path configuration**

The characteristics of the LSM are indicated in Fig. 10, and it is possible to improve that performance by deriving the magnetic circuit [3, 4]. Figure 10a, b show a permanent surface magnet facing the gap, Fig. 10c, d show an embedded yoke magnet, and Fig. 10e, f show the reluctance type that generates thrust from reluctance change without a permanent magnet field. Figure 10a shows the permanent magnet array

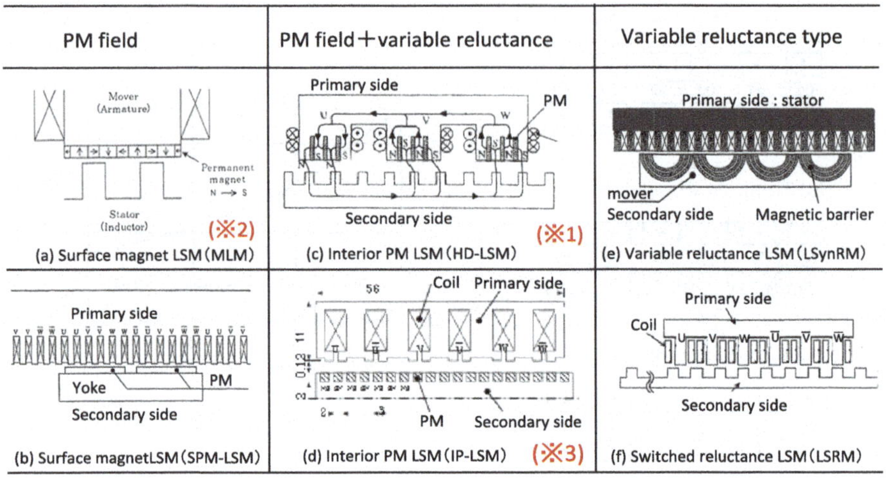

Fig. 10 LSM classification by magnetic path configuration [3, 4]. Modified from [3], Copyright 2003, and [4], Copyright 2009, with permission from IEEJ

placed at the tip of the armature, and the magnet field is a Halbach array. Figure 10d IP-LSM shows a repulsion arrangement. In the case of the configurations shown in Fig. 10a, c, the thrust can be increased by decreasing the magnetic pole pitch. Figure 10c, d show that thrust effectively increases not only from tangent stress from permanent magnetic field, but also from the force coming from reluctance change in the yoke position.

Figure 10e shows a linear motor in a synchronous reluctance motor (SynRM) arrangement for rotating the machine linearly. SynRM does not need to worry about demagnetization due to temperature change. Since SynRM does not use magnets, cost reduction can be expected. However, for positioning control, closed-loop control is required using a positioning sensor. The thrust of the LSynRM F is

$$F = \pi\left(L_d - L_q\right)i_d i_q/\tau \text{ [N]}$$

where as indicated L_d, L_q: *d, q* axial inductance [H], i_d, i_q: *d, q* axial current [A], τ: pole pitch [m], it is important for the design to make (L_r _ L_q) larger.

Figure 10f shows the motor moving by reluctance thrust occurring from magnetic saliency just like a LSynRM, so that the design must have concentrated winding.

(2) **Cylindrical LSM**

Figure 11 is an example of a cylindrical permanent magnet LSM (PM-CLSM) [8]. The permanent magnet is arranged in a Halbach array aiming to reduce use of the magnet while minimizing the cylindrical gap to reduce the load of the supporting mechanism.

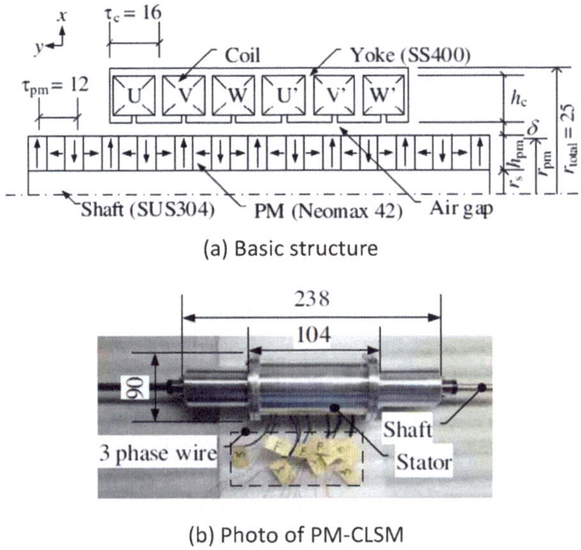

Fig. 11 Cylindrical LSM (PM-CLSM) [8]. Reprinted from [8], Copyright 2015, with permission from IEEJ

(3) High thrust double-sided LSM

Figure 12 is an example of a double-sided LSM, and its armature winding wire processed with a concentrated winding [9]. Because of the configuration in Fig. 12, high-performance LSMs can be made for conventional PM-LSM even with an ordinary magnet arrangement, but here, higher performance has been achieved because of the following differences: the magnet was a Halbach array, a soft magnetic material was added to the gap, a directional silicon steel plate was used as the armature, and the shape of the armature core was optimized. The result was a high-performance TI-LSM.

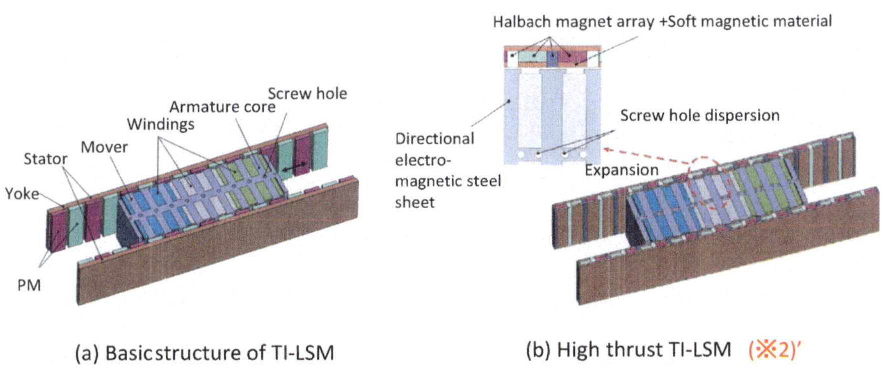

Fig. 12 Double-sided hight thrust LSM [9]. Reprinted from [9], Copyright 2012, with permission from IEEJ

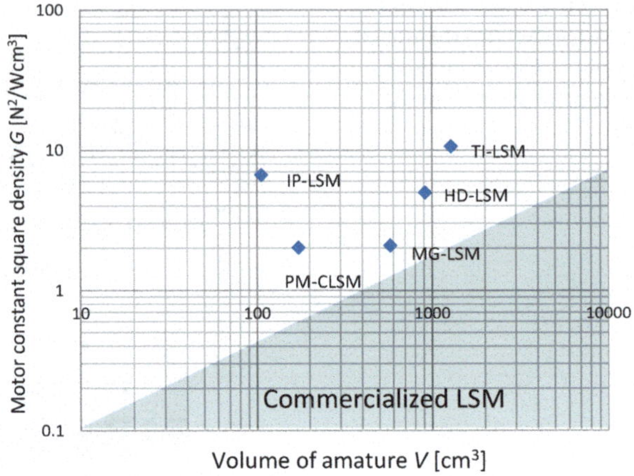

Fig. 13 Comparison of motor constant square density

Table 2 Comparison of five LSMs' characteristics

	Unit	HD-LSM	MLM	IP-LSM	TI-LSM	PM-CLSM
Volume of amature	cm^3	916	577	106	1286	173
Thrust	N	880	930	119	1600 peak	214
Thrust constant	N/A	52	45.2	300	125	143
Motor constant	$N/W^{0.5}$	68	34.6	26.6	117.2	16
Motor constant square density	N^2/Wcm^3	5	2.1	6.7	10.7	2.03
		(※1)	(※2)	(※3)	(※4)	(※5)

4.4 Comparative Evaluation of High Performance

The motor constant square density of LSMs with a general magnetic circuit structure already in commercial use was investigated. It was found to be distributed in the lower right portion of Fig. 13. Also, as shown in Table 2, the five high-performance LSMs ※1–4 shown in Figs. 10 and 11 exceed the characteristics of a conventional LSM as shown in Table 2.

References

1. IEEJ: Linear motors and their applications. IEEJ (1984). (in Japanese)
2. Masada, E. (ed.): Linear drive technologies and their applications. Ohmsha, Ltd. (1991). (in Japanese)
3. IEEJ: Technical terminology for linear drive technologies and their applications. IEEJ Tech. Rep. **911** (2003) (in Japanese)

4. IEEJ: Elemental technology use for industrial linear electromagnetic drive systems and their trend. IEEJ Tech. Rep. **1154**(2009). (in Japanese)
5. IEEJ. Elemental technology use for industrial linear electromagnetic drive systems and their applications. IEEJ Tech. Rep. **1195** (2010). (in Japanese)
6. Yajima, H., Wakiwaka, H., Seno, S., Minegishi, K., Fujiwara, N., Tamura, K.: Relation between thrust constant and size for design of moving coil type linear DC motor. IEEJ Trans. Ind. Appl. **117-D**,863–869 (1997). (in Japanese)
7. IEEJ: Industrial linear motor, its characteristic measurement and evaluation methods: IEEJ Tech. Rep. **1024** (2005) (in Japanese)
8. Azhar, F., Wakiwaka, H., Tashiro, K., Nirei, M.: Design and performance index comparison of the permanent magnet linear motor. Prog. Electromagn. Res. M **43**, 101–108 (2015)
9. Kakihara, M., Hoshi, T., Shikayama, T., Ohto, M.: Development of high-force-density iron-core synchronous motor. IEEJ Trans. Ind. Appl. **132**(4), 480–486 (2012). (in Japanese)

Final Remark

In the present when an application of a motor drive system is still expanding to an overall moving body such as not only a car but also an airplane and a ship, the role of the magnetic material which is one of the main materials in the motor drive system becomes more and more important until the future. On the other hand, since it is said that a number of lectures at universities or colleges about the magnetism decrease, an opportunity to learn the basics of the magnetism is thought to increase as much as possible. Meanwhile, since this book can summarize a whole science and technology from the basics of the magnetism and magnetic material to the application to the motor drive system widely, it is thought to have a significant existence. I am very grateful heartily for the authors of each chapter and the related members as for making this book.

It is tried on writing the book in this time to have the same notation such as A/m for the unit of the magnetic field due to the support of all the authors. However, some disagreement still remains for a term, a sign or a definition such as magnetic polarization, magnetization, J, M, saturated magnetization, saturated magnetic flux density. A difference occurs by each situation, significance, how to use, evaluation, when they are seen well. A lot of discussions will be expected to be carried out with the related electric and magnetic researchers and engineers well.

The application of the electrical motor advances, and the researches, develops, design and production are performed widely in each field, and the social requirement for the motor drive system increases. However, it is a present situation that the number of related laboratories at university not so enough increases. Therefore, the situation still continues that almost a lot of numbers of engineers, researchers go to a business circumstance without obtaining an enough opportunity to learn a motor, power electronics, and magnetic material and must engage in this field.

This book intends to extracts the basics matter as much as possible and to explain it carefully in consideration for an engineer and a researcher who learned a different field in the university or so, but some parts may be misunderstanding or not enough explanation. It will be happy if some readers could indicate for that matter.

K. Fujisaki (ed.), *Magnetic Material for Motor Drive Systems*,
Engineering Materials, https://doi.org/10.1007/978-981-32-9906-1

Finally, it is the last message to expect that each reader will deepen his/her specialized field more and more and contribute the development of his/her concerned field by reading this book.

Keisuke Fujisaki
Toyota Technological Institute

The manufacturer's authorised representative in the EU is Springer Nature Customer Service Centre GmbH, Europaplatz 3, 69115 Heidelberg, Germany. If you have any concerns regarding our products, please contact ProductSafety@springernature.com

Printed and bound by CPI Group (UK) Ltd, Croydon, CR0 4YY
25/09/2026
02222884-0010